Lecture Notes in Computer Science   9107

Commenced Publication in 1973
Founding and Former Series Editors:
Gerhard Goos, Juris Hartmanis, and Jan van Leeuwen

More information about this series at http://www.springer.com/series/7407

José Manuel Ferrández Vicente
José Ramón Álvarez-Sánchez
Félix de la Paz López
Fco. Javier Toledo-Moreo
Hojjat Adeli (Eds.)

Artificial Computation in Biology and Medicine

International Work-Conference on the Interplay
Between Natural and Artificial Computation, IWINAC 2015
Elche, Spain, June 1–5, 2015
Proceedings, Part I

 Springer

Editors

José Manuel Ferrández Vicente
Universidad Politécnica de Cartagena
Cartagena
Spain

José Ramón Álvarez-Sánchez
Universidad Nacional de Educación
a Distancia
Madrid
Spain

Félix de la Paz López
Universidad Nacional de Educación
a Distancia
Madrid
Spain

Fco. Javier Toledo-Moreo
Universidad Politécnica de Cartagena
Cartagena
Spain

Hojjat Adeli
Ohio State University
Columbus, Ohio
USA

ISSN 0302-9743   ISSN 1611-3349   (electronic)
Lecture Notes in Computer Science
ISBN 978-3-319-18913-0   ISBN 978-3-319-18914-7   (eBook)
DOI 10.1007/978-3-319-18914-7

Library of Congress Control Number: 2015939151

LNCS Sublibrary: SL1 – Theoretical Computer Science and General Issues

Springer Cham Heidelberg New York Dordrecht London
© Springer International Publishing Switzerland 2015

Printed on acid-free paper

Springer International Publishing AG Switzerland is part of Springer Science+Business Media
(www.springer.com)

Preface

The computational paradigm considered here is a conceptual, theoretical, and formal framework situated above machines and living creatures (two instantiations), sufficiently solid, and still non exclusive, that allows us:

1. to help neuroscientists to formulate intentions, questions, experiments, methods, and explanation mechanisms assuming that neural circuits are the psychological support of calculus;
2. to help scientists and engineers from the fields of artificial intelligence (AI) and knowledge engineering (KE) to model, formalize, and program the computable part of human knowledge;
3. to establish an interaction framework between natural system computation and artificial system computation in both directions, from Artificial to Natural and from Natural to Artificial.

With these global purposes, Prof. José Mira organized the 1st International Work Conference on the Interplay between Natural and Artificial Computation, which took place in Las Palmas de Gran Canaria, Canary Islands (Spain), 10 years ago, trying to contribute to both directions of the interplay.

Today, the hybridization between social sciences and social behaviors with robotics, neurobiology and computing, ethics and neuroprosthetics, cognitive sciences and neurocomputing, neurophysiology and marketing is giving rise to new concepts and tools that can be applied to ICT systems, as well as to natural science fields. Through IWINAC we provide a forum in which research in different fields can converge to create new computational paradigms that are on the frontier between Natural sciences and Information technologies.

As a multidisciplinary forum, IWINAC is open to any established institutions and research laboratories actively working in the field of this interplay. But beyond achieving cooperation between different research realms, we wish to actively encourage cooperation with the private sector, particularly SMEs, as a way of bridging the gap between frontier science and societal impact, and young researchers in order to promote this scientific field.

In this edition, four main themes outline the conference topics: gerontechnology and e-therapy, Brain–Computer Interfaces, Biomedical imaging applications for health, and artificial vision and robotics.

Gerontechnology is an interdisciplinary field combining gerontology and technology. Gerontechnology aims at matching systems to health, housing, mobility, communication, leisure, and work of the elderly. The development of computing systems for gerontechnology has turned into a challenging activity requiring disciplines as diverse as artificial intelligence, human–computer interaction, and wireless sensor networks to work together in order to provide solutions able to satisfy this growing societal demand.

Brain–Computer Interfaces implement a new paradigm in communication networks, namely Brain Area Networks. In this paradigm, our brain inputs data (external stimuli), performs multiple media-access control by means of cognitive tasks (selective attention), processes the information (perception), takes a decision (cognition) and, eventually, transmits data back to the source (by means of a BCI), thus closing the communication loop. The objectives include neuro-technologies (e.g. innovative EEG/ECG/fNIRS headsets, integrated stimulation-acquisition devices, etc.), Tele-services (e.g. applications in Telemedicine, tele-rehabilitation programs, tele-control, mobile applications, etc.), innovative biosignal processing algorithms, training techniques, and novel emerging paradigms.

Image understanding is a research area involving both feature extraction and object identification within images from a scene, and a posterior treatment of this information in order to establish relationships between these objects with a specific goal. In biomedical and industrial scenarios, the main purpose of this discipline is, given a visual problem, to manage all aspects of prior knowledge, from study start-up and initiation through data collection, quality control, expert independent interpretation, to design and development of systems involving image processing capable of tackling with these tasks. Brain imaging using EEG techniques or different MRI systems can help in some neural disorders, like epilepsy, Alzheimer, etc.

Over the last decades there has been an increasing interest in using machine learning methods combined with computer vision techniques to create autonomous systems that solve vision problems in different fields. This research involves algorithms and architectures for real-time applications in the areas of computer vision, image processing, biometrics, virtual and augmented reality, neural networks, intelligent interfaces, and biomimetic object-vision recognition. Autonomous robot navigation sets out enormous theoretical and applied challenges to advanced robotic systems using these techniques.

Ten years after the birth of IWINAC meetings these ideas maintain the visionary objectives of Prof. Mira. This wider view of the computational paradigm gives us more elbow room to accommodate the results of the interplay between nature and computation. The IWINAC forum thus becomes a methodological approximation (set of intentions, questions, experiments, models, algorithms, mechanisms, explanation procedures, and engineering and computational methods) to the natural and artificial perspectives of the mind embodiment problem, both in humans and in artifacts. This is the philosophy that continues in IWINAC meetings, the "interplay" movement between the natural and the artificial, facing this same problem every two years. This synergistic approach will permit us not only to build new computational systems based on the natural measurable phenomena, but also to understand many of the observable behaviors inherent to natural systems.

The difficulty of building bridges between natural and artificial computation is one of the main motivations for the organization of IWINAC 2015. The IWINAC 2015 proceedings contain the works selected by the Scientific Committee from more than 190 submissions, after the refereeing process. The first volume, entitled Artificial Computation in Biology and Medicine, includes all the contributions mainly related to the methodological, conceptual, formal, and experimental developments in the fields of neural sciences and health. The second volume, entitled Bioinspired Computation in Artificial Systems, contains the papers related to bioinspired programming strategies and all the contributions related to the computational solutions to engineering problems in different application domains.

An event of the nature of IWINAC 2015 cannot be organized without the collaboration of a group of institutions and people who we would like to thank now, starting with UNED and Universidad Politécnica de Cartagena. The collaboration of the UNED Associated Center in Elche was crucial, as was the efficient work of the Local Organizing Committee, chaired by Eduardo Fernández with the close collaboration of the Universidad Miguel Hernández de Elche. In addition to our universities, we received financial support from the Spanish CYTED, Red Nacional en Computación Natural y Artificial and Apliquem Microones 21 s.l.

We want to express our gratefulness to our invited speakers Prof. Hojjat Adeli, Ohio State University (USA), Prof. Marc de Kamps, University of Leeds (UK), Prof. Richard Duro, University of A Coruña (Spain), and Prof. Luis Miguel Martínez Otero, University Miguel Hernández (Spain) for accepting our invitation and for their magnificent plenary talks.

We would also like to thank the authors for their interest in our call and the effort in preparing the papers, condition sine qua non for these proceedings. We thank the Scientific and Organizing Committees, in particular the members of these committees who acted as effective and efficient referees and as promoters and managers of preorganized sessions on autonomous and relevant topics under the IWINAC global scope.

Our sincere gratitude goes also to Springer and to Alfred Hofmann and his collaborators, Anna Kramer and Christine Reiss, for the continuous receptivity, help efforts, and collaboration in all our joint editorial ventures on the interplay between neuroscience and computation.

Finally, we want to express our special thanks to Viajes Hispania, our technical secretariat, and to Chari García and Beatriz Baeza, for making this meeting possible, and for arranging all the details that comprise the organization of this kind of event. We want to dedicate these two volumes of the IWINAC proceedings to the memory of Professor Mira, whose challenging and inquiring spirit is in all of us. We greatly miss him.

June 2015   José Manuel Ferrández Vicente
                                               José Ramón Álvarez-Sánchez
                                               Félix de la Paz López
                                               Fco. Javier Toledo-Moreo
                                               Hojjat Adeli

Organization

General Chairman

José Manuel Ferrández Vicente   Universidad Politécnica de Cartagena, Spain

Organizing Committee

José Ramón Álvarez-Sánchez   Universidad Nacional de Educación a Distancia, Spain

Félix de la Paz López   Universidad Nacional de Educación a Distancia, Spain

Fco. Javier Toledo-Moreo   Universidad Politécnica de Cartagena, Spain

Honorary Chairs

Rodolfo Llinás   New York University, USA
Hojjat Adeli   Ohio State University, USA
Zhou Changjiu   Singapore Polytechnic, Singapore

Local Organizing Committee

Eduardo Fernández Jover   Universidad Miguel Hernández, Spain
Arantxa Alfaro Sáez   Universidad Miguel Hernández, Spain
Ariadna Díaz Tahoces   Universidad Miguel Hernández, Spain
Nicolás García Aracil   Universidad Miguel Hernández, Spain
Alejandro García Moll   Universidad Miguel Hernández, Spain
Lawrence Humphreys   Universidad Miguel Hernández, Spain
Carlos Pérez Vidal   Universidad Miguel Hernández, Spain
José María Sabater Navarro   Universidad Miguel Hernández, Spain
Cristina Soto-Sánchez   Universidad Miguel Hernández, Spain

Invited Speakers

Hojjat Adeli   Ohio State University, USA
Marc de Kamps   University of Leeds, UK
Richard Duro   University of A Coruña, Spain
Luis Miguel Martínez Otero   University Miguel Hernández, Spain

Field Editors

Diego Andina	Spain
Jorge Azorín-López	Spain
Rafael Berenguer Vidal	Spain
Germán Castellanos-Dominguez	Spain
Miguel Cazorla	Spain
Antonio Fernández-Caballero	Spain
José Garcia-Rodriguez	Spain
Pascual González	Spain
Álvar Ginés Legaz Aparicio	Spain
Javier de Lope Asiaín	Spain
Miguel Ángel López Gordo	Spain
Darío Maravall Gómez-Allende	Spain
Rafael Martínez Tomás	Spain
Elena Navarro	Spain
Pablo Padilla de la Torre	Spain
Daniel Ruiz Fernández	Spain
Antonio J. Tallón-Ballesteros	Spain
Hujun Yin	UK

International Scientific Committee

Andy Adamatzky	UK
Michael Affenzeller	Austria
Abraham Ajith	Norway
José Ramón Álvarez-Sánchez	Spain
Antonio Anaya	Spain
Diego Andina	Spain
Davide Anguita	Italy
Margarita Bachiller Mayoral	Spain
Dana Ballard	USA
Emilia I. Barakova	The Netherlands
Francisco Bellas	Spain
Guido Bologna	Switzerland
María Paula Bonomini	Argentina
François Bremond	France
Giorgio Cannata	Italy
Enrique J. Carmona Suarez	Spain
German Castellanos-Dominguez	Colombia
Jose Carlos Castillo	Spain
Antonio Chella	Italy

Rafael Martínez Tomás	Spain
Antonio Martínez-Álvarez	Spain
Jose Javier Martinez-Alvarez	Spain
Jose del R. Millan	Switzerland
Taishin Y. Nishida	Japan
Richard A. Normann	USA
Lucas Paletta	Austria
Juan Pantrigo	Spain
Alvaro Pascual-Leone	USA
Gheorghe Paun	Spain
Francisco Peláez	Brazil
Franz Pichler	Austria
Maria Pinninghoff	Chile
Andonie Razvan	USA
Mariano Rincon Zamorano	Spain
Victoria Rodellar	Spain
Camino Rodriguez Vela	Spain
Daniel Ruiz	Spain
Ramon Ruiz Merino	Spain
José María Sabater Navarro	Spain
Diego Salas-Gonzalez	Spain
Pedro Salcedo Lagos	Chile
Angel Sanchez	Spain
Eduardo Sánchez Vila	Spain
Jose Santos Reyes	Spain
Shunsuke Sato	Japan
Andreas Schierwagen	Germany
Guido Sciavicco	Spain
Radu Serban	The Netherlands
Igor A. Shevelev	Russia
Shun-ichi Amari	Japan
Settimo Termini	Italy
Javier Toledo-Moreo	Spain
Jan Treur	The Netherlands
Ramiro Varela Arias	Spain
Marley Vellasco	Brazil
Lipo Wang	Singapore
Stefan Wermter	UK
Hujun Yin	UK
Juan Zapata	Spain
Changjiu Zhou	Singapore

Contents – Part I

Contents – Part II

Automated Diagnosis of Alzheimer's Disease by Integrating Genetic Biomarkers and Tissue Density Information

Andrés Ortiz[1(✉)], Miguel Moreno-Estévez[1], Juan M. Górriz[2], Javier Ramírez[2],
María J. García-Tarifa[1], Jorge Munilla[1], and Nuria Haba[1], for the Alzheimer's
Disease Neuroimaging Initiative

[1] Communications Engineering Department
University of Málaga, 29004 Málaga, Spain
[2] Department of Signal Theory, Communications and Networking
University of Granada, 18060 Granada, Spain
aortiz@ic.uma.es

Abstract Computer aided diagnosis (CAD) constitutes an important
tool for the early diagnosis of Alzheimer's Disease (AD), which, in turn,
allows the application of treatments that can be simpler and more likely
to be effective. This paper presents a straightfoward approach to determine the most discrimanative brain regions, defined by the Automated
Anatomical Labelling (AAL), based on the measurements of the tissue
density at the different brain areas. Statistical analysis of GM and WM
densities reveal significant differences between controls (CN) and AD at
specific brain areas associated to tissue density diminishing due to neurodegeneration. The proposed method has been evaluated using a large
dataset from the Alzheimer's disease Neuroimaging Initiative (ADNI).
Classification results assessed by cross-validation proved that computed
WM/GM densities are discriminative enough to differentiate between
CN/AD. Moreover, fusing density measurements with ApoE genetic information help to increase the diagnosis accuracy.

1   Introduction

Alzheimer's Diasese (AD) is the most common cause of dementia among older
people and a third of young people with dementia have AD, affecting 30 million people worldwide. Due to the increasing life expectancy and the ageing
of the population in developed nations, it is expected AD affects 60 million
people worldwide over the next 50 years. It is a slow neurodegenerative disease
associated to the production of β-amyloid peptide (Aβ) and its extracellular

Data used in preparation of this article were obtained from the Alzheimer's Disease Neuroimaging Initiative (ADNI) database (adni.loni.ucla.edu). As such, the
investigators within the ADNI contributed to the design and implementation of
ADNI and/or provided data but did not participate in analysis or writing of this
report. A complete listing of ADNI investigators can be found at: http://adni.loni.
ucla.edu/wpcontent/uploads/how_to_apply/ADNI_Acknowledgement_List.pdf.

© Springer International Publishing Switzerland 2015
J.M. Ferrández Vicente et al. (Eds.): IWINAC 2015, Part I, LNCS 9107, pp. 1–8, 2015.
DOI: 10.1007/978-3-319-18914-7_1

deposition as well as the flame-shaped neurofibrillary tangles of the microtubule binding protein tau [9]. This progressively causes the loss of nerve cells, whose symptoms usually start with mild memory problems, turning into severe brain damage in several years. There is no cure for AD, and currently developed drugs can only help to temporarily slow down the progression of the disease [14]. However, treatment success depends on the early diagnosis that can allow an early treatment.

Since the AD neurodegeneration process progresively affects different brain functions, functional images such as Single Emission Computerized Tomography (SPECT) [12,6,8] or Positron Emission Tomography (PET) [1,13] have been extensively used in Computer Aided Diagnosis systems. AD also causes structural changes in the brain and thus structural differences between controls and AD patients can be revealed by analysis of Magnetic Resonance Images (MRI). In fact, MRI has been used in many previous works for automatic diagnosis [11,10,4,5]. However, these works use GM or WM images on whole brain volume to classify controls and AD patients [4,5] or to compute Regions of Interest (ROI). By contrast, we use here the 116-regions Automated Anatomical Labeling Atlas (AAL) to extract brain patches corresponding to these areas, allowing to compute GM and WM tissue densities at each brain region. Moreover, clinical information such as gender and APOE genetic biomarkers have been included to 1) determine the discriminative power of each biomarker and 2) improve the classification accuracy by combining biomarkers and image data.

The rest of the paper is organized as follows. Section 2 describes the database used in this work. Image preprocessing, densty computation and the methods used to select the most relevant brain regions are shown in subsections 2.1, 2.2 and 2.3, respectively. Section 3 shows details on the experiments performed and the results obtained using patient data from the ADNI database. Finally, the conclusions are drawn in Section 4.

2   Database

The database used in this work contains multimodal PET/MRI image data from 138 subjects, comprising 68 Controls (CN), 70 AD and 111 MCI patients from the ADNI database [2]. This repository was created to study the advance of the Alzheimer disease, collecting a vast amount of MRI and Positron Emission Tomography (PET) images as well as blood biomarkers and cerebrospinal fluid analyses. The main goal of this database is to provide a maens for the early diagnosis of the Alzheimer's disease. Patient's demographics are shown in Table 1. However, in this work only MRI data are used.

2.1   Image Preprocessing

MRI images from the ADNI database have been spatially normalized according to the VBM-T1 template and segmented into White Matter (WM) and Grey Matter (GM) tissues using the VBM toolbox for SPM [3,16]. This ensures each

Table 1. Patient Demographics

Evaluation	Sex (M/F)	Mean Age ± Std	Mean MMSE ± Std
NC	43/25	75.81 ± 4.93	29.06 ± 1.08
AD	46/24	75.33 ± 7.17	22.84 ± 2.91

image voxel corresponds to the same anatomical position. After image registration, all the images from ADNI database were resized to 121x145x121 voxels with voxel-sizes of 1.5 mm (Sagittal) x 1.5 mm (coronal) x 1.5 mm (axial). MRIs are further segmented to obtain information about GM and WM tissue distributions, which can be used to differentiate AD from CN patients [10,7,11]. This process is guided by means of tissue probability maps of grey matter, white matter or cerebro-spinal fluid. A nonlinear deformation field is estimated that best overlays the tissue probability maps on the individual sujects' image. The tissue probability maps provided by the International Consortium for Brain Mapping (ICBM) are derived from 452 T1-weighted scans, which were aligned with an atlas space, corrected for scan inhomogeneities, and classified into grey matter, white matter and cerebro-spinal fluid. Segmentation through SPM/VBM provides values in the range [0, 1] which denote the membership probability to a specific tissue.

2.2 Density Computation

Features used in this work are based on GM and WM volumes in specific brain regions. The brain atlas delimitates 116 regions that can be used to mask the brain and extract information from different areas. Thus, the GM volume in the i-region can be computed using the following expression,

$$\text{Vol}_i = \frac{\#voxels_i > thr}{1000} * \text{voxel size} \tag{1}$$

where thr indicates the probability threshold that determines whether a voxel belongs to a specific tissue. Similarly, tissue density for each region can be computed as follows,

$$D_i = \frac{\#voxels_i > thr}{Vol_i} \tag{2}$$

The threshold thr, in Equations 1 and 2, indicates how the partial volume effect is taken into account. That is, if $thr = 0.5$, information from GM and WM are equally mixed at the same voxel; the lower the thr ($thr < 0.5$) the less the relative importance of one tissue over the other, and for thr=1 no volume effect is taken into account. Since thr determines the classification accuracy, it has been determined experimentally through classification experiments for different threshold values, as shown in Section 3.

2.3   Selection of Most Relevant Brain Regions

In this work, feature selection is accomplished in order to select the most discriminant brain regions for AD. This is performed by the Student's two sample t-Test, with pooled variance estimate, which allows to test the difference in the means of two populations. t-test defines the statistic t which is a significance measurement on the means difference. Indeed, greater t-values correspond to lower p-values that allows to reject the null hyphotesis. t-statistic can be computed by the following expression

$$I^t = \frac{|I^\mu_{CN} - I^\mu_{AD}|}{\sqrt{\frac{I^{\sigma^2}_{CN}}{N_{CN}} + \frac{I^{\sigma^2}_{AD}}{N_{CN}}}} \tag{3}$$

where I^μ_{CN} and I^μ_{AD} are the mean images for CN and AD respectively, I^σ_{CN} and I^σ_{AD} are the variance images for CN and AD, respectively and N_{CN}, N_{AD} are the number of CN and AD images, respectively.

3   Experimental Results

Firstly, experiments to determine the optimal threshold value indicated in Section 2.2 were carried out. This was addressed by classifying using a Support Vector Machine (SVM) with selected brain regions (p-value < 0.05). As a result, the value 0.7, see Figure 1, is obtained as the optimal threshold that provides the best accuracy outcome. Results shown in Figure 1 were obtained by k-fold (k=10) cross-validation, performing the Welch's test on each training partition.

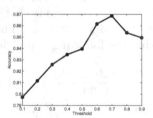

Fig. 1. Classification accuracy obtained for different threshold values for CN/AD subjects

Hereafter, this optimal threshold is used for the rest of our experiments. Thus, density computation is addressed with $thr = 0.7$ for both GM and WM tissues.

3.1   Diagnostic Relevance of Brain Regions

The most relevant regions at each cross-validation iteration are computed by statistical hypothesis test as shown in Section 2.3. A set of p-values (and t-statistic values) corresponding to the discriminative power of each brain region is calculated at each iteration. Thus, p-values coming from each test using the training samples are combined to provide a summarized view of the relevance of different brain areas. Hence, p-value combination has been addressed by the *Stouffer* method [15] that transforms the p-values via the standard normal distribution. This way, p-values combination is defined as

$$p_{comb} = \Phi\left(\frac{1}{\sqrt{n}}\sum_{i=1}^{n}\Phi^{-1}(p_i)\right) \tag{4}$$

where Φ is the cumulative distribution function (c.d.f.), and $p_i, i = \{1,..,n\}$ are independent p-values.

Although all regions with p-value below 0.05 (5% significance level) have been considered in the classification task, Figure 2 shows the most ten relevant regions according to the combined p-value.

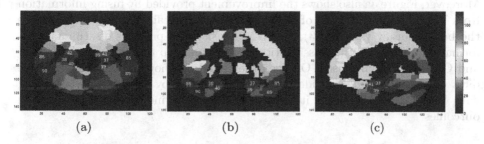

(a)   (b)   (c)

Fig. 2. Top ten selected brain regions in axial (a), coronal (b) and sagittal (c) planes. Sorted by significance order, (38) Right Hippocampus, (90) Right Inferior temporal Gyrus, (38) Left Hippocampus, (85) Left Middle Temporal gyrus, (86) Right Middle Temporal gyrus, (89) Inferior Temporal gyrus, (40) Right ParaHippocampal gyrus, (39) Left ParaHippocampal gyrus, (42) Right Amygdala, (41) Left Amygdala.

3.2   Integrating ApoE Genetic Data

The ApoE gene contains the information necessary to make the apolipoprotein E, which is combined with fats (lipids) in the body to lipoproteinsis. There are four genotypes, depending on the alleles present in the gene. Specifically, genotypes that have been found to be related to AD are ApoE 4,4 and ApoE 3,3. However, subjects with two copies of the allele 4 (ApoE-ε4) have a higher risk of contracting AD than ApoE 3,3 subjects. Conversely, subjects with genotype ApoE 2,3 are considered to be protected against AD. In this paper, ApoE genetic information from each subject has been included in the feature space to be considered in the classification task.

3.3   Classification

In this work, WM and GM density information has been computed from each
brain region according to the AAL atlas, consisting of 116 brain areas. These
densities have been used to compose the feature space and then to assess their
discriminant capabilities. Hence, statistical tests performed on density data revealed discriminant regions corresponding to those found in the medical literature. Moreover, the discriminant power of ApoE genetic information extracted
from the ADNI database has been also assessed, showing less prediction capability than MRI data. A series of classification experiments were performed using
both GM/WM density information and genetic data. Subsequently, density and
genetic information is fused to improve the prediction capability using a SVM
classifier for CN/AD subjects. In the first experiment, once the most relevant
regions have been identified, only density information from these regions are
taken into account for classification. Figure 3a shows the accuracy, sensitivity
and specificity values obtained when WM and GM densities are used as features.
Moreover, Figure 3a also shows the improvement provided by fusing information
from WM and GM. As most of the information is contained in GM, it produces
the best accuracy value (0.83), but WM density also provides discriminant information, increasing the classification performance up to 0.86 when both WM
and GM densities are fused. On the other hand, as shown in Figure 3b, ApoE
data does not provide provide enough information by itself, but when fused
with WM/GM density data, it sightly contributes to increase the classification
outcomes.

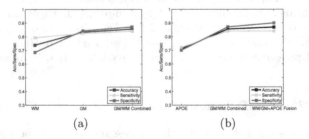

(a)   (b)

Fig. 3. Classification results assessed by k-fold (k=10) cross-validation for (a) WM/GM
density and (b) WM/GM density + ApoE genetic information

Additionally, Receiver Operating Curves (ROC) curves have been computed
to prove the discriminant power of density and genetic features used in this work.
We obtain that fusing WM and GM density information provides greater Aurea
Under ROC Curve (AUC) values than using those values separately, see Figure
4a. Moreover, Figure 4b shows a sightly improvement provided by incorporating
ApoE data to the feature space.

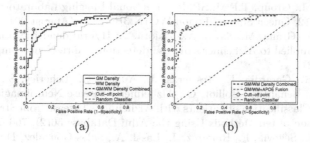

(a)   (b)

Fig. 4. ROC curves for (a) WM/GM density and (b) WM/GM density + ApoE genetic information

4   Conclusions

This paper proposes a straightforward method to compute tissue densities from different brain regions defined by the AAL atlas. Different experiments were carried out using these densities as features to assess their capability to distinguish between CN and AD subjects. Thus, statistical tests performed on density data obtain discriminant regions corresponding to those found in medical literature, such as the hippocampus in both hemispheres. Moreover, artrophy in other regions such as inferior temporal gyrus have been also revealed. Classification experiments performed using GM and WM density values from most discriminant brain regions show an accuracy up to 85%, and AUC up to 0.90.

Finally, although genetic ApoE information has not shown a high discriminative capability, it helps to increase the classification accuracy when incorporated to the WM/GM density information. Indeed, fusion of WM/GM density data and ApoE genetic biomarkers provides accuracy values up to 87% and an AUC of 0.92.

Acknowledgments. This work has been partly supported by the MICINN under the TEC2012-34306 project (DIAGNOSIS), the Consejería de Innovación, Ciencia y Empresa (Junta de Andalucía, Spain) under the Excellence Projects P09-TIC-4530 and P11-TIC-7103, and the Universidad de Málaga, Programa de fortalecimiento de las capacidades de I+D+I en las Universidades 2014-2015, de la Consejería de Economía, Innovación, Ciencia y Empleo, cofinanciado por el fondo europeo de desarrollo regional (FEDER) under Project No. FC14-SAF-30.

References

1. Álvarez, I., Gorriz, J.M., Ramirez, J., Salas-Gonzalez, D., Lopez, M.M., Segovia, F., Chaves, R., Gomez-Rio, M., Garcia-Puntonet, C.: 18f-fdg pet imaging analysis for computer aided Alzheimer's diagnosis. Information Sciences 184(4), 196–903 (2011)
2. Alzheimer's Disease Neuroimaging Initiative (2014), http://adni.loni.ucla.edu/ (accessed March 10, 2014)

3. Ashburner, J., Group, T.F.M: SPM8. Functional Imaging Laboratory, Institute of Neurology, 12, Queen Square, Lonon WC1N 3BG, UK (August 2011)
4. Chyzhyk, D., Graña, M., Savio, A., Maiora, J.: Hybrid dendritic computing with kernel-lica applied to Alzheimer's disease detection in mri. Neurocomputing 75(1), 72–77 (2012)
5. Cuingnet, R., Gerardin, E., Tessieras, J., Auzias, G., Lehéricy, S., Habert, M., Chupin, M., Benali, H., Colliot, O.: Alzheimer's Disease Neuroimaging Initiative. Automatic Classification of patients with Alzheimer's Disease from Structural MRI: A Comparison of ten Methods Using the Adni Database 56(2), 766–781 (2010)
6. Górriz, J.M., Segovia, F., Ramírez, J., Lassl, A., Salas-González, D.: Gmm based spect image classification for the diagnosis of Alzheimer's disease. Applied Soft Computing 11, 2313–2325 (2011)
7. Liu, M., Zhang, D., Shen, D.: Disease Neuroimaging Initiative. Ensemble sparse classification of alzheimer's disease. Ensemble sparse classification of alzheimer's disease 60(2), 1106–1116 (2012)
8. López, M., Ramírez, J., Górriz, J.M., Álvarez, I., Salas-González, D., Segovia, F., Chaves, R., Padilla, P., Gómez-Río, M.: Principal component analysis-based techniques and supervised classification schemes for the early detection of Alzheimer's disease. Neurocomputing 74(8), 1260–1271 (2011)
9. Paul Murphy, M., LeVine, H.: Alzheimer's disease and the β-amyloid peptide. Journal of Alzheimer's Disease 19(1), 311–318 (2010)
10. Ortiz, A., Górriz, J.M., Ramírez, J., Martínez-Murcia, F.J.: LVQ-SVM based CAD tool applied to structural MRI for the diagnosis of the Alzheimer's disease. Pattern Recognition Letters 34(14), 1725–1733 (2013)
11. Ortiz, A., Górriz, J.M., Ramírez, J., Martínez-Murcia, F.J.: Automatic roi selection in structural brain mri using som 3d projection. PLOS One 9(4) (2014)
12. Ramirez, J., Chaves, R., Gorriz, J.M., Lopez, M., Alvarez, I.A., Salas-Gonzalez, D., Segovia, F., Padilla, P.: Computer aided diagnosis of the Alzheimer's disease combining spect-based feature selection and random forest classifiers. In: Proc. IEEE Nuclear Science Symp. Conf. Record (NSS/MIC), pp. 2738–2742 (2009)
13. Segovia, F., Górriz, J.M., Ramírez, J., Salas-González, D., Álvarez, I., López, M., Chaves, R.: The Alzheimer's Disease Neuroimaging Initiative. A comparative study of the feature extraction methods for the diagnosis of Alzheimer's disease using the adni database. Neurocomputing 75, 64–71 (2012)
14. Alzheimer's Disease Society. Factsheet: Drug treatments for alzheimer's disease (2014)
15. Stouffer, S.A., Suchman, E.A., DeVinney, L.C., Star, S.A., Williams Jr., R.M.: Adjustment During Army Life, vol. 1. Princeton University Press, Princeton (1949)
16. Structural Brain Mapping Group. Department of Psychiatry (2014), http://dbm.neuro.uni-jena.de/vbm8/VBM8-Manual.pdf (accessed March 10, 2014)

A Neural Model of Number Interval Position Effect (NIPE) in Children

Michela Ponticorvo[1](✉), Francesca Rotondaro[2], Fabrizio Doricchi[2], and Orazio Miglino[1]

[1] Department of Humanistic Studies, University of Naples "Federico II", Naples, Italy
[2] Department of Psychology, Sapienza, University of Rome, Rome, Italy
michela.ponticorvo@unina.it

Abstract. In the present paper we describe an artificial neural model of the Number Interval Position Effect (NIPE;[5]) that has been observed in the mental bisection of number intervals both in adults and in children. In this task a systematic error bias in the mental setting of the subjective midpoint of number intervals is found, so that for intervals of equal size there is a shift of the subjective midpoint towards numbers higher than the true midpoint for intervals at the beginning of decades while for intervals at the end of decades the error bias is directionally reversed towards numbers lower than the true midpoint. This trend of the bisection error is recursively present across consecutive decades.
Here we show that a neural-computational model based on information spread by energy gradients towards accumulation points based on the logarithmic compressed representation of number magnitudes that has been observed at the single cell level in rhesus monkeys [9] effectively simulates the performance of adults and children in the mental bisection of number intervals, in particular replicating the data observed in children.

Keywords: Artificial Neural Models · Numerical Cognition · Mental Number Line · Bisection of Number Intervals · NIPE effect

1 Introduction

Numbers are everywhere around us and dealing with them covers an important part of our cognitive activity throughout our life. A number of studies have suggested that when left/right response codes must be associated to number magnitudes, healthy participants belonging to western cultures with left-to-right reading habits map numbers upon a mental number line (MNL) with small integers positioned to the left of larger ones. This is reflected in the SNARC effect, (Spatial-Numerical Association of Response Codes) first demonstrated by Dehaene, Bossini, and Giraux [4] who argued that A representation of number magnitude is automatically accessed during parity judgments of Arabic digits. This representation may be likened to a mental number line, because it bears a natural and seemingly irrepressible correspondence with the left/right

© Springer International Publishing Switzerland 2015
J.M. Ferrández Vicente et al. (Eds.): IWINAC 2015, Part I, LNCS 9107, pp. 9–18, 2015.
DOI: 10.1007/978-3-319-18914-7_2

coordinates of external space (p. 394). More recently a inherent spatial and spatial-response-code independent nature of the MNL was suggested by the finding that during the mental bisection of number intervals right brain damaged patients with attentional neglect for the left side of space shift the subjective midpoint of number intervals toward numbers higher than the true midpoint, i.e. supposedly to the right of the true midpoint [14].

However, several ensuing studies have demonstrated that this numerical bias is unrelated to left spatial neglect and that it is rather linked to a deficit in the abstract representation of small numerical magnitudes [1,2]; for a review see Rossetti and collegues [11]. This conclusion was suggested by the finding that in right brain damaged patient the pathological bias toward numbers higher than the midpoint in the mental bisection of number interval is correlated to a similar bias in the bisection of time intervals on an imagined clock face where higher number are positioned to the left, rather than to the right, of the mental display [2,11]. In a recent study, Doricchi and colleagues [5] have discovered an new interesting psychophisical property of the number interval bisection task. It was found that in this task, human participants show a systematic error bias which is linked to the position occupied by the number interval in a decade (Number Interval Position Effect, NIPE). The subjective midpoint of number intervals of the same length is placed on numbers higher than the true midpoint the closer the interval is to the beginning of a decade and on numbers lower than the midpoint the closer the interval is to the end of the same decade. For example, in case of 7 units intervals the bias is positive for the intervals at the beginning of the decade (1-7) and negative for the intervals at the end of the decade (3-9). This effect has been observed in healthy adults [1,5], right brain damaged patients [1,5] and in pre-school children [12] thus suggesting that it is not related to learning of formal arithmetics and that it could be linked to some fundamental properties of the neural representation of number magnitudes.

Neurophysiological studies have demonstrated a neuronal representations of numerosity in the prefrontal and parietal cortex of rhesus monkeys [9]. In these areas different neuronal populations code for different numerosities. For small numerosities, the neural discharge is narrowly tuned, according to a gaussian function, to the preferred numerosity of the neuron so that the discharge is weak for adjacent numerosities. This gaussian tuning becomes progressively larger, i.e. less selective, for increasing numerosities, so that neurons tuned to larger numerosities show some discharge also for numerosities that are immediately adjacent to the preferred one. The organisation of the gaussian curves linked to the different and progressively increasing numerosities is best described by a nonlinearly logarithmic compressed scaling of numerical information.

In what follows we shall propose that the NIPE observed in the mental bisection of number intervals can be simulated by a neural-computational model based on infomation spread by energy gradients towards accumulation points based on the logarithimic compressed representation of number magnitudes that has been observed at the single cell level in the rhesus monkey [9].

2   Materials and Method

2.1   The Task

In order to investigate neuro-cognitive structures and mechanisms underlying basic arithmetics, namely neural coding of natural numbers and simple arithmetics operations, it is often proposed a task in which the participant has to identify the natural number that divides equally, bisects, a numerical series that is delimited by two natural numbers. For example, if we consider only the series of the first natural ten (1-10), the partecipant can be asked to identify the middle number between 1 (lower bound) and 7 (upper bound) or between 2 (lower bound) and 6 (upper bound) and so on. This task includes various forms: some of them permit one single solution, the ones whose limits sum is an even number, some others, the ones whose limits sum is an odd number, permit two solutions. This latter case is exemplified by the identification of the middle number between 1 and 8: the solutions are 4 and 5. To reply univocally the partecipant must choose the number that is closer to the lower bound, rounding down, or the upper, rounding up. For this reason, it is preferred to propose the task form with even sum.

2.2   The Model

We propose a neural model where no linear spatial representation is present. For this reason we start from two general principles about neural mechanisms which are strongly funded:

a. **Natural numbers neural coding**: basic numbers in a certain notation are coded in an amodal way by distinct neural groups. In other words, if we consider the decimal notation, there is a neural group whose activation is more probable when the number 1 is presented regardless of the presentation form, another one for number 2 and so on up to 10.

b. **Neural accumulation mechanisms**: neural elaboration takes place by energy transfer between neural groups and arrives to its conclusion when some neural group accumulates a certain energy level.

Fig. 1. The neural network architecture with nodes connections

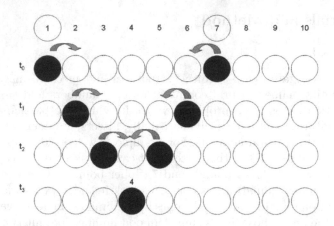

Fig. 2. The bisection problem: the even sum case

Let us consider that m neural groups code n natural numbers: $m=n$. In decimal notation, we will therefore have ten neural groups a,b,c,d,e,f,g,h,i,l which code natural numbers from 1 to 10. Let us imagine that neural groups are communicating vessels that transfer from one to another their energy level and the transfer dynamic ends when one neural group goes beyond a certain accumulation threshold. The various groups are connected in such a way that the neural group who presents its biggest activation probability when the number n is presented, is connected with the groups representing $n-1$ and $n+2$. The groups representing the bounds 1 and 10 are an exception. Number 1 is connected only with $n+1$ group and 10 is connected only with $n-1$. This architecture is represented in figure 1. Please note that each node does not represent a single neuron, but a group of neurons, a network.

We dictate the following dynamic to our network:

1. A neural group is univocally associated to a natural number. It therefore activates when this number is presented. The node a,b,c,d,e,f,g,h,i,l are associated with numbers 1, 2, 3, 4, 5, 6, 7, 8, 9, 10.
2. At time t_0, nodes coding upper and lower bound have 1.0 activation whereas other nodes have 0.0.
3. An highly active node transfers its energy to a node to whom it is connected with lower energy.
4. At time t_1 energy flows between nodes according to the following constraint: the node that codes the lower limit transfers its energy to the node representing the number immediately superior; the node coding the upper bound transfers its energy to the node representing the immediately inferior number.
5. If two contiguous nodes have the same energy level the energy flow interrupts.

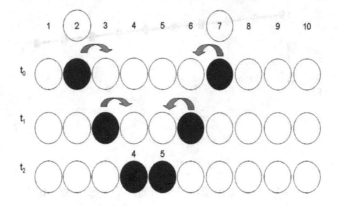

Fig. 3. The bisection problem: the odd sum case

6. One single node activation is the sum of all energies that collects.
7. Every node has an accumulation threshold that, when overcome, interrupts the network dynamic. Let us dictate that this threshold is equal to 1.5.

Such a neural network is able to calculate the intermediate number between two limits whose sum is an even number and the two intermediate numbers when the sum is an odd number. Figure 2 and 3 represent two examples: in figure 2 we have the intermediate number between 1-7 (even sum), figure 3 represents the odd sum (2-7).

To obtain in every case an univocal result, even if the limit numbers sum is odd, it is necessary hypothesize that various neural groups at moment t_0 present different energetic levels in order to have a convergence toward a single neural group. In other words, it is necessary to hypothesize that to each neural group is associated an activation coefficient or energy gradient that reinforce or soften stimulation coming from outside. For example, at time t_0, the energy level associated is inversely proportional to the numerical value to which it is associated. This relation is shown in figure 4. It is necessary to underline that this relation is arbitrary.

Let us modify the point 2. of the above described neural dynamic in the following way:

2. At time t_0, nodes coding upper and lower bound have an activation that varies in function of an activation coefficient that is specific for each neuronal group. All the other nodes have activation 0.0.

This condition is the equivalent of defining a neuro-cognitive bias that, depending on the activation coefficient associated to lower and upper bound produces a solution of rounding up or down. Obviously this bias is valid only for "odd sum" problems. The example in figure 3, with this change in neural dynamic and of parameters defined in figure 4, produces a new solution, as illustrated in figure 5.

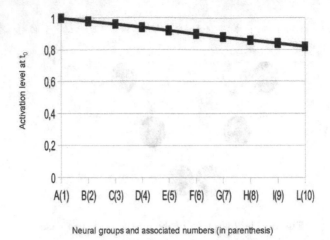

Fig. 4. The bisection problem: the odd sum case

The relation defined in figure 4 imposes to the network to select an univocal intermediate point also in cases where the limit numbers is an odd number (see figg. 3 and 5). It is worth noting that the relation between initial activation level of neural groups reported in fig.4 produces rounding up solutions, the intermediate point of "odd sum" problem is placed toward the upper bound. Obviuosly if we change the relation, the rounding changes too. For example, if the relation is directly proportional we have a rounding down. If the relation is non-linear we would observe sometimes a rounding up, some others down.

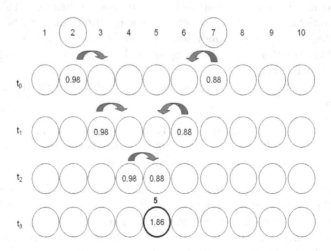

Fig. 5. The bisection problem: the odd sum case with the modified neural dynamic

This network is able to calculate the natural number between two limit numbers. In even sum case, it produces solutions without any mistake. On the contrary, human beings committ systematic errors, as shown in the introduction. They select, in some conditions, the intermediate number toward the lower limit (rounding down) and, in some others, toward the upper bound (rounding up).

This means that the initial activation gradient, shown in figure 4, is non linear, as shown by already cited studies. One possible explanation can be that when a certain number is presented a particular neural group activates selectively. Small numbers (1, 2, 3) are always associated to the same neural group, whereas bigger numbers are associated with neural groups with a certain probabilty. In other words a given neural group can activate more probably than others, but other groups, devoted to coding numbers that are close to the presented one, can activate too. For example, ift number 8 is presented, neural groups for 7 and 9 can activate too. If we introduce in our network dynamic this phenomen, the item 1 becomes:

1. A neural group is **probabilistically** associated to a natural numbers. Each neural group has a probability distribution where the association between the neural group and the natural number is defined.

Behavioral and neurophysiological evidence show that representations of increasing numerosities increasingly overlap, thereby becoming progressively less discriminable from adjacent ones [9]. Let us now imagine to adopt the probability distribution as reported in figure 6, deduced from cited studies.

Applying the relation illustrated in figure 6, a subject to whom is proposed to identify the intermediate number between 1 and 7 can accomplish the task sometimes as the demand is with 1 and 6 limits and some others as it is with 1 and 8 limits. Data indicate that the systematic error is non-linear: sometimes we observe a overestimation, in other cases an underestimation. For these reasons we modify figure 4 giving it a non-linear trend, as shown in figure 7.

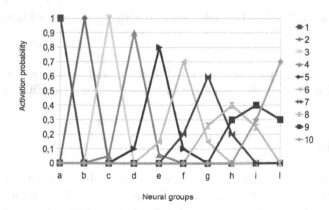

Fig. 6. Probability distribution of each neural group activation for number from 1 to 10

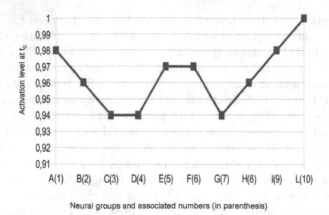

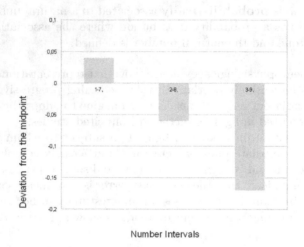

Neural groups and associated numbers (in parenthesis)

Fig. 7. Probability distribution of each neural group activation for number from 1 to 10

Number Intervals

Fig. 8. Data about Number Intervals Bisection with 7 units derived from the neural model

Please not that the parameters related to the model have been obtained using a genetic algorithm [3]. For more detailsl, please look at additional materials section.

3 Results

The proposed model is able to replicate the data observed with children [12] indicating the presence of NIPE effect. In figure 8 the data about the 7 units task are reported. The results reported in figure 8 indicate that the model shows the NIPE effect too, displayng the same trend as the children. The artificial neural network in fact commits a systematic error that is consistent with the NIPE effect, in fact the closer one boundary of the interval was to the border of a ten, the more its midpoint was shifted.

4   Conclusions and Future Directions

In the present paper we have described a neural model that replicates data observed in mental bisection of numerical intervals in children. This model is exclusively based on energy transfer and accumulation and, despite of this, it can replicate data observed in children.

These results support the idea that the mental number line does not represents numbers in a spatial guise and the arithmetics module can, at least in principle, work on energy transfer rather than on number spatial representation.

Numbers are a fundamental part of our cognitive environment and it is worth interrogating on how they are represented in the brain. Feigenson and collegues [6] underline that there are two core systems that underlie the ability to think and reason about number: one system that is devoted to represent large, approximate numerical magnitudes, and another system that precisely represent small numbers of individual objects. These systems are shared across different developmental stages and different species and represent the basis on which the sophisticated human numerical ability is built.

The reported studies and the presented results indicate that the NIPE effect is widespread too. It can be observed in human adult and children and in our artificial model too. What does this tell us about number representation? The NIPE effect can mirror the logarithmic central representation of numerical magnitudo that is independent from school education and that is shared by non-human species too.

In this context a computational model can be an interesting way to approach cognitive issues [7,10]. Artificial models, in fact, can give us the chance to produce an artifact to be included in the list of species to be studied. If comparative sciences can give us insights about cognition, artificial models can give further insights in reproducing a certain phenomenon. In this case the scientific challenge is buiding a new artificial species with its own specific features. These artificial networks can reproduce phenomena at various levels: behavioural, physiological, neural with different granularity from the single neuron to whole structures. This approach has been already used in modelling neuropsychological phenomena,[13] linking these phenomena with neural representation as well as organisms interaction with the environment [8] giving useful insights to this research field.

The next step will be to build an extended model with more different layers to reproduce not only the behavioural side of NIPE effect but also the supposed corresponding neural circuitry.

Additional Materials

More details about the model and the related code can be provided to whom is interested by emailing the author Orazio Miglino (orazio.miglino@unina.it) or Michela Ponticorvo (michela.ponticorvo@unina.it).

References

1. Aiello, M., Merola, S., Doricchi, F.: Small numbers in the right brain: evidence from patients without and with spatial neglect. Cortex 49(1), 348–351 (2013)
2. Aiello, M., Jacquin-Courtois, S., Merola, S., Ottaviani, T., Tomaiuolo, F., Bueti, D., Doricchi, F.: No inherent left and right side in human mental number line: evidence from right brain damage. Brain, aws114 (2012)
3. Davis, L. (ed.): Handbook of genetic algorithms, vol. 115. Van Nostrand Reinhold, New York (1991)
4. Dehaene, S., Bossini, S., Giraux, P.: The mental representation of parity and number magnitude. Journal of Experimental Psychology: General 122(3), 371–396 (1993)
5. Doricchi, F., Merola, S., Aiello, M., Guariglia, P., Bruschini, M., Gevers, W., Tomaiuolo, F.: Spatial orienting biases in the decimal numeral system. Current Biology 19(8), 682–687 (2009)
6. Feigenson, L., Dehaene, S., Spelke, E.: Core systems of number. Trends in Cognitive Sciences 8(7), 307–314 (2004)
7. Miglino, O., Ponticorvo, M.: Exploring the Roots of (Spatial) Cognition in Artificial and Natural Organisms. The Evolutionary Robotics Approach. The Horizons of Evolutionary Robotics, 93–123 (2014)
8. Miglino, O., Ponticorvo, M., Bartolomeo, P.: Place cognition and active perception: a study with evolved robots. Connection Science 21(1), 3–14 (2009)
9. Nieder, A., Miller, E.K.: Coding of cognitive magnitude: compressed scaling of numerical information in the primate prefrontal cortex. Neuron 37(1), 149–157 (2003)
10. Ponticorvo, M., Walker, R., Miglino, O.: Evolutionary Robotics as a tool to investigate spatial cognition in artificial and natural systems. Artificial Cognition Systems, 210–237 (2007)
11. Rossetti, Y., Jacquin-Courtois, S., Aiello, M., Ishihara, M., Brozzoli, C., Doricchi, F.: Neglect around the Clock: Dissociating number and spatial neglect in right brain damage. In: Dehaene, S., Brannon, E.M. (eds.) Space, Time and Number in the Brain: Searching for the Foundations of Mathematical Thought, pp. 149–173. Elsevier, Amsterdam (2011)
12. Rotondaro, F., Gazzellini, S., Peris, M., Doricchi, F.: The Mental Number Line in children: number interval bisection and number to position performance. Poster presented at the European Workshop Cognitive Neuroscience 2012, Bressanone, Italy (2012)
13. Urbanski, M., Angeli, V., Bourlon, C., Cristinzio, C., Ponticorvo, M., Rastelli, F., Bartolomeo, P.: Negligence spatiale unilaterale: une consequence dramatique mais souvent negligee des lesions de lhemisphere droit. Revue Neurologique 163(3), 305–322 (2007)
14. Zorzi, M., Priftis, K., Umilt, C.: Brain damage: neglect disrupts the mental number line. Nature 417(6885), 138–139 (2002)

A Volumetric Radial LBP Projection of MRI Brain Images for the Diagnosis of Alzheimer's Disease

F.J. Martinez-Murcia[1], Andrés Ortiz[2], J. Manuel Górriz[1], Javier Ramírez[1], and I.A. Illán[1] and the Alzheimer's Disease Neuroimaging Initiative

[1] Department of Signal Theory, Networking and Communications,
Universidad of Granada, Granada, Spain
gorriz@ugr.es
[2] Department of Communications Engineering, University of Málaga, Málaga, Spain

Abstract. Alzheimer's Disease (AD) is nowadays the most common type of dementia, with more than 35.6 million people affected, and 7.7 million new cases every year. Magnetic Resonance Imaging (MRI) is a fairly widespread tool used in clinical practice, and has repeatedly proven its utility in the diagnosis of AD. Therefore a number of automatic methods have been developed for the processing of MR images. In this work, a new algorithm that projects the three-dimensional image onto two-dimensional maps using Local Binary Patterns (LBP) is presented. The algorithm yields visually-assessable maps that contain the textural information and achieves up to a 90.5% accuracy in a differential diagnosis task (AD vs controls), which proves that the textural information retrieved by our methodology is significantly linked to the disease.

Keywords: LBP · SVM · MRI · Alzheimer's Disease · Projection

1 Introduction

Neurodegenerative diseases have attracted considerable research attention, specially in developed countries where the ageing population is a major concern. According to the World Health Organization, the most common type of dementia is Alzheimer's Disease (AD), with more than 35.6 million people affected, and 7.7 million new cases every year.

Currently, new imaging techniques such as Magnetic Resonance Imaging (MRI) or Single Photon Emission Computed Tomography (SPECT) are being

Data used in preparation of this article were obtained from the Alzheimers Disease Neuroimaging Initiative (ADNI) database (adni.loni.ucla.edu). As such, the investigators within the ADNI contributed to the design and implementation of ADNI and/or provided data but did not participate in analysis or writing of this report. A complete listing of ADNI investigators can be found at: http://adni.loni.ucla.edu/wpcontent/uploads/how_to_apply/ADNI_Acknowledgement_List.pdf.

© Springer International Publishing Switzerland 2015
J.M. Ferrández Vicente et al. (Eds.): IWINAC 2015, Part I, LNCS 9107, pp. 19–28, 2015.
DOI: 10.1007/978-3-319-18914-7_3

intensively used in the diagnosis. These techniques allow not only the manual processing of the images to obtain helpful data related to the neurodegeneration that occurs, but the prediction of the conversion from prodromal stages (Mild Cognitive Impairment, or MCI) to AD [1].

Given its availability and non-invasive nature, MRI is often recommended as a first diagnosis tool, and therefore, a huge variety of methods have been developed for the medical imaging processing in what is known as the Computer Aided Diagnosis (CAD) paradigm. Since MRI provides high tissue contrast, the most common brain image analysis methods focus on the intensity information: analysis of Regions of Interest (ROIs) [2], Voxel-Based Morphometry (VBM) [3] or Cortical Thickness [4], among others. However, there is an increasing interest in multivariate approaches that are able to handle regional patterns, texture features and voxel and region-wise relationships [5,6,7,8]. These approaches could reveal other information than the volumetric, complementing and providing new insights into the disease.

Recently, some CAD systems based on texture features have been proposed [5,9,10]. Particularly, in [9], an easily computed texture descriptor, Local Binary Pattern (LBP) has demonstrated its utility in various high-level brain MR image analysis. At the same time, a new algorithm was proposed in [11], that allow the projection of three-dimensional MR images to a two-dimensional map of some radial texture features including average, entropy or kurtosis, with promising results in the diagnosis of AD vs normal individuals.

In this work, a new projection system based on a combination of the two aforementioned methods is proposed, where, instead of using direct texture features from the projection vector, a LBP is computed around it, to avoid the spatial information loss in the vicinity of the vector.

The article is organised as follows. First, in Section 2 the methodology is presented, and the usage of LBP, volumetric LBP (VLBP) and projection algorithms and in a projection environment is explained. Later, in Section 3, the database and the results are presented and analysed. Finally, some conclusions are drawn in Section 4.

2   Methodology

2.1   Local Binary Patterns

Local Binary Patterns (LBP) were first introduced in [12] to describe the texture of an image with application to face recognition. In its first version, LBP generates a value that describe the local texture, using 3x3 pixel blocks. Roughly, LBP operator computes binary values by comparing the grey values in a neighbourhood of radius 1 to the central pixel value. Then, 8 binary values are sampled counterclockwise in each block, generating a binary histogram and finally, the LBP operator transforms the binary histogram in a unique descriptor by converting it to a decimal value. In contrast to the first version of LBP which only considers the use of 8 sampling points in each 3x3 pixels block, [13] generalizes the LBP operator to any radius and any number of sampling sampling points.

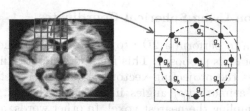

Fig. 1. Computation of the generalized $LBP_{8,2}$ descriptor from a 5x5 image block

Figure 1 shows the sampling procedure on a 5x5 block using a 8-pixel neighbourhood and 2 pixel radius ($LBP_{(8,2)}$). This process can be formally described as follows. Let I be an image and $g_0 = I(x, y)$ the grey level of a pixel located at position (x, y) in the image. P samples can be extracted going counterclockwise in the r-neighbourhood around p_0 taking the pixel values at positions:

$$x_{p,r} = x + r\cos(2\pi p/P)$$
$$y_{p,r} = y + r\sin(2\pi p/P) \tag{1}$$

with $p = \{0, ..., P - 1\}$.

Thus, the 2D-LBP operator can be computed as:

$$LBP_{P,r} = \sum_{p=1}^{P} s(g_{p,r} - g_0)2^{p-1} \tag{2}$$

where $g_{p,r}$ is the p-th sampling point at a distance r from the central pixel, and s is the sign function defined as:

$$s(x) = \begin{cases} 1 & x \geq 0 \\ 0 & x < 0 \end{cases} \tag{3}$$

where $g_{p,r}$ is the grey level sampled at position $(x_{p,r}, y_{p,r})$ and g_0 is the central pixel in the block. Taking $g_{p,r} - g_0$ differences aims to obtain grey scale invariance, as signed differences are not affected by changes in mean luminance [13].

However, the described LBP descriptor is defined for 2D images. Extensions of LBP to volumes have been proposed to compute image texture in the spatiotemporal domain, specifically developed to extract textural features from video sequences [14,15], namely Volume Local Binary Patterns (VLBP). Other proposals such as [16] use a different approach directed to characterize 3D textures, substituting the 2D neighbourhoods uses in classical LBP to spherical ones. Moreover in [17], the authors propose a different approach to compute LBP in 3D images, by sampling the 6 nearest voxels around the central one. In this case, the developed method is used to obtain 3D textural patterns in PET images for AD diagnosis. The method proposed in this work uses a different sampling algorithm and computes a unique descriptor for a group of layers across the image.

2.2 2D Projection Using Spherical Coordinates

In [11], an algorithm to perform a 2D projection based on the use of spherical coordinates in the brain is proposed. This method computes different statistical measurements across the projecting vector $\mathbf{p}_{\theta,\varphi}$, which is defined in a pair of inclination (θ) and azimuth (φ) angles in the range $[-\pi/2, \pi/2]$ and $[-\pi, \pi]$ respectively, and sampling the nearest voxel. In other words, a 2D projection is obtained by computing an unique value R from the projecting vector $\mathbf{p}_{\theta,\varphi}$ for each (θ, φ) pair of coordinates, what yields a two-dimensional map in the $\theta - \varphi$ plane. As an example, 2D projection of different statistics is shown in Figure 2.

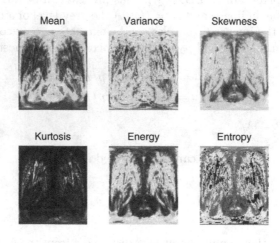

Fig. 2. Example of 2D projections of different statistics using the method proposed in [11].

This smart representation of the brain is able to provide insight into different and useful properties of the anatomical structures such as brain folds or the cortical thickness in a 2D representation. However, the mapping process of reducing from 2D to 3D losses textural information in the vicinity of the vector $\mathbf{p}_{\theta,\varphi}$ that may be important from a discriminative point of view. The method proposed in this work preserves part of the 3D textural information by means of the $VLBP^{\theta,\varphi}$ descriptor which takes into account not only the voxels in the radius of a specific direction (θ, φ) but also the voxels in a predefined neighbourhood.

2.3 Projecting 3D LBP Features

Although the discriminative properties of the maps computed by projecting statistics across a $\mathbf{p}_{\theta,\varphi}$ vectors has been proved in [11], these statistic do not preserve per layer textural information, as only voxels that are crossed by $\mathbf{p}_{\theta,\varphi}$

are considered. In this work, a 3D LBP descriptor is defined by computing the LBP in the neighbourhood of different layers across the $\mathbf{p}_{\theta,\varphi}$ vector. This way, each $\mathbf{p}_{\theta,\varphi}$ vector is the axis of a cylinder oriented in the direction indicated by (θ, φ) and whose radius define the neighbourhood used to compute the 3D-LBP descriptor. Figure 3 shows the sampling method devised in this work, which is based on the VLBP descriptors developed for characterization of dynamic textures in video sequences by using three consecutive frames. In this case, voxels in each layer across the radius $\mathbf{p}_{\theta,\varphi}$ of the sphere that contains the brain are sampled following helical coordinates as shown in Figure 3.

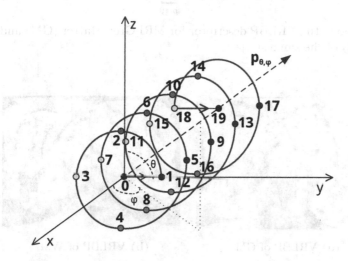

Fig. 3. Sampling voxels in VLBP model. Note that in this case the vector $\mathbf{p}_{\theta,\varphi}$ is placed along the x axis, which corresponds to $\mathbf{p}_{0,0}$ in our coordinate system.

This method allows to extend the LBP to 3D and eventually defining a 3D texture in a local neighbourhood by means of a texture sequence in a similar way that [15] defined the basic Volume LBP (VLBP).

Formally, the sequence of image voxels in the 3D neighbourhood taken by helical sampling as depicted in Figure 3 can be expressed as:

$$V = v_{P,r}^{\theta,\varphi}(g_0, g_{1,r}, g_{2,r}, \cdots g_{P-1,r}) \tag{4}$$

where $g_{p,r}$ indicates the voxel sampled in p-th place at a sampling radius r. The coordinates of $g_{p,r}$, for a projection vector $\mathbf{p}_{\theta,\varphi}$ orientated in the (θ, φ) direction, are given by:

$$g_{p,r} = \begin{cases} x_{p,r} = p\sin(\varphi)\cos(\theta) - r\sin(2\pi n \frac{p}{P}) \\ y_{p,r} = p\sin(\varphi)\sin(\theta) + r\cos(2\pi n \frac{p}{P}) \qquad p = \{0, ..., P-1\}, P \in \mathbb{N} \\ z_{p,r} = p\cos(\varphi) \end{cases} \tag{5}$$

where r is the neighbourhood radius, P is the total number of sampling points and n is the number of layers ($n = 5$ in the example of Fig. 3).

As proposed in [15], the voxels that do not fall exactly at coordinates computed by Equations 5 are estimated by interpolation of closest points. It is interesting to note that first and end sampling points correspond to points in the $\mathbf{p}_{\theta,\varphi}$ vector. Specifically, the first and final sampling point would be $(0, 0, 0)$ and $(P\sin(\varphi)\cos(\theta), P\sin(\varphi)\sin(\theta), P\cos(\varphi))$ respectively. Hence, our VLBP, namely Volumetric Radial LBP (VRLBP), defined across each $\mathbf{p}_{\theta,\varphi}$ vector can be computed using the following expression:

$$VRLBP_{P,r}^{\theta,\varphi} = \sum_{p=0}^{P} v_{p,r}^{\theta,\varphi} 2^p \tag{6}$$

Figure 4 shows the VRLBP descriptor for MRI Grey Matter (GM) and White Matter (WM) of the same subject.

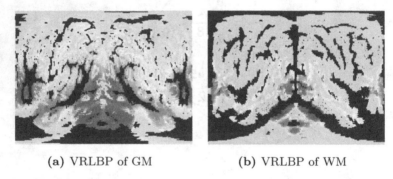

(a) VRLBP of GM   (b) VRLBP of WM

Fig. 4. Volumetric Radial LBP computed for GM (a) and WM (b) MRI images of the same subject

3   Experimental Results

3.1   Database

Data used in the preparation of this article were obtained from the Alzheimer's Disease Neuroimaging Initiative (ADNI) database (adni.loni.usc.edu). The ADNI was launched in 2003 by the National Institute on Aging (NIA), the National Institute of Biomedical Imaging and Bioengineering (NIBIB), the Food and Drug Administration (FDA), private pharmaceutical companies and non-profit organizations, as a $60 million, 5-year public-private partnership. The primary goal of ADNI has been to test whether serial magnetic resonance imaging (MRI), positron emission tomography (PET), other biological markers, and clinical and neuropsychological assessment can be combined to measure the progression of MCI and early Alzheimers Disease. Determination of sensitive and specific markers of very early AD progression is intended to aid researchers and clinicians to develop new treatments and monitor their effectiveness, as well as less en the time and cost of clinical trials.

The Principal Investigator of this initiative is Michael W. Weiner, MD, VA Medical Center and University of California-San Francisco. ADNI is the result of efforts of many co-investigators from a broad range of academic institutions and private corporations, and up to 1500 adults (ages 55 to 90) have been recruited from over 50 sites across the U.S. and Canada in ADNI and its following initiatives ADNI-GO and ADNI-2.

In this article, a subset of the ADNI 1075-T1 database (subjects who have a screening data) has been used. The database contains 1075 T1-weighted MRI images, containing a total amount of 229 normal controls (NOR), 401 MCI (312 stable MCI and 86 progressive MCI) and 188 AD images (see www.adni-info.org). Our methodology has been tested against a subset of 360 randomnly selected patients (180 AD and 180 NOR). The MR images have been spatially normalized and then segmented in Grey Matter (GM) and White Matter (WM) tissues using SPM software [18], after a skull removing procedure.

3.2   Classification Experiments

In this section, VRLBP projections computed for all the GM and WM MRI images are used as features to classify between controls and AD patients. To this end, VRLBP features are first selected by the t-statistic criteria, computed as

$$t = \frac{|I_{CN}^{\mu} - I_{AD}^{\mu}|}{\sqrt{\frac{I_{CN}^{\sigma^2}}{N_{CN}} + \frac{I_{AD}^{\sigma^2}}{N_{CN}}}} \tag{7}$$

where I_{CN}^{μ} and I_{AD}^{μ} are the mean images for CN and AD respectively, I_{CN}^{σ} and I_{AD}^{σ} are the variance images for CN and AD, respectively and N_{CN}, N_{AD} are the number of CN and AD images, respectively. Then, a linear SVC [19] is used

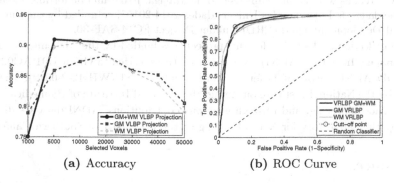

(a) Accuracy   (b) ROC Curve

Fig. 5. Performance results of our proposed VRLBP methodology. 5a displays the accuracy of the system when using GM, WM and both combined, and 5b shows the ROC curve for the same cases.

to classify the samples. the results have been assessed by k-*fold* cross-validation with k=10. Figure 5a shows the classification accuracy obtained for different number of selected voxels by means of varying the threshold in the t-statistic.

This figure states the discriminative capabilities of the VRLBP projections computed for WM and GM, obtaining an accuracy of 88% and 86.5% respectively. Moreover, accuracy up to 90.5% is obtained when fusing the projection of VRLBP descriptors for WM and GM. In the same way, ROC curves are depicted in Figure 5b, showing AUC values of 0.930, 0.940 and 0.945 for GM, WM and combined GM+WM VRLBP projections.

4   Conclusions and Future Works

In this paper, a volumetric descriptor based on LBP has been proposed to retain part of the textural information loosed during the 2D projection process. This way, classification experiments have been performed to show the discriminative power of that descriptor using the ADNI 1075-T1 database, and assessed by k-fold cross-validation. These experiments shown that 1) the proposed descriptor is discriminative enough to be used with the projections, 2) it improves the results obtained by classical first order statistics such as average, variance, kurtosis, etc., and 3) fusing GM and WM projections increases the discriminative power compared to the use of single WM or GM projections, providing an accuracy up to 90% for CN/AD classification using a linear SVC classifier.

As future work, we plan to extend the VRLBP to MCI patients to realize its capabilities for early AD diagnosis. Moreover, also plan to use different methods to fuse WM and GM information as in this work that fusion has been implemented by simply concatenating WM and GM descriptors considering them as different feature dimensions.

Acknowledgments. This work was partly supported by the MICINN under the TEC2012-34306 project and the Consejería de Economía, Innovación, Ciencia y Empleo (Junta de Andalucía, Spain) under the Excellence Projects P09-TIC-4530 and P11-TIC-7103, as well as the "Universidad de Málaga. Programa de fortalecimiento de las capacidades de I+D+I en las Universidades 2014-2015", cofunded by the European Regional Development Fund (ERDF) under Project FC14-SAF-30.

Data collection and sharing for this project was funded by the Alzheimer's Disease Neuroimaging Initiative (ADNI) (National Institutes of Health Grant U01 AG024904) and DOD ADNI (Dept. of Defense award number W81XWH-12-2-0012). ADNI is funded by the National Institute on Aging, the National Institute of Biomedical Imaging and Bioengineering, and through generous contributions. ADNI data are disseminated by the Laboratory for Neuro Imaging at the University of Southern California.

References

1. Schroeter, M.L., Stein, T., Maslowski, N., Neumann, J.: Neural correlates of alzheimers disease and mild cognitive impairment: A systematic and quantitative meta-analysis involving 1351 patients. NeuroImage 47(4), 1196–1206 (2009)

2. Ayache, N.: Analyzing 3D Images of the Brain. NeuroImage 4(3), S34–S35 (1996)
3. Shiino, A., Watanabe, T., Maeda, K., Kotani, E., Akiguchi, I., Matsuda, M.: Four subgroups of Alzheimer's disease based on patterns of atrophy using VBM and a unique pattern for early onset disease. NeuroImage 33(1), 17–26 (2006)
4. Han, X., Jovicich, J., Salat, D., van der Kouwe, A., Quinn, B., Czanner, S., Busa, E., Pacheco, J., Albert, M., Killiany, R., et al.: Reliability of mri-derived measurements of human cerebral cortical thickness: the effects of field strength, scanner upgrade and manufacturer. Neuroimage 32(1), 180–194 (2006)
5. Kovalev, V.A., Kruggel, F., Gertz, H.J., von Cramon, D.Y.: Three-dimensional texture analysis of mri brain datasets. IEEE Transactions on Medical Imaging 20(5), 424–433 (2001)
6. Fan, Y., Rao, H., Hurt, H., Giannetta, J., Korczykowski, M., Shera, D., Avants, B.B., Gee, J.C., Wang, J., Shen, D.: Multivariate examination of brain abnormality using both structural and functional MRI. NeuroImage 36(4), 1189–1199 (2007)
7. Ortiz, A., Górriz, J.M., Ramírez, J., Martínez-Murcia, F.: Lvq-SVM based CAD tool applied to structural MRI for the diagnosis of the alzheimer's disease. Pattern Recognition Letters 34(14), 1725–1733 (2013)
8. Yoon, U., Lee, J.M., Im, K., Shin, Y.W., Cho, B.H., Kim, I.Y., Kwon, J.S., Kim, S.I.: Pattern classification using principal components of cortical thickness and its discriminative pattern in schizophrenia. NeuroImage 34(4), 1405–1415 (2007)
9. Unay, D., Ekin, A., Cetin, M., Jasinschi, R., Ercil, A.: Robustness of local binary patterns in brain mr image analysis. In: 2007 29th Annual International Conference of the IEEE Engineering in Medicine and Biology Society (August 2007)
10. Martinez-Murcia, F., Górriz, J., Ramírez, J., Moreno-Caballero, M., Gómez-Río, M., Initiative, P.P.M.: Parametrization of textural patterns in 123i-ioflupane imaging for the automatic detection of parkinsonism. Medical Physics 41(1), 012502 (2014)
11. Martínez-Murcia, F.J., Górriz, J.M., Ramírez, J., Alvarez Illán, I., Salas-González, D., Segovia, F.: Alzheimer's Disease Neuroimaging Initiative. Projecting mri brain images for the detection of alzheimer's disease. Stud. Health Technol. Inform. 207, 225–233 (2015)
12. Ojala, T., Pietikäinen, M., Harwood, D.: A comparative study of texture measures with classification based on featured distributions. Pattern Recognition 29(1), 51–59 (1996)
13. Ojala, T., Pietikäinen, M., Mäenpää, T.: Multiresolution gray-scale and rotation invariant texture classification with local binary patterns. IEEE Trans. Pattern Anal. Mach. Intell. 24(7), 971–987 (2002)
14. Chetverikov, D., Peteri, R.: A brief survey of dynamic texture description and recognition. In: Proc. Intl. Conf. Computer Recognition Systems, pp. 17–26. Springer (2005)
15. Zhao, G., Pietikainen, M.: Dynamic texture recognition using local binary patterns with an application to facial expressions. IEEE Trans. Pattern Anal. Mach. Intell. 29(6), 915–928 (2007)
16. Paulhac, L., Makris, P., Ramel, J.-Y.: Comparison between 2d and 3d local binary pattern methods for characterisation of three-dimensional textures. In: Campilho, A., Kamel, M.S. (eds.) ICIAR 2008. LNCS, vol. 5112, pp. 670–679. Springer, Heidelberg (2008)

17. Montagne, C., Kodewitz, A., Vigneron, V., Giraud, V., Lelandais, S.: 3D Local Binary Pattern for PET image classification by SVM, Application to early Alzheimer disease diagnosis. In: 6th International Conference on Bio-Inspired Systems and Signal Processing (BIOSIGNALS 2013), Barcelona, Spain, pp. 145–150 (February 2013)

18. Friston, K., Ashburner, J., Kiebel, S., Nichols, T., Penny, W.: Statistical Parametric Mapping: The Analysis of Functional Brain Images. Academic Press (2007)

19. Vapnik, V.N.: Statistical Learning Theory. John Wiley and Sons, Inc., New York (1998)

Telemetry System for Cochlear Implant Using ASK Modulation and FPGA

Ernesto A. Martínez–Rams[1], Vicente Garcerán–Hernández[2(✉)],
and Duarte Juan Sánchez[3]

[1] Universidad de Oriente, Avenida de la América s/n, Santiago de Cuba, Cuba
eamr@fie.uo.edu.cu
[2] Universidad Politécnica de Cartagena, Antiguo Cuartel de Antiguones
(Campus de la Muralla), 30202, Cartagena, Murcia, España
vicente.garceran@upct.es
[3] Instituto Superior de Ciencias Médicas, Avenida de la América s/n,
Santiago de Cuba, Cuba
duarte.juan@sierra.scu.sld.cu

Abstract. This paper presents the design, development, simulation and
test of a directional telemetry system for cochlear implants using FPGA.
We used Manchester codification and ASK modulation in order to achieve
a high transmission speed. The design was simulated using the System
Generator for FPGA by Xilinx and Simulink developed by Mathworks.
Also, the design was emulated using the ISE design software by Xilinx.
The design has been tested under noisy environment. The design was
optimised so as to obtain a power consumption equal or less than the
maximum allowed in the receiver. We achieved the use fewer components of the FPGA. As a result, the telemetry system has been designed
to meet with specifications for use it in the development of a prototype
of cochlear implant for research purposes.

1   Introduction to the Telemetry System

Cochlear implants are high-technology electronic devices. They transform acoustic signals into electrical stimuli of the auditory nerve. Signals are processed
through different elements that composed of the cochlear implant. Cochlear implants have basically two units: the external and internal units. The external
unit is formed by a sound processor and by a data and energy transmitter. The
internal unit is surgically inserted in the internal ear, this is formed by a data
and energy receiver, an internal processor and an electrode array that stimulate
the auditory nerve [1].

The external unit encodes/codifies and processes the input voice signal and
converts it into digital data. These data are send to a transmitter of the telemetry
system in order to be codified, modulated and transmitted to the internal unit by
means of an inductive link. A receiver of the telemetry system gets the modulated
signal. This signal is demodulated and decodified in order to obtain the original
data. These recovery data carry information about the stimulation level as well as

© Springer International Publishing Switzerland 2015
J.M. Ferrández Vicente et al. (Eds.): IWINAC 2015, Part I, LNCS 9107, pp. 29–38, 2015.
DOI: 10.1007/978-3-319-18914-7_4

of the electrode to stimulate. The electrode array stimulates electrically different zones of the cochlea, and excites the auditory nerve that transports the neural impulses to the acoustic area of the brain. Finally, the brain does the most complex cognitive function like speech perception.

A telemetry system can be conceptualised through a model of abstract layers, Figure 1. It has a physical layer, a communication layer and an application layer in both processors external and internal. These layers are connected by direct and virtual form. The virtual connection is established between the layers at the same level, it requires a direct way through the lower level in order to establish the connection between the external and the internal unit. The application

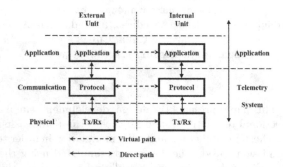

Fig. 1. A telemetry system for biomedical applications

layer corresponds to the control of the whole functionality of the system. This layer includes the control of both units, external and internal. These layers are virtually and physically interconnected by a communication system of the lower layers. The communication layer is formed by a set of elements that transport the information from one point to another of the system. In the same way as the communication layer, the physical layer permits bidirectional communication between the external and internal units. In this layer are included all the components needed to transmit and to receive the information, either through radio, light, inductive link or galvanic connection. Physical layer is the lowest layer of the virtual connection and at the same time, it the units connects directly. The Figure 1 shows a telemetry system for biomedical applications, it is defined by two layers, the communication layer and the physical layer.

The physical layer connects directly the external unit (transmitter) with the internal unit (receiver) of the cochlear implant. This layer is defined by inductive links due to the need to transfer energy and information. The telemetry system is based in a weakly-coupled inductive link. Resonant circuits are used in order to increase the coupling between both units.

In order to provide to the external unit the highest amount of energy from the power supply, the primary coil must reflect a low impedance to the source, i.e. a performance equal to a pure resistance. This condition is reached using a series resonant circuit in the transmitter, tuned in the working frequency. In the secondary, or receiver, there are two possibilities. In the first, the charge is in parallel with the resonant circuit. On the other hand, the charge is in series.

Modulations used frequently in the telemetry system have to have few complexes, because a highly complex modulation consumes more energy. These conditions are very difficult to obtain in the receiver due to the low quantity of energy it can be transferred for fulfill the requirements of the IEEE [2].

ASK modulation is used in the telemetry system. In this modulation, the variations of the amplitude of the carrier represent the values of the digital data. Modulation and demodulation are process that carry out in a relatively easy form. This characteristic is used in the receiver due to the need to design a small circuit and low complexity.

Amplitude Shift Keying (ASK) corresponds to a Double-Sideband with Suppressed Carrier (DSSC or DSB-SC). This type of modulation can be managed using a balanced modulator if the modulating signal is bipolar or using an amplitude modulator, with a modulation rate equal to 100 %, if the modulating signal is unipolar.

With the present technology, the design of a digital modulation by means Field Programmable Gate Arrays (FPGA) can be an easy task. The FPGA are programmed via Matlab/Simulink and Xilinx System Generator (XSG). XSG is a high-level tool developed by Xilinx for designing high-performance Digital Signal Processing (DSP) in systems targeting FPGAs. It produces the same code if we had used a hardware description language (HDL) [4,5].

XSG permits to model the telemetry system into a specific hardware platform. This potentiality is due to its flexibility, robustness and facility to employ high performance DSPs. The design developed with this tool are composed for a wide diversity of elements: XSG specific blocks, HDL code and functions produced by Matlab. Therefore, all of these elements can be used concurrently and they are simulated and synthesised in order to obtain signals processing under FPGA.

This paper has the goal of designing a telemetry system for cochlear implants using a FPGA by Xilinx.

2 Design and Simulation of the Telemetry System

The following parameters have been taken into account in the design of the telemetry system: the number of channels of stimulation 18; the resolution equal to 12 bits/symbol; the control bits 6; the data frame 222 bits; the maximum frequency of stimulation per channel 2000 pps; the maximum transfer rate 24 kbps per channel; the maximum transfer rate of data to the implant 432 kbps; the maximum transfer rate 444 kbps.

With all these considerations, the following technical specifications have been chosen: the signalling rate 444 kbps; Manchester coding; ASK modulation; the carrier frequency 5 MHz; the attenuation of the transmission channel 25 dB; power consumption of the receiver less than 20 mW; maximum power of the cochlear stimulator was 72.5 mW (18.60 dBm); maximum power of the receiver 100 mW (20 dBm).

Given the above considerations and [3,6,7,8], we suggest the telemetry system shown in the Figure 2. Figure 2 a) shows the transmitter block diagram. It is formed by a Manchester encoder and an ASK modulator. Figure 2 b) shows the block diagram of the receiver. It is formed by an ASK demodulator and a Manchester decoder.

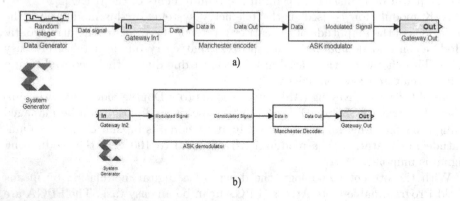

Fig. 2. Block diagram of a) the transmitter, b) the receiver

2.1   Design and Simulation of the Telemetry System Using System Generator

Manchester Encoder/Decoder. Manchester encoder is a line code very used for the codification of bit frames that provides applications of a simple way. This codifier avoids loss of synchronicity in the transmission with inductive links and even errors in the bit frames. These losses of synchronicity are due to very long bit frames without level transitions [9]. Figure 3 shows the XSG block diagram of the Manchester encoder. In this, data are put by means of an input port. The data signal is obtained using a random number generator with a period of $2\,\mu s$, equivalent to a signalling rate of 500 kbps which is a value greater than the 444 kpbs specified. The clock signal of the encoder is obtained from a counter block. This block generates a symmetric square waveform and period equal to $1\,\mu s$, which is half of the data period. Both signals data and clock are connected to a XNOR block, then to a NOT block. Figure 4 shows the Manchester decoder. The input data flow through two paths. In the upper branch, each sample is subtracted mathematically to the previous sample. This approach would obtain impulses in all the transitions of the encoded signal. Afterwards, these impulses are rectified in order to obtain positive values of the samples. On the other hand, in the lower path, the data are delayed with a value equal to half of the number of data bits, then they are subtracted and rectified; hence we obtain positive impulses in the initial instants of each bit without transition in the original data signal. In order to get the clock signal, first the data of each path

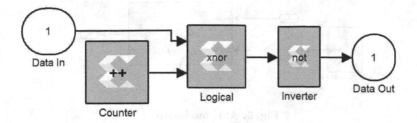

Fig. 3. Manchester Encoder

are logically added by the block OR, so impulses in the start of each bit are generate. With the XOR block, we obtain small impulses in the moments when the signal change its state. The CAST block converts the signal to Boolean values before of the COUNTER block. Finally, the COUNTER block generates the data signal encoded in Manchester format.

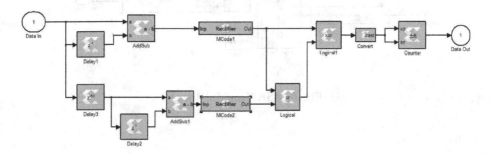

Fig. 4. Manchester Decoder

Modulator/Demodulator. ASK modulation is obtained from a balanced modulator with a unipolar modulating signal, Figure 5. The modulating signal is the data output of the Manchester encoder. The frequency of the carrier signal is 5 MHz and the sampling interval is equal to 20 ns. This carrier signal is generated using a frequency synthesizer (DDS compiler 4.0). The carrier signal is fitted to the suitable levels of the modulator through the CMult block. Both signal, modulating and carrier are inserted to the ASK modulator (Mcode). The ASK demodulator is designed using an enveloping detection noncoherent, Figure 6. The noncoherent detection just picks up data without the necessity of recover the carrier signal in the receiver. This type of detection can be carried out in the XSG in an easy approach and with a very low hardware resources of the FPGA.

The received modulated signal is full-wave rectified. The consequent signal is processed through a finite impulse response (FIR) low-pass filter (FIR Compiler

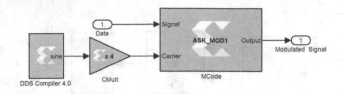

Fig. 5. ASK modulator

5.0). The Mcode block works as a decision threshold and the output signal is sent to the Mancherter decoder. We would highlight that we first had chosen a 79-order filter, a cutoff frequency of 0.5 MHz and a minimum of attenuation equal to 60 dB at 1.0 MHz. Afterwards, we searched the minimum filter order for a minimum power consumption in the receiver and we took into account the integrity of data even in a noisy environment.

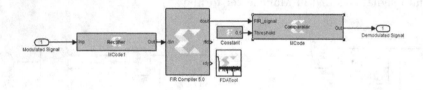

Fig. 6. ASK demodulator

2.2   Emulation Using the HDL Coder and Xilinx ISE Design Suite

We have done the emulation of the design using System Generator. For this process, first it is necessary to choose the FPGA more suitable to implement the design. System Generator automatically synthesizes the design on the selected FPGA, and HDL Coder generates the VHDL code from Simulink models. With EDA ISE tool, the internal interconnection of the FPGA hardware is made. This tool reports on the percentage of resources used. If the quantity of resources are greater than the FPGA has got, we must change the design and/or choose another FPGA. Power Analyzer tool works out the overall power consumption of the design. If the overall power consumption is greater than that expected, we must change the design until to accomplish that target. Our implementation was made on a Xilinx Virtex-6 xc6vsx315t-3ff1156 FPGA.

Transmitter and Receiver Emulation. We took into account the Mancheter encoder, the ASK modulator and PWM generator for the transmitter emulation. In addition, the ASK demodulator and Manchester decoder for the receiver emulation. In both cases, all of those elements were implemented on the same FPGA.

3   Results

3.1   Verification of the Telemetry System Design Using System Generator

Figures 7 a), b) and c) show signals that are involved in the encoding process. Figure 7 a) shows the data signal to encode. Figure 7 b) shows the clock signal that is necessary in the Manchester encoding process. Figure c) shows the data signal encoded. We can notice that each data encode with a high level, Figure 7 a), it corresponds with a unipolar pulse of the type ON-OFF in the encoded data, Figure 7 c). In the same way, each data encode with a low level, and corresponds with a unipolar pulse of the type OFF-ON in the encoded data. Figures 7 d),

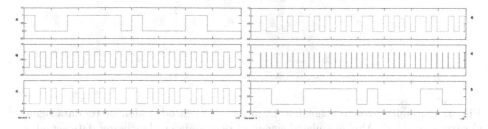

Fig. 7. Signals in the Manchester encoder: a) Data signal; b) Clock signal; c) Encoding signal. Signals in the Manchester decoder: d) Input signal; e) Clock pulses; f) Decoding signal.

e) and f) show signals that are involved in the Manchester decoding process. Figure d) depicts the input signal and Figure e) shows the clock pulses. Finally, in Figure f) it shows the decoded data. In the last figure, we can observe that pulses with an ON-OFF relation of the encoded signal correspond to a high level symbol in the decoded signal. However, pulses with an OFF-ON relation of the encoded signal correspond to a low level symbol in the decoded signal.

ASK Modulator/Demodulator. Figures 8 a), b) and c) show signals corresponding to the ASK modulator. Figure 8 a) depicts Manchester encoded data, which is the modulating signal. Figure 8 b) shows the carrier signal, 5 MHz. Figure 8 c) shows the modulated ASK signal. In this figure we can observe that sinusoidal signals correspond with high values of the modulating signal. Therefore, the implemented modulator works perfectly. Figures 8 d), e) and f) shows signals of the ASK demodulator. Figure 8 d) shows the modulated signal. Figure 8 e) depicts the filter output signal. Figure 8 f) shows the recovered modulating signal. In this figure we can observe that the filter output signal is the envelope of the modulating signal but delayed. Furthermore, a high level of the recovered modulating signal corresponds with the instant of a high level of the filter output signal.

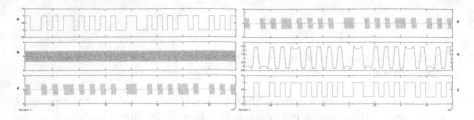

Fig. 8. ASK modulator (left) / demodulator (right): a) modulating signal; b) carrier signal; c) modulated ASK signal; d) modulated signal; e) filter ouput signal (envelope); f) recovered m

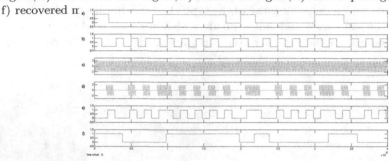

Fig. 9. Transmitter-receiver signal: a) Data signal; b) encoded Manchester data signal; c) carrier signal; d) ASK modulated signal; e) demodulater Manchester data signal; f) recovered data signal

Telemetry system (Transmitter/Receiver). In the Figure 9 it can be appreciated that corresponding signals to input and output of each block which builds the transmitter and the receiver (Figures 2 and 3, respectively). Figures 9 a) and b) show both data signal and encoded Manchester data signal. The encoded data signal is the modulating signal of the ASK modulator. Figure 9 c) shows the carrier signal, 5 MHz. Figure 9 d) shows the ASK modulated signal to the output of the transmitter (and input of receiver). Figure 9 e) shows the demodulated data signal which is the same as the encoded data signal depicted in the Figure 9 b). Finally, Figure 9 f) shows the recovered data signal which is the same as the original signal but it is delayed for the own demodulation and decodification process. Figure 10 shows the spectrum of the transmitted signal for the designed telemetry system. In this we can observe that the maximum power comes to the carrier frequency, 5 MHz. And its bandwidth is 1 MHz. We can also see a very low direct component, which is good in a transmission system with inductive link, therefore avoiding loss of data.

3.2   Emulation Using Xilinx ISE Design Tools, Results

Transmitter/Receiver. In the emulation of the transmitter, we took special interest in achieving a design that would get the default maximum dynamic power. In this case, it was possible that the maximum power consumption was the required 12 mW.

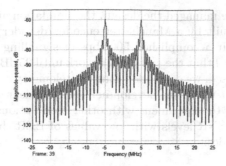

Fig. 10. Spectrum of the transmitted signal by the telemetry system

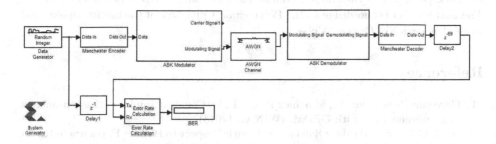

Fig. 11. Block diagram to analyse the reception in noyse environment

In the emulation of the receiver, results of the first emulation gave a power consumption higher than the required maximum. This high consumption is due to the high computational cost of the high level order of the filter, initially equal to 79. We proceeded to calculate the minimum order of the filter capable to keep a suitable Bit Error Rate (BER) and an adequately Signal to Noise Ratio (SNR). It was possible to reduce the power consumption using a 12-order filter, from 48 mW to 12 mW, the value of BER=10^{-5} and SNR=25 dB (or BER=0 and SNR=26 dB). For all the cases, the measurement of the BER was realized using a random sequence of data of 100 ms of duration that correspond to 5×10^4 bits. We can conclude that the SNR values are acceptable considering the relatively high power in the transcutaneous transmission across and inductive link and the proximity of both transmitter and receiver in a cochlear implant. Furthermore, it is possible to emphasize that energetic levels needed for the electronics meet the design parameters suggested in the section 2. Figure 11 shows the Simulink diagram used for this purpose.

4   Conclusion and Recommendations

With the present work it was possible to design a telemetry system for cochlear implants using a FPGA device, it complies with a low power consumption and low hardware cost. We managed the design, the simulation and the emulation of a telemetry system built by a Manchester encoder/decoder and an ASK modulator/demodulator. It accomplished 500 kbps of signalling rate and 1 MHz of bandwidth. We checked that the system can work until 25 dB of SNR. For everything previously exposed, we concluded that the telemetry system complies with the request need for its use in the development of a cochlear implant for research aims. We recommend to experiment with other modulations less complex that meet equal or better SNR levels with low values of BER that maintaining the characteristics of the design.

Acknowledgements. This research has been funded by the project 9786 Procesamiento de Seales Biomdicas, de la Universidad de Oriente. To the finished Thematic Network Computacin Natural y Artificial para la Salud (CANS) of CYTED, Spain. The authors also acknowledge to the Polytechnic University of Cartagena, Spain, and to the University of Oriente, Cuba.

References

1. Garcerán Hernández, V., Martíncz Rams, E.A.: Cochlear Implant: Transcutaneous Transmission Link with OFDM. IWINAC (2013)
2. IEEE C95.1. Standard for Safety Levels with Respect to Human Exposure to Radio Frequency Electromagnetic Fields, 3 kHz to 300 GHz (2005)
3. Jordi, S.R.: Sistema implantable para la estimulación y registro de nervio periférico. Universitat Autónoma de Barcelona. Departament d'Enginyeria Electrónica, Barcelona (2006)
4. Xilinx Inc. System Generator for DSP Reference Guide. EEUU (2011)
5. Xilinx Inc. Xilinx System Generator v2.1 for Simulink Basic Tutorial. EEUU (2013)
6. Ghovanloo, M., Atluri, S.: A Wide-Band Power-Efficient Inductive Wireless Link for Implantable Microelectronic Devices Using Multiple Carriers. IEEE Transactions on Circuits and Systems 54(10), 2211–2221 (2007)
7. Elamare, G.A.: Investigation of High Bandwidth Biodevices for Transcutaneous Wireless Telemetry. Newcastle University (2010)
8. Svensson, A.: Design of inductive coupling for powering and communication of implantable medical devices. Royal Institute of Technology, Stockholm (2012)
9. Alexandru, N.D.: Improved encoder circuit for inverse differential Manchester code. In: 8th International Conference on Development Application System, pp. 181–183 (2006)

MBMEDA: An Application of Estimation of Distribution Algorithms to the Problem of Finding Biological Motifs

Carlos I. Jordán[✉] and Carlos. J. Jordán

Facultad de Ingeniería en Electricidad y Computación,
Escuela Superior Politécnica del Litoral (ESPOL), Guayaquil, Ecuador
cjordan@espol.edu.ec

Abstract. In this work we examine the problem of finding biological motifs in DNA databases. The problem was solved by applying MBMEDA, which is a evolutionary method based on the Estimation of Distribution Algorithm (EDA). Though it assumes statistical independence between the main variables of the problem, results were quite satisfactory when compared with those obtained by other methods; in some cases even better. Its performance was measured by using two metrics: precision and recall, both taken from the field of information retrieval. The comparison involved searching a motif on two types of DNA datasets: synthetic and real. On a set a five real databases the average values of precision and recall were 0.866 and 0.798, respectively.

Keywords: DNA dataset · Estimation of distribution algorithms · Molecular biology · Transcription factor · Motifs

1 Introduction

The search for biological motifs is an important problem in molecular biology. A motif or transcription factor binding site (TFBS) is the sequence of nucleotides in the promoting zone of a gene, where a transcription factor (TF) binds and controls the process of transcription of that gene into an mRNA molecule [1]. This molecule eventually will be translated into a protein at a cells ribosome; all this happens according to the central dogma of molecular biology.

Basically the problem can be formulated as follows: given a DNA base consisting of n promoting zones of size m, with one TFBS per sequence, find a pattern of length l that constitutes a motif. No doubt this problem is rather difficult, because we dont know a priori the length of the motif or its location in the promoting zone, neither the specific sequence of nucleotides we are looking for. To make matters even worse, the TFBS may mutate from one instance to another. Fig. 1 shows how difficult is to find a pattern of nucleotides on a real DNA base.

There exists, however, a key to break this code: the motif is a sequence of nucleotides of length l that repeats with the highest frequency in the DNA

© Springer International Publishing Switzerland 2015
J.M. Ferrández Vicente et al. (Eds.): IWINAC 2015, Part I, LNCS 9107, pp. 39–46, 2015.
DOI: 10.1007/978-3-319-18914-7_5

```
taatgtttgtgctggtttttgtggcatcgggcgagaatagcgcgtggtgtgaaagactgtttttttgatcgttttcacaaaaatggaagtccacagtcttgacag
gacaaaaacgcgtaacaaaagtgtctataatcacggcagaaaagtccacattgattatttgcacggcgtcacactttgctatgccatagcatttttatccataag
acaaatcccaataacttaattattgggatttgttatatataactttataaattcctaaaattacacaaagttaataactgtgagcatggtcatattttatcaat
cacaaagcgaaagctatgctaaaacagtcaggatgctacagtaatacattgatgtactgcatgtatgcaaaggacgtcacattaccgtgcagtacagttgatagc
acggtgctacacttgtatgtagcgcatctttctttacggtcaatcagcaaggtgttaaattgatcacgttttagaccattttttcgtcgtgaaactaaaaaaacc
agtgaattatttgaaccagatcgcattacagtgatgcaaacttgtaagtagatttccttaattgtgatgtgtatcgaagtgtgttgcggagtagatgttagaata
gcgcataaaaaacggctaaattcttgtgtaaacgattccactaatttattccatgtcacactttcgcatctttgttatgctatggttatttcataccataagcc
gctccggcggggttttttgttatctgcaattcagtacaaaacgtgatcaacccctcaattttcccttgctgaaaaattttccattgtctcccctgtaaagctgt
aacgcaattaatgtgagttagctcactcattaggcaccccaggctttacactttatgcttccggctcgtatgttgtgtggaattgtgagcggataacaatttcac
acattaccgccaattctgtaacagagatcacacaaagcgacggtggggcgtaggggcaaggaggatggaaagaggttgccgtataaagaaactagagtccgttta
ggaggagcgcgggaggatgagaacacggcttctgtgaactaaaccgaggtcatgtaaggaatttcgtgatgttgcttgcaaaaatcgtggcgattttatgtgcgca
gatcagcgtcgttttaggtgagttgttaataaagatttggaattgtgacacagtgcaaattcagcacacataaaaaaacgtcatcgcttgcattagaaaggtttct
gctgacaaaaagattaaacataccttatacaagactttttttttcatatgcctgacggagttcaacacttgtaagttttcaactacgttgtagactttacatcgcc
ttttttaaacattaaaattcttacgtaatttataatctttaaaaaaagcatttaatattgctccccgaacgattgtgattcgattcacatttaaacaatttcaga
cccatgagagtgaaattgttgtgatgtggttaacccaattagaattcgggattgacatgtcttaccaaaaggtagaacttatacgccatctcatccgatgcaagc
ctggcttaactatgcggcatcagagcagattgtactgagagtgcaccatatgcggtgtgaaataccgcacagatgcgtaaggagaaaataccgcatcaggcgctc
ctgtgacggaagatcacttcgcagaataaataaatcctggtgtccctgttgataccgggaagccctgggccaacttttggcgaaaatgagacgttgatcggcacg
gattttatactttaacttgttgatatttaaaggtatttaattgtaataacgatactctggaaagtattgaaagttaatttgtgagtggtcgcacatatcctgtt
```

Fig. 1. DNA base for searching the TFBS of CRP in Escherichia Coli

dataset. This clue reduces the problem to a mathematical one, i.e., an optimization problem. To solve it a number of different methods have been devised; among others: MEME (Multiple Expectation Maximization for Motif Elicitation) and BioProspector [2].

It is well known that optimization problems can be solved efficiently by evolutionary methods [3]. For instance: genetic algorithms are a good option; but in this case we are required to guess appropriate values for the rates of crossover and mutation, which are its classical operators [4]. We could avoid guessing these values if we use the method Estimation of Distribution Algorithm (EDA) [5]. However, in this case the challenge is to construct a good estimator. For the problem of finding biological motifs, it has been proposed in [6] to use a multivariate Gaussian estimator in order to capture possible correlations among the positions in the motif instances.

However, looking for simplicity and better processing times, we assume here that the nucleotides on a motif instance are statistically independent. Then, four univariate Gaussian Estimators (GE) will be required instead of a multivariate one to generate a new individual, where each estimator represents the distribution of a particular nucleotide estimated from the best individuals in the population. Our method will be called MBMEDA (Mtodo de Bsqueda de Motivos con base en un Algoritmo por Estimacin de Distribuciones) and its results will be compared systematically with those of EDAMD (Estimation of Distribution Algorithms for Motifs Discovery) published in [6]; this will allow us to explore two questions: 1) whether our method gives better or similar results compared with those of the multivariate approach, and 2) whether an EDA based motif search algorithm is more efficient than other computational motif search methods.

2   Materials and Methods

To test MBMEDA we used two types of DNA datasets: synthetic and real. A synthetic base is generated artificially following criteria used in other similar works: the length of the motif, the size of the promoting zones and the presence of noise [7]; in this bases a motif is implanted at known sites. On the other hand, in the real or biological DNA datasets, the sequences of nucleotides of the motif were determined experimentally by analyzing a number of promoting regions for each particular organism. Each biological dataset is labeled with the name of the TF that binds on its motif. Here we use five databases: CRP, E2F, ERF, ME2F and MYOD.

MBMEDA is a method that does a global search on the problem space of possible solutions, where a solution -also known as an individual- is defined by a vector VIP of the initial positions of a candidate motif on the n rows of the DNA dataset. Therefore, with each individual we associate a vector S of n sequences of length l that starts at the initial positions specified in VIP; we also associate with each individual a matrix of positional weights, denoted as PWM m x l, where l is the length of the motif sequence and m the cardinality of the nucleotide alphabet, in this case m = 4. The PWM has one row for each symbol of the alphabet: 4 rows in our case; it also has one column for each position in the pattern of sequences [8]. Each entry on the PWM represents the relative frequency of a nucleotide on its correspondent column in S. See Fig. 2.

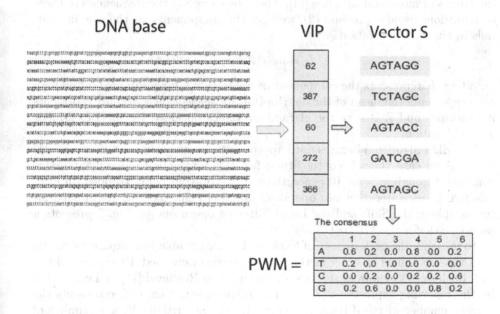

Fig. 2. Representation of an individual or candidate solution in MBMEDA

The initial population of individuals is usually generated randomly. The quality of a solution is evaluated by the fitness function, which in this case is the information content (IC) of the individual as defined by expression (1) [9] where fb is the frequency that nucleotide b appears at position i on the PWM and pb is the frequency of b on the entire DNA base. At each iteration, a number of the best individuals in the current population will be chosen by a tournament selection operator, in order to model with them how solutions will distribute in the next generation.

$$IC = \sum_{i=1}^{L} \sum_{b} f_b(i) \, log \left(\frac{f_b(i)}{p_b} \right) \tag{1}$$

The information content of an individual is a measure of the difference between the distribution of the nucleotides in the PWM -which represents a solution- and the distribution of the nucleotides on the entire DNA dataset. The larger is this difference, the more information content the solution has and, therefore, the larger is the possibility that it to be a motif. This concept of IC is crucial to the process of getting a subset of the best individuals in a population [10]; with them well estimate the four univariate gaussian models that will be used to calculate the next population.

Since we work here with the assumption of statistical independence of nucleotides on the motif instances, we have to estimate a set of four Univariate Gaussian distributions, one for each nucleotide, by calculating their corresponding values of mean and variance [11]. Then, by sampling the frequencies of these distributions using expression (2), well get the components for the new individuals in the next generation.

$$I_b = \mu_b + Z * \sigma_b^2. \tag{2}$$

Where I_b represents the component of nucleotide b for a sampled individual, mu_b represents the mean of the distribution for nucleotide b, $sigma_b^2$ represents its variance and Z is a vector of random values obtained by the Box Muller Transformation.

The EDA algorithm iterates until appropriate termination conditions are satisfied; in our case, the value of the fitness function for the best individual remains constant through at least 10 generations [12]. To avoid being trapped on local minima, at each iteration two operators unique to this method are applied after sampling: the Shift and the Local Filtering operators [6]. Fig. 3 presents a pseudo-code for the MBMEDA algorithm.

To measure the performance of EDA so that we are able to compare its results with those obtained by other methods, two metrics were used: Precision and Recall; both were taken from the field of Information Retrieval [13] and calculated by the following expressions (3) and (4), respectively; where N_c represents the correct number of motif instances found by the algorithm, N_p the number of promoter regions in the DNA database and N_t represents the total number of real instances of the motif.

$$Precision = \frac{N_c}{N_p}. \tag{3}$$

$$Recall = \frac{N_c}{N_t}. \tag{4}$$

MBMEDA (DNA database B):
  P   // Population Set
  M   // Estimated Gaussian Model
  Parents   // Set of best individuals, chosen to estimate M
  Children   // Set of new individuals generated from sampling M
  initialize P {randomly generated from B}
  repeat:
    for each individual pi in P do:
      fitness (pi)
    Best_Individual <- Best (P)   // Best individual in P
    Parents <- **Tournament Selection** (P)
    M <- **Estimate Gaussian Model** (Parents)
    Children <- **Sample Gaussian Model** (M)
    Every 10 Generations do:
      Local Filtering (Children)
      Shift (Best (Children))
    P <- P U Children
  until (termination condition)
  return Best_Individual

Fig. 3. MBMEDA algorithm

Table 1. Results of MBMEDA applied on synthetic bases

Number of Sequences	Motif Size	Noise			
		Noiseless		With-Noise	
		Pr	Rc	Pr	Rc
100	16	1.00	1.00	0.99	0.97
20	16	0.99	0.99	0.98	0.95
100	8	1.00	1.00	0.99	0.93
20	8	0.99	0.99	0.98	0.92
100	16	0.97	0.97	0.84	0.79
20	16	0.95	0.95	0.93	0.86

3   Results

Table 1 shows the MBMEDAs performance with different synthetic DNA bases
that corresponds to each row in the table. When we include noise for each base
-which represents a more realistic situation-, the average values for both metrics
were above 0.90; this is certainly promising

Table 2. Results of applying MBMEDA and EDAMD on real DNA bases

Base	MBMEDA		EDAMD	
	Pr	Rc	Pr	Rc
CRP	0.83	0.65	0.94	0.74
ERE	0.80	0.80	0.76	0.76
E2F	0.80	0.74	0.71	0.80
MYOD	1.00	0.80	0.86	0.90
ME2F	1.00	1.00	1.00	1.00

Table 3. Results when MBMEDA and other methods are applied on real DNA bases

Base	MBMEDA		MBMAG		MEME		BioProspector	
	Pr	Rc	Pr	Rc	Pr	Rc	Pr	Rc
CRP	0.83	0.65	0.88	0.69	0.92	0.52	1.00	0.35
E2F	0.80	0.75	0.76	0.70	0.80	0.70	0.52	0.41
ERE	0.80	0.80	0.76	0.76	0.88	0.60	0.30	0.56
ME2F	1.00	1.00	0.94	0.94	0.93	0.82	0.71	0.71
MYOD	1.00	1.00	0.94	0.76	0.00	0.00	0.00	0.00

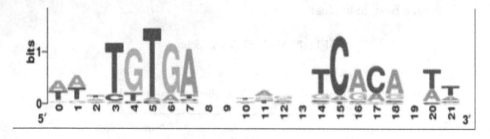

Fig. 4. Sequence logo of CRP motif consensus found experimentally

Table 2 shows the performance of MBMEDA and EDAMD on real datasets; this methods are both based on estimation of distribution algorithms. The average value for Precision for MBMEDA on these bases was 0.886, better than 0.854 for the reference method EDAMD, while for the other metric, Recall, the average value for MBMEDA was 0.798, a bit smaller than the 0.846 obtained for the reference method.

Table 3 presents results for the same real DNA bases as those of Table 2, obtained by applying our method MBMEDA and three others: MBMAG (Mtodo de Bsqueda basado en Algoritmos Genticos), which is a method based on genetic algorithms [9], and two non-evolutionary ones: MEME and BioProspector. The average values for Precision and Recall were 0.886 and 0.798 respectively, higher for the method we propose than for the other three.

Figure 4 shows a sequence logo [14], which is a graphic representation of the consensus word for the TFBS of transcription factor CRP, found experimentally

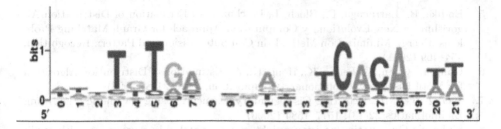

Fig. 5. Sequence logo of CRP motif consensus found by MBMEDA

[6]. Figure 5 on the other hand presents the sequence logo for the same motif as it was found by MBMEDA, the method proposed in this work.

4   Discussion and Conclusion

From the tables and figures above, its clear that the results of applying MBMEDA on DNA synthetic and real databases are quite satisfactory; they are similar and in some cases better than those obtained by other methods, like EDAMD for example.

Comparing Fig. 4 and Fig. 5, we observe that logo sequences for the motif consensus of the TFBS of protein CRP resemble each other quite well, which confirms the good results obtained with MBMEDA when searching for a motif. All this would imply that the assumption of statistical independence among the positions of the nucleotides in the motif instances is a reasonable one. However, we still consider necessary to make a more rigorous analysis of this assumption, which is fundamental to the performance of EDA based methods, since it simplifies the modeling of distributions and the process of sampling new individual for the next population.

Acknowledgements. The authors would like to thank Dr. Daniel Ochoa, Director of the Artificial Vision and Robotics Laboratory at ESPOL, for its constructive comments that helped shape the present form of this work.

References

1. Stormo, G.: DNA binding sites: representation and discovery. Bioinformatics 16(1), 16–23 (2000)
2. Liu, X.: Bioprospector: Discovering Conserved DNa Motifs in Upstream Regulatory Regions of Co-expressed Genes. In: Pacific Symposium on Biocomputing, vol. 6, pp. 127–138 (2001)
3. Hertz, Z., Stormo, G.: Identifying DNA and Protein Patterns with Statistically Significant Aligments of Multiple Sequences. Bioinformatics 15(7), 563–577 (1999)
4. Eiben, E. , Smith, J. : What Is an Evolutionary Algorithm. Introduction to Evolutionary Computing. Springer, New York (2003)

5. Endika, B., Larrañaga, P., Bloch, I., Perchant, A.: Estimation of Distribution Algorithms: a New Evolutionary Computation Approach for Graph Matching Problems. Energy Minimization Methods in Computer Vision and Pattern Recognition, 454–469 (2001)

6. Gang, L., Chan, T., Leung, K., Hong, K.: An Estimation of Distribution Algorithm for Motif Discovery. Evolutionary Computation, 2411–2418 (2008)

7. Wei, Z.: GAME: Detecting Cis-regulatory Elements Using a Genetic Algorithm. Bioinformatics 22(13), 1577–1584 (2006)

8. Sinha, S.: On counting position weight matrix matches in a sequence, with application to discriminative motif finding. Bioinformatics 22(14), 454–463 (2006)

9. Schneider, T., Stormo, G., Gold, L., Ehrenfeucht, A.: Information Content of Binding Sites on Nucleotide Sequences. Journal of Molecular Biology 188(3), 415–431 (1986)

10. Shannon, C.: A Mathematical Theory of Communication. Bell Syst., Techn. J. 27, 379–423 (1948)

11. Jordán, I., Jordán, C.: Aplicación de Algoritmos Evolutivos a la búsqueda de motivos biológicos en bases de regiones promotoras de ADN. Revista Matemática ICM, 33–42 (2012)

12. Fogel, D.: Evolutionary Computation: Toward a new Philosophy in Machine Intelligence. IEEE Press (1995)

13. Manning, D., Raghavan, P., Schutze, H.: Introduction to Information Retrieval, pp. 151–158. Cambridge UP, New York (2008)

14. Schneider, T., Stephens, R.: Sequence Logos: A New Way to Display Consensus Sequences. Nucleic Acids Res. 18(20), 6097–6100 (1990)

Towards a Generic Simulation Tool of Retina Models

P. Martínez-Cañada[✉], Christian Morillas, Begoña Pino,
and Francisco Pelayo

CITIC and Department of Computer Architecture and Technology,
University of Granada, Granada, Spain
{pablomc,cmg,bpino,fpelayo}@ugr.es

Abstract. The retina is one of the most extensively studied neural circuits in the Visual System. Numerous models have been proposed to predict its neural behavior on the response to artificial and natural visual patterns. These models can be considered an important tool for understanding the underlying biophysical and anatomical mechanisms. This paper describes a general-purpose simulation environment that fits to different retina models and provides a set of elementary simulation modules at multiple abstraction levels. The platform can simulate many of the biological mechanisms found in retinal cells, such as signal gathering though chemical synapses and gap junctions, variations in the receptive field size with eccentricity, membrane integration by linear and single-compartment models and short-term synaptic plasticity. A built-in interface with neural network simulators reproduces the spiking output of some specific cells, such as ganglion cells, and allows integration of the platform with models of higher visual areas. We used this software to implement whole retina models, from photoreceptors up to ganglion cells, that reproduce contrast adaptation and color opponency mechanisms in the retina. These models were fitted to published electro-physiological data to show the potential of this tool to generalize and adapt itself to a wide range of retina models.

Keywords: Retina simulator · Contrast adaptation · Color opponency · Neural network · Spikes

1 Introduction

The retina is the visual sensory input of the brain. Photons arriving at photoreceptors are first translated into a biochemical message and then into an electrical message that can stimulate all of neurons in the retina. Analog electrical signals are finally transformed into patterns of spikes for transmission along the optic nerve to various downstream brain regions. The retina is not merely a simple spatiotemporal prefilter [1]. On the contrary, retinal cells connect in different and complex neural structures that provide a wide visual behavioral repertoire. However, many aspects of retinal connectivity are still controversial and certain functional mechanisms are not entirely clear [2]. Therefore, new efforts and approaches are required to improve and advance in both neurophysiological studies and the modeling of the retina.

© Springer International Publishing Switzerland 2015
J.M. Ferrández Vicente et al. (Eds.): IWINAC 2015, Part I, LNCS 9107, pp. 47–57, 2015.
DOI: 10.1007/978-3-319-18914-7_6

Initially, computational models of neurons were programmed by researchers with the purpose of fitting the results of specific experiments conducted in their labs. There was no effort made to disseminate the software or to generalize it beyond a particular model [3]. Neural simulation systems, such as NEST [4] or NEURON [5], exploited common properties of neurons (e.g., their ionic-selective channels) to provide researchers with general and efficient simulation environments. They have become reliable research tools that facilitate and accelerate neural modeling and provide options for extensibility, interoperability, and model sharing between laboratories.

A remarkable amount of research also pursued a generalization of common features of retina processing and a unification of different biophysical retina models [6–13]. Computations performed by these systems reproduce those retina behaviors that they have been intentionally designed for, but lack the configurability to modify their simulation circuitry and adapt to new experiments. In agreement with other authors [1], we consider that there are sufficient examples of single neural structures that serve quite different roles in retina processing to motivate the generalization of basic retinal circuits. Moreover, many physiological experiments can be approximated by simple linear temporal filters and static nonlinearities, such as the widely used linear-nonlinear (LN) model for measuring contrast adaptation [14–17]; single-compartment models for the description of adaptation phenomena (e.g., neural gain control) [7, 18, 19]; and spatial Gaussian receptive fields when the spatial processing of the signal is relevant for the experiment [20, 21]. These models have led to a quantitative understanding of many dynamic phenomena occurring both at the cellular and conceptual level in the retina and they have been commonly used in neural modeling of the retina.

We have implemented a software platform that provides researchers with a general and efficient simulation environment of neural mechanisms in the retina. Computations performed by the platform are based on generic microcircuits at different abstraction levels whose interconnection schema can be fully modified to configure different retina architectures. The platform combines the efficient filtering scheme of retina simulators based on image-processing techniques and some biological concepts used to simulate the dynamics and structure of neural networks. Simulation of electrophysiological experiments, covering some significant phenomena observed in the retina (i.e., contrast adaptation and chromatic opponency), show the potential of this tool to adapt to a wide range of retina models. The software can be easily used as an efficient benchmark to simulate and understand the visual processing at low-level.

The rest of the paper is organized as follows. In section 2 we detail the neural models implemented by the platform. We present simulation results of the physiological experiments in section 3. Finally, in section 4, we discuss the conclusions and summarize the contributions of this software.

2   Neural Models of Retinal Cells

Neural models are constructed to explain the biophysical mechanisms responsible for generating neural activity. Such models range from highly detailed descriptions involving thousands of coupled differential equations to greatly simplified systems that are used to explain results of specific physiological experiments. The simulation platform implements a series of neural models that have been recurrently used over the last few decades to explain certain phenomena such as contrast adaptation or synaptic integration within the receptive field. One of the main contribution of our work has been to summarize and generalize these models to provide a set of general neural tools that can be used to construct a wide range of retina architectures. In this section we discuss the mathematical framework of these models and how they are integrated in the simulation platform.

Membrane potential of cells can be defined using a single-compartment model. The basic equation that explains the temporal evolution of a single-compartment model is [22]:

$$C_m \frac{dV(t)}{dt} = \sum_i I_i(t) + \sum_j g_j(E_j(t) - V(t))g \tag{1}$$

where the index j indicates the input ionic channel, C_m is the membrane capacitance, V the membrane potential, g_j is the conductance of the channel, E_j the reversal potential of the channel and the term $\sum_i I_i$ denotes the sum of external input currents. Channel conductances can be modified by other neural modules of the simulator to reproduce the shunting inhibition effect. Conductances with reversal potentials near the membrane potential conduct little current. Instead, their primary impact is to change the membrane resistance of the cell. Such conductances are called shunting because their main effect is to increase the total conductance of a neuron. Shunting inhibition has been used to reproduce nonlinear mechanisms of the retina, such as contrast and luminance gain control [7, 18] or directional selectivity to motion [23, 24], and normalization of the linear response in the primary visual cortex [25]. Voltage-gated and calcium-dependent conductances can be easily implemented following a similar mathematical formalism. However, instead of a conductance controlled by another neural module the gating variable would be the membrane potential or the concentration of calcium-binding molecules.

In some specific physiological experiments the membrane potential is approximated by linear models. A linear approximation of the neural response of a cell, $L(t)$, can be defined based on the linear kernel $K(x, y, \tau)$ [22, 26]:

$$L(t) = \int_0^\infty d\tau \int_{(x,y) \in RF} K(x, y, \tau)s(x_0 - x, y_0 - y, t - \tau)dxdy \tag{2}$$

where $s(x, y, t)$ is the visual stimulus and RF the receptive field of the cell. The neural response of the cell depends linearly on all past values of the input stimulus located in the cells receptive field RF. This integral corresponds to the well-defined convolution operation:

$$L(t) = (s * K)(x_0, y_0, \tau) \tag{3}$$

For some neurons $K(x, y, \tau)$ can be broken down as a product of two functions, one that accounts for the spatial receptive field and the other one for the temporal receptive field:

$$K(x, y, \tau) = K_s(x, y)K_t(\tau) \tag{4}$$

The temporal linear kernel is the basis of a well-known mathematical tool, the linear-nonlinear analysis (LN) [16, 17, 27], used to describe, for example, contrast adaptation mechanisms. The LN analysis separates the temporal behavior of the cell from nonlinear response components (e.g., synaptic rectification or saturation). The neural response is first correlated with the input pattern to obtain the temporal filter. This filter is convolved afterwards with the stimulus to generate a linear model of the response; this linear model (i.e. filtered stimulus) serves as the input passed through a static nonlinearity, which works like a lookup table, and translates linear model values into output values (nA, mV or spikes/s) [28]. The simulation platform offers the possibility of creating these two modules, the temporal linear filter and the static nonlinearity, to reproduce this type of experiments. Low-pass filters are based on recursive implementations of exponential and exponential cascade functions. Linear subtraction of low-pass filters (e.g., between photoreceptors and horizontal cells at the outer plexiform layer) produce the typical biphasic shape observed in bipolar and subsequent neural layers [16, 27].

The spatial receptive field is modeled as a Gaussian function, similarly to the kernels used in the receptive field model proposed by Rodieck [20] and Enroth-Cugell and Robson [21]. The software also reproduces morphological and physiological variations associated with eccentricity. It is possible to simulate the spread of neuron dendrites with eccentricity and its consequent increase of the receptive field size. To this end, Gaussian kernels vary with eccentricity. Space-variant filters are based on the Deriche's recursive approach [29–32]. The main advantage of these filters is that the number of operations per pixel is constant and does not depend on the size of the kernel. Moreover, kernel coefficients can be modified at every pixel to simulate a foveated retina [31]. We have improved the performance of the spatial filtering in a multi-core processor by taking advantage of the fact that every row and every column of the image are processed independently according to the Deriche's recursive algorithm and can be executed in different threads.

The history of activity at a synapse influences the probability of transmitter release and the magnitude of the resulting conductance change. This phenomenon is known as short-term plasticity when it lasts anywhere from milliseconds to tens of seconds[22]. Short term plasticity is also present in the retina; Contrast adaptation originates in bipolar cells and neither photoreceptors nor horizontal cells are involved in the process [16, 27]. Recent experiments have shown that contrast adaptation effects are still present under physiological blockade of amacrine synapses, ruling out a critical role for amacrine cells in driving contrast adaptation [27, 28, 33]. Slow adaptation mechanisms are apparently driven by prolonged depression of glutamate release at bipolar cell synapses [14, 34–36], whereas inactivation of voltage-dependent Na+ channels in ganglion cells [17, 19] and calcium-related mechanism in bipolar cells [33] may be responsible for the fast component. In addition, a large fraction of adaptation has been observed at the synapse bipolar-to-ganglion [27, 37].

Polarization and hyperpolarization offsets of the nonlinearity are implemented by a model of short-term plasticity. It was suggested that opposing mechanisms of plasticity (i.e., depression and facilitation) could be combined together to compensate the mutual information loss [38]. Following this idea, the model includes a short-term plasticity module that correlates synaptic weight with the neural input to simulate a depolarizing offset of the ganglion membrane for high contrast steps [14, 16]. On the other hand, synaptic depression occurs for maintained values of contrast with the synaptic offset decaying exponentially back to its resting value. This module is defined by:

$$P = P + k_f(k_m(t)abs(input) - P) \tag{5}$$

where P is the offset of the synapse, the parameter k_f controls the degree of facilitation, and the factor $(k_m(t)abs(input) - P)$ prevents the offset from growing indefinitely. A rectification of the input is applied by the term of absolute value. The variable k_m is responsible for the slow depression of the synapse. Its exponential decay is approximated by:

$$k_m(t + 1) = k_{mInf} + (k_m(t) - k_{mInf})\exp(-step/tau) \tag{6}$$

with a temporal constant defined by the quotient of the simulation $step$ and the parameter tau. k_{mInf} fixes the resting value and is inversely proportional to the input using a depression factor k_d:

$$k_{mInf} = \frac{k_d}{abs(input)} \tag{7}$$

A summarize of these neural modules and the software connection with a neural network simulator is represented in figure 1. When interfacing the platform with models of higher brain areas the neural network simulator integrates our software efficiently and the retina module can be easily loaded in the neural network script. Configuration parameters are passed through the retina script, whose programming syntax is similar to that used in other neural simulators. Other modules can be also instantiated in the retina script to generate synthetic input stimuli (e.g., spatially drifting gratings or white noise) or to read video sequences. The user can also create monitoring panels to visualize intermediate outputs and the temporal and spatial activity of simulated cells.

3   Simulations of Physiological Experiments

We fitted four different retina models to published physiological data. Different combinations of the neural modules, included in the simulation platform, were used to reproduce chromatic opponency mechanisms, red-green and blue-yellow, and contrast adaptation phenomenon in the retina. Chromatic models

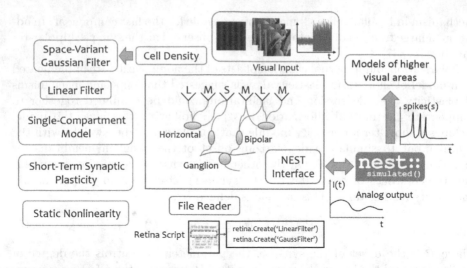

Fig. 1. Schematic view of the simulation platform connected to a spiking neural network simulator (e.g., NEST [4]). Green boxes indicate those modules that can be configured and interconnected by the user through a retina script (such as neural models and visual input sources). The analog outputs of the platform correspond to the ganglion synaptic currents, which are processed by the neural network simulator to produce the spiking output. This spiking output can feed afterwards models of higher brain areas implemented in the neural network simulator. The neural network simulator drives the simulation time and synchronizes the update process of spatiotemporal equations in the retina model. In the figure we show a possible retina configuration, where L, M and S correspond to L-, M-, and S-cones, respectively. Other different retina architectures can be easily configured by creating new neural layers and modifying the connection scheme in the retina script.

were inspired by the retina structures suggested by different authors [39–41]. We would like to remark that authors of these chromatic experiments used simple mathematical models, based on difference of Gaussian functions that depend on spatial frequency, to fit the physiological recordings. However, our simulated models correspond to whole retina models that provide a neural basis at each retinal stage to explain measured data.

A summary of the simulation experiments conducted for two of these models is shown in figure 2. The model on the right was inspired by the blue-yellow retina circuitry proposed by Crook et al. [39]. The spatially coextensive receptive field of the blue-yellow pathway is explained by a retina architecture of parallel ON and OFF cone bipolar inputs to ganglion cells in primate retina. The S-ON bipolar (left branch of the circuit) inherits a LM-OFF surround created by H2 horizontal cell feedback to the S-cone. However, this LM-OFF surround is canceled out by the LM-ON surround created by H1 horizontal cell feedback to LM-OFF bipolar (right branch). Thus, no net surround appears at the ganglion cell resulting in a cone-opponent receptive field that lacks center-surround spatial opponency.

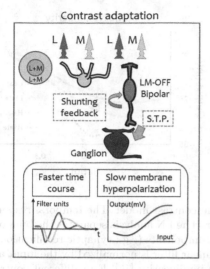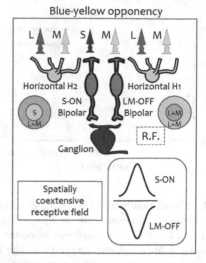

Fig. 2. Two retina experiments reproduced by the simulation tool. The retina architecture on the left reproduces both fast and slow contrast adaptation by a combined model of shunting feedback loop at bipolar cells [7] and short-term plasticity (S.T.P.) at the bipolar-to-ganglion synapse [14, 34–36]. A representation of the type of results that are obtained for the contrast experiment is shown below the retina circuit. Further details of this experiment are included in the text and in figure 3. The model on the right corresponds to the spatially coextensive receptive field of the blue-yellow pathway [39]. Excitatory contributions to receptive fields (R.F.) of each bipolar cell are marked in orange and inhibitory in blue. The S-ON bipolar (left branch of the circuit) is fed by a LM-OFF surround created by H2 horizontal cell feedback to the S-cone. However, this LM-OFF surround is counterbalanced by the LM-ON surround created by H1 horizontal cell feedback to LM-OFF bipolar (right branch). Therefore, the center and the periphery of the ganglionar receptive field are similar in size and it does not present the common center-surround spatial opponency of other retinal cells.

Fast and slow contrast adaptation mechanisms have been also characterized. The simulation tool captures both forms of adaptation by a combined model of shunting feedback at the bipolar level [7] and short-term plasticity at the bipolar-to-ganglion synapse [14, 34–36]. Results of the LN analysis for this experiment are shown in figure (figure 3). A contrast increase of the visual input accelerates kinetics of the linear filter, reduces sensitivity (defined as the average slope of the nonlinearity) and depolarizes the membrane potential (reflected in the increase of the mean value of the nonlinearity). Slow adaptation does not affect the temporal response but produces a progressive hyperpolarization of membrane potential (figure 3). Upon a decrease in contrast, all these changes reverse direction but with asymmetric time constants for slow adaptation.

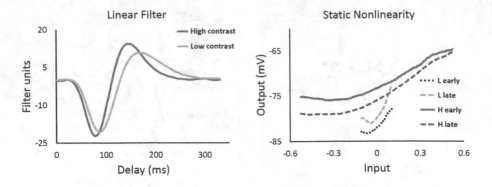

Fig. 3. Linear-nonlinear (LN) analysis of the contrast model. The temporal response of the retina is represented by the linear filter. The LN analysis separates the temporal behavior of the cell from nonlinear response components (e.g., synaptic rectification or membrane depolarization). In addition, the linear filter is normalized by the input and any variation of gain is reflected only in the nonlinearity [16]. Four different contrast intervals are considered to calculate the nonlinearity: 'L early' corresponds to the first 10 seconds after a low contrast step and 'L late' to the period from 10 to 20 seconds after a low contrast step. 'H early' and 'H late' are defined similarly for a high contrast step. Only two contrast periods are considered for the linear filter because the temporal behavior does not vary within a contrast period.

4   Discussion

A great number of retina models have been proposed to explain the biophysical and anatomical mechanisms of the retina. Neural response observed in different physiological experiments have been recurrently explained by a set of basic mathematical tools, such as the well-known Gaussian receptive field [20, 21]. There are sufficient examples of the same neural structures that serve quite different roles in retina processing to motivate the generalization of basic retinal circuits [1]. In this paper we presented a general-purpose simulation environment that adapts to different retina models and provides a set of elementary simulation modules at multiple abstraction levels. The platform can simulate many of the biological mechanisms found in retinal cells, such as signal gathering though chemical synapses and gap junctions, variations in the receptive field size with eccentricity, membrane integration by linear and single-compartment models and short-term synaptic plasticity. The interconnection scheme of these modules can be modified to create different retina architectures. The platform is easily integrated with neural network simulators to generate a spiking output that can interface models of higher visual areas.

Four different retina models were fitted to published physiological data. These retina models simulate chromatic opponency, blue-yellow and red-green, and contrast adaptation mechanisms in the retina. The chromatic models we propose provide a neural basis for the retina architectures suggested by different authors

[39–41]. The simulation tool was configured to reproduce both theories of the red-green pathway (the cone-type selective surround [40, 42, 43] and the random-wiring or mixed surround [41, 44]) and the spatially coextensive receptive field of the blue-yellow pathway [39]. A detailed description of these experiments can be found in previous publications [45]. The simulation platform also captures both form of contrast adaptation observed in the retina, fast and slow, by a combined model of shunting feedback loop [7, 18] in bipolar cells and short-term plasticity at the bipolar-to-ganglion synapse [14, 34–36]. Simulation results of these physiological experiments show the potential of this tool to generalize and adapt to a wide range of retina configurations.

Acknowledgments. This work has been supported by the Human Brain Project (SP11 - Future Neuroscience), project P11-TIC-7983, Junta of Andalucia (Spain), Spanish National Grant TIN2012-32039, co-financed by the European Regional Development Fund (ERDF), and the Spanish Government PhD scholarship FPU13/01487.

References

1. Gollisch, T., Meister, M.: Eye smarter than scientists believed: neural computations in circuits of the retina. Neuron 65(2), 150–164 (2010)
2. Lee, B.B., Martin, P.R., Grünert, U.: Retinal connectivity and primate vision. Progress in Retinal and Eye Research 29(6), 622–639 (2010)
3. Beeman, D.: History of neural simulation software. In: 20 Years of Computational Neuroscience, pp. 33–71. Springer (2013)
4. Gewaltig, M.-O., Diesmann, M.: Nest (neural simulation tool). Scholarpedia 2(4), 1430 (2007)
5. Hines, M.L., Carnevale, N.T.: The neuron simulation environment. Neural Computation 9(6), 1179–1209 (1997)
6. Benoit, A., Caplier, A., Durette, B., Hérault, J.: Using human visual system modeling for bio-inspired low level image processing. Computer Vision and Image Understanding 114(7), 758–773 (2010)
7. Wohrer, A., Kornprobst, P.: Virtual retina: a biological retina model and simulator, with contrast gain control. Journal of Computational Neuroscience 26(2), 219–249 (2009)
8. Hérault, J., Durette, B.: Modeling visual perception for image processing. In: Sandoval, F., Prieto, A.G., Cabestany, J., Graña, M. (eds.) IWANN 2007. LNCS, vol. 4507, pp. 662–675. Springer, Heidelberg (2007)
9. Morillas, C.A., Romero, S.F., Martínez, A., Pelayo, F.J., Ros, E., Fernández, E.: A design framework to model retinas. Biosystems 87(2), 156–163 (2007)
10. Hérault, J.: A model of colour processing in the retina of vertebrates: From photoreceptors to colour opposition and colour constancy phenomena. Neurocomputing 12(2), 113–129 (1996)
11. De Valois, R.L., De Valois, K.K.: A multi-stage color model. Vision research 33(8), 1053–1065 (1993)
12. Andreou, A.G., Boahen, K.A.: A contrast sensitive silicon retina with reciprocal synapses. Advances in Neural Information Processing Systems (NIPS) 4, 764–772 (1991)
13. Mead, C.: Neuromorphic electronic systems. Proceedings of the IEEE 78(10), 1629–1636 (1990)

14. Ozuysal, Y., Baccus, S.A.: Linking the computational structure of variance adaptation to biophysical mechanisms. Neuron 73(5), 1002–1015 (2012)
15. Mante, V., Frazor, R.A., Bonin, V., Geisler, W.S., Carandini, M.: Independence of luminance and contrast in natural scenes and in the early visual system. Nature Neuroscience 8(12), 1690–1697 (2005)
16. Baccus, S.A., Meister, M.: Fast and slow contrast adaptation in retinal circuitry. Neuron 36(5), 909–919 (2002)
17. Kim, K.J., Rieke, F.: Temporal contrast adaptation in the input and output signals of salamander retinal ganglion cells. The Journal of Neuroscience 21(1), 287–299 (2001)
18. Mante, V., Bonin, V., Carandini, M.: Functional mechanisms shaping lateral geniculate responses to artificial and natural stimuli. Neuron 58(4), 625–638 (2008)
19. Kim, K.J., Rieke, F.: Slow na+ inactivation and variance adaptation in salamander retinal ganglion cells. The Journal of Neuroscience 23(4), 1506–1516 (2003)
20. Rodieck, R.W.: Quantitative analysis of cat retinal ganglion cell response to visual stimuli. Vision Research 5(12), 583–601 (1965)
21. Enroth-Cugell, C., Robson, J.G.: The contrast sensitivity of retinal ganglion cells of the cat. The Journal of Physiology 187(3), 517–552 (1966)
22. Dayan, P., Abbott, L.F.: Theoretical neuroscience: computational and mathematical modeling of neural systems. Journal of Cognitive Neuroscience 15(1), 154–155 (2003)
23. Torre, V., Poggio, T.: A synaptic mechanism possibly underlying directional selectivity to motion. Proceedings of the Royal Society of London. Series B. Biological Sciences 202(1148), 409–416 (1978)
24. Amthor, F.R., Grzywacz, N.M.: Nonlinearity of the inhibition underlying retinal directional selectivity. Visual Neuroscience 66(03), 197–206 (1991)
25. Carandini, M., Heeger, D.J., Movshon, J.A.: Linearity and normalization in simple cells of the macaque primary visual cortex. The Journal of Neuroscience 17(21), 8621–8644 (1997)
26. Wohrer, A.: Model and large-scale simulator of a biological retina, with contrast gain control. PhD thesis, Nice (2008)
27. Beaudoin, D.L., Borghuis, B.G., Demb, J.B.: Cellular basis for contrast gain control over the receptive field center of mammalian retinal ganglion cells. The Journal of Neuroscience 27(10), 2636–2645 (2007)
28. Demb, J.B.: Functional circuitry of visual adaptation in the retina. The Journal of Physiology 586(18), 4377–4384 (2008)
29. Deriche, R.: Recursively implementating the gaussian and its derivatives (1993)
30. Deriche, R.: Fast algorithms for low-level vision. IEEE Transactions on Pattern Analysis and Machine Intelligence 12(1), 78–87 (1990)
31. Tan, S., Dale, J.L., Johnston, A.: Performance of three recursive algorithms for fast space-variant gaussian filtering. Real-Time Imaging 99(3), 215–228 (2003)
32. Triggs, B., Sdika, M.: Boundary conditions for young-van vliet recursive filtering. IEEE Transactions on Signal Processing 54(6), 2365–2367 (2006)
33. Rieke, F.: Temporal contrast adaptation in salamander bipolar cells. The Journal of Neuroscience 21(23), 9445–9454 (2001)
34. Jarsky, T., Cembrowski, M., Logan, S.M., Kath, W.L., Riecke, H., Demb, J.B., Singer, J.H.: A synaptic mechanism for retinal adaptation to luminance and contrast. The Journal of Neuroscience 31(30), 11003–11015 (2011)
35. Dunn, F.A., Rieke, F.: Single-photon absorptions evoke synaptic depression in the retina to extend the operational range of rod vision. Neuron 57(6), 894–904 (2008)

36. Manookin, M.B., Demb, J.B.: Presynaptic mechanism for slow contrast adaptation in mammalian retinal ganglion cells. Neuron 50(3), 453–464 (2006)
37. Zaghloul, K.A., Boahen, K., Demb, J.B.: Contrast adaptation in subthreshold and spiking responses of mammalian y-type retinal ganglion cells. The Journal of Neuroscience 25(4), 860–868 (2005)
38. Kastner, D.B., Baccus, S.A.: Coordinated dynamic encoding in the retina using opposing forms of plasticity. Nature Neuroscience 14(10), 1317–1322 (2011)
39. Crook, J.D., Davenport, C.M., Peterson, B.B., Packer, O.S., Detwiler, P.B., Dacey, D.M.: Parallel on and off cone bipolar inputs establish spatially coextensive receptive field structure of blue-yellow ganglion cells in primate retina. The Journal of Neuroscience 29(26), 8372–8387 (2009)
40. Lee, B.B., Shapley, R.M., Hawken, M.J., Sun, H.: Spatial distributions of cone inputs to cells of the parvocellular pathway investigated with cone-isolating gratings. JOSA A 29(2), A223–A232 (2012)
41. Crook, J.D., Manookin, M.B., Packer, O.S., Dacey, D.M.: Horizontal cell feedback without cone type-selective inhibition mediates red–green color opponency in midget ganglion cells of the primate retina. The Journal of Neuroscience 31(5), 1762–1772 (2011)
42. Lee, B.B., Kremers, J., Yeh, T.: Receptive fields of primate retinal ganglion cells studied with a novel technique. Visual Neuroscience 15(01), 161–175 (1998)
43. Lee, B.B., Dacey, D.M., Smith, V.C., Pokorny, J.: Horizontal cells reveal cone type-specific adaptation in primate retina. Proceedings of the National Academy of Sciences 96(25), 14611–14616 (1999)
44. Verweij, J., Hornstein, E.P., Schnapf, J.L.: Surround antagonism in macaque cone photoreceptors. The Journal of Neuroscience 23(32), 10249–10257 (2003)
45. Martínez-Cañada, P., Morillas, C., Nieves, J.L., Pino, B., Pelayo, F.: First stage of a human visual system simulator: The retina. In: Trémeau, A., Schettini, R., Tominaga, S. (eds.) CCIW 2015. LNCS, vol. 9016, pp. 118–127. Springer, Heidelberg (2015)

Specialist Neurons in Feature Extraction Are Responsible for Pattern Recognition Process in Insect Olfaction

Aaron Montero[1]([⊠]), Ramon Huerta[1,2], and Francisco B. Rodriguez[1]

[1] Grupo de Neurocomputación Biológica, Dpto. de Ingeniería Informática,
Escuela Politécnica Superior, Universidad Autónoma de Madrid,
28049 Madrid, Spain
[2] BioCircuits Institute, University of California, La Jolla, San Diego,
CA 92093-0402, USA
{aaron.montero,f.rodriguez}@uam.es, rhuerta@ucsd.edu

Abstract. In the olfactory system we can observe two types of neurons based on their responses to odorants. Specialist neurons react to a few odorants, while generalist neurons respond to a wide range of them. These kinds of neurons can be observed in different parts of the olfactory system. In the antennal lobe (AL), these neurons encode odorant information and in the extrinsic neurons (ENs) of the mushroom bodies (MB) they can learn and identify different kind of odorants based on the selective and generalist response. The classification of specialists and generalists neurons in Kenyon cells (KCs), which serve as a bridge between AL and ENs, may seem arbitrary. However KCs have the unique mission of increasing the separability between different odorants, to achieve a better information processing performance. To carry out this function, the connections between the antennal lobe and Kenyon cells do not require a specific connectivity pattern. Since KCs can be specialists or generalists by chance and olfactory learning performance relies on their feature extraction capabilities, we analyze the role of generalist and specialist neurons in an olfactory discrimination task. Role that we studied by varying the percentage of these two kind of neurons in KC layer. We determined that specialist neurons are a decisive factor to perform optimal odorant classification.

Keywords: Pattern recognition · Specialist neuron · Generalist neuron · Olfactory system · Neural variability · Supervised learning · Heterogeneous threshold · Lateral inhibition

1 Introduction

Insects possess an olfactory system that identifies a large number of odorants using a simple structural organization. This neural network allows pattern recognition under different environmental conditions, gas concentrations, and mixtures. Inside this network, there are two kind of neurons based on their

J.M. Ferrández Vicente et al. (Eds.): IWINAC 2015, Part I, LNCS 9107, pp. 58–67, 2015.
DOI: 10.1007/978-3-319-18914-7_7

response to odorants: specialists and generalists. Specialist neurons have selective responses to stimuli and generalist neurons code for multiple stimuli [19,4,20,26,21]. The role of both classes of neurons in the olfactory system is still under debate [10,4]. However, it is suggested that specialist neurons are crucial for discrimination, while generalist neurons play a key role in extracting and discovering common features [25]. This Hypothesis has been supported, in the case of the role of specialists neurons using experimental studies [7,2] and computational models [16]. In this paper we investigate the impact that changing the ratio of specialists and generalists has in the performance in a pattern recognition task.

To answer this question, we focused on Kenyon cells of the mushroom body. While projection neurons of the antennal lobe and extrinsic neurons of the mushroom body have a reason to react to certain odorants, the first ones encode the odorants and the second ones identify them, the response to different odorants of Kenyon cells is circumstantial. The role of Kenyon cells in the olfactory systems is primarily to increase the separability of different odorants to facilitate subsequent learning and identification. It has been observed that for this task the Kenyon cells do not require specific connections with the antennal lobe [14,24]. In fact, these connections vary between individuals of the same species. Thus, it seems that there is not a criterion by which a neuron is defined as a generalist or specialist. Therefore, aspects of KCs as:

- arbitrary creation of their connections from AL and, therefore, their specialist and generalist neurons,
- large number of neurons (50,000 in locust), and
- being the final stage of feature extraction started in the olfactory receptor neurons (ORNs),

are the reasons why we consider them the best ones to analyze the implications of varying the number of generalist and specialist neurons for pattern recognition process.

In order to classify Kenyon cells as specialist or generalist neurons, we use neural sensitivity. This can be estimated from the distribution of neurons that respond to n out of N different stimuli [20,21]. However, because the boundary between specialists is arbitrary in a continuous distribution of sensitivity, a systematic analysis is required for a proper differentiation of these neurons. We will define, therefore, the minimum percentage of reaction to an odorant for which a neuron can be considered sensitive to this, as well as the sensitivity degree that makes a neuron be specialist or generalist.

To perform this study, we used a single-hidden-layer neural network that represents a computational model focused on the AL and MB, which we can see in Fig. 1. The input of this neural network is the AL activity, which is connected to MB through a non-specific connectivity matrix [14,24,6]. The other layers, hidden and output, are made of KCs and ENs respectively. These neurons are connected by a connectivity matrix subjected to learning that is modulated by

Hebbian learning [5,1]. ENs give us the final result of the classification, once the lateral inhibition process between them has finished [13,3].

We will show in results that the most specialist neurons as responsible for odorant classification.

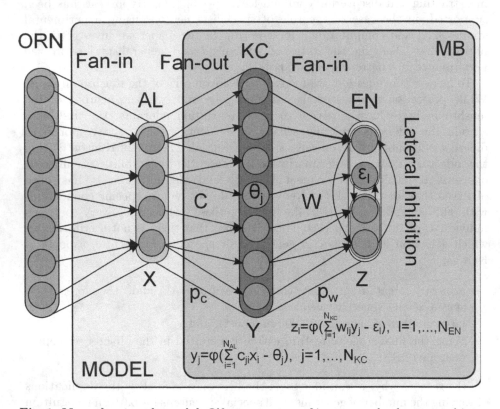

Fig. 1. Neural network model. Olfactory system of insects can be decomposed into 3 parts: olfactory receptor neurons (ORN), antennal lobe (AL) and mushroom body (MB). The ORNs sent olfactory information in a fan-in phase to AL that transmits this in a fan-out phase to MB. The olfactory information is received by Kenyon cells (KC), which is responsible for increasing the separability of information and transmit it, in a fan-in phase, to extrinsic neurons (EN), responsible for its learning and classification. Our model is a single-hidden-layer neural network with AL as input (X) connected by a random matrix (C) to KCs, the hidden layer (Y). KCs are connected by a matrix with Hebbian learning (W) to ENs, the output layer (Z). The thresholds for the hidden and output layer are θ_j and ε_l respectively. The Heaviside activation function φ is 0 when its argument is negative or 0 and 1 otherwise.

2   Neural and Network Model

The model focuses on the AL and MB, dividing the MB into KCs and ENs (Fig. 1). Therefore, the network model is a single-hidden-layer neural network with an input layer of 1,568 neurons (due to our patterns), a hidden layer with 50,000 neurons (locust [12] has a ratio of 1:50 between neurons of the AL, input layer, and KCs, hidden layer, and similar dimensions to those we selected) and an output layer with 100 neurons. These neurons of the output layer are divided into populations of 10 neurons, a population for each pattern class, and there are lateral inhibitions between these populations [1]. This facilitates that only a specific population of neurons reacts to a particular pattern class.

The KC neurons of the MB display very low activity [19]. These neurons are inactive most of the time, with a mean firing frequency lower than 1 Hz. But when they are activated, their neuronal response is produced by the coincidence of concurrent spikes followed by a reset. Bearing in mind this behavior, we chose the McCulloch-Pitts model in all neurons of the hidden and output layers.

The connectivity matrices, C and W, are initialized at the beginning of each learning process. These matrices are created by using the connection probabilities, p_c and p_w, as a threshold on matrices with random values uniformly distributed. The connectivity matrix W is updated using Hebbian learning [8,9], which is subjected to a target t of the output layer (supervised learning), while matrix C remains fixed.

The synaptic model is binary. Therefore, activation states and weights can only take values 0 or 1.

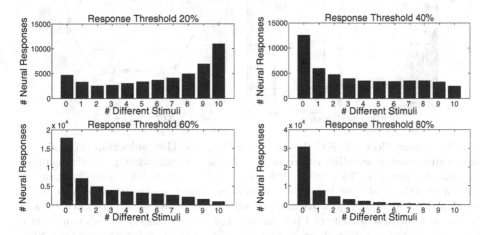

Fig. 2. Neural sensitivity depending on the response threshold. Response threshold is the percentage of odorants from a class that a neuron needs to respond to be considered sensitive to it. When response threshold rises, the neurons that respond to few stimuli increase in number compared to those that respond to many of them.

3   Neural Sensitivity

To define KCs as specialist or generalist neurons, we use a criterion based on neural sensitivity [20,21]. Neural sensitivity represents the number of neural responses of a neuron to different stimuli. However, it is necessary to define what is the minimum response degree to a pattern class so that a neuron can be considered sensitive to this. To analyze this response degree, we used different response thresholds and we present in Fig. 2 some cases: the neurons have to respond to 20/40/60/80% of patterns for a specific odorant to be considered sensitive to it.

As we can see in Fig. 2, when response threshold is higher, the number of neurons that respond to no or few pattern classes increases, while the number of those respond to all or many of them decreases. Since specialist neurons let odorant discrimination, we are interested to have more neurons of this kind. Also a neuron should be considered sensitive to a pattern class when it responds to this in most cases. For that, we consider that response threshold of 80% as desirable response percentage.

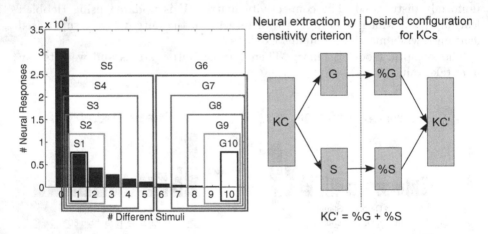

Fig. 3. Generation of KC layer according to the selection criteria of generalist and specialist neurons. We define specialist and generalist neurons by neural sensitivity. To perform this, we initially define that specialist neurons have a sensitivity of 5 or less and generalist ones have a sensitivity of 6 or more. Subsequently, we start to remove neurons with intermediate sensitivity until we only have neurons with sensitivity of 1 (specialists) and 10 (generalists), see left panel. Once these sensitivities are defined, we extract generalist and specialist neurons from KCs, excluding the neurons with sensitivity 0, and we create two sets with them. Then, a new KC' layer is generated, with the same dimensions than the original one and the desired percentage of generalist and specialist neurons, see right panel.

4   Selection Criteria of Generalist and Specialist Neuron

At the beginning of each simulation, we determine which kinds of sensitivities define neurons as generalist or specialist. To perform this, we excluded neurons with sensitivity 0. Once these sensitivities are defined, we extract these neurons and create with them two sets, one generalist and other specialist, as we can see in right panel of Fig. 3. We create a new KC layer with the same dimensions than the original one by extracting neurons of these sets, which let us to control the percentages of these two kinds of neurons. The neural network starts with all generalist neurons and we gradually replace these by specialist neurons and observe how the classification error varies during this process.

Because we used 10 kinds of patterns, we established different definitions of specialist and generalist neurons in terms of neural sensitivity. First, we defined that specialist neurons have a sensitivity of 5 or less and generalist ones have a sensitivity of 6 or more. Subsequently, we start to remove neurons with intermediate sensitivity until we only have neurons with sensitivity of 1 (specialists) and 10 (generalists), see left panel of Fig. 3. This way of defining neurons as specialists and generalists allow us to assess the relevance of neurons with intermediate sensitivities.

5   Odor Patterns

We used the MNIST digits [11], which have dimensions of 28 × 28 pixels, and binarized their information on values of 0 and 1. Since antennal lobe has a gain control mechanism [18,22,23] that keeps a constant neuronal activity for all odorants and their alterations (concentrations, mixtures, etc.), each MNIST pattern is subjected to gain control too. The method for performing this gain control is simple, we duplicated the information of each pattern to use their positive and negative image [9], see Fig. 4.

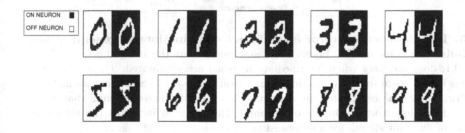

Fig. 4. Gain control in the patterns. Example of MNIST digit patterns with gain control where two populations of neurons respond inversely to each other.

We used 100 patterns, 10 patterns for each class. In the learning process these patterns are divided into 5 parts, taking one as test set and the other four as training set. This process is repeated 5 times, in order to each part can be used as test set (5-cross-validation). Thus, the training data set has 80 patterns and test set has 20 patterns.

6   Results

To analyze what happens if we vary the proportion of generalist and specialist neurons in KC layer, we used a computational model focused on AL and MB where we introduced 100 MNIST digits, 10 for each pattern class, as input. The following averaged results for the test set were obtained by supervised learning. We ran 10 simulations with 5-cross-validation. This assumes an average of 50 results for each of the system configurations shown below.

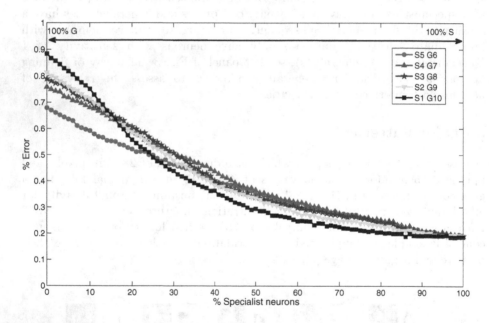

Fig. 5. Test classification error for different combinations of generalist and specialist neurons in KC layer. We can observe that the minimum error, for a $p_c = 0.1$, does not vary when intermediate neurons are eliminated. Therefore, these neurons do not seem to participate in classification success, as generalist ones increase its error. The S value indicates the maximum number of different stimuli that make firing a specialist neuron. On the other hand, the G value indicates the minimum number of different stimuli for a generalist ones, see left panel of Fig. 3.

We have used a p_c connection probability of 0.1 [15,17], for C matrix that connects AL to MB. We initialize the weights of W matrix, which connects KCs to ENs, with an intermediate p_w value, 0.5, before Hebbian learning [9]. The combination of values for Hebbian probabilities that we selected are $p_+ = 0.1$ and $p_- = 0.05$ by their good performance [8].

6.1 Only the Most Specialist Neurons are Required for a Good Odor Classification

The minimum classification error in Fig. 5 is 18.5% that are consistent with other studies [9]. Since the neural selection process reaches its minimum with 100% of specialist neurons, we can say that these neurons are responsible for odorant classification. This also happens when we do not introduce changes in KC layer. In this case, KC layer neurons are mostly specialists, as we can see in left panel of Fig. 3, and therefore we obtain a similar error to the previous one. Because its percentage of specialist neurons place its result on the right side of the error curve, Fig. 5. We also note that the minimum classification error does not change when we have only the most specialist neurons or we also have some neurons with intermediate sensitivities. This leads us to think that neurons with intermediate sensitivities do not contribute to achieve this minimum error. On the other hand, we can observe that classification error increases when in the KC layer there are more generalist neurons.

7 Discussion and Conclusions

The objective of this work is to investigate what happens if we change the percentage of generalist and specialist neurons during the feature extraction process, to learn about how odorant information is processed at olfactory system. To investigate this point, we used a simple model that retains the most relevant structural properties of the olfactory system. This model focuses on the AL and MB, where the input to single-hidden-layer neural network is the AL activity. The other layers, hidden and output that represent the MB, are composed by KCs and ENs respectively. These latter layers are connected by a connectivity matrix that implements a supervised Hebbian learning. Also ENs possess a lateral inhibition process between the different populations of neurons that are specialized in a particular pattern class. Using MNIST digits as odorant information, we analyze the neural sensitivity of each neuron and define these ones as a specialist or generalist by this information. Once we established generalist and specialist populations, we begin to vary their proportions in the KC layer. This process that not only allows observing the behavior of the generalist and specialist neurons, but also those with intermediate sensitivities.

We show that in the feature extraction phase, pathway from ORNs to KCs, the achieved minimum error in the learning phase, ENs, is obtained by the most specialized KCs. This is clearly seen in the case of neural selection, Fig. 5, since the classification error does not change when we also have neurons with intermediate sensitivities. However, it also happens before this modification of KC layer, left panel of Fig. 3, where almost all neurons are specialist. On the other hand, this error increases for the most generalist neurons and those with intermediate sensitivities. These results are consistent with researches about *Drosophila* [7,2] that measure the KC activity by calcium images, since their experimental results reveal that KCs show high selectivity for a particular odorant.

These results raise questions on the functional role of generalist neurons and neurons with intermediate sensitivities. It is possible that in other kind of problems, with a larger degree of overlapping between patterns, these neurons could have a greater role in classification. However, for this problem, they are not needed. Other aspects such as gain control and lateral inhibition of ENs deserve further analysis, since their impact in odorant classification seem critical as we observed during this work. On the other hand, we need to study the different aspects of the neuronal network for their relationship with the existence of specialist and generalist neurons. For example, the impact on these neural populations by the network dimensionality and fan-in/out phases. Their relationship with p_c and what is the value of this probability that provides an optimal configuration of specialist and generalist neurons. As if the existence of variability between neuronal thresholds affects these populations. All these points reveal details about the role of these kinds of neurons and allow us to understand how olfactory system processes odorant information.

Acknowledgments. This work was supported by the Spanish Government project TIN2010-19607 and predoctoral research grant BES-2011-049274. R.H. acknowledges partial support by NIDCD-R01DC011422-01.

References

1. Bazhenov, M., Huerta, R., Smith, B.H.: A computational framework for understanding decision making through integration of basic learning rules. The Journal of Neuroscience 33(13), 5686–5697 (2013)
2. Campbell, R.A.A., Honegger, K.S., Qin, H., Li, W., Demir, E., Turner, G.C.: Imaging a population code for odor identity in the drosophila mushroom body. The Journal of NeuroscienceSPIE Proc. 33(25), 10568–10581 (2013)
3. Chandra, S.B., Wright, G.A., Smith, B.H.: Latent inhibition in the in the honeybee, apis mellifera: is it a unitary phenomenon? Anim Cogn. 13, 805–815 (2010)
4. Christensen, T.A.: Making scents out of spatial and temporal codes in specialist and generalist olfactory networks. Chem. Senses 30, 283–284 (2005)
5. Dubnau, J., Grady, L., Kitamoto, T., Tully, T.: Disruption of neurotransmission in drosophila mushroom body blocks retrieval but not acquisition of memory. Nature 411(6836), 476–480 (2001)
6. Garcia-Sanchez, M., Huerta, R.: Design parameters of the fan-out phase of sensory systems. J. Comput. Neurosci. 15, 5–17 (2003)
7. Gruntman, E., Turner, G.C.: Integration of the olfactory code across dendritic claws of single mushroom body neurons. Nature Neuroscience 16, 1821–1829 (2013)
8. Huerta, R., Nowotny, T., Garcia-Sanchez, M., Abarbanel, H.D.I., Rabinovich, M.I.: Learning classification in the olfactory system of insects. Neural Comput. 16, 1601–1640 (2004)
9. Huerta, R., Nowotny, T.: Fast and robust learning by reinforcement signals: Explorations in the insect brain. Neural Comput. 21, 2123–2151 (2009)
10. Kaupp, U.B.: Olfactory signalling in vertebrates and insects: differences and commonalities. Nature Reviews Neuroscience 11, 188–200 (2010)

11. LeCun, Y., Cortes, C.: Mnist database (1998),
    http://yann.lecun.com/exdb/mnist/
12. Leitch, B., Laurent, G.: GABAergic synapses in the antennal lobe and mushroom
    body of the locust olfactory system. J. Comp. Neurol. 372, 487–514 (1996)
13. Lubow, R.E.: Latent inhibition. Psychol Bull. 79, 398–407 (1973)
14. Marin, E.C., Jefferis, G.S., Komiyama, T., Zhu, H., Luo, L.: Representation of the
    glomerular olfactory map in the Drosophila brain. Cell 109, 243–255 (2002)
15. Montero, A., Huerta, R., Rodriguez, F.B.: Neuron threshold variability in an
    olfactory model improves odorant discrimination. In: Natural and Artificial Models
    in Computation and Biology - 5th International Work-Conference on the Interplay
    Between Natural and Artificial Computation, IWINAC 2013, Proceedings, Part I,
    Mallorca, Spain, June 10-14, pp. 16–25 (2013)
16. Montero, A., Huerta, R., Rodriguez, F.B.: Neural trade-offs among specialist
    and generalist neurons in pattern recognition. In: Proceedings of the Engineering
    Applications of Neural Networks - 15th International Conference, EANN 2014,
    Sofia, Bulgaria, September 5-7, pp. 71–80 (2014)
17. Montero, A., Huerta, R., Rodriguez, F.B.: Regulation of specialists and
    generalists by neural variability improves pattern recognition performance.
    Neurocomputing 151, 69–77 (2015)
18. Olsen, S.R., Wilson, R.I.: Lateral presynaptic inhibition mediates gain control in
    an olfactory circuit. Nature 452(7190), 956–960 (2008)
19. Perez-Orive, J., Mazor, O., Turner, G.C., Cassenaer, S., Wilson, R.I., Laurent,
    G.: Oscillations and sparsening of odor representations in the mushroom body.
    Science 297(5580), 359–365 (2002)
20. Rodríguez, F.B., Huerta, R.: Techniques for temporal detection of neural sensitivity
    to external stimulation. Biol. Cybern. 100(4), 289–297 (2009)
21. Rodríguez, F.B., Huerta, R., Aylwin, M.: Neural sensitivity to odorants in deprived
    and normal olfactory bulbs. PLoS ONE 8(4) (2013)
22. Salinas, E., Thier, P.: Gain modulation: A major computational principle of the
    central nervous system. Neuron 27, 15–21 (2000)
23. Serrano, E., Nowotny, T., Levi, R., Smith, B.H., Huerta, R.: Gain control network
    conditions in early sensory coding. PLoS Computational Biology 9(7) (2013)
24. Tanaka, N.K., Awasaki, T., Shimada, T., Ito, K.: Integration of chemosensory
    pathways in the Drosophila second-order olfactory centers. Curr. Biol. 14,
    449–457 (2004)
25. Wilson, R.I., Turner, G.C., Laurent, G.: Transformation of olfactory
    representations in the drosophila antennal lobe. Science 303(5656), 366–370
    (2004)
26. Zavada, A., Buckley, C.L., Martinez, D., Rospars, J.-P., Nowotny, T.: Competition-
    based model of pheromone component ratio detection in the moth. PLoS One 6(2),
    e16308 (2011)

Intensity Normalization of ^{123}I-ioflupane-SPECT Brain Images Using a Model-Based Multivariate Linear Regression Approach

A. Brahim$^{(\boxtimes)}$, J. Manuel Górriz, Javier Ramírez, and L. Khedher

Department of Signal Theory, Networking and Communications,
University of Granada, Granada, Spain
brahim@ugr.es

Abstract. The intensity normalization step is essential, as it corresponds to the initial step in any subsequent computer-based analysis. In this work, a proposed intensity normalization approach based on a predictive modeling using multivariate linear regression (MLR) is presented. Different intensity normalization parameters derived from this model will be used in a linear procedure to perform the intensity normalization of ^{123}I-ioflupane-SPECT brain images. This proposed approach is compared to conventional intensity normalization methods, such as specific-to-non-specific binding ratio, integral-based intensity normalization and intensity normalization by minimizing the Kullback-Leibler divergence. For the performance evaluation, a statistical analysis is used by applying the Euclidean distance and the Jeffreys divergence. In addition, a classification task using support vector machine to evaluate the impact of the proposed methodology for the development of a computer aided diagnosis (CAD) system for Parkinsonian syndrome detection.

Keywords: Intensity normalization · DaTSCAN SPECT images · Multivariate Linear Regression · Parkinsonian syndrome · Computer-aided diagnosis system

1 Introduction

There have been a growing number of studies showing the importance of functional imaging studies using single-photon emission tomography (SPECT) or positron emission tomography (PET) tracers in neurology. In particular, Parkinson's disease (PD) and other neurodegenerative disorders are useful disease models to understand the contribution of modern functional neuroimaging techniques. The PD, the most common cause of Parkinsonism, is a neurodegenerative disease that provokes the degeneration of the central nervous system. Early in the course of the disease, the most obvious symptoms are movement-related. These include shaking, rigidity, slowness of movement and difficulty with walking and gait. Later, cognitive and behavioral problems may arise, with dementia commonly occurring in the

© Springer International Publishing Switzerland 2015
J.M. Ferrández Vicente et al. (Eds.): IWINAC 2015, Part I, LNCS 9107, pp. 68–77, 2015.
DOI: 10.1007/978-3-319-18914-7_8

advanced stages of the disease. Other symptoms include sensory, sleep, and emotional problems. The neuropathology of the disease is characterized by the progressive loss of dopaminergic neurons of the nigrostriatal pathway. This leads to a corresponding loss of dopamine transporters (DaTs) in the striatum [1]. The diagnosis of Parkinsonian syndrome (PS) is usually based on the results of clinical assessments and clinical signs have proved to be insufficient for accurate diagnosis especially at an early stage and in elderly subjects. The initial diagnoses of PD made by general neurologists have shown to be incorrect in 24% to 35% of the cases [2]. A reliable diagnostic test, which could be used to differentiate between different tremor disorders, would therefore be of great value. Thus, as a feature of PD is a marked reduction in dopaminergic neurons in the striatal region, brain imaging techniques (SPECT or PET) with specific ligands, derived from cocaine such as I-Ioflupane (better known as DaTSCAN [3] or [123I]FP-CIT [4]) can be used as a valuable tool to evaluate PD patients [3]. These specific radio-ligands bind to the dopamine transporters in the striatum and have evolved as in vivo markers of progressive dopaminergic neuron loss in PD. Previous to any kind of image processing, the functional brain images have to be normalized in terms of intensity, so that it is guaranteed that the differences between images of different subjects are due to physiological reasons and the brain functioning, and not due to the baseline calibration of the Gamma camera used for the acquisition among other factors. The most fruitful way of carrying out the intensity normalization is to consider as a reference for all the images the brain region that is not significantto differentiate between ill subjects and healthy ones. In the case of PD, the discriminant region is the striatum and the occipital region is usually chosen as a reference because it is devoid of DATs and it is usually selected as the background region [5]. However, in previous works, the whole brain area is considered, minus the striatum, as a non-specific region [6, 7]. In this sense, a new normalization technique is applied in this work, based on multivariate linear regression approach. Its performance is validated through a statistical analysis and in the classification performance for improving the diagnostic accuracy in Parkinsonism. This new approach is an extension of the mean-squared error normalization method presented in our previous work in [8]. Thus, this normalization methodology can be applied to the whole medical image, not only in a non-specific regions.

2   Materials and Methods

2.1   DaTSCAN SPECT Dataset

The database consisting of 189 SPECT images (94 Normal Controls (NCs) and 95 Parkinsonian Syndrome (PS)). The brain images were acquired by the "Virgen de la Victoria" hospital (Malaga, Spain). The demographic details are shown in Table 1. All SPECT images were spatially normalized using SPM 8 software yielding a $73 \times 73 \times 45$ three-dimensional functional activity map for each subject. This spatial normalization ensures that any given voxel in different images refers to the same anatomical position across the brains.

Table 1. Demographic details of the DaTSCAN SPECT dataset. μ and σ stand for the average and the standard deviation respectively.

| | # | Sex | | Age | | |
		M	F	μ	σ	range
NCs	94	49	45	69.26	10.16	33-89
PS	95	54	41	68.29	9.62	30-87

2.2 Intensity Normalization Approaches

There are a variety of normalization methods available in the literature for the Parkinson's disease image normalization. Several approaches approaches are described below:

Specific-to-Non-Specific Binding Ratio (BR$_{all}$): This normalization approach is based on the binding potential (specific/non-specific binding ratio (BR)) [9] which can be calculated as:

$$BR = \frac{C_{VOI} - C_N}{C_N} \tag{1}$$

where C_{VOI} is the mean count per voxel in the volume of interest (striatum, putamen or caudate nucleus) and C_N represent the mean count per voxel in the occipital cortex. This binding ratio is used for the normalization of functional brain images. BR$_{all}$ denotes the binding ratio calculated using all the brain voxels except those in the striatum as non-specific region.

Integral-based Intensity Normalization [4]: It It consists of the computation of an intrinsic parameter from the image, I_p. This normalization is performed by the estimation of the binding activity:

$$\hat{I} = \frac{I}{I_p} \tag{2}$$

where I denotes the spatially normalized image, $\hat{I}$ denotes the intensity normalized image and I_p is the integral intensity value. It can be approximated as the sum of all the intensity values of the image, giving an integral value of intensity:

$$I_p = \int I(x, y, z) \approx \sum I(x, y, z) \tag{3}$$

Intensity Normalization by Minimizing the Kullback-Leibler Divergence (MKL): The basic idea of the method presented in [10] is to estimate a multiplicative correction field in order to match a template histogram to a reference model density. The observed image I can be expressed as:

$$I = F\hat{I} + n \tag{4}$$

where F is a multiplicative intensity corruption field, n is the additional acquisition noise and $\hat{I}$ is the desired correct image. After neglecting n for having only little influence on the problem of intensity normalization and solving the eq. 4 for $\hat{I}$, the uncorrupted image is obtained as $\hat{I} \approx F^{-1}I$. The intensity adjustment parameter F^{-1} has to be chosen in a way that the Kullback-Leibler divergence [11] between the adjusted source and target data sets is minimized. The Simultaneous Perturbation Stochastic Approximation (SPSA) is used to generate the gradient estimate and then to adjust the current solution estimate according to the gradient estimate.

2.3   Multivariate Linear Regression (MLR) Model

Multivariate regression analysis is a well-known technique that is widely used in many branches of science and engineering to predict values of D responses from a set of P regressors, where $D \geq 1$ and $P \geq 1$. A MLR is generally based on the following statistical model [12]:

$$\mathbf{Y}_i = \boldsymbol{\beta_0} + \mathbf{B}^T\mathbf{x}_i + \boldsymbol{\epsilon}_i \tag{5}$$

where the symbol i is used to denote a sample unit; $\mathbf{Y}_i = (Y_{i1}, \ldots, Y_{id}, \ldots, Y_{iD})^T$ and $\mathbf{x}_i = (x_{i1}, \ldots, x_{ip}, \ldots, x_{iP})^T$ are the D-dimensional vector of the response variables and the P dimensional vector of the fixed regressor values for the ith unit, respectively; $\boldsymbol{\beta_0}$ is a D-dimensional vector containing the intercepts for the D responses; $\mathbf{B}$ is a matrix of dimension $P \times D$ whose (p, d)th element, β_{pd}, is the regression coefficient of the pth regressor on the dth response; finally, $\boldsymbol{\epsilon}_i$ denotes the D-dimensional random vector of the error terms corresponding to the ith unit.

To simplify the computation, the multiple regression model in terms of the observations can be written using matrix notation. Using matrices allows for a more compact framework in terms of vectors representing the observations, levels of regressor variables, regression coefficients, and random errors. The model is in the form:

$$\mathbf{Y} = \mathbf{X}\boldsymbol{\beta} + \boldsymbol{\epsilon} \tag{6}$$

2.4   Intensity Normalization Using MLR

In this paper, we apply MLR to a specific pre-processing step of image processing application, that is intensity normalization. The following assumption will be used to perform the normalization task of each image subject:

– Let $\mathbf{Y} \sim \bar{\mathbf{I}}$ be an $N \times 1$ vector of observations on the dependent variable, that is the template image.
– Let $\mathbf{X} \sim \mathbf{I}$ be an $N \times 1$ vector where we have observations on 1 independent variables for N observations, that is the raw data.
– Let the number of observations N be the number of image voxels.
– Let β be an 1×1 vector of unknown parameters that we want to estimate and ϵ be an $N \times 1$ vector of errors.

Therefore, by estimating the values of β and ϵ we can easily transform the intensity distribution of $\mathbf{I}$ to $\hat{\mathbf{I}}$ using the following expression:

$$\hat{\mathbf{I}} = \mathbf{I}\,a + b \tag{7}$$

where $\hat{\mathbf{I}}$ is the normalized image, $a = \beta$ and $b = \bar{\epsilon}$ are the intensity normalization parameters, they represent the scale and offset of the intensity transformation [8]. The used criteria for obtaining our estimates of β is to minimize the residual sums of squares (or error sums of squares) (RSS), which can be defined as:

$$\epsilon^T \epsilon = (\bar{\mathbf{I}} - \mathbf{I}\beta)^T (\bar{\mathbf{I}} - \mathbf{I}\beta) = \bar{\mathbf{I}}^T\bar{\mathbf{I}} - \bar{\mathbf{I}}^T\mathbf{I}\beta - \mathbf{I}^T\beta^T\bar{\mathbf{I}} + \beta^T\mathbf{I}^T\mathbf{I}\beta$$
$$= \bar{\mathbf{I}}^T\bar{\mathbf{I}} - 2\beta^T\mathbf{I}^T\bar{\mathbf{I}} + \beta^T\mathbf{I}^T\mathbf{I}\beta \tag{8}$$

To find the β that minimizes the RSS, we need to take the derivative of eq. 8 with respect to β. This gives us the following equation:

$$\frac{\partial \epsilon^T \epsilon}{\partial \beta} = -2\mathbf{I}^T\bar{\mathbf{I}} + 2\mathbf{I}^T\mathbf{I}\beta = 0 \tag{9}$$

From eq. 9, the "normal equations" are:

$$(\mathbf{I}^T\mathbf{I})\beta = \mathbf{I}^T\bar{\mathbf{I}} \tag{10}$$

Multiplying both sides of the eq. 10 by the inverse $(\mathbf{I}^T\mathbf{I})^{-1}$ gives us the following equation:

$$(\mathbf{I}^T\mathbf{I})^{-1}(\mathbf{I}^T\mathbf{I})\beta = (\mathbf{I}^T\mathbf{I})^{-1}\mathbf{I}^T\bar{\mathbf{I}} \tag{11}$$

By definition, $(\mathbf{I}^T\mathbf{I})^{-1}(\mathbf{I}^T\mathbf{I}) = \mathbf{I}_N$, where $\mathbf{I}_N$ is the identity matrix and N is the number of image voxels. As a result, the least square solution of β is :

$$\beta = (\mathbf{I}^T\mathbf{I})^{-1}\mathbf{I}^T\bar{\mathbf{I}} \tag{12}$$

The residuals are computed as:

$$\epsilon = \bar{\mathbf{I}} - \mathbf{I}\beta = \bar{\mathbf{I}} - \mathbf{I}(\mathbf{I}^T\mathbf{I})^{-1}\mathbf{I}^T\bar{\mathbf{I}} = (\mathbf{I}_n - \mathbf{I}(\mathbf{I}^T\mathbf{I})^{-1}\mathbf{I}^T)\bar{\mathbf{I}} \tag{13}$$

The goal in this work is to transform linearly all the intensity values for different image subjects using predictive modeling based on MLR. The procedure to perform the intensity normalization is summarized as follows:

- Firstly, the template $\bar{\mathbf{I}}$ is computed as :

$$\bar{\mathbf{I}} = \frac{1}{N_c} \sum_{i \in X_c} (\mathbf{I}_i(x, y, z) + \mathbf{I}_i(-x, y, z)) \tag{14}$$

  where X_c denotes the subset of control images, N_c the number of control images, $\mathbf{I}_i(x, y, z)$ is the ith image and $\mathbf{I}_i(-x, y, z)$ is its reflected image in the $x = 0$ hemisphere midplane.
- Secondly, the different parameters β and ϵ of the MLR model are estimated using the eqs. 12 and 13.
- Lastly, the normalized images are computed by linearly transforming the voxel intensity of each image subject using the model in eq. 7.

3 Results and Discussion

3.1 Image Analysis

The proposed methodology has been tested on the database described in section 2.1 which presents a high degree of variability in the intensity level for the specific/non-specific area (see Fig. 1.a). Furthermore, those images present a relatively poor signal-to noise ratio (SNR) in the so-defined non-specific region provided by the image acquisition system at the nuclear medicine department. The visual assessment of normalized subjects using the compared approaches in Figs. 1.b, 1.d and 1.e shows that they are not enough for an accurate intensity normalization procedure. Since the intensity heterogeneity in the non-specific between subjects is quite reduced. After intensity normalization using our proposed methodology detailed in section 2.4, the intensity heterogeneity in the non-specific region is reduced and the inter-subject intensity differences in the non-specific region due to noise and artifacts are clearly reduced. These qualitative effects can be seen more clearly at the image level in the results of Fig. 1.c. Moreover, this figure proves that, post-normalization, the contrasts of relevant features in the striatum are improved. Hence, the separation between the striatum and the non-specific region is increased. Thus, the proposed post-normalization approach allows us as well to guarantee that the differences between the two classes (NC and PS subjects) are due only to the uptake of the tracer in the discriminant region (striatum) and not due to the baseline calibration of the gamma camera applied for the acquisition. Therefore, this normalization method is suitable for preprocessing of ^{123} I-ioflupane brain images for diagnosis purposes.

3.2 Statistical Analysis

The proper adjustment of the resulting normalized set of images can be evaluated by means of some defined metrics that provide a measure of the difference between two sets of samples, such as Euclidean distance (ED) and Jeffreys divergence (JD) measure [13]. In this particular case, the set of samples to be evaluated with these metrics is the histogram of the normalized 3D image, referred to the histogram of a reference target 3D image (typically, the mean image of the subjects) in the non-specific region. Lower values of these divergences represent less difference between the two distributions of histograms.

Thus, the inter-subject variability is quantitatively computed in Tables 2 and 3, both before and after normalization and according to class belonging. Note that, the lowest ED, JD values and the lowest error are obtained (in terms of the standard deviation) by the proposed normalization method based on linear intensity normalization using a model-based MLR. Compared to raw data and normalized images using the comparative approaches, its ED and JD represent less difference between adjusted images and target model, both in the same class or in different classes. Thus, more intensity homogeneity in the reference region.

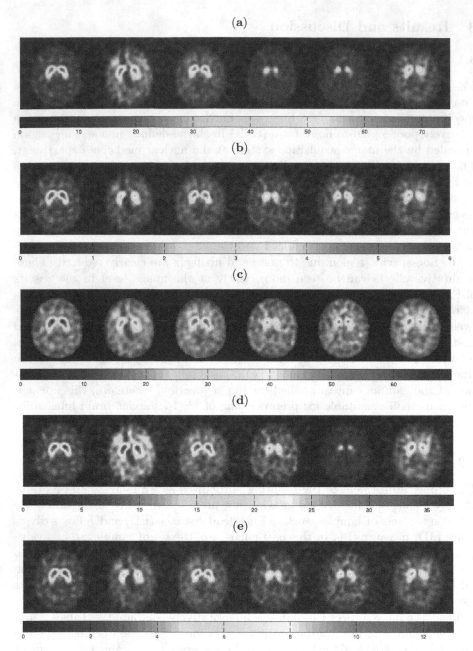

Fig. 1. A given trans-axial slices of 6 selected brain images; 3 healthy subjects (left) and 3 PS patients (right): a) raw DaTSCAN brain images b) BR_{all} brain images c) MLR brain images d) MKL brain images and e) integral-based intensity normalized brain images.

Table 2. Mean Euclidean distance and standard deviation for original images and intensity normalized images in the non-specific region

Normalization approach	class	Euclidean distance
Raw data	NCs	0.5761±0.2567
(spatial normalization)	PS	0.2180±0.1023
	NCs+PS	0.5601±0.2479
BR_{all}	NCs	0.5058±0.2220
	PS	0.4916±0.1886
	NCs+PS	0.5430±0.2303
MLR	NCs	**0.4477±0.1909**
	PS	**0.1805±0.0727**
	NCs+PS	**0.4830±0.1982**
MKL	NCs	0.5097±0.2233
	PS	0.3559±0.1816
	NCs+PS	0.5394±1.1314
Integral	NCs	0.4998±0.2068
	PS	0.4856±0.1777
	NCs+PS	0.4928±0.2163

Table 3. Mean Jeffreys Divergence and standard deviation for original images and intensity normalized images in the non-specific region

Normalization approach	class	Jeffreys Divergence
Raw data	NCs	0.7408±0.5743
(spatial normalization)	PS	0.1427±0.1459
	NCs+PS	0.7422±0.5943
BR_{all}	NCs	0.1634±0.0927
	PS	0.1348±0.0953
	NCs+PS	0.1860±0.1432
MLR	NCs	**0.1511±0.0690**
	PS	**0.0607±0.0340**
	NCs+PS	**0.1657±0.0974**
MKL	NCs	0.5781±0.4166
	PS	0.2406±0.2596
	NCs+PS	0.5700±0.4094
Integral	NCs	0.1655±0.0867
	PS	0.1303±0.0865
	NCs+PS	0.1873±0.1276

3.3   Quantitative Classification Performance of Parkinsonism

The classification performance of the proposed intensity normalization approaches is tested using the classical multivariate approach Voxel-As-Feature approximation (VAF) and Support Vector Machines (SVM) with linear kernel and Leave-One-Out cross-validation strategy in order to extract several performance parameters: accuracy, sensitivity, specificity. This approximation uses all voxels in each image as a feature vector, which is used as an input to the classifiers. Only linear SVM has been used to compute the results, due to the large number of input features to the classifier, to obtain more generalizable results and to avoid the small sample size problem. The performances of the raw images and the intensity normalization methods are presented in Table 4. A significant improvement of the performance results is carried out by the proposed approach. For instance, the accuracy gain is 8.61%, the sensitivity and the specificity gains are 11.99% and 6.30% compared to unnormalized intensity images (raw data). Compared to the other intensity normalization methods, the accuracy gain is up to 7.09%, the sensitivity and the specificity gains are up to 7.72% and 3.64%. The underlying reason for these improvements is the reduction of inter-subject intensity variability between different images of the same class, and between images of different classes as shown in the previous sections. The behavior of the VAF system with this strategy of pre-processing highlights the benefits of using an intensity normalization, as well shows its ability and robustness in PS pattern detection.

Table 4. Comparison between the performance (%) achieved with the proposed intensity normalization methodology, the raw data and the other normalization approaches using VAF approach and linear SVM classifier

Normalization approach	Accuracy	Sensitivity	Specificity
Raw data	79.58%	78.95%	80%
BR_{all}	85.83%	88.68%	83.78%
MLR	**88.19%**	**90.74%**	**86.30%**
MKL	81.10%	83.02%	79.73%
Integral	85.04%	88.46%	82.66%

4   Conclussion

Multivariate regression has been widely used for years applying to almost all the areas in our lives. The present work proposes a new intensity normalization method based on predictive modeling using MLR. This methodology has the advantage of automatically normalizing the 3D functional brain images without using anatomical information. In addition, this approach could be used for visual assessment in clinical practice; since it is not dependent on any pathological information about the specific disease. It might also be applied to other image modalities, such as positron emission tomography (PET). We have proved that, using this approach, we are able to obtain brain images with very similar intensity distribution. We compare our method with a widely used approaches for the Parkinson's disease image normalization. Further analysis reveals that, the intensity normalization is improved by the proposed methodology, the inter-subjects intensity differences are reduced and the artifacts and noise affecting the source images are removed. In addition, the proposed normalization method demonstrates also its ability and robustness in PS pattern detection as it provides good value of accuracy compared to other approaches.

Acknowledgments. This work was partly supported by the Ministerio de Economía y Competitividad (MEC) under the TEC2012-34306 project and the Consejería de Innovación, Ciencia y Empresa (Junta de Andalucía, Spain) under the Excellence Projects P09-TIC-4530 and P11-TIC-7103.

References

1. Booij, J., Habraken, J., Bergmans, P., Tissingh, G., Winogrodzka, A., Wolters, E., Janssen, A., Stoof, J., Van Royen, E.: Imaging of dopamine transporters with Iodine-123-FP-CIT SPECT in healthy controls and patients with parkinson's disease. Journal of Nuclear Medicine 39(11), 1879–1884 (1998)
2. Jankovic, J., Rajput, A., McDermott, M., Perl, D.: The evolution of diagnosis in early parkinson disease. Archives of Neurology 57(3), 369–372 (2000)
3. Seifert, K.D., Wiener, J.I.: The impact of DaTscan on the diagnosis and management of movement disorders: A retrospective study. American Journal of Neurodegenerative Disease 2(1), 29–34 (2013)
4. Illán, I.A., Górriz, J.M., Ramírez, J., Segovia, F., Jimenez-Hoyuela, J.M., Lozano, S.J.O.: Automatic assistance to parkinson's disease diagnosis in DaTSCAN SPECT imaging. Medical Physics 39(10), 5971–5980 (2012)
5. Benamer, H.T.S., Patterson, J., Grosset, D.G.: Accurate Differentiation of Parkinsonism and Essential Tremor Using Visual Assessment of ^{123}I-FP-CIT SPECT Imaging: The ^{123}I-FP-CIT Study Group. Movement Disorders 15(3), 503–510 (2000)
6. Brahim, A., Górriz, J., Ramírez, J., Khedher, L.: Linear intensity normalization of DaTSCAN images using Mean Square Error and a model-based clustering approach. Studies in Health Technology and Informatics 207, 251–260 (2014)
7. Padilla, P., Górriz, J., Ramírez, J., Salas-González, D.D., Illn, I.: Intensity normalization in the analysis of functional DaTSCAN SPECT images: The α-stable distribution-based normalization method vs other approaches. Neurocomputing 150, 4–15 (2015)
8. Brahim, A., Górriz, J., Ramírez, J., Khedher, L.: Applications of gaussian mixture models and mean squared error within datscan spect imaging. In: 2014 IEEE International Conference on Image Processing (ICIP), pp. 3617–3621 (2014)
9. Scherfler, C., Seppi, K., Donnemiller, E., Goebel, G., Brenneis, C., Virgolini, I., Wenning, G., Poewe, W.: Voxel-wise analysis of [123]Iβ-CIT SPECT differentiates the Parkinson variant of multiple system atrophy from idiopathic Parkinson's disease. Brain 128(7), 1605–1612 (2005)
10. Weisenfeld, N., Warfteld, S.: Normalization of joint image-intensity statistics in MRI using the Kullback-Leibler divergence. In: IEEE International Symposium on Biomedical Imaging: Nano to Macro, 2004, vol. 1, pp. 101–104 (2004)
11. Kullback, S.: Information Theory and Statistics. Dover Books on Mathematics. John Wiley & Sons, New York (1959)
12. Galimberti, G., Soffritti, G.: A multivariate linear regression analysis using finite mixtures of t distributions. Computational Statistics and Data Analysis 71, 138–150 (2014)
13. Jeffreys, H.: An invariant form for the prior probability in estimation problems. Proceedings of the Royal Society of London Series A 186, 453–461 (1946)

Independent Component Analysis-Based Classification of Alzheimer's Disease from Segmented MRI Data

L. Khedher, Javier Ramírez, J. Manuel Górriz, A. Brahim, and I.A. Illán

Department of Signal Theory, Networking and Communications,
University of Granada, Granada, Spain
lailaa@ugr.es

Abstract. An accurate and early diagnosis of the Alzheimer's disease (AD) is of fundamental importance to improve diagnosis techniques, to better understand this neurodegenerative process and to develop effective treatments. In this work, a novel classification method based on independent component analysis (ICA) and supervised learning methods is proposed to be applied on segmented brain magnetic resonance imaging (MRI) from Alzheimer's disease neuroimaging initiative (ADNI) participants for automatic classification task. The ICA-based method is composed of three step. First, MRI are normalized and segmented by the Statistical Parametric Mapping (SPM8) software. After that, average image of normal (NC), mild cognitive impairment (MCI) or AD subjects are computed. Then, FastICA is applied to these different average images for extracting a set of independent components (IC) which symbolized each class characteristics. Finally, each brain image from the database was projected onto the space spanned by this independent components basis for feature extraction, a support vector machine (SVM) is used to manage the classification task. A 87.5% accuracy in identifying AD from NC, with 90.4% specificity and 84.6% sensitivity is obtained. According to the experimental results, we can see that this proposed method can successfully differentiate AD, MCI and NC subjects. So, it is suitable for automatic classification of sMRI images.

Keywords: Alzheimer's disease · Mild cognitive impairment · Magnetic resonance imaging · Computer aided diagnosis · Independent component analysis · Support vector machine · Supervised learning

1   Introduction

The progression of AD is of great interest in medical research as 1 in 3 seniors in the United States dies with dementia [1]. Based on signs and symptoms, physicians usually track AD using the Clinical Dementia Rating (CDR) system. Using CDR, subjects are classified in three states such as NC, MCI and AD patient. The progression of AD can be characterized by atrophy of gray matter (GM) and white matter (WM) brain tissues. These structural changes in brain

© Springer International Publishing Switzerland 2015
J.M. Ferrández Vicente et al. (Eds.): IWINAC 2015, Part I, LNCS 9107, pp. 78–87, 2015.
DOI: 10.1007/978-3-319-18914-7_9

facilitate the distinction of an AD brain from a NC brain; however, the distinction between MCI and NC is subtle [2]. To facilitate the distinction between the different classes, the proposed work used the segmentation technique in the structural Magnetic Resonance Imaging (sMRI) datasets in order to consider only the relevant brain regions which are significant for AD detection. Then, different algorithms such as Independent Component Analysis (ICA) [3,4], have been used to perform the feature extraction task and to solve the problem of small sample size [5].

In this study, ICA is used to extract maximally spatially independent sources revealing patterns of variation that occur in segmented sMRI images, in order to identify segmented sMRI difference between AD patients, MCI and NC subjects [6]. Once a significant features were selected, we build a SVM to manage the classification task. Support vector machine (SVM) [7,8], a kind of machine learning techniques, plays an important role to perform classification of sMRI after features extraction by ICA which is novel and innovative. It has been successfully used in diagnosing medical images.

This paper shows a computer aided diagnosis (CAD) system for the early detection of AD using SVM classifiers applied to the projection of each average brain image of each classes into the Independent Component (IC) brain images space (Fig. 1). This process reduced the dimensionality of the feature space, thus facing the small sample size problem. The present work based on the segmentation of brain MRI images, separating white matter (WM) from gray matter (GM) and using either or both of them to extract relevant informations. In addition, the proposed method grows a CAD system for the early detection of the Alzheimer's disease (AD) and developed with the aim of reducing the subjectivity in visual interpretation of these scans by clinicians, thus improving the accuracy of diagnosing Alzheimer's disease in its early stage.

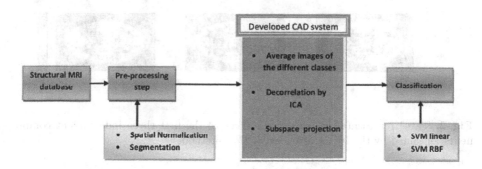

Fig. 1. Detailed schema of the proposed CAD system

1.1 ICA Application to Segmented MRI Images

Independent Component Analysis (ICA) [3, 4], is a statistical technique that represents a multidimensional random vector as a lineal combination of non-gaussian random variables (the so-called "independent components") to be as independent as possible. In brain images, the dataset is an ensemble of 3D brain images Γ_i, whose size M is typically $121 \times 145 \times 121$ voxels. Let the full 3D brain image set be $\Gamma_1, \Gamma_2, ..., \Gamma_N$, each understood as a vector of dimension M, each pertaining to a class ζ_k , $k = 1, 2, ..., K$, where K is the total number of classes. The average brain image of the dataset is defined as:

$$\mathbf{X}_k = \frac{1}{N_k} \sum_{\Gamma_i \subset \zeta_k} \Gamma_i, \qquad k = 1, 2, ..., K. \tag{1}$$

where N_k denotes the number of images in the class ζ_k. These average images $\mathbf{X} = [\mathbf{X}_1, \mathbf{X}_2, ..., \mathbf{X}_K]$, which number is the same as the number of classes.

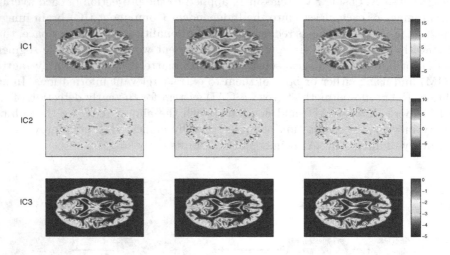

Fig. 2. Three representative transversal slices of the first three independent components, obtained by the ICA feature extraction method

ICA algorithm is expected to separate them into a independent set of sources through $\mathbf{WX} = \mathbf{Y}$, where $\mathbf{W}$ is the estimated separating matrix, and $\mathbf{Y}$ is the estimated set of K independent sources $\mathbf{Y} = [\mathbf{Y}_1, \mathbf{Y}_2, ..., \mathbf{Y}_K]$ (see Fig 2). These latent variables are the essence of the different classes. Furthermore, these independent sources $\mathbf{Y}_k$ define a orthogonal basis which span a independent components subspace of the "brain images space". For the classification task, $\mathbf{Y}$ can be used to project each image onto the independent component (IC) space.

Each projected image produced a vector of weights so that a matrix of weights can be constructed with each brain tissues database. this matrix Ω is giving by:

$$\Omega_{ki} = \mathbf{Y}_k.\mathbf{\Gamma}_i, \qquad i = 1, 2, ..., N, \quad k = 1, 2, ..., K. \tag{2}$$

and describes the contribution of each IC in representing the input brain image $\mathbf{\Gamma}_i$, treating the IC as a basis set for brain images (See Fig. 3).

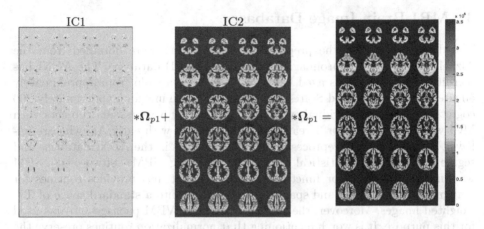

Fig. 3. Reconstruction of a random AD subject using the independent brain source basis. The representation is encoded in the coefficients (Ω_{p1} and Ω_{p2}).

The matrix Ω contains the most relevant information extracted from ICA. We used this matrix Ω for the following classification task, that is N k-dimensional patterns: $\mathbf{x}_i = [\Omega_{1i}, \Omega_{2i}, ..., \Omega_{ki}]$, $i = 1, 2, ..., N$. each of them with its corresponding class labels $y_i \in \times \{\pm 1\}$.

2   Classification Using Support Vector Machines (SVM)

Support Vector Machines (SVM) are a state-of-the-art classification method introduced in 1992 by Boser, Guyon, and Vapnik [9], widely used and applied to lots of different problems, specially in pattern recognition [9]. Being a maximum-margin classifier, and having excellent generalization ability in the linear case, they have been successfully used in a number of CAD systems [10,11]. The principal function of SVM classifier, in this work, is to separate a set of binary labeled training data by means of a hyperplane that is maximally distant from the two classes (known as the maximal margin hyperplane). The goal is to build a function $f \colon \mathbb{R}^K \to \{\pm 1\}$ using training data, consisting of k-dimensional patterns x_i and class labels y_i:

$$(\mathbf{x}_1, y_1), (\mathbf{x}_2, y_2), ..., (\mathbf{x}_N, y_N) \in (\mathbb{R}^K \times \{\pm 1\}) \tag{3}$$

so that f will correctly classify new examples $(\mathbf{x}, y)$. For nonlinearly separable data, the optimization process needs to be modified to work in combination with kernel techniques, so that the hyperplane defining the SVM corresponds to a non linear decision boundary in the input space [12]. The use of kernels enables to map the data into some other dot product space (feature space) through a non linear transformation. We used a Radial Basis Function (RBF) $F(x, y) = exp(-\gamma||x - y||^2)$ as kernel function.

3   MRI Brain Image Database

The database used in the preparation of this work were obtained from the Alzheimer's Disease Neuroimaging Initiative (ADNI) database. The ADNI has recruited over 800 adults aged between 55 to 90 years old from approximately 50 sites across the United States and Canada. These include approximately 200 cognitively normal individuals who are followed for 3 years, 400 subjects with MCI who are followed for 3 years, and 200 patients with early AD who are followed for 2 years. We preprocessed the MRI images in the ADNI database and segmented using the Statistical Parametric Mapping (SPM) software [13]. SPM was initially designed for functional images, but it also provides routines for realignment, smoothing and spatial normalization into a standard space of T1-weighted images. Moreover, the template from the VBM package [14] was used for this purpose. It is worth mentioning that normalization routines preserve the amount of tissues and not the intensities [15]. Thus, images from ADNI database were resized to 121×145×121 voxels with voxel sizes of 1.5 mm (sagital) x 1.5 mm (coronal) x 1.5 mm (axial). After normalization, the whole brain MRI data was automatically segmented using the Statistical Parametric Mapping (SPM8) software. This process partitions brain into gray matter (GM), white matter (WM) and cerebrospinal fluid (CSF) regions. Therefore, the database used in this work contains 1075 T1-weighted MRI images separated into three different classes: 229 NC subjects, 401 MCI (312 stable MCI and 86 progressive MCI) and 188 AD. As only the first exam for each patient has been used in this work, 818 images were used for assessing the proposed approach. The demographic of the subjects who compose the dataset used in this work described in detail in [11]. To prevent this study from being prevalence-dependent, 188 individuals have been randomly selected from each of the classes AD, MCI and normal controls. The same set of patients has been used in the rest of the paper.

4   Experiments and Results

We have developed a computer aided diagnosis (CAD) system using the ICA feature extraction method described above and two SVM classifiers; linear and nonlinear (RBF kernel). Since our purpose is to distinguish between NC, MCI subjects and AD patients, first we have trained the CAD system with only NC and AD images (group 1). Second, we have used MCI images as NC (group 2) and , third, MCI images as AD (group 3). Thus, we can measure the ability of the

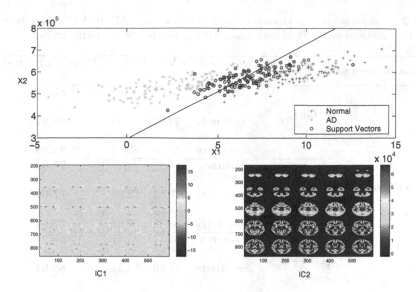

Fig. 4. Projecting of each GM image from group 1 onto the IC images space

Table 1. Statistical performance measures of the proposed model with different SVM classifiers, for the three sample groups

		Group 1	Group 2	Group 3
Brain tissues	Kernel	Acc/Sens/Spec(%)	Acc/Sens/Spec(%)	Acc/Sens/Spec(%)
GM	Linear:	87.11/89.91/84.09	77.59/80.31/75.11	84.79/84.61/85.09
	RBF:	87.61/90.42/84.61	77.61/80.89/74.51	83.51/86.22/80.09
WM	Linear:	77.61/77.61/77.61	72.91/77.11/68.61	80.31/83.09/77.61
	RBF:	77.91/77.59/78.21	72.11/73.42/71.09	81.11/84.59/77.61
GM+WM	Linear:	86.21/88.89/83.49	76.61/80.31/72.89	85.41/85.61/85.11
	RBF:	84.61/86.72/82.41	76.31/81.03/72.01	83.81/85.61/81.91

proposed method to distinguish between NC, MCI and AD patients. The classification performance of our approach is tested using the k−fold cross validation method with the feature vectors dimension k=2.

Fig. 4 shows the distribution of the x_1 training vector versus the x_2 in the feature space. From this plot, it can be read that the separation of the training samples in two classes is giving by the combination of both x_1 and x_2. This result is in correspondence with the IC1 and the IC2 obtained. In the first independent component, there are a high intensity values in the posterior cingulate gyri and precunei, and the second independent component complements the IC1 by giving a high intensity level in temporo-parietal regions. Both of these regions affected by glucose hypometabolism in the AD.

Table 2. Comparaison of performance parameters using VAF, ICA [18] and the proposed ICA method using SVM with linear kernel for the three sample groups

Type of groups	Brain tissues	Parameter(%)	VAF	ICA [18]	Proposed method
		Accuracy	65.61	84.65	**87.12**
	GM	Sensitivity	72.91	86.46	**89.92**
		Specificity	58.29	82.45	**83.98**
		Accuracy	64.49	70.26	**77.61**
NOR.vs.AD	WM	Sensitivity	70.81	72.93	**77.61**
		Specificity	58.29	67.02	**77.61**
		Accuracy	65.71	86.37	**86.21**
	(GM+WM)	Sensitivity	75.09	88.34	**88.91**
		Specificity	56.21	83.98	**83.48**
		Accuracy	55.21	69.46	**77.62**
	GM	Sensitivity	55.11	69.03	**80.27**
		Specificity	58.29	69.96	**74.49**
		Accuracy	44.82	63.51	**72.89**
NOR.vs.MCI	WM	Sensitivity	51.11	63.24	**77.09**
		Specificity	39.55	63.78	**68.48**
		Accuracy	52.2	70.19	**76.62**
	(GM+WM)	Sensitivity	56.3	72.89	**80.27**
		Specificity	48.15	67.49	**72.89**
		Accuracy	48.91	69.19	**84.81**
	GM	Sensitivity	45.82	70.27	**84.62**
		Specificity	52.08	68.11	**85.07**
		Accuracy	59.51	59.46	**80.31**
MCI.vs.AD	WM	Sensitivity	64.98	62.16	**82.98**
		Specificity	52.51	56.76	**77.59**
		Accuracy	61.11	69.83	**85.41**
	(GM+WM)	Sensitivity	66.69	73.43	**85.59**
		Specificity	53.31	66.24	**85.11**

5 Analysis

The results summarized in table 1 show that the idea of compressing a large amount of brain image data to a small element image basis for characterizing the AD is useful to develop the CAD system. The proposed method exhibit interesting performance results , however the best performance result is obtained for group 1 (NC. vs. AD) when combined with a RBF kernel, reaching 87.61% accuracy. Specificity took the value 84.61% and sensitivity 90.42% in the case of using only GM brain tissue images. From table results, we find that within the experiments established to differentiate between NC and MCI, the value of accuracy decreases significantly (77.6% for GM images). However, group 3 (MCI. vs. AD) present a high accuracy rate. According to the results obtained, the MCI subjects can be more similar to NC. It can be noted from this table that the accuracy value with the WM brain tissue images decreases wildly. It means that there are not a remarkable change in the WM brain tissue of the AD patient. Thus, in order to not lose the information from the WM images and to earn the maximum relevant informations related with the disease, we used the combination of features extracted from both GM and WM tissue distributions to improve the classification and to include the relevant informations from the both tissue brains.

In table 2, the method presented in this work was compared with other existing in the literature. The voxel-as-features (VAF) approach results are reported as reference, since different studies have concluded that this method is, at least, comparable to visual assessments performed by experts [16,17]. The use of SVM classifier in combination with ICA leads to better performance, due to the small dimensionality of the feature space and the use of independent features from the brain images. In all the groups, the proposed method outperform the VAF approximation [16,17], as well as the ICA method proposed in [18]. As a conclusion, the proposed CAD system yields better classification results with smaller computational time than the ICA method used in [18]. Furthermore, our methodology can produce a valid approach to perform a CAD system for early diagnosis of AD.

6 Conclussion

For early AD diagnosis, we proposed in this work a new CAD system based on ICA, which makes to extract the highly representative features from each average brain image, related to typical AD patterns, for classification task. The principal aim of ICA application in this work was to find a set of independent component sources that present each Alzheimer's disease stage. Besides, it is proven to be an opportune method of reducing the dimension of the feature with projection into a discriminative subspace , and also, selection the most relevant informations. After that, a SVM classifier is trained to detect these AD patterns, and its performance is evaluated. The resultant system performs significantly well with the segmented MRI database, and demonstrates its ability and robustness in AD detection as its provides high accuracy values. It outperforms several proposed methods in bibliography, specially the baseline VAF approach.

Acknowledgments.. This work was partly supported by the Ministerio de Economía y Competitividad (MEC) under the TEC2012-34306 project and the Consejería de Innovación, Ciencia y Empresa (Junta de Andalucía, Spain) under the Excellence Projects P09-TIC-4530 and P11-TIC-7103.

References

1. Alzheimer's Association, Alzheimer's News (2013), http://www.alz.org/news and events facts and figures report.asp
2. Cuingnet, R., Gerardin, E., Tessieras, J., Auzias, G., Lehricy, S., Habert, M.O., Chupin, M., Benali, H., Colliot, O.: Automatic classification of patients with Alzheimer's disease from structural MRI: A comparison of ten methods using the ADNI database. Neuroimage 56, 766–781 (2011)
3. Illán, I.A., Górriz, J.M., Ramírez, J., Salas-Gonzalez, D., López, M.M., Segovia, F., Chaves, R., Gómez-Rio, M., Puntonet, C.G.: 18F-FDG PET imaging analysis for computer aided Alzheimer's diagnosis. Information Sciences 181, 903–916 (2011)
4. Illán, I.A., Górriz, J.M., Ramírez, J., Salas-Gonzalez, D., López, M.M., Segovia, F., Padilla, P., Puntonet, C.G.: Projecting independent components of SPECT images for computer aided diagnosis of Alzheimers disease. Pattern Recognition Letters 31, 1342–1347 (2010)
5. Duin, R.P.W.: Classifiers in almost empty spaces. In: Proceedings of the 15th International Conference on Pattern Recognition, vol. 2, pp. 1–7 (2000)
6. Magnin, B., Mesrob, L., Kinkingnehun, S., Pelegrini-Issac, M., Calliot, O., Sarazin, M., Dubais, B., Lehericy, S., Benali, H.: Support vector machine-based classification of alzheimer's disease from whole-brain anatomical mri. Neuroradiology 51, 73–83 (2009)
7. Jaramillo, D., Rojas, I., Valenzuela, O., Garcia, I., Prieto, A.: Advanced systems in medical decision-making using intelligent computing. Application to magnetic resonance imaging. In: International Joint Conference on Neural Networks (IJCNN) (2012)
8. Padilla, P., Lopez, M., Gorriz, J.M., Ramirez, J., Salas-Gonzalez, D., Alvarez, I.: NMF-SVM based CAD tool applied to functional brain images for the diagnosis of Alzheimer's disease. IEEE Trans. Med. Imaging 31, 207–216 (2012)
9. Boser, B.E., Guyon, I.M., Vapnik, V.N.: A training algorithm for optimal margin classifiers. In: Haussler, D. (ed.) Proceedings of the 5th Annual ACM Workshop on Computational Learning Theory, pp. 144–152. ACM Press, Pittsburgh (1992)
10. Martinez-Murcia, F.J., Grriz, J.M., Ramrez, J., Moreno-Caballero, M., Gomez-Rio, M.: Parkinson's Progression Markers Initiative. Parametrization of textural patterns in 123i-ioflupane imaging for the automatic detection of parkinsonism. Medical Physics 41, 012502 (2013)
11. Khedher, L., Ramrez, J., Grriz, J.M., Brahim, A., Segovia, F.: Early diagnosis of Alzheimer's disease based on Partial Least Squares, Principal Component Analysis and Support Vector Machine using segmented MRI images. Neurocomputing 151, 139–150 (2015)

12. Chaves, R., Ramrez, J., Grriz, J.M., Lpez, M., Salas-Gonzalez, D., Alvarez, I., Segovia, F.: SVM-based computer-aided diagnosis of the Alzheimer's disease using t-test NMSE feature selection with feature correlation weighting. Neurosci. Lett. 461, 293–297 (2009)
13. Ashburner, J., Friston, K.: Human Brain Function (2003)
14. Psychiatry SBMGD, Vbm toolboxes. University of Jena (2013), http://dbm.neuro.uni-jena.de/vbm8/VBM8-Manual.pdf
15. Ashburner, J., Barnes, G., Chen, C., Daunizeau, J., Flandin, G., Friston, K.: SPM8 manual. In: Functional Imaging Laboratory. Institute of Neurology, London (2012)
16. Stoeckel, J., Ayache, N., Malandain, G., Malick Koulibaly, P., Ebmeier, K.P., Darcourt, J.: Automatic classification of spect images of Alzheimer's disease patients and control subjects. In: Barillot, C., Haynor, D.R., Hellier, P. (eds.) MICCAI 2004. LNCS, vol. 3217, pp. 654–662. Springer, Heidelberg (2004)
17. Stoeckel, J., Malandain, G., Migneco, O., Malick Koulibaly, P., Robert, P., Ayache, N., Darcourt, J.: Classification of SPECT images of normal subjects versus images of alzheimer's disease patients. In: Niessen, W.J., Viergever, M.A. (eds.) MICCAI 2001. LNCS, vol. 2208, pp. 666–674. Springer, Heidelberg (2001)
18. Khedher, L., Ramrez, J., Grriz, J.M., Brahim, A.: Automatic classification of segmented MRI data combining Independent Component Analysis and Support Vector Machines. In: Innovation in Medicine and Healthcare, InMed, vol. 207. Lecture notes in IOS Press (2014)

Trajectories-State: A New Neural Mechanism to Interpretate Cerebral Dynamics

Sergio Miguel-Tomé[1](✉)

Grupo de Investigación en Minería de Datos (MiDa),
Universidad de Salamanca, Salamanca, Spain
`sergiom@usal.es`

Abstract. With regard to neural networks, there are two different areas which have generated two lines of research. One research interest comes from the field of computer science which seeks to create and design neural networks capable of performing computational tasks. In this line of research, any neural network is relevant because the important issue is the problems which they are capable of resolving. Thus, neural networks are computational devices and computational power and the computational process which they perform are researched. The other interest of research is related to neuroscience. This focuses on both neural and brain activity. The big difference between these two lines of research can be observed from the outset. In the first, the neural network is designed and its performance on computational tasks is then researched. In the second, performance on computational tasks is known but the neural mechanism is not and neuroscience seeks to identify it. An interaction between these two lines of research is very positive because it produces synergies which generate important advances in both lines of research e.g. Hopfield's networks. This article enunciates a neural mechanism to interpret neural dynamics based on some of the results produced by computer science. This mechanism identifies an internal or external state s with a formal language L. Independently, if the mechanism exist or not in the human brain, this mechanism can be used to design new architectures for neural networks.

1 Introduction

Human conscious thinking appears to us as a serial symbolic process. This fact was the inspiration for the Turing's work concerning computation. At the same time, the work of Turing was a key element in the emergence of cognitive psychology. However, the computer metaphor does not help us to understand how the brain works because the Turing Machine is an architecture very different from human brain architecture. The human brain has hundreds of billions of processors and performs a massive parallel processing. Allan Newell suggested that a computational process should be described by a multilevel description to reach a full understanding[20]. However, Newell's proposal was focused on his concept of the symbol system. David Marr proposed applying a multilevel description to the visual system[13]. Concretely, He proposed three levels and he

© Springer International Publishing Switzerland 2015
J.M. Ferrández Vicente et al. (Eds.): IWINAC 2015, Part I, LNCS 9107, pp. 88–97, 2015.
DOI: 10.1007/978-3-319-18914-7_10

developed algorithms concerning these three levels. The idea of explaining the nervous system using a multilevel description has also been defended by Mira and Delgado[19][18].

Brain activity is measured at different levels using different techniques EEG, MEG and fMRI. Those techniques are related to the synchrony of a huge number of neurons. Therefore, although EEG, MEG, fMRI are useful to research cognitive behavior, they do not permit discovery or research of the neural mechanism which allows the flow of information nor explain how the information is processed. Since the second half of the 20th century, electrophysiology has researched the processing done by a neuron in depth using microelectrodes. Thus, from the '80s onwards we have known the features of the units of processing of the brain very well[12]. However, there is much theoretical and experimental work which has indicated that information can be coded in sparse patterns of activity[15]. Thus, if we wish to understand the function of a population of neurons and we try to inference that function from independent recordings of individual neurons we will have probably overlooked important aspects of the information coded in changing patterns of activity that are distributed throughout the populations of neurons.

For a long time, there were no techniques available to perform recordings of each neuron in a population simultaneously. Nevertheless, in order to fill the gap between neuron activity and brain activity, mathematical models about the mechanism which the neural populations of the brain may be using to be successful have been proposed and researched. Two possible kinds of neural mechanisms underlying the neural dynamics have been stated: vector-state and chaotic itinerancy. Each one of these kinds of mechanisms groups many methods together. In the 21st century, the time has come for discussion to pass from the merely theoretical to the experimental given that it is now possible to record each neuron of a population simultaneously. One of the techniques which permits this is calcium imaging[25]. Calcium imaging allows studies of neuronal activity in hundreds of neurons within neuronal circuits. These advances have recently permitted the observation of the neural dynamics of neural circuits[3]. The work of Brice Bathellier et al. [3] has shown that the superficial layers of the auditory cortex contain a small number of attractor-like neuronal assemblies which produce categorization of sounds. The issue of how the dynamics of the brain operate appears to be reaching an interesting new point where experiments may allow us to better understand how it works.

This article is structured as follows. The second section of this article explains the problem about understanding neural dynamics from the point of view of multilevel descriptions. The third section summarizes the two kinds of mechanism which currently exist for neural activity. In the fourth section, a short overview of formal languages, neural networks and deterministic finite automata is given. The fifth section proposes a new mechanism to categorization which could be responsible for cerebral dynamics. The last section contains a summary and a short discussion of this article.

2   The Multilevel Description

The idea of a multilevel description to describe computational process was proposed initially by Allan Newell [20]. Newell explained that a level can be defined in two ways :

"First, it can be defined autonomously, without reference to any other level to an amazing degree, programmers need not know logic circuits, logic designers need not know electrical circuits, managers can operate at the configuration level with no knowledge of programming, and so forth. Second, each level can be reduced to the level below. Each aspect of a level - medium, components, laws of composition and behavior - can be defined in terms of systems at the level next below. The architecture is the name we give to the register-transfer level system that defines a symbol (programming) level creating a machine language and making it run as described in the programmers manual for the machine. Neither of these two definitions of a level is the more fundamental. It is essential that they both exist and agree. "[20]

Human conscious thinking is at the symbol level. How the symbol level emerges from the level below in computers is completely understood as a consequence of the fact that computers are designed by humans. However, how the symbol level emerges from the level below in the human brain is an open problem. David Marr proposed that a multilevel description should also be applied to understand visual system but his proposal could also be applied to any neural system. He also proclaimed that there were only three main levels. From that point, Jose Mira Mira and Ana Delgado have defined a hierarchy of three main levels: the physical level, the symbol level and the knowledge level[19][18][17]. They have defended the use of this framework to understand the neural function. One of the issues which they have discussed in their works is about how the symbol level could emerge from the physical level. They argue that knowing all about the processing in the physical level in neural networks is not enough to know what is being calculated at the symbol level. They write the next:

"Let us assume for one moment that we have a complete theory of the physical level. In other words, that we know everything about individual signals, circuits and local operators, similar to our knowledge about digital electronics and computer architecture. Would we know what the brain is calculating? Would we know the 'program' and the emerging cognitive processes? Of course, not."[19]p. 224

Therefore, to move from one level to the next we need a set of rules which permit a translation. One important concept which Mira and Delago introduce in their work is the concept of a 'dynamic symbol' [19]. Mira and Delgado define their concept of a 'symbol' very clearly providing a definition of symbols which includes neurophysiological systems.

" The symbols[in neurophysiological systems] are active and dynamic entities associated with specific patterns of spatio-temporal signals(electrical, chemical and electronics) that are presented repeatedly in a stable, independent and autonomous way and associated with specific references in the organism's internal and external environment."[19].

They expand explicitly the concept of a symbol employed by Newell because they wish to apply the multilevel description to all kinds of computational systems, including conexionist systems.

" *we should remember that the concept of symbol proposed here is not the one proposed by Newell and Simon (1976) in their Physical Symbol System Hypothesis because in our proposal the symbols are evolutive, dynamic, connectionist and grounded in specific physiological mechanisms and they are not programmable in the conventional sense, but via dynamic adjustment processes of synaptic efficiency, controlled by the states of activity in the network. Per contra, symbols in conventional programmable computers are static, descriptive, and arbitrary semantics.*" [19].

Mira and Delgado stated the problematic situation which remains in understanding the global function of the brain:

"*We know how to construct models at the physical level, in terms of mathematical relationships between measured physical quantities as function of time, but we still do not know how to construct formal models of symbols and relations between these symbols with the exception of those peripheral situations in which the neural networks are situated at end of the effector and/or receptor sides of the nervous system, with accessible points of physical measurement. It seems clear to us that we cannot use the conventional programming strategies as used in computers, because our symbols are embodied in the neural dynamics.*"[19].

The research currently being undertaken about biological neural activity is confined basically to the physical level because the register of each neuron of a neural population is really complex and it is only now that some experiments are beginning to be done where each neuron of a population is recorded. However, from the theoretical field of neural networks two main proposals have been to translate neural activity from the physical level to the symbol level. These proposals are summarized in the next section.

3   Neural Activity and Encoding States

There is no knowledge of how the brain encodes human conscious thinking. However, many researchers have proposed hypothetical mechanisms of how populations of neurons could encode symbols states which build the human conscious thinking. Two concepts have been proposed up to now: vector-state and chaotic itinerancy. The concept of vector-state is behind algorithms of neural networks for the associative memories developed by Kohonen, Anderson, Amari, Hopfield, and others[9][6] and the attractor dynamics developed by Amari, Hirsch, Hopfield, Amit, and others[6] [1] [5] [2]. From vector-state concept an internal or external state is represented by a vector which is defined by the state of the units of the neural network. The other concept which is proposed is chaotic itinerancy[26][27], which has been proposed to give a dynamical interpretation of cortical transitory behaviors. Some experiments have shown activity which contains a set of dynamically switching cortical states[8] and irregular transitions in cat visual cortex[4]. The proposers of chaotic itinerancy argue against dynamics related with concept of vector-state (e.g. the vector is an attractor) because they consider it cannot explain the transitory phenomena mentioned[27]. Actually, there is not any experiment which has resolved the contradiction between these two concepts. Therefore, it remains unknown how discrete dynamics reflect perceptual categories in the neural activity. Furthermore, there is no reason to think that vector-state and chaotic itinerancy are unique mechanisms to produce represent internal or external states in a neural network.

4   Formal Languages, Neural Networks and Deterministic Finite Automata

The work of Warren McCulloch and Walter Pitts in 1943 showed that the nervous system could be understood as a computational system[14]. McCulloch–Pitts model is equivalent to other computational model, the Deterministic Finite Automata (DFA)[16]. A useful way to classify the power of a computational model is to compare it with the kind of languages defined by the theory of generative grammars developed by Noam Chomsky. Thus, the McCulloch–Pitts model and the DFA can be seen as acceptors of languages. One of the great advantages of neural networks is the capacity to learn complex task related with classification or pattern recognition but one of the disadvantages is that the knowledge learned by the neural network is difficult to extract and understand. This can be seen in the problem of formal language learning. A formal language L is a set of strings of symbols that may be constrained by a concrete set of rules. A string, w, of a language is denominated a word. The symbols which can contain a word conform a set, Σ named alphabet. The length of a word is the number of symbols which the word contains, denoted $|w|$. Therefore, if $w = 101$ on the alphabet $\Sigma = \{0, 1\}$, then $|w| = 3$ We distinguish languages with finite words and with infinite words. The generative grammars of Chomsky permit the definition of languages with finite words. The language with all the words on the alphabet Σ is denoted by Σ^*. Thus, a formal language L over an alphabet Σ is a subset of Σ^*. Artificial intelligence and cognitive science has been concerned about formal language learning. One of the techniques to resolve this problem has been the use of recurrent networks. Some researchers have proposed that the recurrent networks which recognize formal languages are simulating a DFA. Thus, we could achieve a understanding of the recurrent network generating a DFA from the own neural network. This process is denominated DFA extraction. The success of the DFA extraction from the networks is a matter of debate[10]. However, from the McCulloch–Pitts model to dynamical recurrent networks there is a connection with formal languages.

5   A New Kind of Neural Mechanism: Trajectories-State

In this section a new neural mechanism is proposed for the categorization of inputs. The basis of the new mechanism is the following:

1. Categorizing is a fundamental process of cognition. There are different levels of categorization; the more abstract the categories which an animal can create the greater its cognitive skills. Forgetting the qualitative differences of the processes of categorization, it is a fundamental feature which all animals possess ranging from animals which need this feature for survive[23] to animals which use it to play high cognitive behaviors[22].

2. Existence of discrete dynamics in animal brain for categorizing. Experimental research has shown that firing in the superficial layers of auditory cortex is organized into a small number of attractor-like neuronal assemblies, whose responses can predict an animal's sound discrimination performance[3].

3. A neural population can be seen as a dynamic system. Neural networks have been used to model dynamical systems[28] but a neural network may be considered a dynamical system by itself. Thus, the continuous input which an animal receives from its environment and recurrent connections can convert a neural population into a dynamical system.

4. The time sequence of states produced by a dynamical system can be considered a word. A word is a sequence of symbols; therefore, if the states of the dynamical system are considered an alphabet, then the sequence of states produced by a dynamical system can be seen as words.

5. Conscious states are composed of a finite time sequence of inputs. Several investigations have shown that there is a threshold of time which is needed to overcome to perceive a stimulus[7] [24].

6. A formal language can be interpreted as a category whose objects are words.

7. Neural networks can enable the recognition of formal languages[21]. This is relevant because a formal language is considered as an object at symbol level. Thus, this connection can be used to move from the physical level to the symbol level.

Given the facts (1) and (2) narrated above, we should find a mechanism which will be able to categorize if we wish to explain some brain dynamics. Using facts (3) and (4) we see that the activity of a neural population may be interpreted as a mechanism which generates words. Fact (5) implies that these words should have a finite length. However, the words generated by the dynamical network would not be very interesting if all the words generated belong to the same formal language. Thus, taking fact (6) and if we suppose that the dynamical system generates words of different formal languages, a network can assign a category to a input sequence if the trajectory of the dynamical system interpreted as a word belongs to a formal language which corresponds to the category. Fact (7) shows that a neural network can accept or reject words in order to identify if they belong to a formal language. Thus, the mechanism of the image 1, denominated trajectories-state, is proposed as a one of the possible mechanisms which may produce neural dynamics.

The mechanism consists in two sub-networks connected consecutively which can be interpreted at the symbol level in the following way. The first sub-network will transform the input into a word. The second sub-network takes the word and recognizes to which formal language it belongs. Thus, while usually DFA

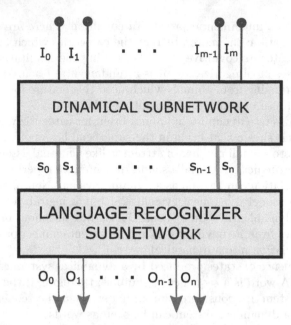

Fig. 1. Scheme of the trajectories-state mechanism proposed to explain neural dynamics

or neural networks are used to define a unique formal language by accepting or rejecting words, here the objective is a neural network which recognizes to which formal language the word belongs. Therefore, the second sub-network could consist of a set of networks where each one is dedicated to recognizing a formal language and the output of each recognizer sub-network conforms to a vector state which determines one formal language among a set of them. Thus, a category is defined as a set of trajectories but when the category must be used it is represented by a state which is interpreted as a formal language at the symbol level. In addition to categorization the mechanism proposed here may filter noise. A sensor can have several variations when it is recording due to noise or fast fluctuations of the environment. Thus, it may be necessary to filter that noise or its fluctuations. If the number of languages which can be recognized is similar to the size of the alphabet then a language can be used to represent a filtered input.

If this mechanism were to exist in the brain, then the states which represent the formal language would be the human thought symbols. How this activity becomes to be consciousness is outside the scope of this article and probably is related to physical issues which are not understood yet[11]. One of the interesting issues is that this mechanism is scalable because it can be reapplied. Thus, the temporal sequence of languages recognized also conform to a string which can be considered a word if we use the languages as the alphabet of formal languages.

Thus, a hierarchy of languages can be defined where the categories are more abstract in a new level than the previous level.

6 Discussion

The brain needs to encode external and internal states to make decisions and perform behaviors. Up to now, two kinds of mechanism have been proposed to explain how a neural network can encode states: vector-state and chaotic itinerancy. Also, these two kinds of mechanism have been proposed to explain cerebral dynamics. Ichiro Tsura considers that the way in which the brain encodes information is chaotic itinerancy.

"*Furthermore, another misleading theory in conventional brain theory is the theory based on the description of non-stationary and transitory processes by a geometric attractor.*"[27]

However, the work of Brice Bathellier et al. [3] has shown that the superficial layers of auditory cortex is organized into a small number of attractor-like neuronal assemblies. Thus, the nature of the neural mechanism which produces cerebral dynamics is still an open question.

This article has proposed a new neural mechanism which may underlie some high-level brain functioning: trajectories-state. The idea expressed by this new concept is that a set of trajectories are grouped to define a category. This mechanism can also be translated to the symbolic level. Thus, at the symbol level the trajectories are words of a formal language and the category is formalized as the formal language. Therefore, a formal language becomes a symbol for an internal or external state.

Neuroscientific experiments have demonstrated two facts:

- Activity which contains a set of dynamically switching cortical states[8] and irregular transitions visual cortex[4].
- Firing in the superficial layers of auditory cortex is organized into a small number of attractor-like neuronal assemblies[3].

These two facts, although apparently appearing contradictory from the discussion between vector-state and chaotic itinerancy, can be reconciled using the trajectories-state mechanism. The performance of a bottom-up analysis of neural dynamics based on the trajectories-state mechanism, it may explain the results obtained in neuroscientific experiments in the following way. The continuous generation of words by the dynamical sub-network may explain the dynamically switching or irregular transitions and the attractors would be the language recognizer sub-network. Of course, this explanation is only a hypothesis and complex experiments will have to be carried out in order to validate it and it is not clear if neuroscientists can perform experiments which demonstrate whether this interpretation is right or wrong. However, there is an interesting issue which must be noted about the mechanism trajectories-state. It is that this mechanism can incorporate the mechanisms of vector-state and chaotic itinerancy because these mechanisms can be used in the sub-networks. From my point of view, it is not that one kind of neural mechanism is true and the others are false in their attempt to explaining cerebral dynamics. It may be that the brain is using more than one mechanism to code information if it allows the brain to optimize its global performance and operate in a more efficient manner. Furthermore, it may be possible that different species of animals have nervous systems which use different neural mechanisms to represent environmental or internal features. Evolution could have produced different mechanisms although all nervous systems are connectionist systems. In any way, independently whether this new proposed mechanism exist in the biological realm or not, this new kind of mechanism is interesting because it can be used in artificial neural networks to do tasks of classification or filtering noise. Finally, what can be deduced about the existence of several neural mechanism to represent external or internal states is that the neural network has a high computational versatility which from an evolutive point of view (where the capacity for variation is very important) it is a very interesting mechanism to produce behavior.

References

1. Amari, S.-I.: Neural theory of association and concept-formation. Biological Cybernetics 26(3), 175–185 (1977)
2. Amit, D.J.: Modeling Brain Function: The World of Attractor Neural Networks. Cambridge University Press (1992)
3. Bathellier, B., et al.: Discrete neocortical dynamics predict behavioral categorization of sounds. Neuron 76(2), 435–449 (2012)
4. Gray, C.M., et al.: Synchronization of oscillatory neuronal responses in cat striate cortex: Temporal properties. Visual Neuroscience 8, 337–347 (1992)
5. Hirsch, M.W.: Convergent activation dynamics in continuous time networks. Neural Networks 2(5), 331–349 (1989)
6. Hopfield, J.J.: Neural networks and physical systems with emergent collective computational abilities. Proceedings of the National Academy of Sciences 79(8), 2554–2558 (1982)
7. Joliot, M., Ribary, U., Llinás, R.: Human oscillatory brain activity near 40 hz coexists with cognitive temporal binding. Proceedings of the National Academy of Sciences 91(24), 11748–11751 (1994)
8. Kenet, T., et al.: Spontaneously emerging cortical representations of visual attributes. Nature 425, 954–956 (2003)
9. Kohonen, T.: Associative Memory-A System Theoretical Approach. Springer (1978)
10. Kolen, J.F.: Fool's gold: Extracting finite state machines from recurrent network dynamics. In: Advances in Neural Information Processing Systems, vol. 6, pp. 501–508. Morgan Kaufmann (1994)
11. Llinás, R.R., et al.: Gamma-band deficiency and abnormal thalamocortical activity in p/q-type channel mutant mice. Proceedings of the National Academy of Sciences 104(45), 17819–17824 (2007)
12. Llinás, R.: The intrinsic electrophysiological properties of mammalian neurons: insights into central nervous system function. Science 242(4886), 1654–1664 (1988)
13. Marr, D.: Vision: A Computational Investigation into the Human Representation and Processing of Visual Information. Henry Holt and Co., Inc., New York (1982)

14. McCulloch, W., Pitts, W.: A logical calculus of the ideas immanent in nervous activity. Bulletin of Mathematical Biophysics (5), 115–133 (1943)
15. Meyers, E.M., Freedman, D.J., Kreiman, G., Miller, E.K., Poggio, T.: Dynamic population coding of category information in inferior temporal and prefrontal cortex. Journal of Neurophysiology 100(3), 1407–1419 (2008)
16. Minsky, M.L.: Computation: Finite and Infinite Machines. Prentice-Hall, Inc. (1967)
17. Mira, J., Delgado, A.E.: Where is knowledge in robotics? some methodological issues on symbolic and connectionist perspectives of AI. In: Zhou, C., Maravall, D., Ruan, D., Kacprzyk, J. (eds.) Autonomous Robotic Systems, pp. 3–34 (2003)
18. Mira, J., Delgado, A.: Neural modeling in cerebral dynamics. Biosystems 71(1-2), 133–144 (2003)
19. Mira, J.M., García, A.E.: On how the computational paradigm can help us to model and interpret the neural function. Natural Computing 6(3), 211–240 (2007)
20. Newell, A.: The knowledge level. AI Magazine 2(2), 1–33 (1981)
21. Omlin, C.: Understanding and Explaining DRN Behavior. In: Field Guide to Dynamical Recurrent Networks, pp. 207–227. Wiley-IEEE Press (2001)
22. Pepperberg, I.: Talking with alex: Logic and speech in parrots. Scientific American 9(4), 60–65 (1998)
23. Polack, C., McConnell, B., Miller, R.: Associative foundation of causal learning in rats. Learning and Behavior 41(1), 25–41 (2013)
24. Sekar, K., et al.: Cortical response tracking the conscious experience of threshold duration visual stimuli indicates visual perception is all or none. Proceedings of the National Academy of Sciences 110(14), 5642–5647 (2013)
25. Stosiek, C., et al.: In vivo two-photon calcium imaging of neuronal networks. Proceedings of the National Academy of Sciences 100(12), 7319–7324 (2003)
26. Tsuda, I.: Toward an interpretation of dynamic neural activity in terms of chaotic dynamical systems. Behavioral and Brain Sciences 24(5), 793–810 (2001)
27. Tsuda, I.: Hypotheses on the functional roles of chaotic transitory dynamics. Chaos: An Interdisciplinary Journal of Nonlinear Science 19(1), 15113 (2009)
28. Zimmerman, H., Neuneier, R.: Neural Network Architectures for the Modeling of Dynamic Systems. In: A Field Guide to Dynamical Recurrent Networks, pp. 311–350. Wiley-IEEE Press (2001)

Global and Local Features
for Char Image Classification

Deisy Chaves[1(✉)], Maria Trujillo[1], and Juan Barraza[2]

[1] Multimedia and Computer Vision Group, Universidad del Valle,
Ciudadela Universitaria Meléndez, Cali, Colombia
{deisy.chaves,maria.trujillo}@correounivalle.edu.co
[2] Coal Science and Technology Group, Universidad del Valle,
Ciudadela Universitaria Meléndez, Cali, Colombia

Abstract. The use of image analysis in understanding how powdered coal burns during the combustion plays a significant role in setting combustion parameters. During the pulverised coal combustion, char particles are produced by devolatising coal and represent the dominant stage in the combustion process. The pyrolysis produces different char morphologies that determine coal reactivity affecting the performance of coal combustion in power plants and the emissions of carbon dioxide, $CO2$. In this paper, an automatic char classification model is proposed using supervised learning. A general classification model is trained given a set of char particles classified by an expert. In particular, Support Vector Machine (SVM) and Random Forest are the trained classifiers. Two types of features are evaluated to built classification models: local and global. Local features are calculated using the Scale-Invariant Transform Feature (SIFT). Global features are defined based on the morphology classification by the International Committee for Coal and Organic Petrology (ICCP). Each classifier is trained by SVM or Random Forest and evaluated using a 10-fold cross-validation. The 70% of data is used as training set and the rest as testing set. A total of 2928 char-particle images are used for evaluating performance of classification models. Additionally, evaluation of model generalisation capability is done using a test set of 732 char particle images. Results showed that global features – defined by the application domain – increase significantly the accuracy of classifiers. Also, global features have more generalisation power than local features. Local features lack of meaning in the application domain and classifiers build with local features – such as SIFT – depend crucially on the training set.

Keywords: Char classification · Global features · Local features · Bag-of-features · Support Vector Machine · Random Forest

1 Introduction

Char is produced by devolatising coal and represents the dominant stage in a combustion process. Char particles have different morphologies that determine coal reactivity which affects the performance of coal combustion in power

© Springer International Publishing Switzerland 2015
J.M. Ferrández Vicente et al. (Eds.): IWINAC 2015, Part I, LNCS 9107, pp. 98–107, 2015.
DOI: 10.1007/978-3-319-18914-7_11

plants [6] [14]. Char morphologies have been studied in order to propose a classification into meaningful groups [13] [2] [1]. The classification proposed by Alvarez and Lester [1] is recommended by the International Committee for Coal and Organic Petrology (ICCP). This classification consists of eight char-types. However, it can be summarised into two groups: *char reactive* with morphology of thin-walled, high porosity and large superficial area and *char no reactive* with morphology of thick-walled, low porosity and small superficial area. Char particle with reactive morphologies are more desirable for coal combustion. Figure 1 illustrates char morphologies and classifications.

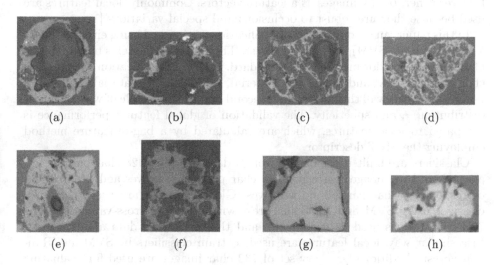

(a)   (b)   (c)   (d)

(e)   (f)   (g)   (h)

Fig. 1. Illustration of char morphologies and classificaction. *Char reactive*: *(a)* Crassisphere; *(b)* Teniusphere: *(c)* Tenuinetwork; *(d)* Crassisnetwork. *Char no reactive*: *(e)* Mixed Porus; *(f)* Mixed Dense; *(g)* Solid; *(h)* Inertoid.

The char classification problem has been addressed using manual and automatic approaches. Manual classification is performed by an expert. Char particles – in a char-block – are observed through a microscope and classified based on observed morphological characteristics [13] [1]. This process is subjective and requires a significant amount of time since a coal sample classification requires observing at least 400 char particles [14]. On the other hand, automatic classification is based on a set of char digital images taken by a camera attached to a microscope. Images are processed to automatically identify char particles and quantify morphological characteristics, following the ICCP standard, such as: area of particle, undevolatilised material, wall thickness and porosity of particle [2] [17] [5]. However, small changes during image acquisition, such as illumination, may produce incorrect classifications. For instance, gray intensities may be erroneously measured as undevolatilised material affecting the particle classification. It is due to the ill-posed characteristics of this problem.

Image classification – in a general context – has been addressed using supervised learning. A classification model may be trained using machine learning algorithms from a given set of labeled images by an expert [9] [11] [3] [15]. A bag-of-feature method may be used to represent image content. The classification process may be summarised in four stages [9] [3]: 1) Feature extraction by partitioning images into patches and describe image content using local descriptors; 2) Image codification by assigning descriptors from each image patch to a predetermined vocabulary; 3) Calculation of aggregated statistics using coded descriptors (pooling) to generate a bag-of-features by image; 4) Classification by applying a classification model. The classifier is trained using the bag-of-features generated by images as a feature vectors. Commonly, local features are used because they are robust to occlusions and special variations [11].

In this paper, an automatic char classification is proposed using either Support Vector Machine (SVM) or Random Forest. The proposed model is based on nine global features, following the ICCP standard. Selected classification features are: char particle area, undevolatilised material, devolatilised material, number of porous, porosity, wall thickness – first, second and third quartile of wall thickness distribution –, and sphericity. The validation of global features performance is compared to local features, which are calculated by a bag-of-feature method employing the SIFT descriptor.

Classifiers are built using images from a dataset with 2928 char particle images where 1464 images correspond to char particle reactives and 1464 images correspond to char particle no reactives. Global features are used to train a classifier using SVM and Random Forest with a 10-fold cross-validation. The 70% of data was used as training set and the rest of the data as testing set. In a similar way, local features are used to train classifiers by SVM and Randon Forest. Additionally, a new set of 732 char images are used for evaluating generalisation capability of the classification models. Results showed that global features – defined by the problem domain – increase significantly the accuracy of classifiers compared to bag of local features. Apparently, Random Forest exhibits better generalisation capabilities. Local features lack of meaning in the application context and classification models depend of the training set.

The remaining of the paper is organised as follows: Section 2 describes global and local features used to represent the char images; Section 3 presents briefly the supervised learning algorithms used to train the char classification models; Section 4 is focused on the experimental evaluation; and Section 5 includes final remarks.

2   Feature Extraction

A feature vector is obtained from a char particle image. The feature vector is formed by characteristics extracted from an image and is used to train classification models. In this paper, two type of features are considered, global and local. Global features are calculated over the whole image. Local features are calculated over image patches or small regions and characteristics are obtained using a bag-of-feature method.

2.1 Global Features

Nine features are selected to represent char particles based on the char classification proposed by the ICCP [1] along with morphological characteristics observed in Colombian chars [5]. A description of global characteristics calculation is presented.

1. **Particle Area:** A binary image is obtained from a char particle image (in gray scale) using the Triangle method [20] . White color pixels correspond to the particle area, in Figure 2b.
2. **Percentage of Undevolatilised Material:** is calculated as the ratio between the total of undevolatilised material and the particle area. The undevolatilised material corresponds to gray intensities in char images with values between 130 and 160 approximate, in Figure 2h.
3. **Percentage of Devolatilised Material:** is defined as the ratio between the total of devolatilised material and the particle area. Undevolatilised materials are recognised using color information, in Figure 2c.
4. **Number of Porous:** identified in a char particle image, in Figure 2d.
5. **Porosity:** is calculated as the ratio between the total area of pores or voids and the total particle area.
6. **Sphericity:** is obtained as the ratio between the minimum and the maximum Feret diameter, in Figure 2e.
7. **First, Second and Third Quartile of Wall Thickness:** Line transects are used for calculating wall thickness, in Figure 2f.

2.2 Local Features

A char particle image is represented by a bag-of-features method which can be summarised in three stages [9] [3], and feature extraction is illustrated in Figure 3.

Firstly, local regions are calculated with a dense regular grid of 16x16 pixels with a spacing of 8 pixels. Each region is described by SIFT descriptor as a vector of 128 values [12]. An unordered vector $X = [x_1, \ldots, x_i, \ldots, x_N] \in \mathbb{R}^{128}$ is obtained from the image, where N is the total of SIFT descriptors extracted. SIFT is choose since it is robust against changes of scale, rotation, and viewpoint [11].

Secondly, given a set of local features, X and a visual codebook, $B = [b_1, \ldots, b_j \ldots, b_M] \in \mathbb{R}^{128}$ with M visual codewords, a codification matrix $H = [\alpha_{1,1}, \ldots, \alpha_{j,i}, \ldots, \alpha_{M,N}] \in \mathbb{R}^{M \times N}$ is obtained. H contains the codification of each x_i image features using the M codewords on the codebook [3]. The codification is done by Locality-constrain Linear Coding (LLC) [16].

The codebook is obtained by the k-means algorithm, with $k = 2000$, to a subset of 300000 local features randomly chosen from all features descriptors of the training set. Each cluster centroid corresponds to a visual codeword.

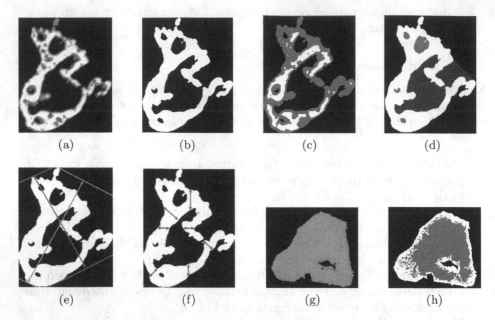

Fig. 2. Illustration of global features. *(a)* Original image of a char particle; *(b)* Area of particle; *(c)* Devolatilised material; *(d)* Identified pores in gray color; *(e)* Illustration of the Feret diameters; *(f)* Line transects used for calculating wall thickness. *(g)* Original image of a char particle; *(h)* Undevolatilised material in grey color using *(f)*.

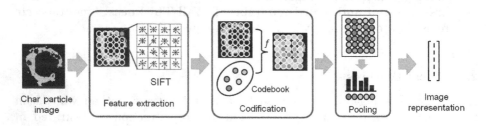

Fig. 3. Extraction of local features

Thirdly, given the codification matrix, H, a single vector $Z = [z_1, \ldots, z_j, \ldots, z_M]$ or bag-of-features to represent the whole image is calculated using a pooling function [3]. In this paper, a max-pooling function is used. The max-pooling function represents an image using the maximum codification values α_i for the codeword b_j by the function [18] [15]:

$$g(H) = z : \forall j, z_j = max(\alpha_{j1}, \alpha_{j2}, \ldots, \alpha_{jN}). \tag{1}$$

The pooling process is illustrated in Figure 4.

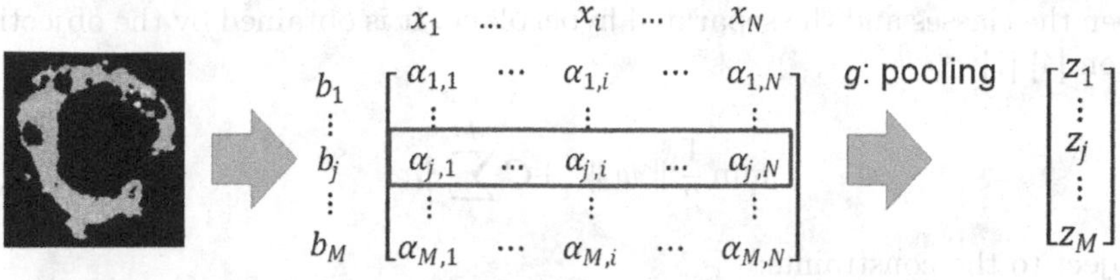

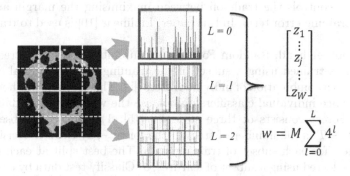

Fig. 4. Pooling of local features

$$w = M \sum_{l=0}^{L} 4^l$$

Fig. 5. Spatial pooling of local features using $L = 2$

Additionally, the spatial pyramid image representation [18] is used to preserve spatial information, since limiting the descriptive ability of pooling representation is disregard in general pooling [18]. In this paper, L pyramid levels are considered to split the image at level into 2^L square regions. At each level region is used the max-pooling function and the final image representation is obtained by concatenating all pooling vectors. the The spatial pooling process is illustrated in Figure 5.

3 Classification Algorithms

Given a training set $\{(Z_i, y_i)\}_{i=1}^n$ where $y_i \in \{-1, +1\}$ corresponds to a label class (char reactive or char no reactive) and Z_i is a feature vector with the obtained representation from a char particle image –in Section 2. A classification model is learnt by a machine learning algorithm. In this paper, Support Vector Machine (SVM) and Random Forest algorithms are used.

SVM learnt the classifier by choosing the best hyperplane that categorises the two considered classes – char reactive and char no reactive. The best hyperplane is defined as one which maximises the margin or distance of separation between the classes and the separated hyperplane. It is obtained by the objective function [4] [7]:

$$\min \frac{1}{2} \parallel w \parallel^2 + C \sum_i^l \xi_i, \tag{2}$$

subject to the constraints:

$$y_i(w \cdot Z_i + b) \geq 1 - \xi_i, \forall Z_i, \xi_i \geq 0, \tag{3}$$

where w is termed the weight vector, b is the bias and C is a regularisation parameter. C controls the trade-off between maximising the margin and minimising the training error term. In this paper, Liblinear [10] is used to train SVM classifiers.

On the another hand, Random Forest is an ensemble of decision trees. Each decision tree is trained using a subset of the training data. The final classifier corresponds to a combination of individual trees [8]. This training method avoids over-fitting since individual classifiers do not use the whole training data [19].

Random Forest consists of three steps [8] [15]: 1) Choose T subsets from training data – T is the number of trees in the forest –; 2) Grow a decision tree, with D nodes, for each subset of training data. The best split at each decision tree node is selected using a subset of features; 3) Classify test data by combining the outputs of the T trees.

4   Building and Evaluating Classification Models

Classification models are built using a dataset composed by 2928 char images – 1464 images correspond to char reactives and 1464 images correspond to char no reactives. Global and local features describe in Section 2 are used to train classifiers. Global features are normalised in order to avoid the effect of different scales. Local features are codified by the LLC algorithm with a spatial pyramidal pooling using the parameters $L = \{0, 1, 2\}$.

Each classifier, learns by SVM or Random Forest algorithm, is evaluated using a 10-fold cross-validation. The 70% of data are used as training set and the rest as testing set. Additionally, the generalisation of the char classification models is evaluated. The best classifier using cross-validation is chosen and used to classify a new set of 732 char images – 366 char reactive images and 366 char no reactive images.

SVM classification models are trained using a regularisation parameter $C = 5$.

Random Forest classifiers are trained using $T = 100$ trees. Each tree is grown to a maximum level size $D = 6$. The number of features selected to learn the split function, at each node, is $\rho = \sqrt{|\tau|}$ where the number of features, τ, depends on the type of feature used (global or local).

4.1   Suppot Vector Machine Evaluation

Table 1 shows the accuracy results using global and local features.

Table 1. Accuracy results of char classification using SVMs

Feature	# of Features	Accuracy I. C. (%)	Accuracy (%)	
			Best classifier	Generalisation
Global	9	99.27 ± 0.37	99.77	99.86
	$2000, L = 0$	88.15 ± 0.71	89.06	90.57
Local	$10000, L = 1$	91.08 ± 0.78	92.25	92.21
	$42000, L = 2$	91.74 ± 0.91	92.94	93.03

Models obtained with SVM classifiers shown that both, global and local features, classified correctly char particles, with accuracies over 88%. However, models built using global features have a higher accuracy than models built with local features – 99.27 ± 0.37 average accuracy using global features and 91.74 ± 0.91 average accuracy using local features with a codification parameter $L = 2$. It appears that global features based on the application domain increase the accuracy of classification models. Moreover, the SIFT descriptor – used in this paper – lacks of meaning in the classification context while the global features are meaningfully since they are measured of char morphological characteristics.

Additionally, it is observed that local features had a higher accuracy using a codification parameter $L > 0$. It shows the importance of taking into account spatial information during char classification. A codification parameter $L = 1$ may be enough for obtaining good classification results – 91.08 ± 0.78 average accuracy – since the used of the parameter $L = 2$ does not increase the accuracy and increase the computational cost of building the classification model.

In general, obtained accuracy using the new set of char images – in order to evaluate generalisation capability – is similar to the average accuracy. Thus, SVM classifiers have generalisation capability.

4.2   Random Forests Evaluation

Table 2 shows the accuracy results using global and local features.

Table 2. Accuracy results of char classification using Random Forest

Feature	# of Features	Accuracy I. C. (%)	Accuracy (%)	
			Best classifier	Generalisation
Global	9	99.96 ± 0.05	100.00	100.00
	$2000, L = 0$	88.05 ± 0.89	89.29	90.44
Local	$10000, L = 1$	90.01 ± 0.96	91.57	92.07
	$42000, L = 2$	90.31 ± 0.97	92.03	91.08

Random Forest classifiers have similar results to the classification models based on SVM. Random Forest models built using global features have a higher accuracy than models built using local features – 99.96 ± 0.005 average accuracy using global features and 90.31 ± 0.97 average accuracy using local features with a codification parameter $L = 2$. These results are consistent with the SVM-based classification models. However, Random Forest classifiers built using local features with a codification parameter $L > 0$ yield an improvement of the classification accuracy. It can be observed a classification accuracy over 88% using a codification parameter $L = 0$ and a classification accuracy over 90% using codification parameters $L = 1$ and $L = 2$.

Regarding the generalisation capability, classification accuracy values are similar to the average accuracy showing Random Forests based classification models have a good generalisation capability. Moreover, Random Forest presents a better generalisation capabilities than SVM.

5   Final Remarks

A correct classification of char images is essential for setting the combustion parameters, such as temperature and residence time. Chars with morphologies no reactives require high temperature, high pressure and long residence time. Whilst chars with morphologies reactives burn faster and produce lower levels of CO_2.

An automatic char classification model was proposed using supervised learning. Two types of features were evaluated for building classification models: local and global. Based on the application domain, nine global features were considered: char particle area, undevolatilised material, devolatilised material, number of porous, porosity, wall thickness (first, second and third quartile), and sphericity. Local features were calculated by a bag-of-feature method. Each local feature was extracted using the SIFT descriptor and codified using the LLC algorithm with a max-pooling function.

Classification models are built based on SVMs and Random Forest classifiers. In general, classification models using global features have higher accuracy, over 99%, than classification models using local features. In char image classification, global features are meaningfully since they are defined taking into account the application domain. Additionally, classification accuracy is affected by the spatial information when local features are used.

Although, classification models using SVM and Random Forest have similar generalisation capabilities, Random Forest shown better classification accuracy using the new test-set.

References

1. Alvarez, D., Lester, E.: Atlas of char occurrences. combustion working group, commission iii. In: Internacional Conference on Coal Petrology, ICCP (2001)

2. Alvarez, D., Borrego, A.G., Menéndez, R.: Unbiased methods for the morphological description of char structures. Fuel 76(13), 1241–1248 (1997)
3. Avila, S., Thome, N., Cord, M., Valle, E., de Araujo, A.: Bossa: Extended bow formalism for image classification. In: 18th IEEE International Conference on Image Processing, ICIP (2011)
4. Boser, B.E., Guyon, I.M., Vapnik, V.N.: A training algorithm for optimal margin classifiers. In: Proceedings of the Fifth Annual Workshop on Computational Learning Theory, COLT 1992 (1992)
5. Chaves, D., García, E., Trujillo, M., Barraza, J.M.: Char morphology from coal blends using images analysis. In: World Conference on Carbon, CARBON (2013)
6. Cloke, M., Lester, E.: Characterization of coals for combustion using petrographic analysis: A review. Fuel 73(3), 315–320 (1994)
7. Cortes, C., Vapnik, V.: Support-vector networks. Mach. Learn. 20(3), 273–297 (1995)
8. Criminisi, A., Shotton, J., Konukoglu, E.: Decision forests: A unified framework for classification, regression, density estimation, manifold learning and semi-supervised learning. Foundations and Trends in Computer Graphics and Vision 7(2), 81–227 (2011)
9. Csurka, G., Bray, C., Dance, C., Fan, L.: Visual categorization with bags of keypoints. In: Workshop on Statistical Learning in Computer Vision, pp. 1–22 (2004)
10. Fan, R.E., Chang, K.W., Hsieh, C.J., Wang, X.R., Lin, C.J.: Liblinear: A library for large linear classification. Journal of Machine Learning Research 9, 1871–1874 (2008)
11. Fei-Fei, L., Perona, P.: A bayesian hierarchical model for learning natural scene categories. In: IEEE Computer Society Conference on Computer Vision and Pattern Recognition, CVPR (2005)
12. Lowe, D.G.: Object recognition from local scale-invariant features. In: The Proceedings of the Seventh IEEE International Conference on Computer Vision, ICCV 1999 (1999)
13. Rojas, A.F., Burgos, J.M.B.: Caracterización morfológica del carbonizado de carbones pulverizados: estado del arte. Revista Facultad de Ingeniería Universidad de Antioquia (41), 84–97 (2007)
14. Rojas, A.F., Burgos, J.M.B.: Caracterización morfológica del carbonizado de carbones pulverizados: determinación experimental. Revista Facultad de Ingeniería Universidad de Antioquia (43), 42–58 (2008)
15. Tang, F., Lu, H., Sun, T., Jiang, X.: Efficient image classification using sparse coding and random forest. In: 5th International Congress on Image and Signal Processing, CISP (2012)
16. Wang, J., Yang, J., Yu, K., Lv, F., Huang, T., Gong, Y.: Locality-constrained linear coding for image classification. In: IEEE Conference on Computer Vision and Pattern Recognition (2010)
17. Wu, T., Lester, E., Cloke, M.: Advanced automated char image analysis techniques. Energy & Fuels 20(3), 1211–1219 (2006)
18. Yang, J., Yu, K., Gong, Y., Huang, T.: Linear spatial pyramid matching using sparse coding for image classification. In: IEEE Conference on Computer Vision and Pattern Recognition, CVPR (2009)
19. Yang, P., Yang, Y.H., Zhou, B.B., Zomaya, A.Y.: A review of ensemble methods in bioinformatics. Current Bioinformatics 5(4), 296–308 (2010)
20. Zack, G.W., Rogers, W.E., Latt, S.A.: Automatic measurement of sister chromatid exchange frequency. J. Histochem. Cytochem. 25(7), 741–753 (1977)

On the Automatic Tuning of a Retina Model by Using a Multi-objective Optimization Genetic Algorithm

Rubén Crespo-Cano[1], Antonio Martínez-Álvarez[1(✉)], Ariadna Díaz-Tahoces[2],
Sergio Cuenca-Asensi[1], J.M. Ferrández[3], and Eduardo Fernández[2]

[1] Department of Computer Technology, University of Alicante, Alicante, Spain
amartinez@dtic.ua.es
[2] Institute of Bioengineering and CIBER BBN,
University Miguel Hernández, Alicante, Spain
e.fernandez@umh.es
[3] Department of Electronics and Computer Technology,
Universidad Politécnica de Cartagena, Cartagena, Spain
jm.ferrandez@upct.es

Abstract. The retina is responsible for transducing visual information into spikes trains which are then sent via the optical nerve to the visual cortex. This is the first step in the visual pathway responsible for the sense of vision. Our research group is working on the design of a cortical visual neuroprosthesis aimed to restore some functional vision to profoundly visual-impaired people. The goal of developing such a bioinspired retinal encoder is not simply to record a high-resolution image, but to process its visual information and transmit it in a meaningful way to the appropriate area on the visual cortex. Retinal models to be implemented have to match as much as possible the output produced by an actual biological retina. The models involve a big search space defined by a set of parameters that have to be appropriately adjusted. This in itself has several problems which need to be addressed. We propose in this paper an automatic evolutionary multi-objective strategy for selecting those parameters which best approximate the outputs by the synthetic retina model and the biological records. A case study is presented where results of a retina model tuned with our method are compared to biological recordings.

Keywords: Retina modeling · Visual neurprostheses · Multi-objective optimization · NSGA-II · Evolutionary search

1 Introduction

The retina is actually a little but important piece of brain capable of being stimulated by direct light focusing in this tissue and coming from the outside world. Indeed, it is responsible for the first stages of the visual processing. The retina integrates a rich set of specialized cells and complex neurostructures which are

© Springer International Publishing Switzerland 2015
J.M. Ferrández Vicente et al. (Eds.): IWINAC 2015, Part I, LNCS 9107, pp. 108–118, 2015.
DOI: 10.1007/978-3-319-18914-7_12

sensitive to color, light intensity, image movements, edges detection, and many others valuable characteristics for the sense of vision. Using these structures, the retina performs chromatic and achromatic spatio-temporal processing of visual information, and finally encodes it into spike trains delivered to the brain visual cortex by the optic nerve. Our research group is working on the design of a cortical visual neuroprosthesis aimed to restore some functional vision to profoundly visual-impaired people. The goal of developing such a bioinspired retinal encoder is not simply recording a high-resolution image, but to process this visual information and transmit it in a meaningful way to the appropriate area on the visual cortex. To achieve this goal we have to take into account the processing and coding features of the biological visual system. In addition, design constraints related to the number and distribution of the microelectrodes where the visual scene is mapped to [1,2] have to be kept in mind. The full description of this problem has been discussed elsewhere [1,3–5] and is beyond the scope of this paper, but Fig.1 summarizes the basic processing blocks of the bioinspired retina model we are currently investigating.

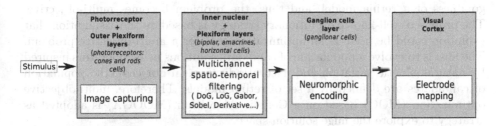

Fig. 1. Functional processing blocks of the bioinspired retina model under study

Before to proceed with acute tests to deliver an stimulation current directly to the visual cortex and thus eliciting an *in-vivo* controlled visual perception, a fine tuning of the involved retina model is needed. In this way, we refer to tuning a retina model as the process of adjusting the parameters and functions defining a retina model to best match its output with the biological records. Appropriate matching metrics are needed to assess the tuning of the synthetic retina.

From a mathematical point of view, the first two processing blocks from Fig. 1 can be modeled as a weighted combination of different well-known convolutive spatio-temporal image filters such as Gaussians, Difference of Gaussians (DoG), Laplacian of Gaussian, Gabor, Sobel, etc. (See Eq. 1) This combination results as an activity matrix feeding an *Integrate & Fire* model that calculates the spike firing of the ganglionar cells to be sent to the visual cortex with a possible remapping for every microelectrode target. Each processing block from the retina model has many parameters to be tuned. Many of them move in a continuous dynamic range (E.g. σ parameter for a Gaussian filter) and some of them can be modeled as natural numbers (E.g. Kernel size ($N \times N$) of a convolutive filter). With such an infinite search space to explore, the process of adjusting those parameters supposes a difficult problem to be solved. We propose in this paper an automatic evolutionary multiobjective strategy for selecting those parameters which best approximate the synthetic retina model outputs with the biological records.

The rest of the paper is structured as follows: Section 2 presents the evolutionary multiobjective strategy for tuning a retina model; Section 3 presents the experimental setup to obtain the biological registers; Section 4 describes and discusses the experimental cases of study, and finally, Section 6 provides the conclusions of this work.

2   Evolutionary Multi-objective Strategy for Tuning a Retina Model

As introduced in Section 1, considering the vast number of possible parameters which must be taken into account, the problem of automatic tuning of a retina model cannot be achieved by exhaustively exploring all the solutions space. In addition, no one but several antagonist objectives come to scene to assess the goodness of a retina model, and thus, the problem becomes multi-objective. The proposed well-known evolutionary strategy is based on the assumption that evolution could be used as an optimization tool for a multi-objective problem. The idea is to evolve a population of candidate solutions using operators inspired by natural genetic variation and natural selection. In our case, the population of candidates are defined by a set of retina models. Therefore, multi-objective optimization (MOO) based on a Genetic Algorithm (MOOGA) is adopted as strategy to explore the huge solution space.

Genetic Algorithms (GAs) are methods for solving search problems, based on sexual reproduction and the mechanics of natural selection. It belongs to the group of techniques known as Evolutionary Algorithms (EAs) which are based on the imitation of evolutionary processes such as natural selection, crossover or mutation. Every individual, which is randomly initialized, represents a possible solution for the problem. The population is evolved through crossover and mutation operators and only those that represents better solutions can be extracted from the surviving population. The selection operator chooses from the population, those individuals that will be allowed to reproduce, being the best individuals those who maximize or minimize the goals, according to their nature. The crossover operation exchanges subparts of two individuals and recombines, since this the way to imitate biological recombination. At last, mutation randomly changes the values of the aleles of the chromosome (Fig. 2).

To solve MOO problems, many different methods of resolution have been proposed in the last couple of years. One of these methods internationally regarded as one of the best is NSGA-II [6], included within of group of elitist algorithms that emphasize computational efficiency as SPEA and SPEA2 [7].

NSGA-II (Non-dominated Sorting Genetic Algorithm II) has been reported to be one of the most successful MOO algorithms [8]. NSGA-II algorithm establishes an order relationship among the individuals of a population mainly based on the concept of non-dominance or Pareto fronts. It is said that one solution

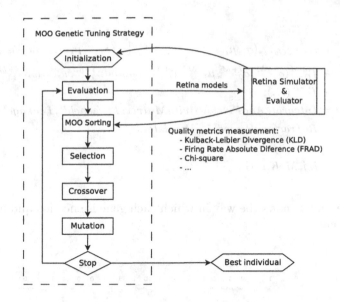

Fig. 2. Steps of the MOO genetic tuning strategy

X_i dominates other X_j if the first one is better or equal than the second in every single objective and, at least, strictly better in one of them (i.e. Pareto fronts are defined by those points in which no improvements in one objective are possible without degrading the rest of objectives). NSGA-II firstly groups individuals in a first front that contains all non-dominated individuals, that is the Pareto front. Then, a second front is built by selecting all those individuals that are non-dominated in the absence of individuals of the first front. This process is repeated iteratively until all individuals are placed in some front. The crowding distance function is used to calculate the diversity of a possible solution, and its purpose is to maintain a good spread of solutions. After that, individuals of the Pareto front are sorted in descending order based on its crowding distance value. As a result, those solutions having more diversity are prioritized.

In summary, the result of the execution of the NSGA-II, will provide us the population sorted by non-dominated fronts and then, by the crowding distance.

Chromosome Codification

In our approach, each individual represents a possible retina model and every gene codifies the values of the parameters of the function which models the retina. As an example, equation 1 represents a general retina model as described in Sec. 1. The parameters of each function will be the candidates to be encoded as genes that make up the chromosome.

$$\begin{aligned}
\frac{Activity}{Matrix} = &\ldots + f^i_{Gauss}(\sigma_i, \mu_i, {}_{KernelSize_i}) + f^{i+1}_{Gauss}(\sigma_{i+1}, \mu_{i+1}, {}_{KernelSize_{i+1}}) + \ldots \\
&+ f^{i+M}_{DoG}(\sigma^{i+M}_1, \sigma^{i+M}_2, \mu^{i+M}_1, \mu^{i+M}_2, {}_{KernelSize^{i+M}_1}, {}_{KernelSize^{i+M}_2}) + \ldots \\
&+ f^{i+K}(\ldots);
\end{aligned}$$

$$\begin{aligned}
RetinaModel = &\ IntegrateAndFire(ActivityMatrix; Threshold^j, Leakage^{j+1}, \quad (1)\\
&RefractoryPeriod^{j+2}, Persistence^{j+3}, \ldots)
\end{aligned}$$

$$i, j, M, K \in \mathbb{N}$$

The figure Fig. 3 shows the way in which each gene is encoded and integrates the chromosome.

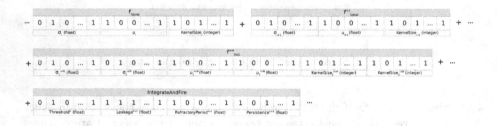

Fig. 3. Example of an individual codification

3   Material and Methods

All experimental procedures were carried out in accordance with the ARVO and European Communities Council Directives (86/609/ECC) for the use of laboratory animals.

Wild-type (C57BL/6J strain) adults mice were bred within a local colony established from purchased breeding pairs (Jackson Laboratories, Bar Harbor, ME). Following inhalational anesthesia with 4% of isoflurane (IsoFlo ®, Esteve vETERINARIA) was carried out the cervical dislocation of animals. Animals were dark-adapted for one hour prior to sacrifice.

After enucleation of the eye, the eyeball was hemisected with a razor blade separating and discarding the cornea and lens. The retinas were then carefully removed from the remaining eyecup, mounted on a glass slide ganglion cell side up and covered with a cut Millipore filter. This preparation was then mounted on a recording chamber and perfused with Ringer medium at physiological temperature. This working process was made under dim red illumination. Extracellular ganglion cell recordings were made in the isolated superfused mice retina using an array of 100 microelectrodes. Simultaneous single- and multi-unit responses were obtained in response to full field flash visual stimuli. Responses were recorded with quadrate array of 100 electrodes, 1.5 mm long (Utah Electrode Array).

The spike trains was recorded with a data acquisition system (Bionic Technologies Inc) and stored on a Pentium-based computer for later analysis. Neural spike events detected when exceed the thresholds established in each electrode using standard procedures described elsewhere [9, 10].

The spike sorting for classify the different units was accomplished with an free open source software based on principal component analysis (PCA) method and different clustering algorithms [11]. Time stamps for each action potentials of the single unit were used to generate peristimulus time histograms and peristimulus spike rasters using NeuroExplorer® Version 4 (Nex Technologies) as well as customized software [12].

Visual stimuli were programmed in Python using an open source library (VisionEgg) for real-time visual stimulus generation [13] and reproduce in a 16-bit ACER TFT monitor. The monitor resolution was 1280×1024 pixels at 60 Hz refresh rate and in this experiment we used an area of 120×154 pixels for the visual stimulation. Pictures drawn on this area were projected through a beam splitter and focused onto the photoreceptor layer with the help of optical lenses.

Finally Fig. 4 summarizes the data acquisition procedure commented bellow.

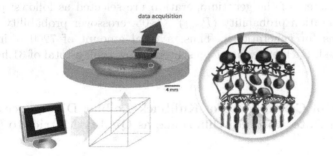

Fig. 4. Data acquisition procedure

4   Case Studies

To assess the feasibility and goodness of the proposed MOOGA strategy for tuning a retina model, two experiments-based case studies have been designed. In both experiments we attempt to demonstrate the importance of a fine tuning for a given retina model to best match the biological records.

In vivo biological records to be synthetically approximated were obtained stimulating a mice retina with a full-field flashing white-black stimulus. The retinal firing spikes response were recorded using the experimental set-up presented in Section 3. The biological stimuli, designed using *VisionEgg*, are defined by a periodic flash stimulus 700ms–white (ON) and 2700ms–black (OFF) repeated 5 times over time, and thus, taking 15 seconds at a frame rate of 60 fps. To feed the synthetic retina a DivX MPEG-4 v4 video file with the exact biological stimulus was prepared using the same *VisionEgg* stimulus.

With such a full black or full white flash stimulus, those retina parameters related with color processing has little or no impact on the biological records. Consequently, for this particular stimulus, we have selected the *Integrate&Fire* (*IaF*) processing block as the most appropriate module to be tuned. This module is based in the *Leaky-Integrate&Fire* model by [14] with the only addition of a new parameter *Persistence time* to model how many times every video frame feeds the retina model. In this way, four parameters within this module were selected to conform the retina model chromosome following the scheme presented in Section 2. The chosen parameters are: *Threshold level, Leakage, Refractory period* and *Persistence time*. The dynamic ranges of variability for these parameters have been chosen as: $[240, 800]$, $[10, 75]$, $[1ms, 10ms]$ and $1, 2, 3, 4, 5, 6$ respectively. Note that, with the exception of the *Persistence time* the remaining parameters move within a continuous range, and thus a infinite search space is presented.

The biological records for every microelectrode are pre-processed to separate different cell responses, and are finally presented as a set of diverse Post-Stimulus Time Histograms (PSTH) for every isolated ganglionar cell response. These PSTHs have to be approximated by our population of retina models.

The parameters of the genetic operation are selected as follows: population size – 60, mutation probability (P_{mut}) – 0.05, crossover probability – 0.3 and 120 iterations (or generations). Thus, a total amount of 7200 retina models were processed. Each experiment presented below took a total of 31 hours to be completed.

4.1 Study of Convergence of Kullback–Leibler Divergence and Firing Rate Absolute Difference as Quality Metrics to Compare PSTHs

To compare synthetic and biological PSTHs, and also test the behavior of the proposed MOOGA strategy, two quality metrics, or fitness functions in the field of genetic algorithms, have been selected. The first one is the Kullback–Leibler Divergence (KLD) between two PSTH vectors, and the second one is their firing rate absolute difference(FRAD).

To test and ensure the convergence of both metrics, two experiments have been done to cope with the challenge of approximating them. Both experiments have been completed with a population of 60 individuals and 120 iterations. The duration was 31 hours each. The fist experiment evaluates the convergence of the Kullback–Leibler divergence, whereas the second one evaluates the FRAD.

As expected and showed by Figure 5, both metrics converge at a plausible time (less than 4 hours) and a plausible number of iterations ($\sim$ 15 in each case). For a better examination of these figures, an appropriate maximization have been added to each one.

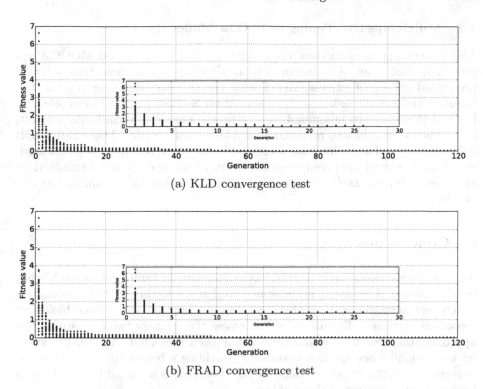

(a) KLD convergence test

(b) FRAD convergence test

Fig. 5. Convergence of Kullback-Leibler divergence (a) and FRAD (b) when comparing biological and synthetic PSTHs

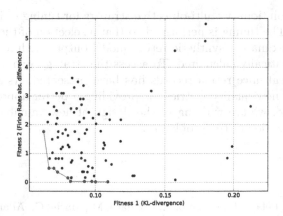

Fig. 6. Kullback-Leibler divergence against FRAD

4.2 Multi-objective Tuning of Retina Model

This experiment aims to show the effectiveness of the proposed MOOGA approach to find a set of retina models maximizing at the same time a set of predefined criteria of interest. To this end, the previous fitness metrics KLD as fitness 1 and FRAD as fitness 2, presented on Section 4.1, have been selected. Both of them have to minimized to best fit the biological records. As a result, Fig. 6 shows the solutions minimizing at a time both metrics. Those individuals belonging to the Pareto front are identified by means of red dots. 10 retina models (from 7200 explored) comprises the Pareto front and are not dominated by any other solution. Each one represents a valid solution for the multi-objective problem.

5 Conclusions

An automatic evolutionary multi-objective strategy for tuning retina models has been presented. The tuning is performed so that a selection of those parameters which best approximate a synthetic retina model output with actual biological records is automatically calculated. To assess the strategy two case studies for approximating real mice retina records has been presented. As a future work, we are designing new quality metrics for providing a better approximation and convergence. Also, we are working in the design of new valuable stimuli for a better characterization of the problem.

6 Conclusions

An automatic evolutionary multi-objective strategy for tuning retina models has been presented. The tuning is performed so that a selection of those parameters which best approximate a synthetic retina model output with actual biological records is automatically calculated. To assess the strategy two case studies for approximating real mice retina records has been presented. As a future work, we are designing new quality metrics for providing a better approximation and convergence. Also, we are working in the design of new valuable stimuli for a better characterization of the problem.

References

1. Fernandez, E., Pelayo, F., Romero, S., Bongard, M., Marin, C., Alfaro, A., Merabet, L.: Development of a cortical visual neuroprosthesis for the blind: the relevance of neuroplasticity. Journal of Neural Engineering 2(4), R1 (2005), doi:10.1088/1741-2560
2. Normann, R.A., Greger, B.A., House, P., Romero, S.F., Pelayo, F., Fernandez, E.: Toward the development of a cortically based visual neuroprosthesis. Journal of Neural Engineering 6(3), 035001 (2009)

3. Morillas, C., Romero, S., Martínez, A., Pelayo, F., Reyneri, L., Bongard, M., Fernández, E.: A neuroengineering suite of computational tools for visual prostheses. Neurocomputing 70(16-8), 2817–2827 (2007), Neural Network Applications in Electrical Engineering Selected papers from the 3rd International Work-Conference on Artificial Neural Networks (IWANN 2005) 3rd International Work-Conference on Artificial Neural Networks (IWANN 2005), http://www.sciencedirect.com/science/article/pii/S0925231207001579, doi:http://dx.doi.org/10.1016/j.neucom.2006.04.017

4. Morillas, C.A., Romero, S.F., Martínez, A., Pelayo, F.J., Fernández, E.: A computational tool to test neuromorphic encoding schemes for visual neuroprostheses. In: Cabestany, J., Prieto, A.G., Sandoval, F. (eds.) IWANN 2005. LNCS, vol. 3512, pp. 510–517. Springer, Heidelberg (2005), http://dx.doi.org/10.1007/11494669_63, doi:10.1007/11494669_63

5. Martínez-Álvarez, A., Olmedo-Payá, A., Cuenca-Asensi, S., Ferrández, J.M., Fernández, E.: Retinastudio: A bioinspired framework to encode visual information. Neurocomputing 114, 45–53 (2013), Searching for the interplay between neuroscience and computation Selected papers from the {IWINAC} 2011 Conference, http://www.sciencedirect.com/science/article/pii/S0925231212007850, doi:http://dx.doi.org/10.1016/S0006-89930003072-9

6. Deb, K., Pratap, A., Agarwal, S., Meyarivan, T.: A fast and elitist multiobjective genetic algorithm: NSGA-II. IEEE Transactions on Evolutionary Computation 6(2), 182–197 (2002)

7. Zitzler, E., Laumanns, M., Thiele, L.: SPEA2: Improving the strength pareto evolutionary algorithm for multiobjective optimization. In: Giannakoglou, K.C., Tsahalis, D.T., Périaux, J., Papailiou, K.D., Fogarty, T. (eds.) Evolutionary Methods for Design Optimization and Control with Applications to Industrial Problems, pp. 95–100. International Center for Numerical Methods in Engineering, Athens (2001)

8. Martínez-Álvarez, A., Cuenca-Asensi, S., Ortiz, A., Calvo-Zaragoza, J., Tejuelo, L.A.V.: Tuning compilations by multi-objective optimization: Application to apache web server. Applied Soft Computing 29, 461–470 (2015), http://www.sciencedirect.com/science/article/pii/S156849461500, doi:http://dx.doi.org/10.1016/j.asoc.2015.01.029

9. Fernández, E., Ferrández, J.-M., Ammermller, J., Normann, R.A.: Population coding in spike trains of simultaneously recorded retinal ganglion cells1. Brain Research 887(1), 222–229 (2000), http://www.sciencedirect.com/science/article/pii/S0006899300030729, doi:http://dx.doi.org/10.1016/S0006-89930003072-9

10. Normann, R.A., Warren, D.J., Ammermuller, J., Fernandez, E., Guillory, S.: High-resolution spatio-temporal mapping of visual pathways using multi-electrode arrays. Vision Research 41(10-11), 1261–1275 (2001), http://www.sciencedirect.com/science/article/pii/S004269890000273X, doi:http://dx.doi.org/10.1016/S0042-69890000273-X

11. Bongard, M., Micol, D., Fernández, E.: Nev2lkit: A new open source tool for handling neuronal event files from multi-electrode recordings. International Journal of Neural Systems 24(04), 1450009, pMID: 24694167 (2014), http://www.worldscientific.com/doi/abs/10.1142/S0129065714500099, doi:10.1142/S0129065714500

12. Ortega, G.J., Bongard, M., Louis, E., Fernández, E.: Conditioned spikes: a simple and fast method to represent rates and temporal patterns in multielectrode recordings. Journal of Neuroscience Methods 133(1-2), 135–141 (2004),   http://www.sciencedirect.com/science/article/pii/S0165027003003418, doi:http://dx.doi.org/10.1016/j.jneumeth.2003.10.005

13. Straw, A.D.: Vision egg: an open-source library for realtime visual stimulus generation. Frontiers in Neuroinformatics 2(4), http://www.frontiersin.org/neuroinformatics/10.3389/neuro.11.004.2008/abstract, doi:10.3389/neuro.11.004.2008

14. Gerstner, W., Kistler, W.M.: Spiking Neuron Models: Single Neurons, Populations, Plasticity. Cambridge University Press (2002)

Creating Robots with Personality: The Effect of Personality on Social Intelligence

Alexandros Mileounis(✉), Raymond H. Cuijpers, and Emilia I. Barakova

Eindhoven University of Technology, P.O. Box 513, Eindhoven,
The Netherlands
amileunis@hotmail.com, R.H.Cuijpers@tue.nl, E.I.Barakova@tue.nl

Abstract. This study investigates the effect of two personality traits, dominance and extroversion, on social intelligence. To test these traits, a NAO robot was used, which was teleoperated through a computer using a Wizard of Oz technique. A within-subject design was conducted with extroversion as within-subject variable and dominance as between-subject. Participants were asked to cooperate with the robot to play "Who wants to be a millionaire". Before the experiment participants filled in a personality questionnaire to measure their dominance and extroversion. After each condition, participants filled in a modified version of the Godspeed questionnaire concerning personality traits of the robot plus 4 extra traits related to social intelligence. The results reveal a significant effect of dominance and extroversion on social intelligence. The extrovert robot was judged as more socially intelligent, likeable, animate, intelligent and emotionally expressive than the introvert robot. Similarly, the submissive robot was characterized as more socially intelligent, likeable and emotionally expressive than the dominant robot. We found no substantial results towards the similarity-attraction hypothesis and therefore we could not make a conclusion about the mediating effect of participant's personality on likeability.

1 Introduction

There is an increasing need for social robots that are able to interact with humans and appear social enough to keep elderly and children company. Making social robots intelligent enough to be perceived as a replacement of a human being in some tasks or at least a trustworthy social agent is extremely difficult. The robot not only has to look social but also behave socially and in an intelligent way[1]. As Fong et al. [1] state in their study, an important trait is missing from social robots nowadays and that is social intelligence.

Research on intelligent robots mainly focuses on the rational part of intelligence, equipping robots with planning, reasoning, navigation, manipulation and other non-social skills [2]. However, this kind of intelligence is not the only skill that social robots need to maintain a long-term acceptance. Social robots need in addition social skills related, but not limited, to persuasion, collaboration, cooperation, emotions, empathy, situational awareness and adaptation [3]. This is called emotional [3] or social intelligence [1]. Social intelligence is still an

© Springer International Publishing Switzerland 2015
J.M. Ferrández Vicente et al. (Eds.): IWINAC 2015, Part I, LNCS 9107, pp. 119–132, 2015.
DOI: 10.1007/978-3-319-18914-7_13

ill-defined concept since it was first presented in 1933 by Vernon [4], but still hasn't reached a unanimous definition. The definition Vernon gives is: "social intelligence is a person's ability to get along with people in general, social technique or ease in society, knowledge of social matters, susceptibility to stimuli from other members of a group, as well as insight into the temporary moods of underlying personality traits of strangers". So Vernon focuses on the emotional aspect of social intelligence that dictates socially intelligent individuals are able to recognize the effect of stimuli on group members and change their behavior accordingly to maintain group harmony.

In another definition by Albrecht [5] social intelligence is defined as "the ability to get along well with others while winning their cooperation." He also posits that social intelligence requires social awareness, sensitivity to the needs and interests of others, an attitude of generosity and consideration, and a set of practical skills for interacting successfully with others in any setting. As it can be noticed, Albrecht focuses more on the cooperative part of social intelligence, which is one of the required skills in social robots.

To summarize, a socially intelligent person is more skilled on judging other people's feelings, thoughts, attitudes and opinions, intentions, or the psychological traits that may determine their behaviour. It is worth noting that social intelligence is directly connected to the social context, between two or more people, and judgment on it necessitates the consideration of contextual factors in play [6].

In this study we follow the cooperation definition by Albrecht, the social skills indicated in Martínez-Miranda & Aldea (except empathy and adaptation) and also keep the social context in play when the robot responds to the participant. In addition, we enhance the persuasion of the robot by incorporating gazing and gestures [7].

The need to find the features and the extent that human social intelligence applies to robots is still an issue that has to be further studied [1]. To do so, we have to analyze the features of the personality of the robots that have already been tested in the past.

Robot personality started concerning scientists in the last twenty years. Since Nass et al. [8] found that humans respond socially to computers and even recognize their personality [9] there was enough fertile ground for studies to develop. Several sources pinpoint the need to determine the connection between human and robot personalities in order to find which traits are critical for HRI to increase satisfaction and enjoyment[10] [11]. Furthermore, Breazeal [12] claims that adaptation of robot's personality to that of the human that interacts with is needed. In addition, she states that the type and the complexity of the personality should be also defined. Personality is considered the key for creating socially interactive robots [10] because it "represents those characteristics of the person that account for consistent patterns of feeling, thinking, and behaving" [13]. Therefore, equipping robots with personality gives people the affordances needed to engage in human-human interaction schemes.

As we said earlier, Nass et al. [8] and Isbister & Nass [9] were the first who studied the effect of adapting an artificial agent's personality to human's. In particular, they investigated whether similarity attraction hypothesis, which supports that humans with similar personality traits are attracted to each other, also applies to artificial agents such as a computer. In their study, they created two different personalities, one dominant and one submissive, and found that participants were in favor of the computer with a similar personality to theirs. The dominant behavior was characterized by the attributes of self-confident, leading, self-assertive, strong and take-charge. On the contrary, submissive behavior was linked to self-doubting, weak, passive, following and obedient.

In another study by Isbister and Nass [9], researchers tested whether previous findings on similarity attraction rules also extend to embodied artificial agents. They used a mannequin avatar on a computer screen which used verbal (text) and non-verbal gestures to interact with participants. In addition, participants interacted with either an extrovert or introvert avatar. In comparison with the previous study [8], results indicated that subjects preferred a complimentary personality to theirs when interacted with an embodied virtual agent.

Similar results were found when participants interacted with the robot dog AIBO by Sony [10]. Subjects interacted with an extrovert (high speech speed, pitch and intensity, facial expressions (LED activity), long moving angles, high moving speed) or introvert (lower intensities than extrovert) robot. They found that a robot with complementary personality is regarded as more intelligent, attractive and socially present than a similar personality robot.

Other findings suggest that when humans interact with non-anthropomorphic robots (interactive closet) they tend to complement the personality of the robot [14]. In particular, when dominant participants interacted with a dominant closet they reported feelings of submissiveness. The same effect was found when a submissive subject interacted with a submissive closet, as they reported feeling of dominance. In the same study, the submissive closet was more favorable than the dominant closet.

However, none of these studies tested this with a humanoid robot. There is currently no evidence that these results will apply to a humanoid robot. Furthermore, these studies did not directly test whether their system appeared more socially intelligent as suggested by Fong et al. [1]. Therefore we combine elements from all these studies that relate to in one experiment and measure the apparent social intelligence of a humanoid robot. Whether people prefer to interact with similar or complementary personality artificial agents is the second question that this study will try to answer.

To identify the personality traits that impact social intelligence, we propose testing two pairs of personality traits that have been empirically tested in previous studies. These pairs are dominance-submissiveness [8] and extroversion-introversion [10]. Based on these theories we expect that:

H1) Participants that interact with the submissive robot will consider it as more socially intelligent and will like it more than the dominant robot. Thus, submissive behaviour will score higher on social intelligence and likeability than dominant behaviour. We also expect the same effect with likeability. The reasoning behind this lies to the fact that the submissive robot will be more cooperative [1] and it will compromise to reach a common final agreement.

H2) Participants will judge the extrovert robot as more socially intelligent and will like it more than the introvert robot. Thus, extrovert behaviour will score higher on social intelligence than introvert behaviour. We believe that an extrovert robot will be able to exhibit more social cues that will be beneficial for the enhancement of the interaction.

H3) We also expect that people prefer a companion of similar personality as they usually pick their human friends. Subsequently, likeability will score higher when the personality of the robot matches the one of the participant than when it is the opposite.

2   Method

2.1   Robot

The robot used in the experiment was a NAO robot from Aldebaran Robotics [15]. NAO is an anthropomorphic robot involving 25 degrees of freedom for movement of the limbs and head, two cameras for movement and face recognition, and two speakers in the head. Furthermore, NAO incorporates two LED-colored eyes that can be used as an additional social cue.

One of the features of the robot that were used in the experiment was the wireless communication. The robot allows to be remotely controlled, something that was beneficial for using the wizard of Oz technique.

2.2   Experimental Design

To test our hypotheses, we used a mixed model 2 between-subject (dominant-submissive) X 2 within-subject (extrovert-introvert) design.

Four different personalities were designed according to each of the conditions. The variations of the behavior of the robot concerned the assertiveness (dominant-submissive) and the expressiveness (extrovert-introvert) of the robot in the following way. A dominant robot uses strong arguments with confident language, while a submissive robot uses arguments, but expresses uncertainty as well. To enhance this feeling, we also varied the pitch of the voice of the robot. The dominant robot sounded more serious and straightforward to its suggestions and was given a lower voice pitch. On the other hand, the submissive robot was given a higher voice pitch to boost its uncertainty and insecurity. The extraversion-introversion factor was designed through changing the intensity of the expressiveness of the robot by using gesticulation and a higher speech speed (see Table 1). For this game, the robot was programmed to give 360 different answers, 90 per condition. In addition, it gave a different answer according to the user's selection (loss or victory).

Table 1. Robot personality differences per condition

	Dominant	Submissive
	Low-pitch speech	High-pitch speech
	Assertive	Insecure
Extrovert	Gestures	Gestures
	Emotions	Emotions
	High-speed speech	High-speed speech
	Talkative	Talkative
	Low-pitch speech	High-pitch speech
	Assertive	Insecure
Introvert	Limited gestures	Limited gestures
	No emotions	No emotions
	Low-speed speech	Low-speed speech
	Less talkative	Less talkative

2.3   Participants

We have recruited 70 subjects for this experiment, 42 male and 28 female. Their ages ranged from 18 to 82 (Mean 28.1) and some of them were recruited from Technical University of Eindhoven and Fontys University of applied sciences, while others from a database of participants all over the Nord Brabant province. The interaction was in English, so only participants that could understand and speak English were recruited. All of the participants' data was used for the analysis, since there were not significant technical difficulties during the experiment or the outliers did not affect the outcome. Only a handful of participants were acquainted with the robot, with few of them having participated in a similar experiment in the past. Participants were randomly placed in one of the two between-subject conditions, so each of them would face either the dominant or the submissive robot. Before the experiment, participants were requested to sign an informed consent form.

2.4   Task

To satisfy the needs of this experiment, we wanted a task that allowed one-way interactions between the robot and the participant, without hindering the social intelligence of the robot. To fulfill this purpose, we redesigned the "Who wants to be a millionaire" TV-show (http://en.wikipedia.org/wiki/Who_Wants_to_Be _a_Millionaire%3F). In the TV-show, the player has to answer fifteen questions of escalating difficulty to win one million euros/pounds/dollars (depending on the currency of the country). After each correct answer the money the person has acquired is doubled, starting from 100. On each question four different answers appear on the screen and the player is requested to select the correct one. Furthermore, the game provides three assistances (lifelines): a) Phone-a-Friend, b) 50:50, c) Ask the Audience. The first lifeline allows the player to call a friend of his/hers in thirty seconds, read them the question and the answers, and receive an input. On the second lifeline, the computer eliminates two of the incorrect answers. Finally, the last lifeline requests the members of the audience to use the touch pads in front of them to vote for the correct answer. After the voting, the outcome is displayed on the player's screen. In the experiment, we kept the main goal of the game, but we changed the way the player played it. More specifically, we asked the participants to cooperate with the robot. Additionally, we removed the lifelines, so the participant requested help from the robot to win the game. The robot's behavior was designed in such a way that the robot utilized the same strategy the lifelines did to assist the player. For instance, the robot in the dominant and extrovert condition responded to the participants like they were taking advice from a friend that was confident about his/her answer supported by several arguments. To make the task more challenging we made the robot respond correctly with a 50% chance. In other words, on each question the robot had 50% chance to provide the correct answer. This arrangement made the interaction to be also based on trust. Last but not least, to increase gaming time so participants can interact more with the robot, we gave participants three lives. When the participant failed to answer a question correctly, they lost a life and started a new set of 15 questions from the beginning. For experimental purposes, we tracked the time spent on each condition, so each participant would spend ten minutes per condition. For the design of the game, we used Axure RP Pro by Axure Software Solutions (www.axure.com) to create the graphics and then connect each of the buttons on the screen with different functionalities. The scripts run on the robot were created in Python 2.6. The gestures of the robot were initially created in Aldebaran's Choregraphe 1.14.5 and then exported as a script to Python.

2.5   Questionnaires

In this study, we used two questionnaires to create the data we needed. The first questionnaire concerned participant's personality and the second robot's personality. Both questionnaires used a 5-point Likert scale with 1 being the lowest and 5 the highest. Participant's personality questionnaire involved 48 questions focusing first on defining the extraversion of the participants and then their dominance (see Table 2). This questionnaire was needed to answer the third hypothesis by testing the effect of robot's extraversion and dominance on likeability when controlling for participant's extraversion and dominance respectively. The items used in this questionnaire were derived from Wiggins (1979) and International Personality Item Pool (IPIP) online database. From the scales provided on the website, we used BIG-FIVE, CAT-PD and CPI and NEO:E3. Participants filled in this questionnaire before engaging the task with the robot. The second questionnaire, robot's personality questionnaire, was a modified version of Godspeed questionnaire by Bartneck et al., (2009) and (Waytz et al., 2010) with the safety and anthropomorphism dimensions replaced by emotion, extraversion and dominance plus four new items that represented social intelligence (cooperative, supportive, persuasive and situation aware). In other words, it consisted of 34 questions describing the following seven factors: animacy, perceived intelligence, emotion, likeability, social intelligence, extraversion and dominance. In more detail, animacy represents the liveliness of the robot. Interaction is a key ingredient of animacy as animate robots tend to interact more with the environment and incorporate less inactive moments. Perceived intelligence is a dimension that represents the general intelligence term described in the introduction, more closely related to the rationality aspect of the robot's decisions. Emotion corresponds to an expressive and empathetic agent. Next, likeability is connected to the extent participants consider the robot as friendly, nice and kind. Social intelligence describes the cooperation, persuasion and the level of social skills of the robot according to the context of the conversation. Last, extraversion is related to the extent the robot is expressing itself by being talkative and outgoing, and dominance to the extent the robot is assertive and competitive. All dimensions and items used can be seen in Table 3. This questionnaire was answered after each condition to receive evaluation on robot's personality.

Table 2. Participant traits addressed by the questionnaire

Extraversion	Dominance
Silent - Talkative	Dominant
Shy – Not shy	Assertive
Introverted- Extraverted	Forceful
Inward - outgoing	Domineering
	Submissive

2.6   Procedure

Participants were picked up from the waiting area and the experimenter led them to the lab. At the entrance of the lab, participants were asked about their familiarity with the lab, the experiment and the robot. Next, they were given an informed consent form to read and sign. The informed consent form involved instructions about the experiment. After signing the form, participants were requested to sit on an armchair next to the robot facing a TV-set. They were asked to fill in a questionnaire about their personality and were given a keyboard (to write their name) and a mouse. The experimenter told them that the purpose of this questionnaire was to check whether they match with the robot. When they were done, the experimenter took the keyboard away and started explaining the task. First, he asked the participants whether they know the TV-show and explained the differences between the TV-show and the task. Only a couple of participants were completely unaware of the TV-show. After the explanation, the experimenter summarized again the process and demonstrated on the screen how the participant should play it. When the experimenter confirmed that participants understood the rules of the game, he set the timer and said

Table 3. Robot behavioral traits addressed by the questionnaire

Animacy	Perceived Intelligence	Emotion	Likeability
Dead-Alive	Irrational-Rational	Insensitive-Compassionate	I dislike-like it
Stagnant-Lively	Incompetent-Competent	Emotionally unstable-Emotionally stable	Unfriendly-Friendly
Mechanical-Lively	Ignorant-Knowledgeable	Passive-Active/Energetic	Unkind-Kind
Artificial-Lifelike	Irresponsible-Responsible	Apathetic-Empathetic	Unpleasant-Pleasant
Inert-Interactive	Unintelligent-Intelligent		Awful-Nice

Social Intelligence	Extraversion	Dominance
Uncooperative-Cooperative	Silent - Talkative	Forceful (have the final word)
Unsupportive-Supportive	Shy – Not shy	Dominant (competitive)
Unpersuasive-Persuasive	Extraverted (play for the team)	Dominant (doesn't like to be outperformed)
Situation aware-unaware	Outgoing (expressive)	Assertive
	Outgoing (share personal experiences)	Domineering (enforcing opinion)

to the participants to start playing when the robot stood up. The experimenter left the room and initiated the robot. On each question, participants read the question and clicked on one of the answers. The experimenter, as soon as the participant selected one of the answers on the screen, made the robot give an answer to that particular question. Participants could decide either to confirm their own answer, or change it to follow the robot's suggestion. After every question, the robot responded according to the condition that it was on. For example, on dominant condition the robot either gave a positive feedback such as "Good Job" or "We can do it", or a neutral feedback "Ok, I don't know everything". On the other hand, on submissive condition, the robot never discouraged participants and was eager to take responsibility of its suggestions, such as "Sorry! I wasn't sure about the answer". The extrovert robot on both dominant and submissive conditions tended to gesticulate using obtuse joint angles and talk more by providing more arguments towards their suggestion. On the contrary, the introvert robot used little to no movement of the body and the responses were short and straightforward. Next, when the participant lost all of his/her lives the

Table 4. Reliability analysis on participant and robot personality questionnaires

Factors	Cronbach's Alpha
Participant personality questionnaire	
Silent - Talkative (4 items)	0.468
Shy – Not shy (4 items)	0.699
Introverted - Extraverted (6 items)	0.535
Inward - Outgoing (4 items)	0.613
Introvert – Extrovert (18 items)	0.783
Dominant (8 items)	0.733
Assertive (7 items)	0.784
Forceful (3 items)	0.650
Domineering (6 items)	0.795
Submissive (6 items)	0.843
Dominant – Submissive (30 items)	0.861
Robot personality questionnaire	
Animacy (6 items)	0.769
Perceived Intelligence (5 items)	0.774
Perceived Emotion (4 items)	0.703
Likeability (5 items)	0.921
Social Intelligence (4 items)	0.699
Extraversion (5 items)	0.564
Dominance (5 items)	0.780

experimenter walked in the room and requested from the participants to fill in the robot's questionnaire. After the participants were done with the questionnaire, the experimenter set the timer again and moved to the next room to initiate the robot again. When the second condition was completed, the participants were asked to fill in the same questionnaire as before to evaluate the new behavior of the robot. A lot of participants became really excited about the experiment and when the experiment finished wanted to learn more things about the study. The experimenter answered all their questions and, after handing their monetary compensation, led the participants out of the room.

3   Results

3.1   Reliability Analysis

For the first part of the analysis, we did a reliability analysis in order to check the internal validity of the questionnaire factors in participant and robot questionnaires. In addition, in the analysis of the robot questionnaires, to increase the reliability of the scale we joined the results of both questionnaires, doubling our sample size to 140. The results of the analysis can be seen in Table 4.

All the items were eventually used in the analysis, since the removal of certain items did not improve the alpha of the factor significantly. The final factor scores were computed by averaging the values of the items that describe each factor in the questionnaire.

3.2   Hypotheses Analysis

To answer our first two hypotheses, we conducted a repeated-measures analysis of variance (ANOVA) with robot's extraversion as the within-subject factor and robot's dominance as the between-subject factor.

The first hypothesis (H1) was supported. The results showed that participants judged the extrovert robot to be more socially intelligent (M=3.4, SD= .71) than the introvert robot (M=2.96, SD= .81). The main effect of robot's extraversion was statistically significant, $F(1,68) = 12.6$, $p < .001$, $\eta2 = .16$ (see Figure 1(left)). In addition, the main effect of robot's extraversion on likeability was also significant, $F(1,68) = 13.27$, $p < .001$, $\eta2 = .16$. The extrovert robot was liked more (M= 3.6, SD= .9) than the introvert (M=3.07, SD= .96), which is also verified by participant's verbal statements after the experiment (see Figure 1(right)).

Similarly, the second hypothesis (H2) was also supported. A significant main effect of dominance was found, $F(1,68) = 4.17$, $p < .05$, $\eta2 = .002$. More specifically, the dominant robot was judged as less socially intelligent (M=3.05, SD= .72) than the submissive robot (M=3.32, SD= .79) (see Figure 1(left)). Moreover, there was a significant main effect of dominance on likeability, $F(1,68) = 16.88$, $p < .001$, $\eta2 = .008$. Participants liked more the submissive robot (M=3.65, SD= .81) than the dominant robot (M=3.03, SD= .94). No interaction effects between extraversion and dominance were found on either social intelligence or likeability (see Figure 1(right)).

Finally, the third hypothesis (H3) was partly supported. The personality of the participants was used as a covariate to test whether similarity-attraction hypothesis applies on this experiment. This part of the analysis gave two opposite results. First, the participant's extraversion dimension was used as a covariate, which did not return significant results. Second, we used the dominance dimension of the participant's personality. Compared to extraversion, controlling for participant's dominance had a statistically significant effect on likeability, $F(1,67) = 5.48$, $p=.02$, $\eta2 = .03$. Likewise, robot's dominance was also found to be significant, while controlling for participant's dominance, $F(1,67) =15.3$, $p<.001$, $\eta2 = .09$. Due to the insignificance of the interaction effect, we cannot make a conclusion about whether the similarity or complementarity attraction rule applies in our case.

Apart from the hypotheses, we also analyzed the rest of the dimensions in the robot questionnaire, namely animacy, perceived intelligence, perceived emotion, extraversion and dominance (see Figure 2(left)).

Robot's extraversion had a significant effect on animacy, $F(1,68) = 6.49$, $p<.05$, $\eta2 = .09$. The extrovert robot appeared more lifelike and alive (MD = 3.48, SD= .6) than the introvert robot (MD = 3.27, SD= .66). There was

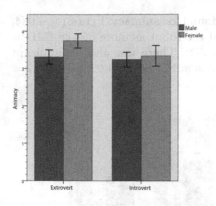 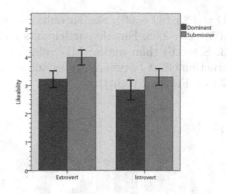

Fig. 1. The effect of personality on social intelligence (left) and likeability (right)

no main effect of dominance on animacy, p=.27, nor an interaction effect of dominance and extraversion.

Next, we found a significant effect of robot's extraversion on perceived intelligence, F(1,68) = 11.49, p<.001, η2 = .14. The extrovert behavior of the robot was rated higher (MD = 3.24, SD= .72) than the introvert behavior (MD = 2.84, SD= .84). No main effect of dominance, p=.45 or an interaction effect between dominance and extraversion was found.

Emotional expression was found to be significantly affected by both robot's extraversion, F(1,68) = 8.29, p<.05, η2 = .11 and dominance, F(1,68) = 13.81, p<.001, η2 = .006. More specifically, the extrovert robot was perceived as more emotionally expressive (MD = 3.12, SD= .69) than the introvert robot (MD = 2.87, SD= .68) and the submissive robot as more emotionally expressive (MD =3.24, SD= .63) than the dominant robot (MD = 2.75, SD= .67). Again no interaction effects were noticed to be significant.

Finally, the last two dimensions of the robot questionnaire extraversion and dominance, were tested for verification purposes over the effectiveness of the manipulation. Indeed, the results indicated that robot expressing extravert behaviors had a significant effect on the perceived extraversion F(1,68) = 22.11, p<.001, η2 = .25. The extrovert robot was correctly perceived as more extrovert (MD = 3.46, SD= .55) than the introvert one (MD = 3.03, SD= .66). However, dominance had, as expected, no effect on perceived extraversion, p=.55. Conversely, when we used dominance as the dependent variable, we got the opposite pattern. The dominant behaviors expressed by the robot significantly affected perceived dominance by the participants F(1,68) = 47.19, p<.001, η2 = .02, but not the extraversion, p=.38. In fact, the dominant robot was perceived as more dominant (MD = 3.41, SD= .74) than the submissive robot (MD = 2.53, SD= .69).

Another interesting finding is the effect of gender on the perceived emotion in the robot behavior. When gender was used as a covariate there was a significant main effect, F(1,66) = 7.17, p<.05, η2 = .016. Female participants found the robot more emotionally expressive (MD = 3.2, SD= .64) than male participants

(MD = 2.85, SD= .69). Similar outcome was found for animacy, $F(1,66) = 4.084$, p<.05, $\eta 2$ = .008. Female participants judged the robot as more lifelike (MD = 3.53, SD= .6) than males (MD =3.26, SD= .62). There was also a significant interaction effect between gender and extraversion $F(1,66) = 4.92$, p<.05, $\eta 2$ = .067 (see Figure 2(right)).

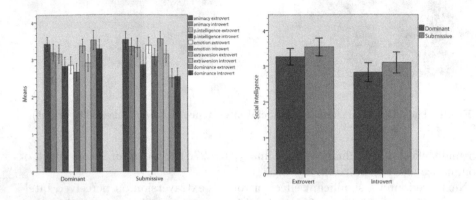

Fig. 2. Overall the remaining dimensions (left), interaction effect extraversion X gender (right)

4   Discussion

Many scientists supported the idea of designing social robots with personalities and social intelligence. However, the connection between these two factors has not been empirically tested. The purpose of this study was to verify whether the theory behind social intelligence and personality applies to the design of robotic agents.

Therefore, we combined in a different way the expression of two different personality traits and this way created four robot personalities. To enrich the interaction, we used robot features based on findings from previous studies on robot personalities such as gestures, voice prosody and emotional expression. Our results showed that the expressed personality does affect the perceived social intelligence of robots.

In particular, our first hypothesis supported that the submissive robot will be judged as more socially intelligent and be liked more as it will be more cooperative than the dominant robot. The results indicated that, indeed, the submissive behavior was seen as more socially intelligent and more likeable than the dominant one. This is in line with the findings of [14] for a non-anthropomorphic robot. We assume that the cooperative attitude of the submissive behavior is mainly responsible for this result. In addition, the verbal feedback of the robot, by exhibiting supportive behavior, was equally responsible for the increased likeability. Independently from the outcome of the in-game decision, the robot verbally supported the participant to continue playing and felt responsible for any erroneous suggestions. On the other hand, the dominant robot blamed the participant after a wrong in-game decision and did not feel responsible for any wrong suggestion. We did not measure the performance of the participants between the conditions so we could not detect the effects of the robot's assertiveness. In future studies the performance difference between these two behaviors should be tested.

The second hypothesis tested the differences between the extrovert and introvert behavior of the robot. The results supported our expectations. The extrovert robot was perceived as more socially intelligent and was liked more than the introvert robot. This result can be attributed to the vivid gestures of the robot and the verbal feedback in the extrovert condition that in the introvert condition were quite limited or absent.

Our third hypothesis tested whether the similarity-attraction hypothesis also applies in anthropomorphic robots. We did not find a significant effect that could indicate that either similarity or complementarity attraction hypothesis is supported. We believe that not screening out the participants by their extraversion or dominance as [8], [10], and [14] did is responsible for the absence of interaction effects. More specifically, most participants scored 3 on the participant questionnaire (middle) and thus there was no clear difference between extrovert and introvert participants.

In general, the extrovert robot was judged as more socially intelligent, likeable, animate, intelligent (this corroborates with [10]), and emotionally expressive than introvert robot. Similarly, the submissive robot was characterized as more socially intelligent, likeable (in consent with [14]) and emotionally expressive than the dominant robot. Last, female participants found the extrovert robot more emotionally expressive and lifelike than the male participants.

Although our results cannot support any of the previous studies about personality matching preferences, they provide substantial evidence on the personality and social intelligence studies. [1] discussed the need of finding the personality features that affect social intelligence and we can say that dominance-submissive and extrovert-introvert are two dimensions that affect perceived social intelligence. Nevertheless, we strongly believe that this is just the beginning of the identification of the critical personality characteristics that affect social intelligence and HRI in general. There is a big list of social traits in [3] that define emotional intelligence and future studies should put these to test.

References

1. Fong, T., Nourbakhsh, I., Dautenhahn, K.: A survey of socially interactive robots. Robotics and Autonomous Systems 42(3-4), 143–166 (2003), doi:10.1016/S0921-8890(02)00372-X
2. Dautenhahn, K.: Socially intelligent robots: dimensions of human-robot interaction. Philosophical Transactions of the Royal Society of London. Series B, Biological Sciences 362(1480), 679–704 (2007), doi:10.1098/rstb.2006.2004
3. Martínez-Miranda, J., Aldea, A.: Emotions in human and artificial intelligence. Computers in Human Behavior 21(2), 323–341 (2005), doi:10.1016/j.chb.2004.02.010

4. Vernon, P.E.: Some Characteristics of the Good Judge of Personality. The Journal of Social Psychology (1933), doi:10.1080/00224545.1933.9921556
5. Albrecht, K.: Social Intelligence: The New Science of Success. John Wiley and Sons (2009)
6. De Ruyter, B., Saini, P., Markopoulos, P., Van Breemen, A.: Assessing the effects of building social intelligence in a robotic interface for the home. Interacting with Computers 17(5), 522–541 (2005)
7. Ham, J., Bokhorst, R., Cuijpers, R.: Making robots persuasive: the influence of combining persuasive strategies (gazing and gestures) by a storytelling robot on its persuasive power. Social Robotics, 71–83 (2011)
8. Nass, C., Moon, Y., Fogg, B.: Can computer personalities be human personalities? International Journal on Human-Computer Studies 43, 223–239 (1995), doi:10.1006/ijhc.1995.1042
9. Isbister, K., Nass, C.: Consistency of personality in interactive characters: verbal cues, non-verbal cues, and user characteristics. International Journal of Human-Computer Studies 53, 251–267 (2000), doi:10.1006/ijhc.2000.0368
10. Lee, K.M., Peng, W., Jin, S.-A., Yan, C.: An Empirical Test of Personality Recognition, Social Responses, and Social Presence in Human?Robot Interaction. Journal of Communication 56(4), 754–772 (2006), doi:10.1111/j.1460-2466.2006.00318.x
11. Woods, S., Dautenhahn, K., Kaouri, C., Boekhorst, R.: Is this robot like me? Links between human and robot personality traits. In: 5th IEEE-RAS International Conference on Humanoid Robots 2005, vol. 2005, pp. 375–380. IEEE (2005), doi:10.1109/ICHR.2005.1573596
12. Breazeal, C.: Social Interactions in HRI: The Robot View. IEEE Transactions on Systems, Man and Cybernetics, Part C (Applications and Reviews) 34(2), 181–186 (2004), doi:10.1109/TSMCC.2004.826268
13. Pervin, L., John, O., Robins, R.: Handbook of Personality, 3rd edn. Theory and Research, p. 862. The Guildford Press, New York (2008)
14. Hiah, L., Beursgens, L., Haex, R., Romero, L.P., Teh, Y.F., Ten Bhomer, M., Barakova, E.I.: Abstract robots with an attitude: Applying interpersonal relation models to human-robot interaction. In: Proceedings - IEEE International Workshop on Robot and Human Interactive Communication (2013), doi:10.1109/ROMAN.2013.6628528
15. Aldebaran Robotics. Aldebaran Robotics. Aldebaran Robotics - SAS, n.d. (February 25, 2015)

Artificial Metaplasticity:
Application to MIT-BIH Arrhythmias Database

Santiago Torres-Alegre[1][✉], Juan Fombellida[1],
Juan Antonio Piñuela-Izquierdo[2], and Diego Andina[1]

[1] Group for Automation in Signals and Communications,
Technical University of Madrid, 28040, Madrid, Spain
{santiago.torres,d.andina}@upm.es, jfv@alumnos.upm.es
[2] Universidad Europea de Madrid, Villaviciosa de Odón, Madrid, Spain
juan.pinuela@uem.es

Abstract. Artificial Metaplasticity are Artificial Learning Algorithms based on modelling higher level properties of biological plasticity: the plasticity of plasticity itself, so called Biological Metaplasticity. Artificial Metaplasticity aims to obtain general improvements in Machine Learning based on the experts generally accepted hypothesis that the Metaplasticity of neurons in Biological Brains is of high relevance in Biological Learning. Artificial Metaplasticity Multilayer Perceptron (AMMLP) is the application of Metaplasticity in MLPs ANNs trying to improve uniform plasticity of the Backpropagation algorithm. In this paper two different AMMLP algorithms are applied to the MIT-BIH electro cardiograms database and results are compared in terms of network performance and error evolution.

Keywords: Metaplasticity · Plasticity · MLP · MMLP · AMP · MIT-BIH ECGs · Feature Extraction · Machine Learning · Artificial Neural Network

1 Introduction

In this research we continue with our previous work [1], [2] applying the artificial metaplasticity multilayer perceptron (AMMLP) algorithm for the classification of cardiac arrhythmias. In this work AMMLP based on the input distribution presented in our previous work is compared with a new AMMLP based on the output of the network using it to modify the learning process of the MLP. AMMLP algorithm is tested using the well-known MIT-BIH (MIT-Beth Israel Hospital) dataset. For assessing this algorithm's accuracy of classification, we used the most common performance measures: specificity, sensitivity and accuracy. The results obtained were validated using the 10-fold cross-validation method.

Different artificial neural networks (ANNs) have been suggested for the detection of cardiac arrhythmias. Some based just on the ANN and others combining a preprocessing technique with the ANN system. For comparison purposes a table with the results of the different approaches will be provided in section 3.

© Springer International Publishing Switzerland 2015
J.M. Ferrández Vicente et al. (Eds.): IWINAC 2015, Part I, LNCS 9107, pp. 133–142, 2015.
DOI: 10.1007/978-3-319-18914-7_14

Regarding to the importance of the problem to which we apply the AMMLP algorithm is important to mention that cardiovascular diseases (CVDs) are one of the mayor cause of death worldwide as stated by the World Health Organization. There are significant cardiovascular abnormal symptoms which appear before the sudden occurrence of a heart attack. Therefore, having an effective method for early detection and treatment would reduce the number of disabilities and deceases caused by heart attack.

The remainder of this paper is organized as follows. Section 2 presents a description of the database and the algorithms. In Section 3 we present the experimental results obtained. A brief discussion of these results is showed in Section 4. Finally section 5 summarizes the main conclusions.

2   Materials and Methods

2.1   MIT-BIH Dataset

The MIT-BIH Arrhythmia Database was the first generally available set of ECGs (Electrocardiograms) test material for evaluation of arrhythmia detectors [3]. Database contains 48 half-hour excerpts of two-channel, 24-hour, studied by the BIH Arrhythmia Laboratory. These 48 half-hour excerpts were split in two groups: 23 (the "100 series") were chosen at random from a collection of over 4000 Holter tapes, and the other 25 (the "200 series") were selected to include examples of uncommon but clinically important arrhythmias that would not be well represented in a small random sample.

2.2   Data Preparation

1000 annotated ECG beats examined by specialists in MIT-BIH were selected for this study, which contain 4 different waveforms related to cardiac arrhythmias target, Normal beat (N); Premature ventricular contraction (PVC); Right bundle branch block (RBBB) and Left bundle branch block (LBBB). In Table 1 the eleven features descriptors that seem to be most important for characterizing the cardiac arrhythmias are presented. This features has been chosen with the help of specialists in cardiology. Componentes of ECG signal could be seen in Figure 1.

The data set formed by 1000 patterns is divided equally in four classes with 250 patterns each (N, PVC, RBBB and LBBB). We denominate respectively these classes as H_1, H_2, H_3, and H_4. To obtain results statistically independent of the distribution of the patterns a 10 fold cross validation evaluation method has been considered. Using this method the possible dependence of the results with the distribution of the samples in the training or performance evaluation sets is eliminated: all the samples are used to train the networks and all the samples are used to evaluate the performance of the results in different executions of the experiment for the same initial neural networks. Mean values are calculated to establish the final performance results.

Table 1. Features descriptors

Attributes	Meaning
Duration P	The width of the P wave.
PR interval	The distance between the beginning of the P wave and the beginning of QRS.
QRS complex	The distance between the beginning of the Q wave and the end of the S wave.
Duration T	The width of the T wave.
ST segment	The distance between the end of the S wave or R and the beginning of the T wave.
QT interval	The distance between the beginning of QRS and the end of the T wave.
RR previous: RRp	The distance between the peak R of the present beat and the peak R of the previous beat.
RR next : RRn	RRn: the distance between the peak R of the present beat and the peak R of the following beat.
RDI (delay of the deflexion)	From the beginning of QRS to the top of the latest wave of positivity R peak.
Beat duration	The distance between the beginning of the P wave and the end of the wave T.
RRp / RRn.	The ratio RRp / RRn

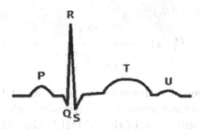

Fig. 1. ECG signal components

2.3   Artificial Metaplasticity Neural Network Model

ANNs, widely used in pattern classification within medical fields, are biologically inspired distributed parallel processing networks based on the neuron organization and decision-making process of the human brain [4]. In this paper we contiue with our prevoius work [1] applying metaplasticity to the MLP for classifying cardiological patterns.

The concept of biological metaplasticity was defined in 1996 by Abraham and now is widely applied in the fields of biology, neuroscience, physiology, neurology and others [5], [6]. Ropero-Pelez [6], Andina [2] and Marcano-Cedeño [7] have introduced and modeled the biological property metaplasticity in the field of

ANNs, obtaining excellent results. The model is applicable to general ANNs [2],[7], although in this paper it has been implemented for a multilayer perceptron (MLP).

The Backpropagation algorithm (BPA) presents some limitations and problems during the MLP training [8]. The artificial metaplasticity on multilayer percep-tron algorithm (AMMLP) tries to improve BPA by including a variable learning rate in the training phase affecting the weights in each iteration step, that is the metaplasticity, instead of the uniform plasticity that applies in the BPA. If s, j, i ∈ N are the MLP layer, node and input counter respectively, for each W(t) com-ponent $\omega_{ij}^{(s)}$(t)∈ R, where W(t)is the weight matrix, we can express the weight reinforcement in each iteration as:

$$w_{ij}^{(s)}(t+1) = w_{ij}^{(s)}(t) - \eta \frac{\partial E^*\left[W(t)\right]}{\partial w_{ij}^{(s)}} = w_{ij}^{(s)}(t) - \eta \frac{1}{f_X^*} \frac{\partial E\left[W(t)\right]}{\partial w_{ij}^{(s)}} \qquad (1)$$

It is up to the designer to find a function f_X^* that improves MLP learning. Several have been already proposed , and we introduce a new one in this study, given by Equation 3. In order to validate AMMLPs performance paper two alternative weighting functions are considered.

Artificial Metaplasticity by Gaussian Weighting Function as Estima-tion of Inputs Distribution. In AMMLP based on the distribution of the input patterns the function f_X^* is:

$$f_X^*(x) = \frac{A}{\sqrt{(2\pi)^N}.e^{B\sum\limits_{i=1}^{N} x_i^2}} \qquad (2)$$

Where N is the number of neurons in the MLP input layer, and parameters A and B ∈ R^+ are algorithm optimization values empirically determined which depend on the specific application of the AMMLP algorithm [1],[2],[7].

Artificial Metaplasticity by Outputs of the Network as an Estima-tion of a Posteriori Probability. In this case the estimation of a *posteriori* probability density function is considered as follows:

$$\widehat{y}_L \cong P(H_i/x) = f_X^*(x) \qquad (3)$$

where $\widehat{y}_L$ is the output of the neuron that estimate the a *posteriori* probability of the class. It can be seen that Equation 3 takes advantage of the inherent a *posteriori* probability estimation for each input class of MLP outputs.

The algorithm presented performs in the following way, if the output is far from the expected output the learning ratio is reinforced in next iteration and if the output is close to the expected output then the learning ratio in slowed down. It has to be noticed that in the first steps of the training the outputs of the MLP do not provide yet any valid estimation of the probabilities, so the training may not converge. In practice there are rarely instability problems of this kind, but if they occur, it is then better in these first steps of training, either to apply ordinary BPA training or to use another valid weighting function till BPA starts to minimize the error objective.

2.4 Network Structure Selection and AMMLP Algorithm

A 11/9/4 network structure is selected, that is 11 input neurons, one for each relevant features selected, 9 neurons in the hidden layer and 4 output neurons to represent the four possible classes. Output layer could also be composed of 3 neurons if output [0, 0, 0] is interpreted as one of the four classes.

Regarding to the AMMLP training phase the following parameters are consider:

- Learning rate $\eta = 1$
- Activation function is sigmoidal with value between (0,1).
- Initialize all weights in weight matrix W randomly between (-0.5,0.5)
- if epochs = 200 stop training
- if Mean Squared Error, MSE = 0.01 stop training
- for Gaussian AMMLP case A=39 and B=0,5 are selected

3 Results

In this section we present the results obtained in this research. All the models used in this study were trained and tested with the same data and validated using 10-fold cross-validation. The MLP and AMMLP proposed as classifiers for cardiac arrhythmias were implemented in MATLAB (software MATLAB version R2013a). The eleven attributes detailed in Table 1 were used as the inputs of the ANNs.

3.1 Measures of Quality

To evaluate the performance of the classifiers three measures are used and defined as follows:

$$Sensitivity(SE) = \frac{TP}{TP + FN}(\%) \tag{4}$$

$$Specifity(SP) = \frac{TN}{TN + FP}(\%) \tag{5}$$

$$Accuracy(AC) = \frac{TP + TN}{TP + TN + FP + FN}(\%) \tag{6}$$

Where TP, TN, FP, and FN stand for true positive, true negative, false positive and false negative, respectively.

3.2   Model Evaluation

For test results to be more valuable, a *k-fold* cross-validation is used among the researchers because it minimizes the bias associated with the random sampling of the training. In this method, the whole data are randomly divided into k mutually exclusive and equal size subsets. The classification algorithm is trained and tested k times. In each case, one of the folds is taken as test data and the remaining folds are added to form training data. Thus k different test results exist for each training-test configuration. The average of these results provides the test accuracy of the algorithm. A 10-fold cross-validation is used in all of our experiments by separating the selected 1000 samples randomly into 10 subsets with 100 records each and then taking each subset as test data in turns.

3.3   Performance Evaluation

In Table 2 results are presented, MLP AMMLP1 and AMMLP2 stands respectively for standard MLP Backpropagation ANN, MLP using gaussian function to modify the weights of the network and MLP using the output of the network to modify the weights. MLP and AMMLP1 results were presented in [1], the models are evaluated based on the accuracy measures discussed above (classification accuracy, sensitivity and specificity). The results were achieved using 10-fold cross-validation for each model, and are based on the average results obtained from the test data set for each fold.

Table 2. Results for 10-fold cross-validation for all folds and AMMLP and MLP models. Bold values highlight the average results obtained in this research.

Fold N	MLP (%)			AMMLP1 (%)			AMMLP2 (%)		
	SP	SE	AC	SP	SE	AC	SP	SE	AC
1	100	100	100	98.94	98.87	98.91	100	97.66	98.75
2	96.12	100	99.03	100	99.43	99.63	100	99.33	99.75
3	84.70	100	96.18	97.89	100	99.26	99.20	98.75	99.25
4	100	98.68	99.01	96.56	98.52	97.59	100	98.66	99.25
5	100	100	100	92.73	91.6	92.81	100	97.33	98.00
6	88.28	97.33	95.07	97.59	100	99.06	98,40	98.99	99.25
7	76.85	94.67	90.21	100	96.29	97.83	100	96.63	97.00
8	80.24	100	95.06	100	100	100	100	97.66	98.50
9	100	86,67	90	99.02	100	99.97	98.40	97.66	98.00
10	100	100	100	95.2	98.29	97.47	100	98,66	99,25
Average	92.62	97.73	96.45	97.79	98.3	98.25	**99.60**	**98.10**	**98.70**

For comparison purposes, Table 3 gives the classification accuracies of our method and previous methods applied to the same database. As can be seen from the results, AMMLP2 method is among the best in classification accuracy.

3.4   Error Evolution

Error evolution is evaluated at the end of every epoch of training. In each epoch 900 patterns are presented to the network. In Figures 2, 3, 4, error evolution is presented for MLP, AMMLP1 and AMMLP2.

Table 3. Classification accuracies obtained with our method and other classifiers from the literature

Authors (year)	Method	Accuracy (%)
Hu Y.et al., [9] (1997)	Expert Approach	94.00
Minami K. et al., [10] (1999)	Fourier-NN	98.00
Owis M.I. et al., [11] (2002)	Blind Source Separation	96.79
Yu S.N. et al., [12] (2008)	ICA-NN	98.71
Benchaib Y. et al., [13] (2009)	MLP BPA	95.12
Gothwal H. et al., [14] (2011)	Fourier-NN	98.48
P. Ghorbanian P. et al., [15] (2011)	Fourier-NN	99.17
Benchaib Y. et al., [1] (2013)	AMMLP1	98.25
in this study	**AMMLP2**	**98.70**

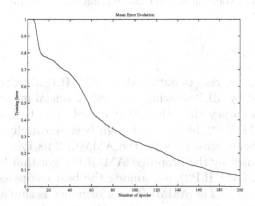

Fig. 2. Evolution of the classification error $\eta = 1$ - Nominal Backpropagation

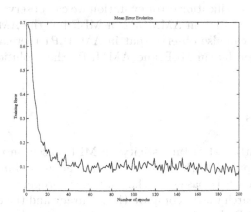

Fig. 3. Evolution of the classification error - AMMLP1

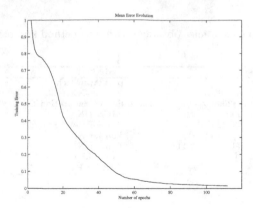

Fig. 4. Evolution of the classification error - AMMLP2

4   Discussion

As seen in Table 2, the results obtained by AMMLP algorithms are superior to the ones obtained by MLP showing that the Artificial metaplasticity models produce a higher accuracy than the MLP model. The best average accuracy is obtained with AMMLP2, 98,70% although best sensitivity is obtained with AMMLP1, 98,3%, being sensitivity for the AMMLP2 98,1%.

The results obtained by the proposed AMMLP algorithm based on the output of the network, (AMMLP2), are among the best compared with the other state-of-the-art methods. The AMMLP2 performance is similar to the solution proposed by Yu S.N et al., [12] and is beat for the solution proposed by [15]. It must be taken into account that in these better methods preprocessing techniques as Independent Component Analysis and Wavelets have been applied. In our future study we will combine these preprocessing techniques with AMMLP.

Regarding to the classification error evolution we can observe that the learning is quicker in AMMLP1 than in AMMLP2 or MLP but the AMMLP2 reach the final value first. We can also observe that in AMMLP1 there are a lot peaks in the learning evolution but in MLP and AMMLP2 the evolution of the error is more natural.

5   Conclusions

In this paper, the artificial metaplasticity on MLP based on the output of the network has been applied to the problem of cardiac arrhythmias classification. The AMMLP approach is based on the biological property of metaplasticity. The goal of this research was to compare the accuracy and the error evolution of the proposed AMMLP2 based on the output of the network with the AMMLP1 based on the distribution of the input patterns, and the classical MLP with BPA. Proposed AMMLP2 algorithm provides better results than AMMLP1, and MLP with BPA and is among the best of the state-of-the- art algorithms applied to the same database. It can be observed also that the evolution of the classification error is quicker in the AMMLP cases than in BPA altough more peaks appear in AMMLP1 being AMMLP2 and BPA evolution more natural. The results indicate that the use of the AMMLP algorithm based on the output of the network, as well as the AMMLP based On the distribution of the input patterns, is an alternative option for cardiac arrhythmias detection and could be used as a computer aided detection system for second opinion by physicians when making their diagnostic decisions.

References

1. Benchaib, Y., Marcano-Cedeño, A., Torres-Alegre, S., Andina, D.: Application of Artificial Metaplasticity Neural Networks to Cardiac Arrhythmias Classification. In: Ferrández Vicente, J.M., Álvarez Sánchez, J.R., de la Paz López, F., Toledo Moreo, F. J. (eds.) IWINAC 2013, Part I. LNCS, vol. 7930, pp. 181–190. Springer, Heidelberg (2013)
2. Andina, D., Alvarez-Vellisco, A., Jevtic, A., Fombellida, J.: Artificial metaplasticity can improve artificial neural network learning. Intelligent Automation and Soft Computing; Special Issue in Signal Processing and Soft Computing 15(4), 681–694 (2009)
3. Moody, G.B., Mark, R.G.: The impact of the MIT-BIH Arrhythmia Databaoc. IEEE Engineering in Medicine and Biology Magazine 20(3), 45–50 (2001)
4. Ropero-Pelaez, J., Andina, D.: Do biological synapses perform probabilistic computations? Neurocomputing (2012), http://dx.doi.org/10.1016/j.neucom.2012.08.042
5. Abraham, W.C.: Activity-dependent regulation of synaptic plasticity (metaplasticity) in the hippocampus. In: The Hippocampus: Functions and Clinical Relevance, pp. 15–26. Elsevier Science, Amsterdam (1996)
6. Kinto, E.A., Del Moral Hernandez, E., Marcano, A., Ropero Peláez, F.J.: A preliminary neural model for movement direction recognition based on biologically plausible plasticity rules. In: Mira, J., Álvarez, J.R. (eds.) IWINAC 2007. LNCS, vol. 4528, pp. 628–636. Springer, Heidelberg (2007)
7. Marcano-Cedeño, A., Quintanilla-Dominguez, J., Andina, D.: Breast cancer classification applying artificial metaplasticity algorithm. Neurocomputing 74(8), 1243–1250 (2011)
8. Leung, H., Haykin, S.: The complex backpropagation algorithm. IEEE Transactions on Signal Processing 39(9), 2101–2104 (1991)
9. Hu, Y.H., Palreddy, S., Tompkins, W.J.: A patient- adaptable ECG beat classifier using a mixture of experts approach. IEEE Transactions on Biomedical Engineering 44(9), 891–900 (1997)
10. Minami, K., Nakajima, H., Toyoshima, T.: Real-time discrimination of ventricular tachyarrhythmia with Fourier-transform neural network. IEEE Transactions on Biomedical Engineering 46(2), 179–185 (1999)
11. Owis, M.I., Youssef, A.B.M., Kadah, Y.M.: Characterization of ECG signals based on blind source separation. Medical and Biological Engineering and Computing 40(5), 557–564 (2002)
12. Yu, S.N., Chou, K.T.: Integration of independent component analysis and neural networks for ECG beat classification. Expert Systems with Applications 34(4), 2841–2846 (2008)

13. Benchaib, Y., Chikh, M.: A Specialized learning for neural classification of cardiac arrhythmias. Journal of Theoretical and Applied Information Technology 6(1), 81–89 (2009)
14. Gothwal, H., Kedawat, S., Kumar, R.: Cardiac arrhythmias detection in an ECG beat signal using fast fourier transform and artificial neural network. Journal of Biomedical Science and Engineering 4, 289–296 (2011)
15. Ghorbanian, P., Jalali, A., Ghaffari, A., Nataraj, C.: An improved procedure for detection of heart arrhythmias with novel pre-processing techniques. Expert systems 29(5), 478–491 (2009)

Toward an Upper-Limb Neurorehabilitation Platform Based on FES-Assisted Bilateral Movement: Decoding User's Intentionality

Andrés Felipe Ruiz-Olaya[1(✉)], Alberto López-Delis[2], and Alexander Cerquera[1]

[1] Facultad de Ingeniería Electrónica y Biomédica, Universidad Antonio Nariño,
Bogotá, Colombia
{andresru,alexander.cerquera}@uan.edu.co
[2] Centro de Biofísica Médica, Universidad de Oriente,
Santiago de Cuba, Cuba
alberto.lopez@cbiomed.cu

Abstract. In the last years there has been a noticeable progress in motor learning, neuroplasticity and functional recovery after the occurrence of brain lesion. Rehabilitation of motor function has been associated to motor learning that occurs during repetitive, frequent and intensive training. Neuro-rehabilitation is based on the assumption that motor learning principles can be applied to motor recovery after injury, and that training can lead to permanent improvements of motor functions in patients with muscle deficits. The emergent research field of Rehabilitation Engineering may provide promise technologies for neuro-rehabilitation therapies, exploiting the motor learning and neural plasticity concepts. Among those technologies, the FES-assisted systems could provide repetitive training-based therapies and have been developed to aid or control the upper and lower limbs movements in response to user's intentionality. Surface electromyography (SEMG) reflects directly the human motion intention, so it can be used as input information to control an active FES-assisted system. The present work describes a neurorehabilitation platform at the upper-limb level, based on bilateral coordination training (i.e. mirror movements with the unaffected arm) using a close-loop active FES system controlled by user. In this way, this work presents a novel myoelectric controller for decoding movements of user to be employed in a neurorehabilitation platform. It was carried out a set of experiments to validate the myoelectric controller in classification of seven human upper-limb movements, obtaining an average classification error of 4.3%. The results suggest that the proposed myoelectric pattern recognition method may be applied to control close-loop FES system.

1 Introduction

Neurorehabilitation is a process that take advantage of the neural plasticity to assist people in recovering motor ability [1], [2], whose most important aspects are based on the assumption that patients can improve with practice through motor learning. Repetitive motor activity in a real-world environment with a

© Springer International Publishing Switzerland 2015
J.M. Ferrández Vicente et al. (Eds.): IWINAC 2015, Part I, LNCS 9107, pp. 143–152, 2015.
DOI: 10.1007/978-3-319-18914-7_15

cognitive effort has been identified in several studies as favorable for motor recovery in stroke patients [3]. In this way, functional electrical stimulation (FES) is a technology developed to generate controlled specific muscle functions using electrical impulses, which facilitate movement of motor impaired people. This is achieved by activating skeletal muscles with constant frequency trains of stimulations [6]. In literature, FES has been evaluated to assist neurorehabilitation processes in people after stroke [7], [4], [5], rehabilitation after spinal cord injury [8], and suppression of pathological tremor [9].

During active FES-assisted therapy in a neurorehabilitation process, muscles work to complete a motor function under control of the treated patient. When the active FES-assisted therapy is delivered, the purpose of that intervention is to restore voluntary function. In other words, active FES-assisted therapy helps the neuromuscular system to relearn the execution of a function impaired due to neurological injury or disorder [6]. Thus, the goal of the FES therapy is helping to recover voluntary function as much as possible.

Implementing an active FES-assisted therapy requires a controller to generate commands in response to user's intention. This controller needs a convenient source of information, such as electromyography signals (EMG), whose characteristics reflect the muscular force level and the intention for movement. EMG signals offer a potential to provide residual control channels between the motor disability and assisted rehabilitative engineering [10], [11]. Accordingly, myoelectric pattern recognition is the process to detect the movement intention of the user from EMG signals [12]. To accomplish an effective myoelectric pattern recognition process, it is required to implement methods for pre-processing, features extraction, classification, and post-processing. Feature extraction is a key step of pattern recognition, taking into account that features represent relevant information from the input data in order to perform the desired task. Several works propose methods to extract features from EMG signals in the time-domain, frequency-domain and time-frequency domain. In the time-domain there are features such as mean absolute value (MAV), mean absolute value slope, zero crossing (ZC), slope sign changes (SSC), root mean square (RMS), waveform length (WL), among others [12]. By the other hand, frequency-domain features include coefficients obtained via short-time Fourier transform (STFT), wavelet transform (WT) and auto-regression (AR) of the spectrum of EMG signals [12].

This study presents a novel myoelectric pattern recognition method for decoding muscle movements to be implemented in an active FES-assisted system. It was implemented a feature extraction stage computing from EMG signals the RMS (Root Mean Square), WL (waveform length) and coefficients of a four-order AR model. Pattern recognition of EMG signals requires the generation of repeatable patterns of contractions, whose changes may result in an erroneous controller. A main source of pattern alterations arise from position of upper limb, where there are muscles whose activity depends on the angles and kinematics in joints [13]. Thus, the proposed myoelectric controller include information of the angular velocity at each upper-limb joint, in such a way that the mean value of angular velocity signals was computed as an additional feature. Finally, a classifier based on the Linear Discriminant Analysis (LDA) was implemented. In the validation process, it was carried out a set of experiments aimed to recognize seven upper-limb movements.

Next section presents a review of EMG-controlled FES-assisted techniques for neurorehabilitation and describes the proposed platform. Section 3 describes experimental methods including the protocol, the feature extraction methods and the classification process. Section 4 presents obtained results in the validation stage and last section presents conclusions and future work.

2 EMG-Controlled FES-Assisted Therapy for Neurorehabilitation

During a motor recovery process in neurorehabilitation two operating modes for FES therapy could be defined: passive and active. During passive FES therapy, a FES cycle is applied without intervention of the user, whereas in active FES therapy, the user initializes the application of FES to allow the execution of a specific motor function in response to user's intention [15].

An active FES therapy such as EMG-triggered electrical stimulation is reported to generate cortical changes and to induce neuroplasticity. Kimberley and collaborators reported a study to evaluate the effectiveness of EMG triggered electrical stimulation therapy in patients with longer-term hand paresis and representation of cerebral activation in the functional NMR [16]. Shin et al. investigated the effect of electromyography (EMG)-triggered neuromuscular electrical stimulation on functional recovery of hemiparetic hand and its related cortical activation pattern in chronic stroke patients. They demonstrated that 10-week EMG-stimulation can induce functional recovery and change of cortical activation pattern in the hemiparetic hand of chronic stroke patients [17]. Hara et al. researched the relation between hemiparetic arm function improvement and brain cortical perfusion (BCP) change during voluntary muscle contraction (VOL), EMG-controlled FES (EMG-FES) and simple electrical muscle stimulation (ES), before and after EMG-FES therapy in chronic stroke patients. They concluded that EMG-FES may have more influence on ipsilesional BCP than VOL or ES alone [4].

Active FES-assistive therapy is congruent with the sensorimotor integration theory that underlies EMG-controlled neuromuscular stimulation, which proposes that non-damaged motor areas can be recruited and trained to plan more effective movements using time-locked movement-related afference [18]. The patients initiate a movement and are assisted to complete it, receiving reafference that can be related to the command and the movement. EMG-triggered neuromuscular stimulation involves initiating a voluntary contraction for a specific movement until the muscle activity reaches a threshold level. When EMG activity reaches the chosen threshold, an assistive electrical stimulus is triggered.

In this way, two motor learning principles can be coupled in one protocol: repetition and sensorimotor integration.

2.1 Bilateral Coordination Training

Bilateral Coordination Training is a new treatment aimed to improve recovery of upper limb function in patients with hemiplegic stroke [19]. Specifically, voluntary movements of the intact limb may facilitate voluntary movements in the paretic limb. Activating the primary motor cortex and supplementary motor area for the intact limb increases the likelihood of voluntary muscle contractions (i.e., motor synergies) in the impaired limb when symmetrical movements are executed [20].

Bilateral coordination produces larger improvements than cyclic neuromuscular electrical stimulation on upper extremity impairment and activity limitation in patients [4]. The paradigm incorporates several rehabilitation principles that are important for motor relearning. The bilateral coordination puts the brain back in control of the affected limb by giving the patient direct control of the stimulation intensity [21], which reinforces the principle of intention-driven movement improving the synchronization between motor intention (central neural activity) and stimulated motor response (peripheral neural activity). This synchronization may promote neural reorganization to improve the central control of the impaired limb [21], so that the patient may not control only the stimulation turning on, but also its duration, intensity and therefore the resultant movement.

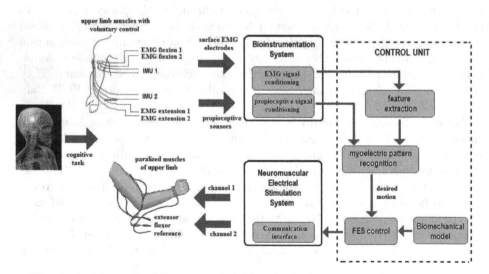

Fig. 1. Architecture of the neurorehabilitation platform at the elbow joint level

2.2   Upper-Limb Motor Recovery Through an EMG-Controlled Bilateral FES-Assisted Therapy

Figure 1 shows a proposed upper-limb closed-loop system based on EMG-controlled FES for impaired patients due to stroke that currently is being implemented. EMG activity of unaffected limb was detected and processed to control the stimulation duration and intensity of muscles in paralyzed limb, in order to generate mirror movements with the unaffected limb. This offers to the patient the ability to control the stimulation involving a cognitive task.

As was described in previous sections, this work describe the implementation and validation of a novel myoelectric pattern recognition method as a controller for decoding user's movements to be used in the neurorehabilitation platform. Next section describes the implementation and validation of the controller.

3   Experimental Methods

3.1   Protocol

Six adult subjects without neurological or musculoskeletal impairments participated under informed consent (four females and two males with ages between 22 and 36 years). The subjects were instrumented at upper limb level with surface EMG electrodes (Ag-AgCl electrodes) following the SENIAM recommendations [22]. Seven surface electromyography records (SEMG) were taken from each subject: two arm muscles (biceps brachii long head and triceps brachii long head) and five muscles around forearm. Subsequently, three 9-axis Inertial Measurements Units - IMU (MPU9150 from InvenSense) were coupled to arm, forearm an hand (see Figure 2). Since IMUs provided absolute angular velocity in its active axis, it was required the combination of two independent IMUs placed distally and proximally to the joint of interest.

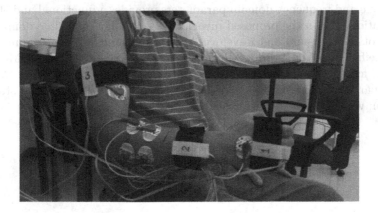

Fig. 2. Surface EMG electrodes and IMUs sensors coupled to upper limb

Signals were acquired using a PowerLab 8/30 system from ADInstruments and a sampling frequency of 1 kHz was used to acquire records from EMG channels and from IMUs. Subjects were seated in front of a computer that indicated movements to execute.

In each trial, subjects repeated each limb motion three times following seven movements: flexion and extension of the wrist, pronation and supination of the forearm, flexion and extension of the elbow, and resting state. The order of these limb motions was randomized. The resting state was kept in both at the beginning and at the end of each test, as well as between movements. A total of three trials were complete in a session.

3.2   Data Analysis and Pre-processing Techniques

The raw EMG signals were normalized respect to the maximum voluntary contraction (MVC) value. EMG signals were also filtered through a band-pass four-pole Butterworth filter, with a range of frequencies of 20-500 Hz. Signals from IMUs was filtered through a low-pass four-pole Butterworth filter with a cut-off frequency of 50 Hz. Electromyographic and angular velocity signals were segmented in windows of 250 ms and overlapped in 50 ms, taking into account that delays in myoelectric control for real-time applications must be inferior to 300 ms [23]. Afterward, each data segment was processed through a feature extraction method given from a combination of parameters in temporal and spectral domains, aimed at extracting information from both sources of signals.

3.3   Feature Extraction and Classification Methods

The work presented by [14] showed that a simple feature extraction method combined with a LDA classifier may provide an suitable performance for real-time myoelectric control. Taking into account its easy implementation and high performance, this system has been widely accepted and was implemented in the present work. Furthermore, recent researches have demonstrated that a combination of time and frequency domain parameters provides a functional and efficient configuration [24]. In the proposed myoelectric controller, two time-domain features (root mean square and waveform length) and four frequency-domain features (coefficients of an 4-order AR model) were extracted from the EMG data. Likewise, for each segment (epochs of 50 ms each), the mean value (MV) of the angular velocity signal was calculated. The RMS value relates to standard deviation, which is expressed in Equation 1.

$$RMS = \sqrt{\frac{1}{n} \sum_{i=1}^{n} x_i^2} \tag{1}$$

where x_i represents the data of the segment i of n samples length. The WL value provides the cumulative length of the waveform over the time segment (see Equation 2):

$$WL = \sum_{i=1}^{n-1} |x_{i+1} - x_i| \tag{2}$$

The MV provides the average amplitude of x_i in the segment i of n samples length (see Equation 3):

$$MV = \frac{1}{n} \sum_{i=1}^{n} x_i \tag{3}$$

AR model described each sample of SEMG signal as a linear combination of previous samples plus a white noise error term. These coefficients were herein used as features describing EMG pattern. The model is described by Equation 4.

$$Xn = -\sum_{x=1}^{p} a_i \cdot x_{n-1} + w_n \tag{4}$$

where Xn is a sample of the model signal, a_i is an AR coefficient, w_n is white noise or error sequence, and p is the order of the AR model. The feature vector, which is then used in the classification stage, results from the concatenation of both time and frequency domains. Thus, the feature vector have 38 features (7 EMG channels x 5 features/channel + 3 angular velocity signal x 1 feature/signal). The classification process was carried out using a LDA classifier, where each movement in the experimental protocol belongs to a class such as follows: class 1-wrist flexion, class 2-wrist extension, class 3-forearm pronation, class 4-forearm supination, class 5-elbow flexion, class 6-elbow extension and class 7-resting. The classifier was trained using data from the first two trials and was tested with data from the last trial, in a individual fashion for each subject.

3.4 Post-processing Techniques

The post-processing method were designed to manage excessive outflows in the classification process and improve the system performance. Errors occur normally during transitional periods, which are expected when the system lays in an undetermined state between contractions. In this way, a transition removal algorithm was implemented to remove errors.

4 Results

Feature extraction and classification methods were implemented using functions in MATLAB. Figure 3 (top) shows the confusion matrix from one working section in the experimental protocol. Rows in the matrix represent the inputs related to classes that are required to obtain, and columns represent obtained patterns as classifier outputs. The main diagonal in both matrices represents the concordance between the true and obtained classes. Figure 3 (bottom) shows the obtained classification error.

Figure 4 presents classification error over all subjects. The average classification error was 4.3%, which suggests that it is a good option for myoelectric pattern recognition. Unlike traditional EMG-based pattern recognition methods found in literature, the proposed myoelectric control method includes EMG-data and kinematics information (angular velocity) of upper-limb joints, in order to compensate EMG pattern alterations that arise from position of upper limb.

Class	wrist flexion	wrist extension	forearm pronation	forearm supination	elbow flexion	elbow extension	rest
wrist flexion	35	0	0	0	0	0	10
wrist extension	0	38	0	0	0	0	7
forearm pronation	0	0	32	0	0	0	13
forearm supination	0	0	0	38	0	0	7
elbow flexion	0	0	0	0	32	0	13
elbow extension	0	0	0	0	0	39	6
rest	0	1	2	1	0	0	84

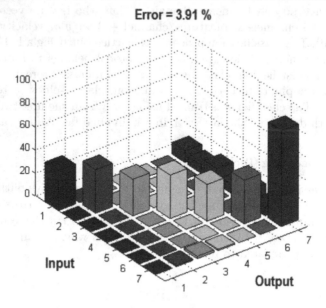

Error = 3.91 %

Fig. 3. Classification errors from one working section. Confusion matrix (top) and obtained errors for all classes (bottom).

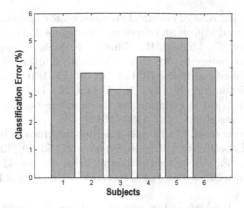

Fig. 4. Classification errors over all subjects. Features extracted from EMG signals and angular velocity information.

5   Conclusions and Future Work

This work presented an experimental protocol to validate a proposed myoelectric pattern recognition method, in order to classify seven upper-limb movements from SEMG information and angular velocity information of upper-limb joints. The results here obtained show a good performance of the method with an average classification error of 4.3%. Likewise, it is presented results for decoding movements executed by a user at the upper-limb level. As future work, this decoding user's intentionality method will be implemented in an upper-limb neurorehabilitation platform based on bilateral movement coordination. In addition, it will be integrated with the FES system to generate a specific real-time motor function in response to users intention at the upper limb level.

Acknowledgement. This work was supported by COLCIENCIAS (Project 415-2011).

References

1. Dietz, V., Nef, T., Rymer, W.Z.: Neurorehabilitation technology. Springer (2012)
2. Sharma, N., Classen, J., Cohen, L.G.: Neural plasticity and its contribution to functional recovery. In: Handbook of Clinical Neurology, vol. 110, pp. 3–12 (2013)
3. Moller, A.R.: Neural plasticity and disorders of the nervous system. Cambridge University Press (2006)
4. Hara, Y., Obayashi, S., Tsujiuchi, K., Muraoka, Y.: The effects of electromyography-controlled functional electrical stimulation on upper extremity function and cortical perfusion in stroke patients. Clinical Neurophysiology 124, 2008–2015 (2013)
5. Sheffler, L., Chae, J.: Neuromuscular electrical stimulation in neurorehabilitation. Muscle Nerve 35, 562–590 (2007)
6. Doucet, B.M., Lamb, A., Griffin, L.: Neuromuscular electrical stimulation for skeletal muscle function. The Yale Journal of Biology and Medicine 85, 201–215 (2012)

7. Hamid, S., Hayek, R.: Role of electrical stimulation for rehabilitation and regeneration after spinal cord injury: an overview. Eur. Spine J. 17, 1256–1269 (2008)
8. Sabut, S.K., Sikdar, C., Mondal, R., Kumar, R., Mahadevappa, M.: Restoration of gait and motor recovery by functional electrical stimulation therapy in persons with stroke. Disability and Rehabilitation 32(19), 1594–1603 (2010)
9. Popovic, L., Jorgovanovic, N., Ilic, V., Dosen, S., Keller, T., Popovic, M.B., Popovic, D.B.: Electrical stimulation for the suppression of pathological tremor. Med. Biol. Eng. Comput. 49, 1187–1193 (2011)
10. Wolczowski, A., Kurzynski, M.: Human-machine interface in bioprosthesis control using EMG signal classification. Expert Systems 27, 53–70 (2010)
11. Farina, D., Negro, F.: Accessing the neural drive to muscle and translation to neurorehabilitation technologies. IEEE Reviews in Biomedical Engineering 5, 3–14 (2012)
12. Oskoei, M.A., Hu, H.: Myoelectric control systems–A survey. Biomedical Signal Processing and Control 2, 275–294 (2007)
13. Jamison, J.C., Caldwell, G.E.: Muscle synergies and isometric torque production: Influence of supination and pronation level on elbow flexion. J. Neurophys. 70(3), 947–960 (1993)
14. Englehart, K., Hudgins, B.: A robust real-time control scheme for multifunction-myoelectric control. IEEE Trans. Biomed. Eng. 50(7), 848–854 (2003)
15. Tarkka, I.M., Pitkanen, K., Popovic, D.J., Vanninen, R., Kononen, M.: Functional Electrical Therapy for Hemiparesis Alleviates Disability and Enhances Neuroplasticity. J. Exp. Med. 225, 71–76 (2011)
16. Kimberley, T.J., Lewis, S.M., Auerbach, E.J., Dorsey, L.L., Lojovich, J.M., Carey, J.R.: Electrical stimulation driving functional improvements and cortical changes in subjects with stroke. Experimental Brain Research 154, 450–460 (2004)
17. Shin, H.K., Cho, S.H., Jeon, H.S., Lee, Y.H., Song, J.C., Jang, S.H., Lee, C.H., Kwon, Y.H.: Cortical effect and functional recovery by the electromyography-triggered neuromuscular stimulation in chronic stroke patients. Neuroscience Letters 442, 174–179 (2008)
18. Sheffler, L., Chae, J.: Technological Advances in Interventions to Enhance Post-stroke Gait. Phys. Med. Rehabil. Clin. N. Am. 24, 305–323 (2013)
19. Cauraugh, J.H., Kim, S.: Two Coupled Motor Recovery Protocols Are Better Than One: Electromyogram-Triggered Neuromuscular Stimulation and Bilateral Movements. Stroke 33, 1589–1594 (2002)
20. Stewart, K.C., Cauraugh, J.H., Summers, J.J.: Bilateral movement training and stroke rehabilitation: A systematic review and meta-analysis. Journal of the Neurological Sciences 244, 89–95 (2006)
21. Knutson, J.S., Harley, M.Y., Hisel, T.Z., Makowski, N.Z., Fu, M.J., Chae, J.: Contralaterally Controlled Functional Electrical Stimulation for Stroke Rehabilitation. In: Conf. Proc. IEEE Eng. Med. Biol. Soc., pp. 314–317 (August 2012)
22. Surface electromyography for the non-invasive assessment of muscles project. web page, http://seniam.org/
23. Huang, Y.H., Englehart, K., Hudgins, B.S., Chan, A.D.C.: A Gaussian mixture model based classification scheme for myoelectric control of powered upper limb prostheses. IEEE Trans. Biomed. Eng. 52(11), 1801–1811 (2005)
24. Phinyomark, A., Phukpattaranont, P., Limsakul, C.: Feature reduction and selection for EMG signal classification. Expert Systems with Applications 39, 7420–7431 (2012)

Decoding of Imaginary Motor Movements of Fists Applying Spatial Filtering in a BCI Simulated Application

Jan Boelts[1], Alexander Cerquera[2], and Andrés Felipe Ruiz-Olaya[3(✉)]

[1] Institute of Cognitive Science, University of Osnabrueck, Osnabrueck, Germany
jboelts@uos.de
[2] Faculty of Biomedical, Electronic and Mechatronics Engineering. Research Group
Complex Systems, Universidad Antonio Nariño. Bogotá, Colombia
alexander.cerquera@uan.edu.co
[3] Faculty of Biomedical, Electronic and Mechatronics Engineering. Research Group
Bioengineering, Universidad Antonio Nariño. Bogotá, Colombia
andresru@uan.edu.co

Abstract. This work presents a study that evaluates different scenarios of preprocessing and processing of EEG registers, with the aim to predict fist imaginary movements utilizing the data of the EEG Motor Movement/Imaginary Dataset. Three types of imaginary fist movements have been decoded: sustained opening and closing of right fist, sustained opening and closing of left fist and rest. Initially, the registers were band-pass filtered to separate frequency ranges given by *mu* rhythms (7.5-12.5 Hz), *beta* rhythms (12.5-30 Hz), *mu&beta* rhythms, and a *broad* range of 0.5-30 Hz. Afterward, the signals of the separated subbands were epoched in time windows of 0-0.5, 0-1, 0-1.5 and 0-2 seconds, as well as preprocessed with two techniques of spatial filtering: common spatial patterns and independent component analysis. In both cases, a set of selected channels was established for feature extraction, by calculation of the logarithms of the variance in the time series corresponding to each preprocessed and selected channel. The classification stage was based on linear discriminant analysis and support vector machines. The results showed that the combination given by common spatial patterns and support vector machines allowed to reach a mean decoding accuracy close to 99.9%, where epoching and filtering to separate subbands did not influence the results in a noticeable way.

1 Introduction

It is well known that mental tasks are encoded in scalp electroencephalographic waves (EEG) that can be used as a source of information to operate devices of brain computer interfaces (BCI). One of the most important challenges in this topic is the application of biomedical signal processing tools to decode motor tasks with high performance of classification. In addition, in BCI applications it is important to extract features from EEG registers to detect characteristics of imaginary movements such as direction, intensity, lateralization and speed [1].

© Springer International Publishing Switzerland 2015
J.M. Ferrández Vicente et al. (Eds.): IWINAC 2015, Part I, LNCS 9107, pp. 153–162, 2015.
DOI: 10.1007/978-3-319-18914-7_16

Several approaches to decode mental tasks have been explored in the last decades to improve the performance of BCI devices, as a way to develop communication technologies between brain and control systems for motor activities. These studies include the evaluation of features extracted via time and time-frequency representations [2], spectral power, coherence, phase locking value [3] and power spectral density [4]. Nevertheless, some years ago the utility of spatial filter techniques began to be explored in BCI applications, as a way to select the most discriminative features of EEG registers in motor imagery tasks and to reduce its huge dimensionality in feature space [5]. In this way, common spatial patterns (CSP) have been applied successfully together with features extracted via frequency-weighted method [6], wavelet packets decomposition [7], local discriminant bases [8] and eigendecomposition [9]. Likewise, ICA has been employed to extract motor-related cortical EEG [10] and hand imaginary movements [11].

These promising studies need to be complemented with an evaluation of different preprocessing and classification methods, whose characteristics can mark the difference between high and low decoding performance of motor tasks from EEG registers. In this way, this work presents a comparison among different EEG preprocessing scenarios to decode a set of imaginary fist movements on a single trial basis. The main preprocessing parameters varied in this study were: 1) EEG subbands: *mu*, *beta*, *mu&beta* and a *broad* spectrum (0.5 to 30 Hz); 2) epoch length: 0-0.5, 0-1, 0-1.5 and 0-2 seconds; 3) spatial filter: *None*, CSP and ICA; and 4) supervised classification method to perform the prediction: linear discriminant analysis (LDA) and support vector machines (SVM). The results allow to infer combinations of these parameters to obtain decoding performances close to 99.9% in a BCI simulated application.

2   Materials and Methods

2.1   Dataset

The database used in this study was the EEG Motor Movement/Imaginary Dataset [12], [13]. This database contains 64-channels EEG data acquired from 109 healthy volunteers performing imaginary movements (sustained opening and closing) of their right and left fists during six minutes. The appearance of the target on the left or right side of a screen in front of each volunteer indicated the onset of imaginary left or right fist movement, which was sustained during four seconds until the target disappeared. Each EEG register contained 87 trials of three events of interest: 43 trials of resting condition (T_0), 22 trials of left fist movement (T_1) and 22 trials of right fist movement (T_2). The files were read using the EEGLAB toolbox [14].

2.2   Data Processing and Experimental Design

Preprocessing. The analysis was carried out in three stages: preprocessing, training and testing. Each stage was performed for every subject individually and in five independent repetitions, where the transformation matrices of the spatial filters and the parameters of the classifiers were determined exclusively during the training stage. This experimental procedure was repeated for every possible combination of 16 preprocessing steps, three spatial filters and two classification methods resulting in 96 experiments.

The preprocessing stage included the combinations of channels selection, re-referencing, temporal filtering and epoching. Whereas the two latter were applied to all combinations, channels selection and re-referencing were applied only in experiments that involved the spatial filtering based on ICA. Firstly, the 64 channels of each EEG register were re-referenced to their average electrode and the analysis was restricted to a subset of eight channels located on the motor cortex (FC3, FCZ, FC4, C3, C1, CZ, C2, and C4), under assumption that this subset transmits most of the relevant information that characterizes imaginary movements from EEG data [15].

The works presented in [16] and [17] showed that significant changes in the power of brain rhythms in response to real or imaginary movements are most prominent in the *mu* (7.5 to 12.5 Hz) and *beta* (12.5 to 30 Hz) EEG subbands. Therefore, four bandwidth scenarios were chosen employing band-pass finite impulse response filters: *mu*, *beta*, *mu&beta* and a *broad* range from 0.5 to 30 Hz. Subsequently, the band-pass filtered EEG data were epoched around the events of interest, labeled as T_0, T_1 and T_2, in four different time windows (the movement onset was set to 0 seconds): 0 to 0.5, 0 to 1, 0 to 1.5 and 0 to 2 seconds. Afterward, the resulting 87 epochs were pseudo-randomly separated into training (80%) and testing (20%) trials.

Training. The training stage started with the application of a spatial filter in order to improve the signal-to-noise ratio of the EEG registers and to optimize the signal for their further analysis. As mentioned, the processing in this work was based on independent component analysis (ICA) and common spatial pattern (CSP). ICA provided a transformation matrix that projected the EEG data from the space of 64 temporal channels into a space of surrogate channels, which are maximally independent from each other. ICA was applied only to the eight EEG channels selected during the preprocessing stage and no independent components were inspected or excluded empirically. Instead, the selection of the independent components with respect to classification task at hand was performed in a data driven way, i.e., the weights obtained during the learning process of the classifiers performed the selection indirectly.

CSP is a spatial filter that was developed to optimize the discriminalibility of a signal with respect to two classes [18]. It has been widely and successfully used for the classification of two movement conditions in BCI research [19], [20]. However, few approaches have been performed to explore its utility in the classification of three movement conditions. Therefore, the CSP technique was extended in this work to the multi-class situation, in such a way that three CSP filters were employed via one-versus-rest approach: $T_0\&T_1$ vs T_2, $T_0\&T_2$ vs T_1 and $T_1\&T_2$ vs T_0. Similar to ICA, one CSP filter yields a transformation matrix that projects the EEG signal to a set of surrogate channels. This transformation is optimized under the constraint that the variances of the resulting surrogate channels are maximally different in two conditions. Usually, the number of surrogate channels used for the further analysis is reduced to the k most discriminant common spatial patterns for each condition, resulting in $2k$ surrogate channels. Here, $k=4$ was used, so that each CSP filter reduced the 64 EEG channels to eight CSP surrogate channels with optimal variances.

The last step in the training stage was the feature extraction from the preprocessed and spatially filtered signals. It was assumed that the power of a certain brain rhythm was approximately equal to the variance of a band-passed filtered signal in the corresponding EEG subband. Accordingly, the features were extracted calculating the logarithm of the variance of each channel resulting from the spatial filtering. This procedure resulted in eight features for the ICA filter, (one for each of the eight independent components), while for the CSP filters 24 features were obtained (eight for each of the three CSP filters). After this task, they were scaled to have zero mean and unit variance.

The classification among the three conditions was carried out with linear discriminant analysis (LDA) and support vector machines (SVM). LDA is a method that finds parameters to construct a hyperplane by separation of the training samples in the feature space. On the other hand, SVM is a more advanced classification method that has proven its usefulness for BCI applications [21], [22]. Similar to LDA, SVM uses the training samples to fit a hyperplane, in order to separate two groups of data points in the feature space. However, by maximizing the margin between the hyperplane and the nearest training samples (the support vectors) it constraints the fitted hyperplane to be optimal and yields a sparse classifier that depends only on a subset of the training samples. As a consequence, it provides a good generalization to data different than the training distribution. Furthermore, by projecting the features in a higher dimensional space, before finding the parameters of the hyperplane, it is capable of dealing with nonlinear structures in the features. In this way, the projection is achieved by applying a kernel function to the features.

In the present study, a polynomial kernel was utilized in the classfication with SVM employing the LIBSVM toolbox [23]. Since the basic SVM is a method to classify between two groups, the classification of the three conditions was performed in three predictive models as well, one for each contrast among the groups: $T_0\&T_1$ vs T_2, $T_0\&T_2$ vs T_1 and $T_1\&T_2$ vs T_0.

Testing. In this stage, preprocessing and feature extraction were performed like it was done during the training stage. The transformation matrices obtained were applied for the spatial filtering, whereas the predictive models obtained from the classification with LDA and SVM were tested on the unseen testing trials using the one-versus-rest approach. For every class contrast j, the corresponding model predicted each trial i and computed the posterior probability, so that given the observation of trial i the underlying class was j (defined as $p(\text{class}_j \mid \text{trial}_i)$). Thus, for each trial the class contrast with the highest posterior probability yielded the final decoded movement.

2.3   Evaluation

The decoding performance was quantified using the classification accuracy, that is, the percentage of correctly classified testing trials. For each subject, the decoding performance was averaged over the five repetitions of the experimental procedure, in such a way that the grand mean over all 109 subjects was calculated in each experiment. In order to evaluate the significant differences among decoding performances, an ANOVA analysis for repeated measures with four within-subject factors was performed on the mean testing accuracies using SPSS. The four within-subject factors were: temporal filter (TF) with four levels (*mu*, *beta*, *mu & beta* and *broad*), epoching (EP) with four levels (0 to 0.5, 0 to 1, 0 to 1.5 and 0 to 2 seconds), spatial filter (SPF) with three levels (*None*, ICA and CSP) and classification method (CM) with two levels (LDA and SVM). This procedure resulted in a total number of $4 \times 4 \times 3 \times 2 = 96$ repeated measures for each subject.

A Mauchly's test of sphericity was performed to assess the compliance of the assumption of sphericity. In case of a significant violation, Greenhouse-Geisser- and Huynh-Feldt-correction of degrees of freedom were applied as proposed in [24]. In addition, post-hoc tests of multiple comparisons on each of the significant main factors and interactions were performed for evaluation of significant differences among the levels within each factor. Here, Bonferroni correction was used to adjust the level of significance for the multiple comparisons. In all tests the initial significance level was established to $p < 0.05$. The classification using each of the 96 combinations of processing methods was repeated five times for each subject, and the performance of each combination for each subject was quantified using mean classification accuracy over the five repetitions.

3   Results

Table 1 shows the grand mean for each combination, i.e. mean over all 109 subjects, giving an overview of the performance of the 96 combinations. A more detailed quantification of the results was performed through a $4 \times 4 \times 3 \times 2$ (TF $\times$ EP $\times$ SPF $\times$ CM) within-subject repeated-measures ANOVA. All four factors showed a significant main effect ($p < 0.01$) and significant interaction between factors were found for TF*EP, TF*SPF, TF*CM, EP*SPF, EP*CM and SPF*CM ($p < 0.01$). Notice that the sign of asterisk denotes the interaction between factors. No other significant effects were found.

The multiple comparison tests on the significant main effects showed small significant differences in some of the factor levels of TF and EP. However, the most interesting and noticeable differences were obtained in the SPF and the CM factor. Here, the marginal mean accuracy of the CSP filter was 13.7% and 11.4% higher than that of *None* and ICA ($p < 0.05$) respectively, and the marginal mean accuracy of the SVM classifier was 39% higher than the obtained with the LDA method ($p < 0.05$). From the multiple comparisons tests within the significant

Table 1. Mean percentages of classification accuracy over all 109 subjects

Epoch in seconds	LDA				SVM			
	mu	beta	mu&beta	broad	mu	beta	mu&beta	broad
None								
0 to 0.5	43.27	42.21	43.80	40.94	78.55	74.91	74.85	74.37
0 to 1	45.03	45.65	46.72	40.23	77.10	78.82	80.22	72.19
0 to 1.5	45.11	47.92	48.44	40.45	77.56	80.11	79.02	72.99
0 to 2	46.14	49.49	49.75	41.01	78.42	80.95	80.96	73.37
ICA								
0 to 0.5	52.24	52.32	56.83	51.06	79.53	79.79	79.27	75.44
0 to 1	56.59	64.00	62.00	51.41	79.46	82.53	81.44	77.57
0 to 1.5	61.41	68.47	67.76	50.24	80.40	82.66	84.64	77.20
0 to 2	64.94	71.53	70.00	49.88	82.90	83.30	84.99	78.42
CSP								
0 to 0.5	49.32	49.50	50.23	48.30	**99.89**	**99.61**	**99.43**	**99.82**
0 to 1	50.48	50.43	52.52	47.06	**99.77**	**99.74**	**99.31**	**99.81**
0 to 1.5	52.07	52.81	37.44	45.61	**99.61**	**99.40**	**99.23**	**99.88**
0 to 2	53.38	54.93	57.01	44.01	**99.46**	**99.45**	**99.15**	**99.81**

interactions, the most interesting results occurred in the interactions TF*SPF and SPF*CM. The corresponding box plots of mean accuracies are presented in Figures 1 and 2, where it can be observed that the marginal mean accuracy for the TF*SPF interaction was maximum in combinations using a temporal mu or beta filter and spatial CSP filter ($p < 0.05$). Regarding spatial filters and classification methods, CSP combined with SVM was significantly better than any other combination.

The relationship between the weights, resulting from spatial filtering and classification with the location of the electrodes, gives a neurobiological meaningful understanding of the involved methods. In this way, Figure 3 shows the topographic maps of the eight most important spatial filters resulting from CSP. The four spatial filters in the first row maximize the variance for the T_0 condition, whereas the other four in the second row do it for the conditions T_1 and T_2. This rationale corresponds to the weights visualized in the topographic maps, showing reverse relations between CSP 2 and CSP 63. Furthermore, CSP 62 and CPS 63 emphasize the channels located over the left and right motor cortex respectively.

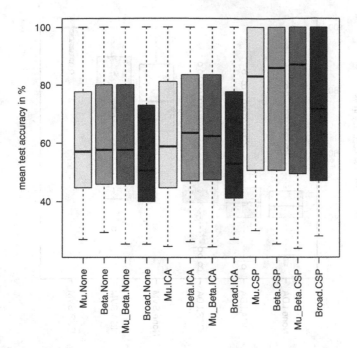

Fig. 1. Box plots diagrams for all combinations of levels in the interaction between the TF and the SPF factors

4   Discussion

The results of this study allow to establish some observations about the influence of preprocessing and classification tasks in EEG registers, in order to predict imaginary movements in a BCI simulated application. Specifically, the significant main effects, interactions and significant differences found in the multiple comparison tests suggest that the combinations including the spatial filter CSP and the classifier SVM resulted in the best classification accuracies. Neither the applied temporal filter nor the extracted epoch seem to have a significant influence on the classification accuracy when CSP and SVM are involved. While CSP yielded the best results in combination with SVM, it was outperformed by ICA when combined with LDA. Regarding the different subbands, similar results for the *mu*, *beta* and *mu&beta* subbands, as well as slightly lower accuracies for the broad band were observed. In any way, the decoding performance was always above the chance level of 33.3%.

Although the works shown in [16] and [17] showed the role of the brain rythms in *mu* and *beta* subbands in response to imaginary movements, in the present study it is observed that the separation of these subbands did not noticeably influence the performance in comparison to those obtained with EEG signals of broad spectrum. The difference in the performance between classifiers was

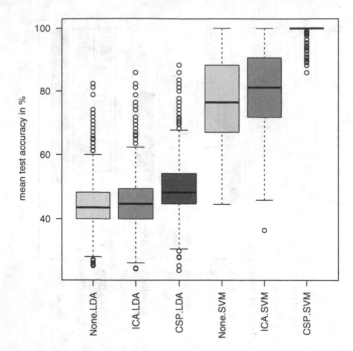

Fig. 2. Box plots diagram for all combinations of levels in the interaction between the SPF and the CM factors

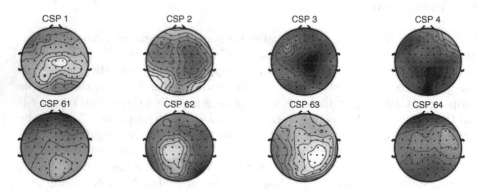

Fig. 3. The weights of the eight most important spatial filters from CSP analysis interpolated in a topographic map. The spatial filters were obtained by contrasting condition T_0 versus rest (99.9% accuracy). The first row shows the four most important commom spatial patterns for condition T_0 and the second row those for conditions T_1 and T_2.

expected, given the utility of the SVM to capture the nonlinear structure of the features by applying a polynomial kernel and by performing intense parameter search. Furthermore, the analysis carried out in this work took advantage of the method of CSP, extending it to the classification of three conditions, optimizing the features for classification, and allowing the prediction to be more successful than using the ICA features.

The computational expensive operations were performed during training, so that it is possible to perform a further implementation of the method in a real-time scenario. Regarding the length of the epochs, although an increase of the time from 0.5 to 2 seconds improve the classification percentages, the preprocessing with CSP allows to obtain performances close to 99.9% with epochs of only 0.5 seconds. These results suggest the feasibility to predict imaginary fist movements from a single-trial EEG record with accuracies, time windows and computational costs sufficient to be realistic in real-time BCI applications.

Acknowledgement. This work was partially supported by COLCIENCIAS (project number 415-2011) and the RISE Worldwide Program of the DAAD.

References

[1] Sanei, S.: Adaptive Processing of Brain Signals, 1st edn., pp. 295–324. John Wiley & Sons (2013)

[2] Hurtado-Rincón, J., Rojas-Jaramillo, S., Ricardo-Céspedes, Y., Alvarez-Meza, A.M., Castellanos-Domínguez, G.: Motor Imagery Classification using Feature Relevance Analysis: An Emotiv-based BCI System. In: XIX Symposium on Image, Signal Processing and Artificial Vision (STSIVA 2014), September 17-19 (2014)

[3] Krusienski, D.J., McFarland, D.J., Wolpaw, J.R.: Value of amplitude, phase, and coherence features for a sensorimotor rhythm-based braincomputer interface. Brain Research Bulletin 87, 130–134 (2012)

[4] Bai, O., Lin, P., Vorbach, S., Li, J., Furlani, S., Hallett, M.: Exploration of computational methods for classification of movement intention during human voluntary movement from single trial EEG. Clin. Neurophysiol. 118, 2637–2655 (2007)

[5] Rejer, I.: EEG feature selection for BCI based on motor imaginary task. Foundations of Computing and Decision Sciences 37, 283–292 (2012)

[6] Liu, G., Huang, G., Meng, J., Zhu, X.: A frequency-weighted method combined with Common Spatial Patterns for electroencephalogram classification in brain-computer interface. Biomedical Signal Processing and Control 5, 174–180 (2010)

[7] Mousavi, E.A., Maller, J.J., Fitzgerald, P.B., Lithgow, B.J.: Wavelet Common Spatial Pattern in asynchronous offline brain computer interfaces. Biomedical Signal Processing and Control 6, 121–128 (2011)

[8] Asensio-Cubero, J., Gan, J.Q., Palaniappan, R.: Extracting optimal tempo-spatial features using local discriminant bases and common spatial patterns for brain computer interfacing. Biomedical Signal Processing and Control 8, 772–778 (2013)

[9] Velásquez-Martínez, L.F., Álvarez-Meza, A.M., Castellanos-Domínguez, C.G.: Motor Imagery Classification for BCI Using Common Spatial Patterns and Feature Relevance Analysis. In: Ferrández Vicente, J.M., Álvarez Sánchez, J.R., de la Paz López, F., Toledo Moreo, F. J. (eds.) IWINAC 2013, Part II. LNCS, vol. 7931, pp. 365–374. Springer, Heidelberg (2013)

162 J. Boelts et al.

[10] Kanoh, S., Miyamoto, K., Yoshinobu, T.: Generation of Spatial Filters by ICA for Detecting Motor-related Oscillatory EEG. In: 34th Annual International Conference of the IEEE EMBS, San Diego, California USA, August 28- September 1, pp. 1703–1706 (2012)

[11] Wang, Y., Wang, Y.T., Jung, T.P.: Translation of EEG Spatial Filters from Resting to Motor Imagery Using Independent Component Analysis. PLoS One 7, e37665 (2012)

[12] Schalk, G., McFarland, D.J., Hinterberger, T., Birbaumer, N., Wolpaw, J.R.: BCI2000: A general-purpose brain-computer interface (BCI) system. IEEE Trans. Biomed. Eng. 51, 1034–1043 (2004)

[13] Goldberger, A.L., Amaral, L.A., Glass, L., Hausdorff, J.M., Ivanov, P.C., Mark, R.G., Mietus, J.E., Moody, G.B., Peng, C.K., Stanley, H.E.: PhysioBank, PhysioToolkit, and PhysioNet: components of a new research resource for complex physiologic signals. Circulation 20, e215–e220 (2000)

[14] Delorme, A., Makeig, S.: EEGLAB: an open source toolbox for analysis of single-trial EEG dynamics including independent component analysis. J. Neurosci. Methods. 134, 9–21 (2004)

[15] Alomari, M.H., Samaha, A., AlKamha, K.: Automated Classification of L/R Hand Movement EEG Signals using Advanced Feature Extraction and Machine Learning (IJACSA) International Journal of Advanced Computer Science and Applications 4, 207–212 (2013)

[16] Pfurtscheller, G., Lopes da Silva, F.H.: Event-related EEG/MEG synchronization and desynchronization: basic principles. Clin. Neurophysiol. 110, 1842–1857 (1999)

[17] Jeona, Y., Namb, C.S., Kimc, Y.J., Whangd, M.C.: Event-related (De)synchronization (ERD/ERS) during motor imagery tasks: Implications for braincomputer interfaces. International Journal of Industrial Ergonomics 41, 428–436 (2011)

[18] Blankertz, B., Tomioka, R., Lemm, S., Kawanabe, M., Mueller, K.R.: Optimizing Spatial Filters for Robust EEG Single-Trial Analysis. IEEE Signal Processing Magazine 25, 41–56 (2008)

[19] Khan, Y.U., Sepulveda, F.: Brain-computer interface for single-trial eeg classification for wrist movement imagery using spatial filtering in the gamma band. IET Signal Process. 4, 510–517 (2010)

[20] Mueller-Gerking, J., Pfurtscheller, G., Flyvbjerg, H.: Designing optimal spatial filters for single-trial EEG classification in a movement task. Clin. Neurophysiol. 110, 787–798 (1999)

[21] Liao, K., Xiao, R., Gonzalez, J., Ding, L.: Decoding Individual Finger Movements from One Hand Using Human EEG Signals. PLOS ONE 9, e85192, 1–12 (2014)

[22] Huang, D., Lin, P., Fei, D.Y., Chen, X., Bai, O.: Decoding human motor activity from EEG single trials for a discrete two-dimensional cursor control. J. Neural. Eng. 6 (2009)

[23] Chang, C.C., Lin, C.J.: LIBSVM: A Library for Support Vector Machines. ACM Trans. Intell. Syst. Technol. 2, 27:1–27:27 (2011)

[24] Girden, E.R.: ANOVA: Repeated Measures, vol. 84. SAGE Publications (1992)

The Koniocortex-Like Network:
A New Biologically Plausible Unsupervised Neural Network

Francisco Javier Ropero Peláez[1](✉) and Diego Andina[2]

[1] Center of Mathematics, Computation and Cognition,
Universidade Federal do ABC, Santo André, Brazil
`francisco.pelaez@ufabc.edu.br`
[2] Group for Automation in Signal and Communications,
Technical University of Madrid, Madrid, Spain
`d.andina@upm.es`

Abstract. In this paper we present a new unsupervised neural network whose architecture resembles the koniocortex, the first cortical layer receiving sensory inputs. For easiness, its properties were incorporated in a step by step manner along successive network versions. In some cases, the version improvement consists in the replacement of a non-biological property by a biologically plausible one. Initially (version 0) the network was merely an scaffold implementing the Bayes Decision Rule. The first network version incorporated metaplasticity and intrinsic plasticity, but neural competition was not biological. In a second version, competition naturally occurred due to the interplay between lateral inhibition and homeostatic properties. Finally, in the koniocortex-like network, competition and pattern classification emerges naturally due to the interplay of inhibitory interneurons and previous version's properties. An example of numerical character recognition is presented for illustrating the main characteristics of the network.

Keywords: koniocortex · Granular cortex · Intrinsic plasticity · Pre-synaptic rule · Competition · Feature extraction · Learning · Neural network

1 Introduction

The koniocortex, also called granular cortex, is the name given to the different regions of the cerebral cortex that exhibit a well-defined inner granular layer (layer IV). Both names (koniocortex and granular) refer to a cortex with a grainy texture (*konia* is a greek word meaning "dust") due to the abundance of spiny stellate neurons in this layer. Brodmann areas 13 of the somatic sensory cortex, area 17 of the visual cortex, and area 41 of the auditory cortex belong to the koniocortex. All these areas behave as topographic maps that change their boundaries and receptive fields according to sensory experience. The visual koniocortex, for example, is the locus of ocular dominance columns and orientation

© Springer International Publishing Switzerland 2015
J.M. Ferrández Vicente et al. (Eds.): IWINAC 2015, Part I, LNCS 9107, pp. 163–174, 2015.
DOI: 10.1007/978-3-319-18914-7_17

columns that are modified when for example one eye is occluded; in the case of the somatosensory cortex, Diamond *et al* [6] discovered that when rodent whiskers are differently stimulated, important somatotopic map modifications occur in this cortex. They attributed these modifications to NMDA receptors in spiny stellate cells [8] . According to Miller *et al.*[11] , map reorganization is a bottom-up process, solely involving the two layered network composed of the thalamo-cortical layer and the IVth layer of koniocortex. They mentioned David Ferster research [7] who found that orientation sensitivity columns developed in the fourth layer of the cortex without the collaboration of upper cortical layers (i.e. layers I,II and III) in which neuron's spiking was prevented by cooling the cortical preparation.

In this paper, an initial complete artificial neural framework is gradually transformed into a new type of neural network, the koniocortex-like network (KLN), that very much resembles the granular layer of the koniocortex. This is done mainly by replacing non-biological properties by biological ones.

Initially (in version 0 of the network), we introduce a non-biological framework, the "Bayesian Decision Framework" (see Fig.1.a) used as a sort of scaffold over which the different biological properties will be placed. This "Bayesian Decision Framework" serves for introducing some algebraic concepts of the following versions of the network.

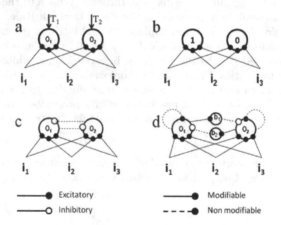

Fig. 1. For allowing a step-by-step presentation of the koniocortex-like network, it was developed along four stages. See details in Section. 1

In the following version, version 1, we added competition to the initial framework. This competitive network is depicted in Fig.1.b. In this version, competition is not yet biological, but externally driven (the winner neuron is obtained by calculation). In this version, biological intrinsic plasticity (see section 2) was incorporated to the neurons.

This network version performs feature extraction and classification tasks as any competitive network. The network in Fig.1.c, is a type of "lateral inhibition network", performing feature extraction and classification tasks. Here, intrinsic plasticity, lateral inhibition and a steep slope in the sigmoidal activation functions allow competition in a biologically plausible way. Finally, the fourth version of the network which is represented in Fig.1.d performs competition and pattern classification similarly to the previous "lateral inhibition network". Here, inhibitory lateral connections are substituted by inhibitory interneurons with non-modifiable negative connections. We have called this network the "Koniocortex-like network" (KLN), due to its similarity to the koniocortex or granular layer of sensory cortex. In this paper, we discuss in more detail some of the findings sketched in a previous work [13] while highlighting the similarities between the KLN and the koniocortex fourth layer. We also apply the network to perform a classification task using alphanumerical input patterns.

2 Homeostatic Properties: Metaplasticity and Intrinsic Plasticity in Rate Code Neurons

All versions of the network use rate-code neurons whose outputs, O_j, (bounded between 0 and 1) represent the probability of occurrence of an action potential. The inner product of neuron's j weights and the normalized input pattern $\vec{i} = \vec{I} / \| \vec{I} \|$ (lower case notation meaning vector normalization) yields the net-input of neuron j. Normalization is performed with the l_1-norm in which:

$$\| \vec{I} \| = \sum_{i=1}^{n} |I_i| \tag{1}$$

Weights can be considered the components of a vector prototype $\vec{T^j}$, so that $\vec{T^j} = \vec{W^j} = [W_{j1}, W_{j2}, ..., W_{jn}]$. Taking this into account, the net-input of neuron j is calculated as $net_j = \| \vec{W^j} \cdot \vec{i} \| = \| \vec{T^j} \cdot \vec{i} \| = \| \vec{T^j_I} \|$, the modulus of the projection of prototype $\vec{T^j}$ over input pattern $\vec{I}$.

For altering synaptic weights, we used the incremental version of the presynaptic rule:

$$\triangle \omega = \xi I(O - \omega) \tag{2}$$

Where O and I are the postsynaptic and presynaptic action potential probabilities, respectively, and ξ, a learning factor.

The presynaptic rule not only yields the empirical plasticity curve obtained by Artola et al. [3] relating postsynaptic voltage to the increment of synaptic weight [12],[14], but also exhibits metaplasticity [1][2], a homeostatic property which elongates the plasticity curves rightwards for higher initial synaptic weights.

The computer simulation [14] of the presynaptic rule yields a family of curves that are similar to biological-plasticity curves exhibiting metaplasticity [12]. For relating the net-input of neuron O^j to its firing probability, O_j a conventional sigmoidal activation function was used.

$$O_j = \frac{1}{1 + e^{-k(net_j + 0.5 - 2s^j)}}$$ (3)

where k is a curve-compressing factor and s^j the horizontal shift of the activation function ranging from zero to one, $0 < s^j < 1$. In our examples we adjust the parameters so that when the sigmoid is completely shifted leftwards $s^j = 0$ and when it is completely shifted rightwards $s^j = 1$.

Related to this horizontal shift s^j, real neuron exhibits intrinsic plasticity [4][5] (see Fig. 2), the homeostatic property that makes very active neurons moderate their spiking rate and inactive neurons increment its firing rate. According to this property [5], the activation function gradually shifts leftwards or rightwards regulating the activation of scarcely or highly activated neurons, respectively.

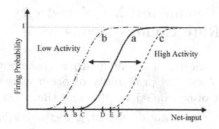

Fig. 2. Intrinsic plasticity allows the neurons' activation function to shift horizontally so that the activation function "follows" the average net-input of the neuron. (a) Initial position of the sigmoidal activation function. (b) In the case of a low regime of net-input values (as in A, B and C), intrinsic plasticity shifts the sigmoid leftwards. (c) In the case of a high regime of net-input values (as in D, E and F), intrinsic plasticity shifts the sigmoid rightwards increasing the sensitivity of the neuron.

For allowing this dynamics, parameter s^j, the horizontal shift of the activation function of neuron O_j is incorporated in the neuron's activation function $f()$ relating the net-input of the neuron to its spiking probability O_j:

$$O^j = f(\| \overrightarrow{T^j_I} \|, s^j)$$ (4)

The following equation is an attempt to model intrinsic plasticity [12]. It calculates the shift of the activation function, s at time t in terms of the shift and output probability of the neuron at time $t - 1$.

$$s^j_t = \frac{v \cdot O_{t-1} + s^j_{t-1}}{v + 1}$$ (5)

Where v, the shifting velocity parameter, is a small arbitrary factor for adjusting the shifting rate of the activation function. s^j shifts rightwards in the case of highly activated neurons so that the neuron will be down-regulated in the future. It shifts left-wards in the case of less active neurons, for allowing the neuron to increase its firing. Notice that when both the shift and the output at time $t-1$ are equal, the shift at time t continues having the same value of the shift at time $t-1$.

3   Step by Step Development of the Koniocortex Like Network

In this section, and for a better understanding of the operation of the KLN, we evolve a network from a Bayesian scaffold to a KLN. In the latter, learning appears as an emergent property, due to the interaction of individual neurons. In the KLN, a "winner takes all" (WTA) kind of operation takes place without the need of applying a function for obtaining the maximally activated neuron.

3.1   The "Bayes Decision Rule Framework"

The objective of presenting this case is to establish a minimal mathematical framework to start dealing with the following network versions. As in the following versions, input patterns are normalised by dividing each pattern by the l_1 norm (the sum of the inputs). In the "Bayes Decision Rule" framework, all patterns belonging to a specific category, let's call it T^j, produce (in a forced manner) the firing O_j of a specific output neuron j, while remaining neurons are forced to be silent. Along the presentation of a certain category of patterns, neuron j set of weights converge through a process of *hebbian* modification to prototype $\overrightarrow{T^j}$. Afterwards, during the testing phase, all prototypes are projected over the testing input pattern, $\overrightarrow{I_{test}}$, (this projection is accomplished by the inner product of the normalized input vector by the prototype's weights). The neuron that fires with highest output, O^* , indicates the category, to which the testing pattern $\overrightarrow{I_{test}}$ belongs to.

$$O^* = O^j \;\; / \;\; \forall k \neq j \quad\quad || \, \overrightarrow{T^j_{I_{test}}} \, || > || \, \overrightarrow{T^k_{I_{test}}} \, || \tag{6}$$

Previous expression is analogous to the Bayes decision rule, used in pattern recognition for determining the class, T^j, to which a certain pattern belongs to.

$$O^* = O^j \;\; / \;\; \forall k \neq j \quad\quad P\left(T^j/\overrightarrow{I_{test}}\right) > P\left(T^k/\overrightarrow{I_{test}}\right) \tag{7}$$

3.2   Forced WTA Network with Intrinsic Plasticity

Differently from previous case in which neuron's outputs are arbitrarily imposed, neuron's outputs in this case results from the network dynamics in which intrinsic and synaptic plasticity are determinant factors. However, there is an external modification performed once neurons' outputs are known: the highest output is forced to be one and remaining outputs are set to zero in a WTA manner.

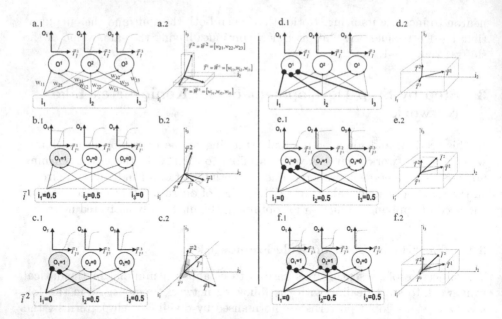

Fig. 3. Example of a WTA network with intrinsic plasticity: see details in section 3.2

This WTA type of network is depicted in Fig.3.a in which input patterns are normalized, by dividing each component by the l_1-norm (we represent normalized inputs with lower-case letters).

The set of weights of each neuron are represented either as a weight vector $\overrightarrow{W^i}$ or a prototype vector $\overrightarrow{T^i}$. The activation functions depicted above each neuron represent the preliminary output O_j of neuron O^j, in terms, $f()$, of its net-input ($\| \overrightarrow{T^j_{\overrightarrow{I}}} \|$, see section 2). Here, the WTA operation takes place through an altered version of eq.6:

$$O^* = O^j \quad / \quad \forall k \neq j \quad f\left(\| \overrightarrow{T^j_{\overrightarrow{I}}} \|, s^j\right) > f(\left(\| \overrightarrow{T^k_{\overrightarrow{I}}} \|, s^k\right) \tag{8}$$

In this equation, activation functions, $f()$, are shifted according to the value of s (see section 2). When pattern $\overrightarrow{I^1}$ is input to the network (Fig.3.b.1), the projections of prototypes $\overrightarrow{T^1}$, $\overrightarrow{T^2}$ and $\overrightarrow{T^3}$ over $\overrightarrow{I^1}$ (Fig.3.b.2) are calculated.

After applying the activation function to these projections, and after calculating the greatest output, we obtain that the winning neuron is O^1. According to the WTA, O_1 output is set to one while remaining neurons' outputs are set to zero. This kind of operation is far from biological because competition is produced, not as the result of the networks dynamics but due to an external algorithm that evaluates which is the higher output neuron.

Before presenting the second pattern $\overrightarrow{I^2}$, notice that prototypes $\overrightarrow{T^j}$ have changed (Fig.3.c.2).

This is because the pre-synaptic rule makes weights w_{1j} from active inputs i_j to O^1 (the winning neuron) increase, and weights from active inputs to non-active neurons, O^2 and O^3, be reduced.

Weights from null inputs to non-winning neurons remain the same. The result of this process of weights changing is that vector $\overrightarrow{T^1}$ evolves towards $\overrightarrow{I^1}$ and vectors $\overrightarrow{T^2}$ and $\overrightarrow{T^3}$ towards a plane orthogonal to $\overrightarrow{T^1}$ (in gray).

In a case of having more neurons in the second layer, all non-winning prototypes, $\overrightarrow{T^j}$, move towards a plane that is orthogonal plane to the winning prototype. The situation of the weights just before a second input-pattern $\overrightarrow{I^2}$ is presented to the network is shown in Fig.3.c.1.

Due to intrinsic plasticity, the activation curve of neuron O^1 is shifted rightwards, while the activation curves of neurons O^2 and O^3 are shifted leftwards. Thicker connections correspond to previously reinforced ones.

When a second pattern $\overrightarrow{I^2}$ is input to the network, projection of $\overrightarrow{T^1}$ over $\overrightarrow{I^2}$ is greater than projection of $\overrightarrow{T^2}$ over $\overrightarrow{I^2}$ (see Fig. 3.c.2), due to the higher value of O^1 neuron's weights. In this conditions O^1 (in gray) wins again.

After several presentations of $\overrightarrow{I^1}$ and $\overrightarrow{I^2}$, neuron O^1 weights continue increasing as shown in Figure 3.d.1 and 3.d.2.

On the other hand, neuron O^2 weights decrease. With higher weights, neuron O^1 will be the winner in future competitions unless other neuron property acts in the opposite direction. This problem is solved with intrinsic plasticity that helps remaining neurons to win by making neuron O^1 less sensitive and neurons O^2 and O^3 more sensitive due to the shift of the activation function. This allows that O^2 output, $O_2 = f\left(|| \overrightarrow{T^2_{I_2}} ||, s^2\right)$, becomes greater than any other neuron's output, $O_2 = f\left(|| \overrightarrow{T^j_{I_2}} ||, s^j\right)$, so that O^1 finally fails to win the competition that is won by neuron O^2, as depicted in Figure 3.e.1.

Figures 3.f.1 and 3.f.2 show the neuron's weights after many epochs of patterns $\overrightarrow{I^1}$ and $\overrightarrow{I^2}$.

When pattern $\overrightarrow{I^2}$ is presented and O^2 wins the competition, weights from non-zero inputs to neuron O^2 are reinforced so that $|| \overrightarrow{T^2} ||$ grows, and $\overrightarrow{T^1}$ and $\overrightarrow{T^3}$ become orthogonal to $\overrightarrow{T^2}$. Along this process, each prototype, $\overrightarrow{T^j}$, evolves for representing each category of the input data. This neural network version was previously used for different purposes like mimicking the illusion of movement of printed static images [15], identifying the direction of moving objects [10], or becoming the "nervous system" of a self-learning robot [16].

3.3   Non-forced WTA Network with Lateral Inhibition

In this case (see Fig.4) the WTA operation is performed without neither the need of an "external" calculation for identifying the most activated neuron, nor the need of setting the highest activated neuron to one and remaining ones to zero. Here, both operations result from the intrinsic dynamics of neurons in which lateral inhibition, intrinsic plasticity and the pre-synaptic rule are the main operating factors. This network has the following characteristics: a) intrinsic plasticity performing an homeostatic control of neurons' firing rates, b) activation curves with steep slopes, c) weights that are modified using the pre-synaptic rule (as in previous cases), d) lateral inhibition so that each neuron inhibits its neighbors. Inhibitory weights are kept in fixed value (ranging from -0.8 to -1).

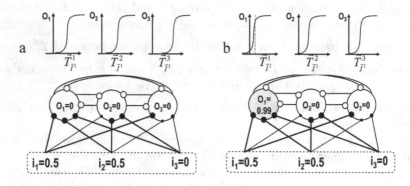

Fig. 4. Intrinsic plasticity, lateral inhibition and steep slope sigmoids orchestrated for a genuine natural competitive process leading to pattern classification: see detailed explanation in section 3.3

This version represents the first stage in our "developing" network in which the WTA algorithm takes place naturally, emerging from the internal dynamic of the interacting neurons, without the aid of any kind of external supervision. The process takes place as follows: Weights start with random, small values (Fig.4.a). Due to these negligible weights, net-inputs $\overrightarrow{T^{j}_{I^1}}$ in all neurons are also negligible and incapable of producing significant outputs. Negligible outputs make all sigmoidal activation-functions shift leftwards (according to the equation $S^{j}_{t} = S^{j}_{t-1}/(v+1)$ in which outputs are negligible and do not appear in the equation). Once, all neurons become prone to fire, due to the leftward shift of sigmoid functions, the most activated neuron (the winner) fires first (O^1 in the example of Fig. 4.b). The firing of this neuron precludes remaining neurons to fire due to lateral inhibition. This winning neuron will certainly stop winning in the future because of intrinsic-plasticity shifting its sigmoid rightwards.

3.4   Koniocortex-like Network, KLN

Biological neurons engaged in competition processes are never inhibitory as in previous version. For this reason, in this version we kept them as excitatory but including among them ancillary inhibitory interneurons. In biology, inhibitory neurons participates in WTA dynamics for allowing main excitatory neurons to win or lose the competition.

Fig.5. is a more complete version of the KLN simplified model presented in Fig.1.d. In the KLN, "B" labeled neurons are inhibitory neurons endowed with intrinsic plasticity. "S" labeled neurons which are the main neurons engaged in competition, also have intrinsic plasticity. Since each S contacts a single B, intrinsic plasticity is concomitantly regulated in both types of neurons. In this way, when S is highly activated, so it happens with B. In consequence, S reduces its excitability and B, the inhibitory field surrounding S. This effect allows other neurons to be the winners in future competitions.

Other neurons with the property of intrinsic plasticity are TC neurons. Single input/single output neurons, like TC neurons, can use intrinsic plasticity to remove the mean of a series of input values. When removing the average, patterns become more uncorrelated and easier to classify.

Fig.5.a shows that each S neuron has a recurrent connection on itself that was initially intended, not for modeling a real connection, but for allowing a sustained activation over time in simple rate-code neurons. Recurrent connections are extremely rare in real neurons. Despite of this, this kind of recurrent connection was indeed present in the koniocortex (Fig.3,[9]). When analyzing this circumstance we noticed that there were many similarities between the "evolved" net and the koniocortex, and, for this reason, the net was called KLN.

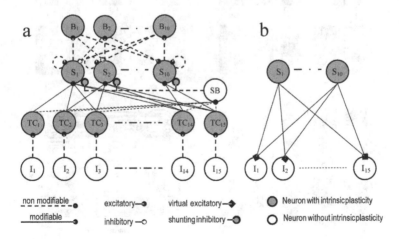

Fig. 5. Architecture of KLN applied to the identification of ten different numerical characters, each one of them represented in a 5x3 pixel matrix: a detailed explanation of the graphs is given in the first part of section 3.4

Another type of neuron with an important role in KLN learning capabilities is the SB neuron that is important in pattern normalization. Similarly to real shunting/dividing inter-neurons, SB neurons perform the arithmetical summation of its inputs (TC outputs), dividing the activation of its target neurons (the S neurons) by this quantity. Although rare, this type of operation is feasible in the real koniocortex [9]. As this operation is far different from other neurons operation, its calculation was performed separately and will not be represented in the bottom-right graph of Fig.6.

The KLN was tested in pattern classification tasks like the one of recognizing numerical characters (Fig.6). Numbers are represented in a 5x3 grid and inputted to the network in each iteration. This KLN (Fig.5) has 15 neurons in its input layer, 15 neurons' in its TC layer, 10 neuron's in the S layer, and 10 neurons in the upper B layer. Once the number is fed to the network, its activation is "propagated" until all layers are activated. One thousand iterations (100 iterations for each pattern) were sufficient for the network to correctly separate all numbers, so that at the end of training, each S neuron strongly fires for a specific numerical pattern, while remaining neurons keep almost inactive. This WTA process occurs naturally as an emergent consequence of the individual computation of each neuron without the need of externally monitoring the network. Besides lateral inhibition, intrinsic plasticity is determinant for this to happen, being this fact the main characteristic of the model.

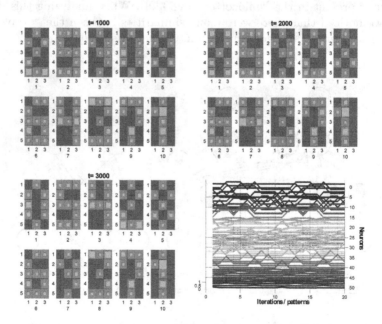

Fig. 6. KLN Matlab simulation, for identifying ten different numerical characters: a detailed explanation of the graphs is given in the second part of section 3.4

Fig.5.b shows an auxiliary network that runs in parallel with the KLN. The main function of this network is to show, at the end of training, each of the numbers that were associated with each S winning neuron. As a unique S neuron fires each time, this auxiliary network is able to associate without crosstalk each S with each input pattern, allowing the visualization of the pattern associated with each S neuron. The "virtual" weights of this "virtual" (or auxiliary) network are represented by the green tiles in each grid of Fig.6. Weights from TC to S neurons are also shown in the same graph (represented as little yellow squares inside the tiles).

The top-left graph of Fig.6 shows that, at the end of 1,000 iterations, each neuron fires for a specific number. In this way neuron S_1 fires for number 2, S_2 for number 5, S_3 for 9, S_4 for 0, S_5 for 8, S_6 for 6, S_7 for 4, S_8 for 3, S_9 for 7 and S_{10} for 1.

We tested the adaptation capacity of the network by substituting a pattern during the training phase (see Fig.6: top-right): pattern "zero" was substituted by pattern "X" from iteration 1,001 to iteration 2,000. At iteration 2,000 a complete reorganization of the network took place. S_1 that encoded number 2, came to encode character X, and S_4 that encoded a zero now came to encode number 2.

From iteration 2,001 ahead (see Fig.6: bottom-left), the initial set of training patterns was replaced back and used again. At iteration 3,000, all neurons correctly encoded the numbers showing the potentiality of the network during continuous learning.

Bottom-right graph shows the output of all neurons during the last two epochs (iterations 2,981 to iteration 3,000). Each ribbon represents the response of each of the 50 neuron along these iterations (the output of the SB neuron is not represented because it was calculated differently from the other neurons). Competition between S neurons is easily seen in this graph (see ribbons 30 to 40) because S neuron's output are uncorrelated.

In this simulation, v was set to 0.025 and ξ to 0.001. The initial sigmoid shift was 0.5. Initial weights from TC to S neurons were negligible and random. Non-modifiable weights were set to $W_{SS} = 0.85$, $W_{SB} = 0.98$, $W_{ITC} = 1.0$ and $W_{Bs} = 0.5$.

4   Conclusions

In this work, an initially non-biological double-layered network was developed through different stages so that non-biological characteristics were gradually substituted by biological homeostatic ones, like metaplasticity and intrinsic plasticity. Lateral inhibition was also introduced to allow competition between second layer neurons: first through direct connections between competing neurons, and, afterwards via inhibitory inter-neurons. Differently from previous WTA models, in the KLN lateral inhibition acts synergistically with synaptic and intrinsic plasticity so that WTA dynamics and learning contribute to make each other possible. With successive improvements, the net-work becomes very similar to the koniocortex. As shown in the simulation, in which the KLN is applied to recognize numerical characters, competition and learning emerge from individual neurons properties, without the need of any external supervisor.

References

1. Abraham, W.C., Bear, M.F.: Metaplasticity: the plasticity of synaptic plasticity. Trends in Neuroscience 19, 126–130 (1996)
2. Abraham, W.C., Tate, W.P.: Metaplasticity: a new vista across the field of synaptic plasticity. Progress in Neurobiology 52, 303–323 (1997)
3. Artola, A., Brocher, S., Singer, W.: Different voltage-dependent threshold for inducing long-term depression and long-term potentiation in slices of rat visual córtex. Nature 347, 69–72 (1990)
4. Desai, N.S.: Homeostatic plasticity in the CNS: synaptic and intrinsic forms. Journal of Physiology 97(4-6), 391–402 (2003)
5. Desai, N.S., Rutherford, L.C., Turrigiano, G.G.: Plasticity in the intrinsic excitability of cortical pyramidal neurons. Nature Neurosciences 2, 515–520 (1999)
6. Diamond, M.E., Armstrong-James, M., Ebner, F.F.: Experience-dependent plasticity in adult rat barrel cortex. Proceedings of the National Academy of Sciences USA 90, 2082–2086 (1993)
7. Ferster, D., Sooyoung Chung, S., Wheat, H.: Orientation selectivity of thalamic input to simple cells of cat visual cortex. Nature 80(6571), 249–252 (1996)
8. Fleidervish, I.A., Binshtok, A.M., Gutnick, M.J.: Functionally Distinct NMDA Receptors Mediate Horizontal Connectivity within Layer IV of Mouse Barrel Cortex. Neuron 21(5), 1055–1065 (1998)
9. Hirsch, J.A.: Synaptic integration in layer IV of the ferret striate cortex. Journal of Physiology 483(1), 183–199 (1995)
10. Kinto, E.A., Del Moral Hernandez, E., Marcano, A., Ropero Peláez, F.J.: A Preliminary Neural Model for Movement Direction Recognition Based on Biologically Plausible Plasticity Rules. In: Mira, J., Álvarez, J.R. (eds.) IWINAC 2007. LNCS, vol. 4528, pp. 628–636. Springer, Heidelberg (2007)
11. Miller, K.D., Pinto, D.J., Simons, D.J.: Processing in layer IV of neocortical circuit: new insights from visual and somatosensory cortex. Current Opinion in Neurobiology 11, 488–497 (2001)
12. Peláez, F.J.R., Andina, D.: Do biological synapses perform probabilistic computations? Neurocomputing 114, 24–31 (2013)
13. Peláez, F.J.R., Godoi, A.C.: From Forced to Natural Competition in a Biologically Plausible Neural Network. Advances in Intelligent Systems and Computing 198, 95–104 (2013)
14. Peláez, F.J.R., Godoy Simoes, M.: A computational model of synaptic metaplasticity. In: Proceedings of the International Joint Conference of Neural Networks 1999, Washington DC (1999)
15. Peláez, F.J.R., Ranvaud, R., Szafir, S., Ramírez-Fernández, F.J.: The illusion of movement in static images analyzed with a biologically plausible unsupervised neural network model. In: Proceedings of Brain Inspired Cognitive Systems, BICS 2008, São Luiz (2008)
16. Ropero Peláez, F.J., Santana, L.G.R.: Doman's inclined floor method for early motor organization simulated with a four neurons robot. In: Ferrández, J.M., Álvarez Sánchez, J.R., de la Paz, F., Toledo, F.J. (eds.) IWINAC 2011, Part I. LNCS, vol. 6686, pp. 109–118. Springer, Heidelberg (2011)

Towards an Integrated Semantic Framework for Neurological Multidimensional Data Analysis

Santiago Timón Reina[1(✉)], M. Rincón Zamorano[1], and Atle Bjørnerud[2]

[1] Departamento de Inteligencia Artificial, UNED, Madrid, Spain
santiagotimon@dia.uned.es
[2] The Intervention Centre, Oslo University Hospital, Oslo, Norway

Abstract. Medical institutions are increasingly aware of the vast amount of available data they have and its potential benefits. These data are being analyzed and shared at institutions all around the world, however, the way the data are stored, managed and secured need for new technological solutions to facilitate its consumption and sharing between institutions. This situation has become a technological challenge for the interoperability, data mining and Big Data fields. Neuroimaging community is one of the most active in looking for effective solutions, like the XNAT project which aims for neuroimaging data acquisition, management and processing. This paper shows the ongoing effort to develop a Semantic Framework to facilitate multidimensional data analysis based on XNAT architecture.

Introduction

Alzheimer's disease (AD) and other neurodegenerative dementias reduce patient and caregiver quality of life and increase health costs to society but are difficult to identify at early stages. Therefore its diagnosis is very important for effective treatment and patient care [1]. Multimodality noninvasive MRI is the principal diagnostic imaging modality for neurological diseases due to superb soft tissue contrast combined with high spatial resolution. Further, novel functional and structural MR based imaging enable new insights into the pathophysiology of the disease. One of the main challenges of neuroimaging is high-level image interpretation, which requires further processing beyond segmentation and object detection [2]. Multi-dimensional biomedical data analysis is a growing topic, not only within clinical research field, but also within clinical practice, because it helps in the process of finding new biomarkers with the aim of facilitating early diagnosis and prognosis for neurological diseases. Also Big Data technologies are allowing to look deeper and more efficiently into the vast amount of research and clinical data [3].

It is common that these projects are carried out in a multi-center and/or international environment, where data is spread over different institutions, with a common need for an efficient mean of data exchange. Sometimes a hub information system or database is deployed for the needs, other times each center has already set up its own information system, which adds complexity and the necessity to build interoperable services for inter-center data exchange. One of

© Springer International Publishing Switzerland 2015
J.M. Ferrández Vicente et al. (Eds.): IWINAC 2015, Part I, LNCS 9107, pp. 175–184, 2015.
DOI: 10.1007/978-3-319-18914-7_18

the challenges of the application of bioinformatics in these heterogeneous data source scenario is the accurate exchange of data, which needs at least some common structure or standardization. This can be accomplished through the use of ontologies.

Ontologies offer a solution to code the domain knowledge so it can be machine-readable. In the biomedical domain the main use is to serve as a way to relate domain knowledge to a great amount of data stored in databases. However, its capabilities go far beyond data relation, it has applications in any complex knowledge intensive task, like radiological assistance [4], surgical planning [5,6] or clinical management [7] and patient care systems [8]. There exists a great quantity of biomedical data and models thanks to international efforts like NCBO [9] Bioportal [10] and Open Biological and Biomedical Ontologies [11]. Many of them based on Semantic Web technologies like OWL[1] for domain models and RDF[2]/SPARQL[3] for the data storing and linking.

Using these resources with Linked Data principles enables to easily access and process biomedical data with many purposes. There are two great examples which are fundamental for the knowledge framework: the Gene Ontology, which have allowed many researchers to store and share valuable genotypic data serving multiple aims [12,13], and the Foundational Model of Anatomy (FMA) [14], an extensive ontological representation of the human anatomy. The FMA has served as knowledge model for many research projects, like the previously mentioned [5,6] and [15]. The Mayo Clinic made another great example of the application of Linked Data principles to its Electronic Health Records [16] with a proof-of-concept case study leveraging publicly available data from the Linked Open Drug Data [17] cloud to federated querying for type 2 diabetes patients. The study highlights several challenges and opportunities in using Semantic Web tools and technologies within a healthcare setting for enabling clinical and translational research.

Semantic Web tools and technologies, and in particular W3C's Linked Open Data project, is providing unprecedented opportunities by harnessing information from publicly available resources, such as Wikipedia and PubMed, and exposing the data as structured RDF that can be queried uniformly via SPARQL. Not only this provides the capabilities for interlinking and federated querying of diverse Web-based resources, but also enables fusion of private/local and public data in very powerful ways.

One of the key aspects for those research projects aiming at data evaluation is how this data will be gathered, comprising a technological challenge for itself. While Semantic Web serves as the foundation of data sharing within the biomedical domain, it is also important to develop solutions implementing these standards, either for open access or inter-center and project internal distribution. Neuroimaging community has invested a vast amount of resources looking into this issue for many years and one solid example is The Extensible Neuroimaging

---

[1] http://www.w3.org/TR/owl2-primer/
[2] http://www.w3.org/TR/rdf-schema/
[3] http://www.w3.org/TR/sparql11-query/

Archive Toolkit (XNAT). It is a software platform designed to facilitate common management and productivity tasks for neuroimaging and associated data. In particular, XNAT enables quality-control procedures and provides secure access to and storage of data [18]. Currently XNAT does not support semantic data-type definition, but is already making efforts to enable a Semantic infraestructure which will allow the use of Triple stores, SPARQL and query federation [19].

XNAT system provides a consistent data consumption/update mean through the use of RESTful Web Services. While this approach is very powerful and fosters flexibility and modularity for external software development and API library designs such as PyXNAT [20], it requires a fair amount of programming to perform complex queries and data retrieval.

Accessing experiment data using the SPARQL endpoint of the semantic framework would allow to write intelligent and semantic aware SPARQL queries, which can be stored in separated independent files and be executed from any programming language, for example, from R script for statistical analysis of demographic and genetic data correlation.

Because the expert evaluation process takes into account not only image-based information, but also the combination of multiple sources of information that could even be implicit, this work is closely related to information systems and automated reasoning over large and diverse information sources, which comprise a set of technological and methodological challenges:

- The creation of a homogenous data access framework built with Semantic Web technologies, using biomedical ontologies and controlled vocabularies for knowledge modelling and data schemas.
- The implementation of new software agents to wrap and decouple image-processing algorithms and expose image feature data as interoperable Web Services.
- Build inference and data mining services that will consume the already normalized and exposed multidimensional data to perform reasoning, clustering and correlation analysis to identify hidden/implicit bio-indicators and extend the WML features to other neurological diseases such as multiple sclerosis and brain tumours.

Methods

The starting point of this work is to create a semantic environment along XNAT. This way it will be possible to use all the storing and management capabilities of XNAT, which will allow us to perform use cases within clinical research projects and also minimize design and development work. Therefore, XNAT will act as a central hub for the system and extra custom software modules will be deployed to be fed from it. We are using Semantic Web technologies (OWL, RDF, SWRL and SPARQL) to develop the Semantic Framework. As this framework will serve as a solid terminology base, it is needed a meticulous selection of domain ontologies and align them for our purposes.

There are three main lines of work to engage to accomplish such a system: 1) Developing and integrating a semantic framework along XNAT, 2) aligning the biomedical domain ontologies and 3) developing pipelines for image processing integration.

Develop and Integrate a Semantic Framework Along the XNAT Environment

As stated before, XNAT barely supports semantic descriptions for the data. We are currently studying the best way to integrate a semantic framework which will be along XNAT environment.

The easiest way to achieve this goal is by using a D2RQ server instance which will be connected to XNAT's PostgreSQL data store, and map each variable to an ontology term using D2RQ's mapping language. However, to facilitate custom extensions and enable reasoning over the triple store is better to look directly into D2RQ engine instead of the server. Using the D2RQ engine with Jena API it will be possible to deploy an external triple store, attach a reasoner to the model, such as Pellet [21], and implement custom web services deployed inside XNAT to allow semantic data consuming from inside and outside XNAT's instance. Figure 1 shows a draft of the modular architecture under development.

This implementation will enable semantic data sharing, using Linked Data and SPARQL for external module querying.

Biomedical Semantic Design and Ontology Alignment

One of the key aspects for the multi-dimensional data analysis is to understand which data is being imported to XNAT and its structure. Currently, the main information sources are a set of research survey forms, modeled as XNAT custom data-types schemas. These forms gather patient's phenotypic, cognitive and genetic data, in particular: patient's basic information, medical history, physical examination, cognitive screening, biochemistry and genetic data. The list of information sources and the set of ontologies used for its semantic modelling are listed in table 1.

Table 1. Information sources and candidate ontologies

Information source	Candidate Ontologies
Patient's basic information	CPRo
Medical history	CPRo, Disease Ontology, Cognitive Atlas
Physical examination	CPRo, FMA
Cognitive screening	Mental State Assessment
Biochemistry	Chemical Entities of Biological Interest Ontology
Genetic data	Gene Ontology

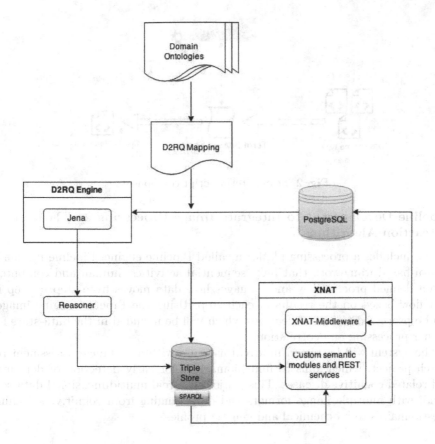

Fig. 1. Semantic extension for XNAT environment

In addition to the subject information, there are different image modalities, mainly PET and MRI. This images are annotated using NIDM (http://nidm.nidash.org/specs/nidm-primer.html) and FMA.

After identifying a set of candidate ontologies, the next step is to select the terms which will represent each of the input variables for the semantic mapping. Instead of performing this task manually, we have developed a semi-automatic method which takes advantage of one of the Bioportal RESTful web services: the search endpoint [10]. Half of the process is therefore automatic, only the final selection is manual.

The method takes two inputs, a XSD schema with the XNAT data type model (one for each form) to extract the variables to be mapped and the list of selected ontologies. For each variable a query is sent to Bioportal's search endpoint with the list of ontologies, the response is a collection of candidate terms for the variable, among other related information, such as the ontology in which the term is defined. The output is a XML file with possible term mappings for each variable.

Fig. 2. Search Term Script components

Pipeline Development to Integrate Image Processing and Feature Extraction Algorithms

XNAT includes a processing platform called Pipeline engine. Pipeline Engine is a Java-based framework that links sequential activities, human and computer, into a defined process flow and manages how data moves from step to step in that flow based on the results of each step. Using the Pipeline engine images can be processed to extract features which will be included in the data-store for further processing and correlation.

The system will be tested in a real use case within a clinical assessment research project, which aims to find biomarkers for early detection of dementia and related cognitive diseases. This project is a real multidimensional data scenario, with multiple image formats and data ranging from cognitive screening questionnaires to biochemical and genetic profiles.

Results

The final goal of the system is to create a semantic-based neuroimaging framework which will facilitate multidimensional clinical data analysis, easing the integration of neuroimaging with other information sources like genetic profiles, biochemistry or cognitive screening.

So far, we have made a revision of current Semantic Web technologies (OWL2, RDF, SPARQL and SWRL) to master the necessary skills to implement scripts, services and systems based on semantic data.

Also, we have made an exhaustive revision of current Biomedical ontologies and vocabularies related to the case of study, such as the Computer-based Patient Record Ontology (CPRo), the Foundational Model of Anatomy (FMA), the Gene Ontology, Chemical Entities of Biological Interest Ontology (CHEBI), Mental State Assessment ontology, Radlex, MESH and UMLNS, the Cognitive Paradigm Ontology, Disease Ontology, Cognitive Atlas, etc. as well as online public services from NCBO's Bioportal.

Regarding the biomedical semantic design and ontology alignment, it is already available a set of ontologies to model and annotate the data. These ontologies are one of the input of the Term Search Script, which is under testing

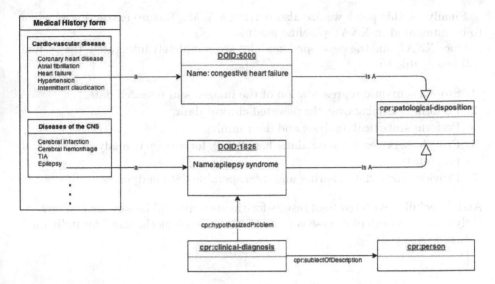

Fig. 3. Form mapping and ontology alignment diagram

before the final term selection is validated. An example of this design process is shown in figure 3, where diseases from the patient's medical history arc mapped to Disease Ontology terms (DOID) which has been aligned with CPR ontology to express the diagnosis relation.

It is under study the semantic framework integration as well, which will be placed along XNAT environment. D2RQ engine is the best option to obtain real-time RDF data, therefore we have made a first set of mappings like the one shown in figure 4.

```
map:XNATDatabase a d2rq:Database;
    d2rq:jdbcDSN "jdbc:postgresql://xnat/xnatdatabase";
    d2rq:username "xnatuser";
    d2rq:password "xnatpassword";
    .

map:SubjectClassMap a d2rq:ClassMap;
    d2rq:uriPattern "subjects/subject_@@xnat_subjectdata.id@@";
    d2rq:class cpr:Patient;
    d2rq:dataStorage map:XNATDatabase;
    .
```

Fig. 4. A sample mapping for subjets in D2RQ mapping language

Currently we are revising the programming techniques to access and manipulate the semantic model mapped from the database, to enable external reasoning and updateable triple store integration.

Finally, at this point we are also porting a WML feature extraction pipeline to be integrated in XNAT pipeline module.

Once XNAT and the developed modules are completely integrated, the system will be capable to:

1. Store a semantic representation of the images and research data.
2. Perform reasoning over the asserted clinical data.
3. Perform statistical analysis and data mining.
4. Provide services for easy data harvesting for external analysis tools and programs.
5. Provide linked data sharing and interoperable data endpoints.

And, hopefully, expected final results for the use case will be new biomarkers for early detection both of disease activity, and of disease mechanisms for individual patients.

Conclusion

Data heterogeneity and integration represent a technological challenge for current biomedical research and clinical evaluation. Aware of this issue, biomedical community has been looking into vocabularies and ontologies for many years. Semantic Web Ontologies offer a way to formally describe the data and its relations using a logic based language, which reduce data ambiguity. Another problem is the need for effective platforms for data gathering, processing and sharing, this is being approached by the neuroimaging community with platforms such as XNAT.

We have started the developing of an Integrated Semantic Framework for multidimensional data analysis which is based on these principles takes advantage of the XNAT software, allowing us to focus on data modelling and consumption. Available OBO ontologies deliver a sound formal and widely accepted model, a properly aligned set of these ontologies is the main data model. The mapping and alignment process is semi-automatic thanks to a method which uses the Bioportal Ontology Term Search endpoint. D2RQ engine is the middleware layer which maps the raw data stored in XNAT's database to ontology classes and properties of our semantic framework. This layer enables reasoning over the data and its sharing using SPARQL and Linked Data. The Semantic Framework will allow to perform intelligent, semantic aware and weak-typed queries, easing the integration of neuroimaging with other information sources like genetic profiles, biochemistry or cognitive screening and the data consumption for multidimensional data analysis.

Acknowledgements. This research has been supported by project No. 018-ABEL-CM-2013 of NILS Science and Sustainability program coordinated by Universidad Complutense (Spain).

References

1. Leifer, B.P.: Early diagnosis of Alzheimer's disease: clinical and economic benefits. Journal of the American Geriatrics Society (2003)
2. Liu, Y., Zhang, D., Lu, G., Ma, W.-Y.: A survey of content-based image retrieval with high-level semantics. Pattern Recognition 40(1), 262–282 (2007)
3. Schadt, E.E., Linderman, M.D., Sorenson, J., Lee, L., Nolan, G.P.: Computational solutions to large-scale data management and analysis. Nature Reviews Genetics 11(9), 647–657 (2010)
4. Mejino, J.L., Rubin, D.L., Brinkley, J.F.: FMA-RadLex: An application ontology of radiological anatomy derived from the foundational model of anatomy reference ontology. In: AMIA Annual Symposium Proceedings, pp. 465–469 (January 2008)
5. Mechouche, A., Golbreich, C., Gibaud, B.: Towards an Hybrid System Using an Ontology Enriched by Rules for the Semantic Annotation of Brain MRI Images, pp. 1–10
6. Mechouche, A., Morandi, X., Golbreich, C., Gibaud, B.: A hybrid system using symbolic and numeric knowledge for the semantic annotation of sulco-gyral anatomy in brain MRI images. IEEE Transactions on Medical Imaging 28(8), 1165–1178 (2009)
7. Sonntag, D.: Towards dialogue-based interactive semantic mediation in the medical domain. In: The 7th International Semantic Web Conference (2008)
8. Su, C.-J., Peng, C.W.: Multi-agent ontology-based Web 2.0 platform for medical rehabilitation. Expert Systems with Applications 39(12), 10311–10323 (2012)
9. Musen, M.A., Noy, N.F.: The national center for biomedical ontology. Journal of the ... (2011)
10. Whetzel, P.L., Noy, N.F., Shah, N.H., Alexander, P.R., Nyulas, C., Tudorache, T., Musen, M.A.: BioPortal: enhanced functionality via new Web services from the National Center for Biomedical Ontology to access and use ontologies in software applications. Nucleic Acids Research 39(Web Server issue), W541–W545 (2011)
11. Smith, B., Ashburner, M., Rosse, C., Bard, J., Bug, W., Ceusters, W., Goldberg, L.J., Eilbeck, K., Ireland, A., Mungall, C.J., Leontis, N., Rocca-Serra, P., Ruttenberg, A., Sansone, S.-A., Scheuermann, R.H., Shah, N., Whetzel, P.L., Lewis, S.: The OBO Foundry: coordinated evolution of ontologies to support biomedical data integration. Nature Biotechnology 25(11), 1251–1255 (2007)
12. Doms, A., Schroeder, M.: GoPubMed: exploring PubMed with the gene ontology. Nucleic Acids Research (2005)
13. Gene Ontology Consortium. The Gene Ontology in 2010: extensions and refinements. Nucleic Acids Research (2010)
14. Rosse, C., Mejino, J.V.: A reference ontology for biomedical informatics: the Foundational Model of Anatomy. Journal of Biomedical Informatics 36(6), 478–500 (2003)
15. Turner, J.A., Mejino, J.V., Brinkley, J.F., Detwiler, L.T., Lee, H.J., Martone, M.E., Rubin, D.L.: Application of neuroanatomical ontologies for neuroimaging data annotation. Frontiers in Neuroinformatics 4, 1–12 (2010)
16. Jyotishman Pathak, R.C.: Kiefer, and CG Chute. Applying linked data principles to represent patient's electronic health records at Mayo clinic: a case report. In: ...Symposium on International Health ..., pp. 455–464 (2012)
17. Samwald, M., Jentzsch, A.: Linked open drug data for pharmaceutical research and development. Journal of ... (2011)

18. Daniel, S., Marcus, T.R.: Olsen, Mohana Ramaratnam, and Randy L Buckner. The Extensible Neuroimaging Archive Toolkit: an informatics platform for managing, exploring, and sharing neuroimaging data. Neuroinformatics 5(1), 11–34 (2007)
19. Herrick, R., McKay, M., Olsen, T., Horton, W., Florida, M., Moore, C.J., Marcus, D.S.: Data dictionary services in XNAT and the Human Connectome Project. Frontiers in Neuroinformatics 8, 65 (2014)
20. Schwartz, Y., Barbot, A., Thyreau, B., Frouin, V., Varoquaux, G., Siram, A., Marcus, D.S., Poline, J.-B.: PyXNAT: XNAT in Python. Frontiers in Neuroinformatics 6, 12 (2012)
21. Parsia, B., Sirin, E.: Pellet: An owl dl reasoner. In: Third International Semantic Web Conference-Poster (2004)

Some Results on Dynamic Causal Modeling of Auditory Hallucinations

Leire Ozaeta[1], Darya Chyzhyk[1], Manuel Graña[1,2(✉)]

[1] Computational Intelligence Group, Universtiy of the Basque Country, UPV/EHU,
Leioa, Spain
[2] ENGINE project, Wroclaw Unversity of Technology (WrUT), Wroclaw, Poland

Abstract. Hallucinations, and more specifically auditory hallucinations (AH), are a perplexing phenomena experienced by many people. Though they are a clinical symptom in some mental diseases, such as Schizophrenia, they are also experienced by normal, healthy persons. There are several models of the mechanics happening in the brain leading to hallucinations, which involve auditory, language and emotion regions. On the other hand, there is not much empirical evidence due to the evanescence of the phenomena, and the difficulty to capture meaningful data. Recent works on resting state functional Magnetic Resonance Imaging (rs-fMRI) data, are providing confirmation of some brain localizations. Dynamic Causal Modeling (DCM) provides estimations of neural effective connectivity parameters from the experimental fMRI data, and recently has been proposed to work on rs-fMRI data. We provide preliminar results on a dataset that recently has been useful to find confirmation of AH model effects.

1 Introduction

Advances in neuroimaging have made the study of the brain more accessible, bringing more extensive and detailed information. New discoveries about brain networks and its behaviors offer an enhanced understanding of several neural conditions that have been quite obscure until now. In this new research environment, the pathophisiological models of neural conditions have been reformulated in order to integrate all the recent knowledge. However, these models are still quite simplistic as there is little room for experimentation in this area due to legal and ethical constraints.

Hallucinations are neural conditions of strong public interest, specially Auditory Hallucinations (AH), whose modeling has experienced a remarkable advance in the late years. Hallucinations are a particularly complex phenomena, involving complex interaction of many brain areas. The specific characteristics of each type of hallucination makes it difficult to achieve a unique model. Hallucinations cannot be observed from outside the patient and, therefore, need some kind of feedback from the studied individual potentially lowering the accuracy of the extracted data due to subjectivity. Computational models offering predictions that can be validated against real data, perhaps constructed following the

© Springer International Publishing Switzerland 2015
J.M. Ferrández Vicente et al. (Eds.): IWINAC 2015, Part I, LNCS 9107, pp. 185–194, 2015.
DOI: 10.1007/978-3-319-18914-7_19

paradigm of multi-agent system modeling [8,15,11] where agent interaction mimics the functional connection between brain regions leading to the generation of hallucinations, are highly desirable. However, there is only one approach right now that may provide some insights into the dynamics of the brain connectivity from empirical data. This approach is the Dynamic Causal Model [3]provided in the SPM package for neuroimage processing. This approach was initially proposed for task oriented fMRI experiments, but recently it has been proposed for rs-fMRI [4].

Intended Contribution. The aim of the work in this article, which is in its initial stages, is to search into the effective connections that can be discovered from rs-fMRI data looking for differences between people with and without AH. The dataset already explored [2] to find discriminant features has been analysed by the DCM approach, with some difficulties because the program is not tailored for dealing with rs-fMRI, despite recent claims [4]. We report results showing some differences between hallucinating and non-hallucinating subjects. The paper contents are as follows: Section 2 provides some background information. Section 3 comments the the abstract functional model. Section 4 presents a detailed anatomical model. Section 5 gives a short review of DCM. Section 6 provides preliminar results of ongoing analysis. Section 7 gives some conclusions.

2   Background

Definition of AH. Hallucinations are defined as any perceptual experience in the absence of external stimuli and sufficiently compelling to resemble a veridical perception. They may involve any sense. They are often regarded as a symptom of mental illness, but they are not necessarily clinical [1]. Auditory (AH) and visual hallucinations (VH) are most prevalent in psychiatric disorders, but auditory verbal hallucinations (AVH) are not uncommon in the general healthy population, with prevalence estimates ranging between 3 and 15% [6,14]. The most widely studied patient group suffering AH are the schizophrenia patients [9,13,7], although some studies have been carried in other clinical patients [12,1] and healthy individuals with hallucinations [14]. It has been proposed that comparison of patients and healthy persons with a history of hallucinations may allow to identify a hallucination brain fingerpring which will help to understand further complex psychiatric illnesses who have hallucinations as a core symptom [6]. In this article we focus our work on AH as they are the most widely present in the variety of mental conditions where hallucinations have been reported, both clinical and non clinical.

Evidences in the Literature. The mechanism of AH generation are not clear yet. Nonetheless, they seem to involve several alterations in grey matter volume, activation, and functional connectivity of a brain regions' network [1]. One of the most widely studied aspects of hallucination prone brains is lateralization. Several studies have reported a reversed lateralization of cerebral activity during AVH, showing right inferior frontal activation when left could be expected, because the left-hemisphere is more relevant than the right in language production in most right-handed subjects. fMRI studies have shown that the stronger the right lateralization was the stronger negative emotional content of the AVH [13]. It has been observed that the right ear advantage (REA) is attenuated in schizophrenia patients, being more predominant in patients with hallucinations [6] implying that left language regions are always "tuned in" to the aberrant signals, and are, therefore, already engaged in processing. Furthermore, patients with AH showed difficulties in shifting the attention to the opposite ear, which implies a difficulty in achieving top-down executive control [14]. Many studies found aberrant activation from emotional attention centers and attenuated activation in areas involved in monitoring processes, such as dorsal anterior cingulate, supplementary motor area, and cerebellum [1].

Hallucination Models. Eearly models contemplate AH as examples of either inner speech misattribution or traumatic memories. These two points of view are especially oriented to explain auditory verbal hallucinations, so that whereas one proposes that AVH are a misattribution of patient's own inner speech, the other looks at AVH as an automatic, or unintentional actualization of memories which the patient fails to inhibit. Nevertheless, both fail to explain all the variations in the hallucination experiences, as some patients can hear a voice they can identify but it is not their own, while others hear benevolent voices narrating their lives. Thereby, new cognitive models present AH as both an aberrant activation of perception and a failure of inhibition control [6,14].

3   Abstract Functional Model

The functional model encompasses six areas, which can be divided in three main groups: emotional regulation/attention and memory related areas, self-monitoring/inhibition areas, and audio/language processing related areas. Comparing the hallucinating and non-hallucinating brains two important differences can be highlighted. On the one hand, aberrant hyper-activation has been found in the emotional regulation/attention and memory related areas, probably related to the common memory triggers and strong emotions contents usually present in the hallucinations. On the other hand, lesser activation has been found in self-monitoring/ inhibition areas, explaining the monitoring error that impedes the brain to recognize thought and sensations as self generated and the weaker capability of many patients to ignore the hallucinations. Audio/language processing related areas, however, have been reported function similar to the response to a real external signal, but with abnormal activations of language related areas of the left hemisphere.

Therefore, the hallucinating signal is expected to originate in the auditory cortex, triggering the language related areas and being amplified by the emotional regulation/memory related areas. As the monitoring/inhibition areas fail to recognize the signal as self-originated due their under-activation, the phrases and voices are sharped, giving the patient the same experience as in the case of

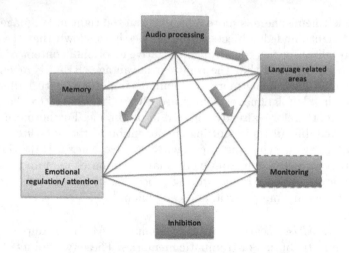

Fig. 1. Abstract functional model of the brain fucntional interactions while experiencing an auditory hallucination. The arrows indicate the general expected hallucinating signal path, starting in the auditory cortex and traveling to emotional regulation/attention, language related and monitoring areas. The areas with thicker border are more activated in the hallucination prone brain, while areas with discontinued border are less activated.

hearing a real voice outside they head. Figure 1 contains a graphic representation of the functional model with the expected signal behavior.

Regarding inter-area connections some differences have been reported between the non-hallucinating and the hallucinating brain, considering bidirectional connections. These differences lay in the strength of the said connection. However, they can be hardly labeled as stronger or weaker as they vary regarding which specific anatomic areas they concern. Strength differences have been observed in the connections from audio processing/language related areas and emotional regulation/attention areas to every other area. Also, from monitoring and inhibition areas to every other area except the memory related one. The model in figure 1 tries to represent the "default" faulty network of a hallucination prone brain, not the event of experiencing the hallucination.

4   Anatomical Model

Some areas appear in most reports and reviews, mostly frontal and subcortical areas. In the frontal areas, the left frontal operculum, both Broca's area and Broca's homologue, dorsolateral gyrus, right orbitofrontal gyrus, left middle frontal gyrus, and right precentral have been highlighted. Whereas in the subcortical areas, special interest has been reported in the right putamen, cingulate (with particular interest to anterior cingulate, and left ventral and dorsal anterior cingulate, separately), hippocampal and right parahippocampal regions, right thalamus, and right amygdala. In other areas, such as parietal, temporal and temporoparietal ones, fewer effects have been observed. These have been

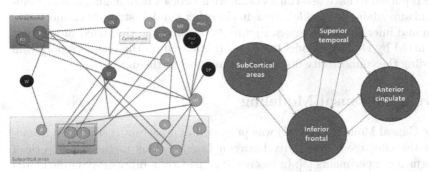

(a) The anatomical model undelying   (b) The simplified model for DCM.
the abstract functional model.

Fig. 2. The detailed and simplified anatomical models of functional connections

superior temporal gyrus; left superior parietal, right postcentral gyrus, and Wer-
nicke's area from the parietal; and Heschl's gyrus, temporoparietal gyrus, and
insula. However, some of them, such as Heschl's gyrus and superior temporal
gyrus have been widely studied, as they are hypothesized to be of importance
in the hallucination mechanisms. More specifically, in studies that capture the
data while patient is experiencing the hallucinations, mostly temporal areas have
been reported to activate, together with abnormal anterior cingulate activation.
Figure 2a shows the graph of connections between anatomical areas there are
shown every considered area and the correspondent connections, with the areas
colour coded as follows: light blue for temporoparietal, dark blue for fontal, black
for parietal, and purple for temporal. Continuous line connections are stronger
in the hallucinating brain, while dotted connections are weaker ones. Obr, orbitofrontal gyrus; DL, frontal dorsolateral gyrus; MF, middle frontal gyrus; PreC,
precentral gyrus, B, Broca's area; FO, frontal operculum; SP, superior parietal;
W, Wernickles area; PostC, postcentral gyrus; A, amygdala; T, Thalamus; P,
putamen; V, ventral anterior anterior cingulate; D, dorsal anterior cingulate;
H, hippocampus; Ph, parahippocamus; I, insula; Hl, Heschl's gyrus; TP, Temporoparietal gyrus; ST, superior temporal.

Regarding functional connectivity, the most studied seed areas have been superior temporal gyrus and Heschl's gyrus (auditive area) [10]. In the hallucinating brain, the superior temporal gyrus has been found to have weaker connections with left frontal operculum, dorsolateral frontal gyrus, left dorso anterior
cingulate, cerebellum, and hippocampus, while being more strongly connected to
Broca's area, ventral anterior cingulate, and Heschl's gyrus. On the other hand,
the Heschl's gyrus of the hallucinating brain has connections of greater strength to
most evaluated frontal areas, save the frontal operculum and dorsolateral gyrus,
and it is more disconnected from some subcortical areas such as hippocampus,
parahippocampus, and thalamus, but with stronger connection with cingulate.
Regarding the rest of the connections it is to notice that temporo parietal gyrus

has been reported to have less connections with Broca's homologue, anterior cingulate and amygdala. Wernickle's area, in the other hand, is strongly connected to putamen and inferior frontal areas. Figure 2b displays the simplified model that is being fitted by DCM to model the data. It is based on the results reported in [2] regarding the connectivity discriminant analysis of the data.

5   Dynamic Causal Modeling

Dynamic Causal Modeling (DCM) was proposed as a bayesian estimation framework for the effective connectivity between brain regions in the framework of fMRI cognitive experiments [3]. In essence it consists of a bilinear dynamic model of the neural dynamics, which is convolved with the hemodynamic response function in the case of fMRI for a better model fitting, but can be applied as such to electroencephalogram (EEG) data. The methodology has evolved during its application to several cases. Recently [5], the model was enlarged to take into account also the phase-delay information, i.e. computing the cross-correlation of spectra (cross-spectra) the result has real and imaginary parts. The real part is the so called coherence, that measures the agreement between the sources. The imaginary part contains information about the time lags between the sources. Formally, the DCM neural dynamics model (before the hemodynamic adjustments) has the form:

$$\dot{\mathbf{x}}(t) = A\mathbf{x}(t) + B\mathbf{v}(t)$$

where $\mathbf{x}(t)$ is the n-dimensional column vector of hidden neuronal states for the n brain regions considered, A is the time invariant matrix of interactions or effective connectivity between brain regions, $\mathbf{v}(t)$ is a vector of exogenous influences and endogenous influences, and B is the time invariant matrix of effects of these influences on the brain regions. The current DCM approach performs a bayesian estimation of the parameters in the spectral representation $Y(\omega) = K(\omega) \cdot V(\omega) + E(\omega)$ according to conventional assumptions, such as the spectral density [4] of the form:

$$g_v(\omega, \theta) = \alpha_v \omega^{-\beta_v} + g_u(\omega, \theta)$$
$$g_e(\omega, \theta) = \alpha_e \omega^{-\beta_e}.$$

Such model covers many forms of noise, including exogenous variates that can be deterministic $g_u(\omega, \theta) = \mathcal{F}(C \cdot u(t))$, where $\mathcal{F}(.)$ represents the fourier transform. The expected signal spectra is

$$g(\omega, \theta) = K(\omega) g_v(\omega, \theta) + K^*(\omega) g_e(\omega, \theta),$$

that is a sampling of the true spectra with Gaussian error $g(\omega) = g(\omega, \theta) + N(\omega)$. The full generative bayesian model

$$p(g(\omega), \theta) = p(g(\omega)|\theta) p(\theta|m),$$

requires the specification of the prior beliefs about the parameter $p(\theta|m)$ distributions. The complex coherence function between two wide-sense stationary can be factorized into the correaltion between the signal amplitudes and the dispersion of the phase-differences [5]:

$$C_{ij} = \frac{\langle \alpha_i \alpha_j \rangle}{\langle \alpha_i^2 \rangle \langle \alpha_j^2 \rangle} \times \phi_{ij},$$

where the first term corresponds to the coherence between signal amplitudes and the second to the phase-delay. We will compare the populations on the basis of the inspection of the these quantities for all brain regions selected.

6   Some Experimental Results

The implementation in SPM[1] of DCM is oriented towards data resulting from some cognitive experimental design. Despite the claim in [4] that it can be applied to rs-fMRI, there actual process is not automatic. SPM proposes a first level analysis to identify the volumes of interest (VOI), which we have already identified in previous works [2]. To find the VOIs, the GLM needs some experimental design wich is lacking in rs-fMRI, so we specified the VOI selection described in figure 2b, running DCM for each subject, performing a population averaging of the results to find population differences. The dataset has been already described in [2,10], as well as its preprocessing.We focus on the differences between subjects with a history of AH versus those without. Figure 3 gives the connectivity results between areas, which are very similar for both populations. There are some slight differences in the connectivity between ST and IF in

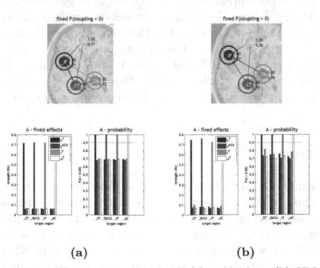

(a)   (b)

Fig. 3. Connectivity results of subjects with AH (a) and without (b). IF Inferior frontal, AC Anterior cingulate, ST Superior temporal, SA Subcortical areas.

---

[1] http://www.fil.ion.ucl.ac.uk/spm/software/spm12/

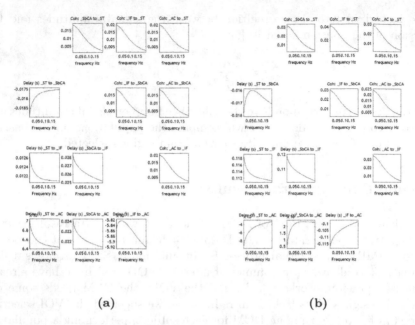

(a)   (b)

Fig. 4. Coherence and delay effects between regions for the AH (a) and nAH (b) subjects. IF Inferior frontal, AC Anterior cingulate, ST Superior temporal, SA Subcortical areas. Upper triangle of plots corresponds to coherence, lower triangle to delays.

both populations, which is stronger in the non hallucinating subjects, implying a stronger control of perception. Figure 4 give the average coherence and delays of AH and nAH populations. Looking at the delays, we find strong differences in all plots between populations, except in the connection from superior temporal and subcortical area with inferior frontal region. The remaining differences point towards the existence of different timing mechanisms in the hallucinating and non-hallucinating brain which deserve further exploration with neurological experts. On the other hand there are no differences in coherence that can be appreciated.

7   Conclusions

Auditory hallucinations (AH) have a high prevalence both in healthy and diseased populations. It is a paradoxical phenomena whose understanding may bring further understanding of the brain mechanisms. There are little empirical neuroimage evidence of the diverse models of AH generation, so that new avenues for research are widely open. In this paper we apply DCM to a rs-fMRI dataset to find effects of effective connection between regions that have been identified by previous research as being involved in the discrimination between hallucinators and non-hallucinators. The application of DCM to rs-fMRI is not immediate, as it has been designed for analysis of cognitive experiment data.

Nevertheless, we have found interesting differences in the the phase-delay between Anterior Cingulate, Superior Temporal and SubCortical Areas suggesting that effective connectivity delays maybe at the root of the difference between the hallucinator and non-hallucinator brain.

Acknowledgements. Thanks to Ann K. Shinn from the McLean Hospital, Belmont, Massachusetts; Harvard Medical School, Boston, Massachusetts, US for providing experimental images. Darya Chyzhyk has been supported by a FPU grant from the Spanish MEC. Support from MICINN through project TIN2011-23823. GIC participates at UIF 11/07 of UPV/EHU. ENGINE project is funded by the European Commission grant 316097.

References

1. Allen, P., Laroi, F., McGuire, P.K., Aleman, A.: The hallucinating brain: A review of structural and functional neuroimaging studies of hallucinations. Neuroscience and Biobehavioral Reviews 32, 175–191 (2008)
2. Chyzhyk, D., Graña, M., Öngür, D., Shinn, A.K.: Discrimination of schizophrenia auditory hallucinators by machine learning of resting-state functional mri. International Journal of Neural Systems (in press, 2015)
3. Friston, K.J., Harrison, L., Penny, W.: Dynamic causal modelling. Neuroimage (2003)
4. Friston, K.J., Kahan, J., Biswal, B., Razi, A.: A dcm for resting state fmri. Neuroimage (94), 396–407 (2014)
5. Friston, K.J., Bastos, A., Litvak, V., Stephan, K.E., Fries, P., Moran, R.J.: {DCM} for complex-valued data: Cross-spectra, coherence and phase-delays. NeuroImage 59(1), 439–455 (2012), Neuroergonomics: The human brain in action and at work
6. Hugdahl, K.: "Hearing voices": Auditory hallucinations as failure of top-down control of bottom-up perceptual processes. Scandinavian Journal of Psychology 50, 553–560 (2009)
7. Mechelli, A., Allen, P., Amaro Jr., E., Fu, C.H.Y., Williams, S.C.R., Brammer, M.J., Johns, L.C., McGuire, P.K.: Misattribution of speech and impaired connectivity in patients with auditory verbal hallucinations. Human Brain Mapping 28, 1213–1222 (2007)
8. Salamon, T.: Design of Agent-Based Models: Developing Computer Simulations for a Better Understanding of Social Processes. Academic series. Bruckner Publishing, Repin (Septermber 2011)
9. Shinn, A.K., T-Baker, J., Cohen, B.M., Öngür, D.: Functional connectivity of left heschl's gyrus in vulnerability to auditory hallucinations in schizophrenia. Schizophrenia Research 143, 260–268 (2013)
10. Shinn, A.K., Baker, J.T., Cohen, B.M., Öngür, D.: Functional connectivity of left heschl's gyrus in vulnerability to auditory hallucinations in schizophrenia. Schizophrenia Research (143), 260–268 (2013)
11. Shoham, Y., Leyton-Brown, K.: Multiagent Systems: Algorithmic, Game-Theoretic, and Logical Foundations. Cambridge Press (2008)
12. Sommer, I.E., Clos, M., Meijering, A.L., Diederen, K.M.J., Eickhoff, S.B.: Resting state functional connectivity in patients with chronic hallucinations. PLoS One (2012)

13. Sommer, I.E.C., Diederen, K.M.J., Blom, J.-D., Willems, A., Kushan, L., Slotema, K., Boks, M.P.M., Daalman, K., Hoek, H.W., Neggers, S.F.W., Khan, R.S.: Auditory verbal hallucination predominantly activate right inferior frontal area. Brain 131, 3169–3177 (2008)
14. Waters, F., Allen, P., Aleman, A., Fernyhough, C., Woodward, T.S., Badcock, J.C., Barkus, E., Johns, L., Varese, F., Menon, M., Vercammen, A., Laroi, F.: Auditory hallucinations in schizophrenia and nonschizophrenia populations: A review and integrated model of cognitive mechanisms. Schizophrenia Bulletin 38(4), 683–692 (2012)
15. Wooldridge, M.: An Introduction to MultiAgent Systems. John Wiley & Sons (2002)

Retinal DOG Filters:
High-pass or High-frequency Enhancing Filters?

Adrián Arias[1], Eduardo Sánchez[1(✉)], and Luis Martínez[2]

[1] Grupo de Sistemas Inteligentes (GSI)
Centro Singular de Investigación en Tecnologías de la Información (CITIUS)
Universidad de Santiago de Compostela, 15782,
Santiago de Compostela, Spain
adrian.arias.abreu@usc.es, eduardo.sanchez.vila@usc.es
[2] Instituto de Neurociencias de Alicante
CSIC-Universidad Miguel Hernández, 03550,
Alicante, Spain
l.martinez@umh.es

Abstract. This paper analyzes the filtering operation carried out by the classical Difference-of-Gaussians model proposed by Rodieck to describe the receptive fields of retinal ganglion cells. Discrete DoG kernels of such functions were developed and compared with High-Pass and High-Frequency Enhancing filters. The results suggest that the DoG Kernels behave as High-Frequency Enhancing filters but in a limited band of frequencies.

Keywords: Retina · Difference of Gaussians · High-pass Filtering · High-Frequency Enhancing Filtering

1 Introduction

The term receptive field (RF) in the visual system was classically defined as a two-dimensional region in visual space where a luminous stimulus triggers a change in response on that neuron [1]. The concept was first applied to the retina to describe the area in which a stimulus drove responses of retinal ganglion cells (RGCs). Later on, Kuffler found that RGCs show RFs with a concentric shape made up of two antagonistic regions: a center, and a surround [2]. Thus, when a bright stimulus is applied to the center region, the RGCs are excited and generate a number of action potentials or spikes; and, conversely, when the same stimulus is applied to the surround, the neuron is inhibited and a weaker or no response is observed. Thereafter, the RFs of RGCs were characterized by this center-surround organization.

The center-surround RF is an empirical model that is useful to understand the spatial organization of the afferent inputs to RGCs but lacks the ability to predict the neuron's response to any given stimulus. An important contribution was later made by Rodieck (1965) by proposing a mathematical model to formally describe the function that maps the input-output relationship of the RFs of

J.M. Ferrández Vicente et al. (Eds.): IWINAC 2015, Part I, LNCS 9107, pp. 195–202, 2015.
DOI: 10.1007/978-3-319-18914-7_20

RGCs. The proposed relationship was the sum of two Gaussian functions: a positive one, representing the center, and a wider negative one representing the surround, both centered at the same point. This model was called the Difference-of-Gaussians (DoG) model and has been used to represent the RFs of RGCs ever since.

The parameters of the DoG model for RGCs were first estimated by Enroth-Cugell and Robson (1966, 1984). They recorded responses of ganglion cells to sinusoidal stimuli and fitted the model against contrast sensitivity curves that were obtained experimentally. For each recorded RGC, a DoG model was fitted and their parameters, the radius and maximum amplitudes of both the center and surround Gaussians, were estimated. As the individual estimated contrast sensitivity curves fitted nicely with the experimental ones, the results provided a strong support for the DoG model as a useful function to describe the center-surround RFs of RGCs.

The issue about the information processing capabilities of a DoG function was subsequently assessed by Marr and Hildreth [6]. They proposed a very influential theory of edge detection on the basis of an analysis of intensity changes which occur in natural images as well as the appropriate filters to signal those changes. Intensity changes or edges could best be detected by finding the zero values of the second derivative, or the Laplacian, of a Gaussian. Moreover, they showed that an accurate approximation of the Laplacian of a Gaussian could be developed by using a DoG function with an appropriate ratio $\sigma_{surround}/\sigma_{center}$ about 1.6. They concluded that neurons in the visual system with RFs being described by such DoG functions could indeed work as edge detectors.

Since the landmark work of Marr and Hildreth, RGCs, which determine the output of the retina, have been understood to perform a kind of edge detection or sharpening filtering. In fact, different authors have reported evidence about the existence of RGCs of the type of Local Edge Detectors in Cats, Rabbits and even Primates [7,8,9]. Moreover, in the field of image processing, discrete kernels derived from DoG functions are usually characterized as high-pass filters [10]. However, it is interesting to note that there was no previous attempt to analyze what is the filtering carried out by the DoG models that were obtained experimentally by Enroth-Cugell and Robson. In this paper we aim at analyzing the information processing capabilities of different versions of the original DoG model and test to what extent it behaves as a canonical edge detector.

2   Methods

2.1   Parameters of the DoG model

The Difference-of-Gaussians model is made up with two gaussians: the first one representing the excitatory center of the RF, the second one the inhibitory surround of the RF. The function as it was used by Enroth-Cugell and Robson [4,5], was formalized as follows:

$$DoG(r) = k_c e^{-(r/r_c)^2} - k_s e^{-(r/r_c)^2} \tag{1}$$

being the relevant parameters: the maximum amplitudes k_c and k_s, and the radius r_c and r_s. For each RGC, 17 cells reported in Enroth-Cugell and Robson (1966) and 6 in Enroth-Cugell and Robson (1984), the theoretical contrast sensitivity function derived from the DoG model was fitted to the empirical contrast sensitivity function measured at different spatial frequencies. As a result, a set of parameters (r_c, r_s, r_s/r_c, and $k_s r_s^2/k_c r_c^2$) were estimated for each cell. For brevity these parameters are not shown here but can be found in the original papers [4,5].

2.2   Discrete DoG kernels

The continuous DoG functions fitted to experimental data has to be converted into discrete DoG kernels to operate with input images. In what follows, the four steps required in that procedure are described.

As the continuous function can take infinite values, the first step consisted on truncating it in order to set a finite range of values. Figure 1 describes the process in one dimension. A variable T was defined to determine the width of the truncation, and an standard rule was followed in order to include the 99, 74% of the area under the curve of either the center or the surround gaussian. Therefore, two possible values were considered: $T = 3 * r_c$, or $T = 3 * r_s$.

The second step consists on sampling the continuous function. By setting the size SxS of the kernel, the number of both the elements of the kernel matrix

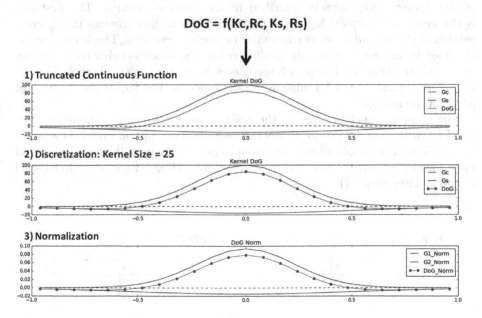

Fig. 1. Discretization of the continuous DoG model: truncation (upper inset), sampling (middle inset), and normalization (lower inset)

as well as the sampling points of the continuous DoG function are set. Both variables T and S define the step of the sampling process:

$$Step = \frac{2T}{S-1} \tag{2}$$

After the convolution of the input image with the discrete kernel, the output image has to preserve the intensity ranges of the input. The third step therefore involves the normalization of the elements of the DoG kernel matrix. Each element or weight w_{ij} of the matrix is normalized as follows:

$$w_{ij}^{norm} = \frac{w_{ij}}{\Sigma_i \Sigma_j w_{ij}} \tag{3}$$

Finally, the goodness of the discretization procedure has been assessed by fitting back the normalized discrete DoG kernel to the original continuous DoG function. The Levenberg-Marquart algorithm was used to solve the non-linear least squares fitting problem. The results (not shown here) confirmed that the parameter set (r_c, r_s, r_s/r_c, and $k_s r_s^2/k_c r_c^2$) is preserved as well as other aspects of the DoG function.

2.3   Baseline filters

Two filter kernels were chosen to analyze the information processing capabilities of the discrete DoG kernels described in the previous section. The first one is the kernel of a typical high-pass filter (HPF), which represents the discrete version of the second-order derivative or Laplacian operator. The kernel can be obtained by means of either the coefficients of the second-derivative operation or by substracting the low-pass filter kernel from the identity kernel (Fig. 2). HPFs are typically used for edge detection tasks as they signal the location of edges in the images.

The second kernel belongs to the class of high-frequency enhancing filter (HFEF), which carries out local contrast enhancement operations. It is a popular HFEF technique known as unsharp masking widely used in the printing and photography industry. The kernel is obtained by adding a high-pass filter kernel to the identity kernel (Fig. 2).

HPF

$$\begin{bmatrix} 0 & 0 & 0 \\ 0 & 1 & 0 \\ 0 & 0 & 0 \end{bmatrix} - \frac{1}{9}\begin{bmatrix} 1 & 1 & 1 \\ 1 & 1 & 1 \\ 1 & 1 & 1 \end{bmatrix} = \frac{1}{9}\begin{bmatrix} -1 & -1 & -1 \\ -1 & 8 & -1 \\ -1 & -1 & -1 \end{bmatrix}$$

Unsharp HFEF

$$\begin{bmatrix} 0 & 0 & 0 \\ 0 & 1 & 0 \\ 0 & 0 & 0 \end{bmatrix} + \alpha\frac{1}{9}\begin{bmatrix} -1 & -1 & -1 \\ -1 & 8 & -1 \\ -1 & -1 & -1 \end{bmatrix} = \frac{1}{9}\begin{bmatrix} -\alpha & -\alpha & -\alpha \\ -\alpha & 9+8\alpha & -\alpha \\ -\alpha & -\alpha & -\alpha \end{bmatrix}$$

Fig. 2. Kernels of baseline filters: high-pass filter (left) and unsharp masking HFEF (right)

2.4   Image processing and kernel analysis

A comprehensive *python* simulation environment was developed to carry out the discretization procedure as well as the analysis of the kernels presented in the Results section. The environment takes advantage of some powerful *python* libraries, such as: *OpenCV*, to convolve an image with a kernel; *Numpy*, to generate 2D kernels and compute 2D discrete Fourier transforms; *Scipy*, to solve the non-linear least squares fitting problem described in section 2.2; and *Matplotlib* to plot graphs and view images.

3   Results

The first task was to analyze the behavior of the discrete DoG kernels obtained for the $23(= 17 + 6)$ cells reported by Enroth-Cugell and Robson (see section 2.1). For each DoG kernel we generated: the kernel representation in both spatial and frequency domain, the Bode diagram, and the output after convolving the kernel with an standard test input (Lena image). It can be concluded that all kernels behave in a similar way regardless of the discretization parameters used (not shown here for the sake of brevity). On the basis of this result, only one of the discrete DoG kernels, which corresponds to cell number 1 of Enroth-Cugell and Robson (1966), was chosen to make the comparison with the baseline filters.

The comparison with the high-pass filter (HPF) is shown in figure 3 for kernels of size 25x25. Both kernels in the spatial domain (red pixels indicating positive values and blue pixels, negative ones) show a positive center and a negative surround, but the extent as well as the structure of these regions are clearly different. These differences are made explicit when the kernels are represented in the frequency domain (Fig. 3, second row). The 2D Fourier spectrum of the DoG kernel in a dB/log scale (red pixels indicating positive coefficients; blue pixels indicating negative ones) shows a region of positive coefficients starting from the center point (DC coefficient) that progressively change into negative coefficients at high frequencies. On the contrary, in the HPF kernel the value of coefficients is negative at low frequencies and positive after a certain threshold frequency. The effect of the kernel on the frequencies of the input image is better analyzed by means of the Bode diagrams (Fig. 3, third and fifth rows), which plots the filter gain for each spatial frequency. It can be seen that (1) the DoG kernel enhances those frequencies found in a band of frequencies; and (2) the HPF Kernel removes low-frequencies as well as keeps the gain of high-frequencies. The image outputs (Fig. 3, fourth and sixth rows), obtained after convolving the input with the kernels, confirm the different nature of DoG and HPF kernels. The first one enhances local edges as well as preserves the intensity levels of the rest of regions, while the second one detects the edges of the image but suppresses all intensity information of constant regions (black pixels). The transformation of the intensities at the output is clarified by looking at the intensity profile of one of the image rows (Fig. 3, fourth and sixth rows). The profile of the DoG kernel output (green line) follows the original profile of the input (black line) and stretches the values at the peaks of the curve. The behavior is different for

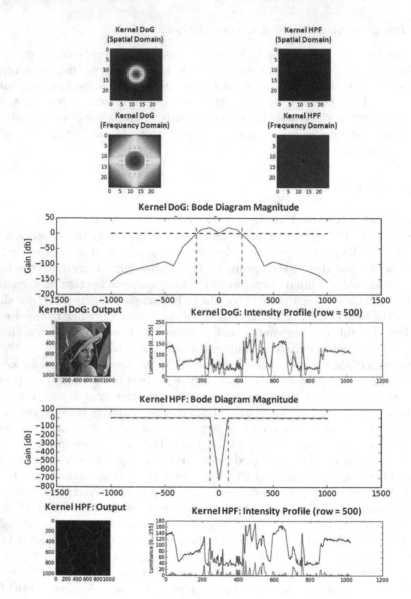

Fig. 3. Comparison of DOG kernel with HPF kernel. The DOG kernel corresponds to cell number 1 of Enroth-Cugell and Robson (1966), parameterized as follows: $k_c = 100$, $k_s = 15.9$, $r_c = 0.32$, $r_s = 0.76$, $SxS = 25x25$, $T = 3 * r_c$, and $Step = 0.04$. The HPF kernel is of the same size as the DoG kernel. The kernels are represented in the spatial domain (first row) and the frequency domain (second row). The Bode diagrams (third and fifth rows) as well as the image outputs and intensity profiles at $row = 500$ (fourth and sixth rows) for both kernels are also plotted.

the HPF kernel output. The baseline is moved down to zero and the positive values of the curve indicate the location of edge points.

The DoG kernel was also compared with the HFEF kernel, as shown in figure 4 for kernels of size 25x25. The main point here is that both the image outputs and intensity profiles (Fig. 4, fourth and sixth rows) are very similar, which might suggest that the behavior of the DoG kernel would belong to the class of HFEF filters. However, the 2D Fourier spectrum as well as the Bode diagram

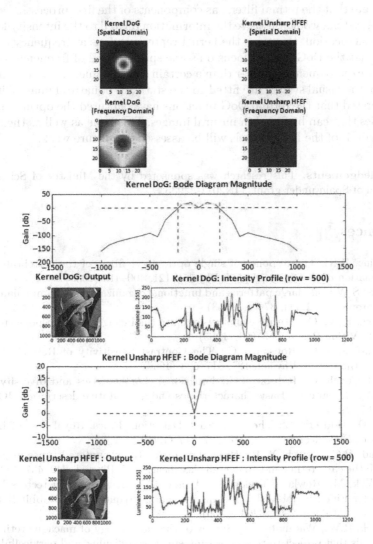

Fig. 4. Comparison of DOG kernel with HFEF kernel. The DoG kernel as well as the analysis plots are the same as shown in 3. The HFEF kernel is of the same size as the DoG kernel.

reveals that the kernel features in the frequency domain are somewhat different. The DoG kernel presents a band-pass behavior whereas the Unsharp HFEF does enhance the high frequencies in the same manner with no upper limit.

4   Discussion

The results shown in section 3 suggests that the DoG kernels would be better classified as High-Frequency Enhancing filters rather than High-Pass filters. Our findings indicate that the retinal filters, as components of the first processing stage of the visual system, would preserve the information related to the intensity levels at each spatial location. However, the kernel representations at frequency domains indicate that the DoG kernels focus on some specific band of frequencies and do not operate on frequencies higher than a certain cut-off value. As the neurons and circuits of the visual system are fitted to the statistics of natural images, it could be interpreted that the retinal DoG functions have captured the optimal band of frequencies that can be found in natural images. This issue as well as the quantitative analysis of the DoG kernels will be assessed in a future work.

Acknowledgements. This research was sponsored by the Ministry of Science and Innovation of Spain under grant TIN2011-22935.

References

1. Hartline, H.K.: The response of single optic nerve fibers of the vertebrate eye to illumination of the retina. Am. J. Physiol. 121, 400–415 (1938)
2. Kuffler, S.W.: Discharge patterns and functional organization of mammalian retina. J. Neurophysiol. 16(1), 37–68 (1953)
3. Rodieck, R.W.: Quantitative analysis of cat retinal ganglion cell response to visual stimuli. Vision Res. 5, 583–601 (1965)
4. Enroth-Cugell, C., Robson, J.G.: The Contrast Sensitivity of Retinal Ganglion Cells of the Cat. J. Physiol. 187, 517–523 (1966)
5. Enroth-Cugell, C., Robson, J.G.: Functional characteristics and diversity of cat retinal ganglion cells. Basic characteristics and quantitative description. IOVS 25, 250–267 (1984)
6. Marr, D., Hildreth, E.: Theory of Edge Detection. Procs. Royal Soc. of London, Series B, Biological Sciences 207, 187–217 (1980)
7. Cleland, B.G., Levick, W.R.: Properties of rarely encountered types of ganglion cells in the cat's retina and an overall classification. J. Physiol. 240, 457–492 (1974)
8. van Wyk, M., Rowland Taylor, W., Vaney, D.I.: Local Edge Detectors: A Substrate for Fine Spatial Vision at Low Temporal Frequencies in Rabbit Retina. J. Neurosci. 26(51), 13250–13263 (2006)
9. Rodieck, R.W., Watanabe, M.: Survey of the morphology of macaque retinal ganglion cells that project to the pretectum, superior colliculus, and parvicellular laminae of the lateral geniculate nucleus. J. Comp. Neurol. 338, 289–303 (1993)
10. Gonzalez, R., Woods, R.: Digital Image Processing. Prentice-Hall (2008)

Spatio-temporal Dynamics of Images with Emotional Bivalence

M.D. Grima Murcia[1(✉)], M.A. Lopez-Gordo[2,3], Maria J. Ortíz[4],
J.M. Ferrández[5], and Eduardo Fernández[1]

[1] Institute of Bioengineering, University Miguel Hernández and CIBER BBN
Avenida de la Universidad, 03202, Elche, Spain
maria.grima@alu.umh.es, e.fernandez@umh.es
[2] Nicolo Association, Churriana de la Vega, Granada, Spain
malg@nicolo.es
[3] Deptartment of Signal Theory, Communications and Networking,
University of Granada, 18071, Granada, Spain
[4] Deptartment of Communication and Social Psycology, University of Alicante,
Alicante, Spain
[5] Deptartment of Electronics and Computer Technologyt, University of Cartagena,
Cartagena, Spain

Abstract. At present there is a growing interest in studying emotions in the brain. However, although in the latest years there have been numerous studies, little is known about their temporal dynamics. Techniques such as fMRI or PET have very good spatial resolution but poor temporal resolution and vice-versa in the case of EEG. In this study we propose to use EEG to gain insight into the spatiotemporal dynamics of emotions processing with a better time resolution. We conducted an experiment in which binary classification (like / dislike) of standardized images was performed. Topographic changes in EEG activity were examined in the time domain. In the spatial dimension, we used a rotating dipole for the spatial location and determination of Cartesian coordinates (x, y and z). Our results showed a temporal window (424-474msec) with a significant difference which involved a lateralization (left to very positive stimuli and right to very negative stimuli) even for neutral stimuli. These results support the lateralization of brain activity during processing of emotions.

Keywords: EEG Teleservices · Brain-computer interface · Brain area networks

1 Introduction

The ability to recognize the emotional states is an important part of natural communication. Emotion plays an important role in human–human communication and interaction. Considering that, in normal live, we all are surrounded by machines; the emotional interaction between humans and machines is one of the most important challenges in advanced human–machine interaction and brain–computer interface [1]. For a robust analysis of the affective human–machine interaction, one

J.M. Ferrández Vicente et al. (Eds.): IWINAC 2015, Part I, LNCS 9107, pp. 203–212, 2015.
DOI: 10.1007/978-3-319-18914-7_21

of the most important requisites is to develop a reliable emotion recognition system capable to guarantee high recognition accuracy, robustness against artifacts and adaptability to applications.

Some researchers support the notion of biphasic emotion, which states that emotion fundamentally stems from varying activation in centrally organized appetitive and defensive motivational systems that have evolved to mediate the wide range of adaptive behaviors necessary for an organism struggling to survive in the physical world [2]. In this framework, neuroscientists have made efforts to determine how the relationship between stimulus input and behavioral output is mediated though specific, neural circuits that have evolved to organize and direct adaptive actions [3].

Relatively little is known about the neural temporal dynamics of emotion processing [4]. The majority of neuroimaging studies are based on methods such as functional Magnetic Resonance Imaging (fMRI) [5] or Positron Emission Tomography (PET) [6] with excellent spatial resolution but a very poor temporal one (in the range of seconds). Conversely, Electroencephalography (EEG) offers excellent temporal resolution (in the range of milliseconds), thus offering a better choice to solve the temporal problem.

Among neuroimaging techniques, EEG has demonstrated it can provide informative characteristics in responses to the emotional states [7]. Since Davidson et al [8] suggested that frontal brain electrical activity was associated with the experience of positive and negative emotions, the studies of associations between EEG asymmetry and emotions has received much attention [9]. In other studies, EEG asymmetry and event-related potentials (indexing a relatively small proportion of mean EEG activity) were also used to study the association with emotion [10].

In this study we investigated the temporal dynamics of neural activity associated to emotions (like/dislike) generated by complex pictures derived from the International Affective Picture System (IAPS) [11]. First, we used EEG to solve the problem of temporal resolution. We evaluated the correspondence between subjective emotional experience induced by the pictures and then the neural signature derived from the temporal profiles associated with their perception. Finally, we estimated with rotating dipole and head reconstruction the underlying neural places in which event-related potentials (ERPs) were generated. The tridimensional location was used for the assessment of changes in the activation of cortical networks involved in emotion processing. We completed the study by analysis of lateralization during emotion identification task in the tridimensional space.

Our results i) provide valuable information to understand the temporal dynamics of emotions, ii) are coherent with other works [12] about hemispheric lateralization and iii) introduce locations in the tridimensional space. Therefore, we suggest that the findings of this study could be useful for the development of effective and reliable neural interfaces.

2   Material and Methods

Participants

Twenty two participants participated in the study (mean age: 24.7; range: 19.7–33; eleven men, eleven women). All participants had no personal history of neurological or psychiatric illness, drug or alcohol abuse, or current medication, and they had normal or corrected to normal vision. All of them were right handed with a laterality quotient of at least + 0.4 (mean 0.8, SD: 0.2) on the Edinburgh Inventory [13]. All subjects were informed about the aim and design of the study and gave their written consent for participation.

Stimuli and Validation

A subset of standardized stimuli (144 pictures in total) was preselected from the IAPS dataset [11]. This is a database that contains a set of normalized emotional stimuli for experimental investigations of emotion and attention. It contains a large set of standardized, emotionally-evocative, internationally accessible, color photographs including contents across a wide range of semantic categories, from pleasant images (e.g. babies and beautiful animals) to unpleasant images (e.g. scenes of violence and injuries). Each image was presented with a score (9-1) concerning their affective valence. Stimuli were presented in color, with equal luminance and contrast.

The preselected IAPS stimuli were categorized into four groups according to punctuation IAPS, namely very nice pictures ($7<$ punctuation ≤ 9), nice pictures ($5 <$ punctuation ≤ 7), unpleasant images ($2 <$ punctuation ≤ 5) and very unpleasant images ($1 <$ punctuation ≤ 2). Each group was composed of 36 images.

IAPS pictures were previously scored with American population. In order to avoid artifacts due to the cultural issue (the participants were Spanish), we executed a previous study to calibrate the valence of the images with our participants. Stimulus categorization was validated in a study including 30 participants who did not participate in the main experiment (mean age: 23.3; range: 20.6–31.3; seventeen men, thirteen women). The stimuli were presented one by one during 1 second followed by a black screen for 3 sec on a 21 inches screen in random order. Subjects were instructed to give each stimulus a score from 1 to 9 avoiding 5 depending on subjective taste (1: dislike; 9: like). Their verbal response was recorded. Eighty out of the 144 images were selected for the main EEG experiment based on their new subjective score. Half of them (40) corresponded to positive images (score >5, CI = 95%) and the other half were negative images (punctuation <5; CI = 95%).

Procedure

Figure 1 summarizes the serial structure of the study. Each image was presented for 500msec and followed by a black screen for 3500msec. The participants task was to view the images and to rate the arousal and valence of their own emotional

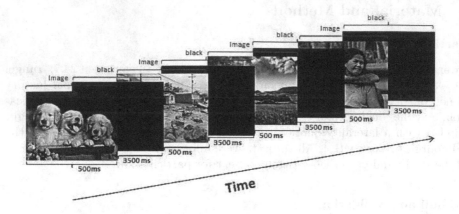

Fig. 1. Experimental design. The sequence of stimuli was presented in continuous mode by using a commercial stimulus presentation software (STIM2, Compumedics, Charlotter, NC, USA).

experience. Pictures score ranged from 1 (very unpleasant) to 9 (very pleasant). The images appeared randomly and only once.

Data Acquisition

We instructed subjects to remain as immobile as possible, avoiding blinking during image exposure and trying to keep the gaze toward the monitor center. EEG data was continuously recorded at a sampling rate of 1000 Hz from 64 locations (FP1, FPZ, FP2, AF3, GND, AF4, F7, F5, F3, F1, FZ, F2, F4, F6, F8, FT7, FC5, FC3, FC1, FCZ, FC2, FC4, FC6, FT8, T7, C5, C3, C1, CZ, C2, C4, C6, T8, REF, TP7, CP5, CP3, CP1, CPZ, CP2, CP4, CP6, TP8, P7, P5, P3, P1, PZ, P2, P4, P6, P8, PO7, PO5, PO3, POZ, PO4, PO6, PO8, CB1, O1, OZ, O2, CB2) using the international 10/20 system [14]. EEG was recorded via cap-mounted Ag-AgCl electrodes. A 64-channel NeuroScan SynAmps EEG amplifier (Compumedics, Charlotte, NC, USA). The impedance of recording electrodes was monitored for each subject prior to data collection and the threshold were kept below 25 KΩ. All the recordings were performed in a silent room with soft lighting.

Signal processing was performed with the help of Curry 7 (Compumedics, Charlotte, NC, USA). Data were re-referenced to a Common Average Reference (CAR) because the statistical and analysis methods required CAR. EEG signals were filtered using a 45 Hz low-pass and a high-pass 0.5 Hz filters.

Electrical artifacts due to motion, eye blinking, etc. were corrected. They were identified as signal levels above 75μV in the 5 frontal electrodes (FP1, FPZ, FP2, AF3 and AF4). These electrodes were chosen because they are the most affected by potential involuntary movements. The time interval for artifact detection was from (-200msec, +500msec) from stimulus onset. The detected artifacts were corrected using Principal Component Analysis (PCA). PCA is a classical technique in statistical data analysis, feature extraction and data reduction [15].

EEG data in the interval (-100, 1000) msec from stimulus onset were analyzed in this study. For each person, records were separated into 8 subgroups according to their given score (9, 8, 7, 6, 4, 3, 2, and 1). In turns, subgroups for dipole analysis were grouped into 4 groups as shown in Table 1.

Table 1. Separation of subjective scores into 4 groups for all people. Dipole separation performed for reconstruction using the mean of all people.

	Dislike				like			
	Group —		Group –		Group ++		Group +++	
opinion punctuation	1	2	3	4	6	7	8	9

Statistical Analyses

To constrain our analysis, we used an approach that has been widely used in psychophysiology: the examination of topographic changes in EEG activity (see [16] for an overview and [17]). This approach considers whole-scalp EEG activity elicited by a stimulus as a finite set of alternating spatially stable activation patterns, which reflect a succession of information processing stages. Differences in topographic patterns of activity between conditions were assessed using the Curry 7 software.

There are two main reasons why we used this analysis rather than the more traditional which is based on the assessment of amplitudes and latencies of a set of predefined ERP components. First, it takes into consideration the entire time course of activity and the entire pattern of activation across the scalp by testing the global field power from all electrodes (see for further explanation [18]). Second, this approach is able to detect not only differences in amplitude, but also differences in underlying sources of activity. The latter is based on the fact that maps that are confirmed to be both spatially and temporally different must necessarily be the product of a different set of generators. However, we emphasize that the analysis of topography changes is not incompatible with the analysis of traditional ERPs.

As recommended, topographical differences were tested through a non-parametric randomization test known as TANOVA (Topographic ANOVA). TANOVA tests for differences in global dissimilarity of EEG activity between two conditions by assessing whether the topographies are significantly different from each other on a time point-by-time point basis. TANOVA were performed to assess differences in activation patterns between different groups of images by subjective scoring. TANOVA is sufficient to indicate the time windows of interest for further analysis dipole. In this study, the significance level is $\alpha=0.01$. As suggested by [19], the corresponding required number of repetitions was chosen to be p> 1000. Map normalization was used for the difference tests, such that the MGFP per map was equal to 1.

The dipole source localization (DSL) solves the EEG inverse problem by using a nonlinear multidimensional minimization procedure that estimates the dipole parameters that best explain the observed scalp potentials in a least-square sense. In this process, we assume that EEG is generated by one or no more than few focal sources. The dipole source model can be further classified as moving, fixing or rotating dipoles depending on the degree of freedom of parameters. In our study we used a rotating dipole, that may be viewed as two independent dipoles whose orientation is allowed to vary with time [20].

Boundary Element Method (BEM) was used in the head reconstruction since it permits to locate the source dipoles. Thus BEM models are superior in non-spherical parts of the head like temporal and frontal lobe or basal parts of the head, where spherical models exhibit systematic localizations of up to 30 mm [21].

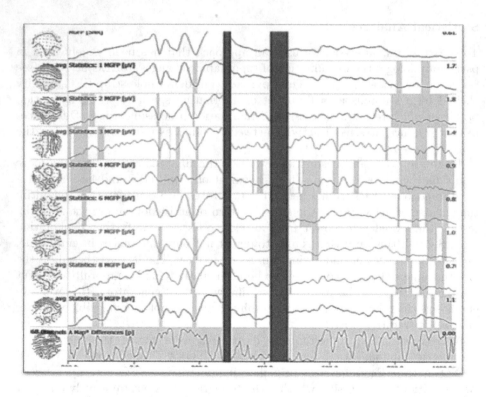

Fig. 2. Time points of significant differences in EEG activity for the 8 contrasts (9, 8, 7, 6, 4, 3, 2 and 1). It is as indicated by the T ANOVA analysis, depicting 1 minus p-value across time. Significant p values are plotted (p<0.01). The two vertical rectangles contain interval with significant differences.

3   Results

Participant Rankings Compression

The participants responded correctly to 1758 images (99.98%) following the instructions before starting the experiment. In only two images volunteer answered incorrectly (score 5) or did not respond. The images followed by incorrect answers were not excluded in the analysis below. The distribution of the new scores (or valences) was 49.4% and 50.6% greater and less than 5 respectively.

EEG

Differences in stimulus-elicited activity are depicted in Figure 2. There were significant differences between pictures with different scores (p<0.01). These differences started approximately 276 msec after stimulus onset. All subgroups were significant different to each other in two time windows, namely [276 - 294] msec and [424 - 474] msec.

Dipoles

One rotating dipole source model was used in the two time windows with significant differences indicated by the TANOVA (see Figure 2). When we focused

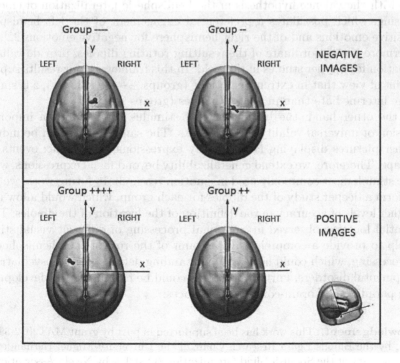

Fig. 3. Head reconstruction by rotating dipole in time window [424-474] msec. Rating was grouped into four groups according to subjective punctuation (see Table 1).

Table 2. Coordinates of dipole in head for window significant [424-474] msec

window 424-474ms											
Dislike (coordinates in mm)						like (coordinates in mm)					
Group ---			Group --			Group ++			Group ++++		
x	y	z	x	y	z	x	y	z	x	y	z
20,7	25,4	8,3	13,9	10,5	20	-0,6	5,3	16,8	-35,8	27,4	7,5

in the larger time window (424-474msec, duration 50msec) we found significant differences in the dipoles for the different types of images (see Figure 3).

The Cartesian coordinates of the rotating dipole for each group are shown in table 2.

4   Conclusions

Our results suggest a strong lateralization in the processing of images with emotive content. Thus, we found an increased activity in the left hemisphere for emotions with a positive valence. In contrast, there was an increased activity in the right hemisphere for emotions with a negative valence. These results are in line with the valence hypothesis in the hemispheric lateralization of emotion processing, which postulates a preferential engagement of the left hemisphere for positive emotions and of the right hemisphere for negative emotions[22],[23]. Furthermore the z coordinate of the resulting rotating dipoles, provide valuable information for further studies in this field. In this framework our results support the point of view that in extreme emotions (groups ++++ and ---), z is smaller or more intermediate than in neutral images (groups ++ and -).

On the other hand, the broad range of stimulus types adds an important dimension of universal validity to the results. The same valence can be induced by either pictures displaying facial, bodily expressions, or complex events and landscape. Therefore, we extend generalizability beyond facial expressions, which are the stimuli most commonly used in emotion research. In future work, we plan to perform a deeper study of the dipoles for each group, which would allow us to get higher levels of accuracy in the definition of the location of the dipoles. Thus, the spatial location observed in emotional processing of different visual stimuli can help to provide a comprehensive account of the role of each hemisphere in this processing, which could help in understanding deficits seen in psychiatric or developmental disorders. Furthermore, this could be helpful for the development of new paradigms of brain-computer interfaces.

Acknowledgement. This work has been supported in part by grant MAT2012-39290-C02-01, by the Bidons Egara Research Chair of the University Miguel Hernández, by a research grant of the Spanish Blind Organization (ONCE), by Nicolo Association for the R&D in Neurotechnologies for disability, by the regional project P11-TIC-7983, by Junta of Andalucia (Spain) and by the Spanish grant TIN2012-32039 (Spain).

References

1. Picard, R.W.: Affective computing. MIT Press
2. Davidson, R.J., Ekman, P., Saron, C.D., Senulis, J.A., Friesen, W.V.: Approach/withdrawal and cerebral asymmetry: Emotional expression and brain physiology(58), 330–341
3. Fanselow, M.S.: Neural organization of the defensive behavior system responsible for fear 1(4), 429–438, http://www.springerlink.com/index/10.3758/BF03210947, doi:10.3758/BF03210947
4. Linden, D.E.J., Habes, I., Johnston, S.J., Linden, S., Tatineni, R., Subramanian, L., Sorger, B., Healy, D., Goebel, R.: Real-time self-regulation of emotion networks in patients with depression 7(6) e38115, http://dx.doi.org/10.1371/journal.pone.0038115, doi:10.1371/journal.pone.0038115
5. Vink, M., Derks, J.M., Hoogendam, J.M., Hillegers, M., Kahn, R.S.: Functional differences in emotion processing during adolescence and early adulthood 91, 70–76, http://linkinghub.elsevier.com/retrieve/pii/S1053811914000561, doi:10.1016/j.neuroimage.2014.01.035
6. Royet, J.P., Zald, D., Versace, R., Costes, N., Lavenne, F., Koenig, O., Gervais, R.: Emotional responses to pleasant and unpleasant olfactory, visual, and auditory stimuli: a positron emission tomography study 20(20) 7752–7759
7. Petrantonakis, P.C., Hadjileontiadis, L.J.: A novel emotion elicitation index using frontal brain asymmetry for enhanced EEG-based emotion recognition 15(5), 737–746, http://ieeexplore.ieee.org/lpdocs/epic03/wrapper.htm?arnumber=5776680, doi:10.1109/TITB.2011.2157933
8. Davidson, R., Fox, N.: Asymmetrical brain activity discriminates between positive and negative affective stimuli in human infants 218(4578), 1235–1237, http://www.sciencemag.org/cgi/doi/10.1126/science.7146906, doi:10.1126/science.7146906
9. Harmon-Jones, E., Allen, J.J.: Anger and frontal brain activity: EEG asymmetry consistent with approach motivation despite negative affective valence 74(5), 1310–1316
10. Schupp, H.T., Cuthbert, B.N., Bradley, M.M., Caccioppo, J.T., Ito, T., Lang, P.J.: Affective picture processing: The late positive potential is modulated by motivational relevance 37(2), 257–261, http://doi.wiley.com/10.1111/1469-8986.3720257, doi.10.1111/1469-8986.3720257
11. Lang, P.J., Bradley, M.M., Cuthbert, B.N.: International affective picture system (IAPS): Technical manual and affective ratings
12. Davidson, R.J.: Anterior electrophysiological asymmetries, emotion, and depression: Conceptual and methodological conundrums 35(5), 607–614, http://doi.wiley.com/10.1017/S0048577298000134, doi:10.1017/S0048577298000134
13. Oldfield, R.: The assessment and analysis of handedness: The edinburgh inventory 9(1), 97–113, http://linkinghub.elsevier.com/retrieve/pii/0028393271900674, doi:10.1016/0028-3932(71)90067-4
14. Klem, G.H., Luders, H.O., Jasper, H.H., Elger, C.: The ten-twenty electrode system of the international federation. the International Federation of Clinical Neurophysiology 52, 3–6

15. Meghdadi, A.H., Fazel-Rezai, R., Aghakhani, Y.: Detecting determinism in EEG signals using principal component analysis and surrogate data testing, pp. 6209–6212. IEEE, http://ieeexplore.ieee.org/lpdocs/epic03/wrapper.htm?arnumber=4463227, doi:10.1109/IEMBS.2006.260679

16. Murray, M.M., Brunet, D., Michel, C.M.: Topographic ERP analyses: A step-by-step tutorial review 20(4) 249–264, http://link.springer.com/10.1007/s10548-008-0054-5,   doi:10.1007/s10548-008-0054-5

17. Martinovic, J., Jones, A., Christiansen, P., Rose, A.K., Hogarth, L., Field, M.: Electrophysiological responses to alcohol cues are not associated with pavlovian-to-instrumental transfer in social drinkers 9(4), e94605, doi:10.1371/journal.pone.0094605

18. Skrandies, W.: Global field power and topographic similarity 3(1) 137–141, http://link.springer.com/10.1007/BF01128870, doi:10.1007/BF01128870

19. Rosenblad, A.: B. f. j. manly: Randomization, bootstrap and monte carlo methods in biology, 3rd edn., 455 p. Chapman & amp; hall/CRC, Boca raton, $79.95 (HB), ISBN: 1-58488-541-6 24 (2) 371372. doi:10.1007/s00180-009-0150-3

20. Fuchs, M., Wagner, M., Wischmann, H.-A., Köhler, A., Theissen, R., Drenckhahn, H.: Improving source reconstructions by combining bioelectric and biomagnetic data 107(2), 93–111, http://linkinghub.elsevier.com/retrieve/pii/S0013469498000467, doi:10.1016/S0013-4694(98)00046-7

21. Vatta, F., Meneghini, F., Esposito, F., Mininel, S., Di Salle, F.: Realistic and spherical head modeling for EEG forward problem solution: A comparative cortex-based analysis, pp. 1–11 (2010), doi:10.1155/2010/972060/

22. Fusar-Poli, P., Placentino, A., Carletti, F., Allen, P., Landi, P., Abbamonte, M., Barale, F., Perez, J., McGuire, P., Politi, P.: Laterality effect on emotional faces processing: ALE meta-analysis of evidence 452(3), 262–267, http://linkinghub.elsevier.com/retrieve/pii/S0304394009001220, doi:10.1016/j.neulet.2009.01.065

23. Costa, T., Cauda, F., Crini, M., Tatu, M.-K., Celeghin, A., de Gelder, B., Tamietto, M.: Temporal and spatial neural dynamics in the perception of basic emotions from complex scenes 9(11), 1690–1703, http://scan.oxfordjournals.org/lookup/doi/10.1093/scan/nst164, doi:10.1093/scan/nst164

Interstimulus Interval Affects Population Response in Visual Cortex *in vivo*

Javier Alegre-Cortés[1], Eduardo Fernández[1,2], and Cristina Soto-Sánchez[1,2(✉)]

[1] Bioengineering Institute, Miguel Hernández University (UMH), Alicante, Spain
jalegre@umh.es
[2] Biomedical Research Networking center in Bioengineering, Biomaterials and Nanomedicine (CIBER-BBN), Malaga, Spain
e.fernandez@umh.es, csoto@umh.es

Abstract. Understanding the underlying properties of neuronal populations over single neurons is a longstanding goal for both basic and applied neurosciences, with a specifically suitable application in the field of neuroprosthesis development, aimed to restore the loss of function of a visual cortex as a result of an injury or disease. We study how the interstimulus interval (ISI) period of a repeated visual stimulus influences the overall activity of rat visual cortex neuronal populations. Our results suggest that certain (3, 5 s) interstimulus intervals do have an increased stimulus response compared to longer or shorter ISIs for a 500 ms grating drifting stimulus. Based on the preliminary results shown in this article, we claim the need of a better understanding of the biological dynamics of the visual cortex neuronal populations in order to properly design suitable brain-machine interfaces for visual neurorehabilitation intracortical neuroprosthetics.

Keywords: Visual cortex · Population analysis · *In vivo* electrophysiology · Neuroprosthesis

1 Introduction

In the aim of understanding the relationship between cortical activity and visual perception, a crucial tool is the ability to analyze neuronal population activity during visual stimulation. Such measurements are increasingly obtained from mice and rats, due to their small size as well as their simpler and further studied cortical micro-structure and brain dynamics [6,11,1], overtaking classical experiments driven in monkeys [7] and cats [3]. In this model, classical single cell studies had focused on the cellular response to a given stimulus [5], concerned only in preventing stimulus specific adaptation (SSA) [12] when considering the interstimulus interval of a task. However, cortical activity is driven by complex interactions between thousands of neurons [4,2,9], rather than the activity profile of single neurons.

Population dynamics are influenced by more complex features than the stimulus representation per se, such as locomotion [1] and attention [10]. Considering

© Springer International Publishing Switzerland 2015
J.M. Ferrández Vicente et al. (Eds.): IWINAC 2015, Part I, LNCS 9107, pp. 213–219, 2015.
DOI: 10.1007/978-3-319-18914-7_22

this landscape, we hypothesize that neuronal populations output to the repetition of a stimulus is sensitive to interstimulus interval, even for longer times than SSA avoidance, and thus, crucial in the design of any brain-machine interface device. Optimal interstimulus periods may improve intracortical neuroprosthetics accuracy if correctly implemented.

To prove this hypothesis, we displayed repetitions of a unique stimulus, to anesthetized Long Evans rats, using interstimulus intervals ranging from 1 to 7 seconds. We recorded visual cortex (V1 and V2) neuronal population activity during this passive task. Finally, we quantified the mean increase in the activity produced by the stimulus, depending on each ISI. Our data suggests the presence of an optimal time frame of visual input, regarding cortical activation.

2   Experimental Methods

2.1   Surgery

Data was obtained from 3 male Long Evans adult rats weighing 450-500gr. Surgical analgesia was induced by buprenorphine (0.025 mg kg^{-1} s.c), and anaesthesia and sedation were induced by ketamine HCl (40 mg kg^{-1} i.p) . The anaesthesia was continued and maintained with a mix of oxygen and 2% of isofluorane during the surgery and afterwards reduced to 1.5% during the electrophysiological recordings. The blinking and the toe pinch reflexes were continuously checked along the experiment to guarantee a proper level of anaesthesia for the animal. The body temperature was maintained with a thermal pad and the heart rate and O$_2$ concentration in blood were monitored throughout the experiment. Animals were pre-treated with dexamethashone (1 mg kg^{-1} i.p) 24 hours and 20 minutes prior to surgery in order to avoid brain edema caused by the electrode insertion. A craniotomy was drilled on top of the visual cortex and the electrode array was inserted 2 mm lateral to the midline and from 0.5 mm anterior to lambda. Then, a Utah array was inserted in the deep layers of the visual cortex with a Blackrock pneumatically-actuated inserter device specifically design for implanting the Utah array through the duramatter with a minimal tissue offense (Blackrock Microsystems, Salt Lake City, USA). The customized microelectrode Utah array consisted of 6 × 6 tungsten microneedles, covering a brain surface of 2 mm × 2 mm millimetres (400 μm spacing). After the insertion, the ipsilateral eyelid to the craniotomy site was closed with cyanocrylate and atropine sulphate 1% was used to dilate the pupil of the contraleral eye.

Ethical Approval. All experimental procedures were performed conformed to the directive 2010/63/EU of the European Parliament and of the Council, and the RD 53/2013 Spanish regulation on the protection of animals use for scientific purposes and approved by the Miguel Hernández University Committee for Animal use in Laboratory.

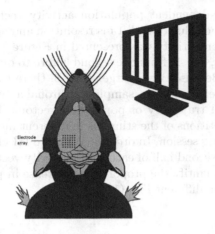

Fig. 1. Schematic representation of the experimental stimulus-recording design

2.2 Visual Stimulation

Visual stimulation consisted on a vertical drifting square-wave grating (90°, light and dark bars, 100% contrast, 6 Hz, 0.6 cycles/degree) interspersed with a dark (uniform) stimulus displayed from 1 to 7 seconds. The stimulus was displayed on a LCD monitor (refresh rate 60 Hz) and a luminance of ~100 cd/m², placed 25 cm in front of the right eye, approximately at 300 from the midline covering a visual field spanning ~100°(Figure 1). The stimulus was generated using the vision egg library and a python script. The room was kept in darkness all along the visual stimulation.

2.3 Extracellular Recording

In vivo neural activity from visual cortex was recorded simultaneously from 16 individual electrodes with a Utah array (Figure 1). The Utah array was connected to a MPA32I amplifier (Multichannel Systems, MCS) and the extracellular recordings were digitized with a MCS analog-to-digital board. The data were sampled at a frequency of 20,000 samples/s and slow waves were digitally filtered out (100-3000Hz) from the raw data. Neural spike events were extracted with a free-tool application for offline spike sorting analysis (Neural Sorter, http://sourceforge.net/projects/neuralsorter/) and the resulting multiunit information obtained from each electrode was storage for further analysis.

2.4 Data Analysis

Population activity analysis was performed using Matlab (MathWorks). Only neuronal clusters with stable waveform and consistent firing rate over the course of a session were considered in the analysis. We appreciated multiunit activity in the majority if the electrodes through the whole recordings sessions.

We constructed time-dependent population activity vectors by temporally binding the activity of each cluster with 1 ms resolution and applied a smoothing function when population activity was presented in Figure 3 (Bin size, B = 50, step size from bin to bin, S = 1) from one second before to one second after each stimulus presentation. Because B was larger than S, there was overlap between time bins. This had the effect of over- sampling neuronal activity and effectively smoothing the temporal trajectory of population vectors. These vectors were averaged for all the repetitions of the stimulus in different mean activity vectors for each ISI and recording session. In order to compare ISI effect on population activity, we divided the second half of each mean activity vector by the first half of the same vectors to quantify the proportional increase in population activity in each session for all the different ISIs (1, 3, 5, 7 s)

3   Results

Data was obtained from 3 Long Evans male adult rats. Multiunit activity from contralateral visual cortex to the stimulated eye was collected simultaneously from 16 individual electrodes. Single or multiunit neural activity was present at least in 13 electrodes consistently throughout each recording session. Spikes

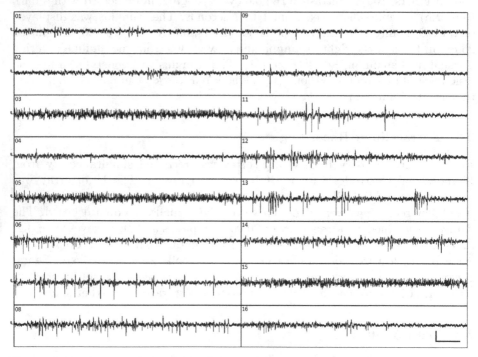

Fig. 2. Screen capture showing the display of the extracellular recording for 16 electrodes simultaneously. Each panel in the image corresponds to an individual electrode of the array. Scale bar in the last bottom panel corresponds to $150\mu V$ in the vertical axis and 50 ms in the horizontal axis.

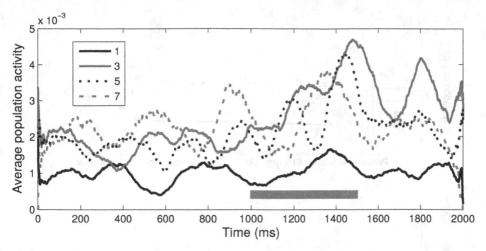

Fig. 3. Mean population activity for ISIs of 1, 3, 5 and 7 seconds (see legend). A peak in the mean population activity level as a response to the stimulus was shown for 3 and 5 ISI seconds. Stimulus duration shown as a horizontal gray bar.

reached amplitudes up to 400 μV in the best signal-to-noise electrode recordings, and both tonic and bursting dynamics were clearly identified (Figure 2). In order to perform the neuronal population activity analysis, both tonic and bursting dynamics were included indifferently for each recording site independently.

We analyzed the mean population activity vectors for the different ISIs (Example session 1 in Figure 3). Two seconds of the activity are shown, centred in the beginning of the grating stimulus (500 ms duration) and averaged for each ISI. A perceptible increase was shown for stimulus response and 3 and 5 seconds of ISI; on the other hand, mean population responses to the same stimulus but with shorter (1 s) or longer (7 s) ISIs were weaker; thus there was not a noticeable increase in population activity in these vectors. The reduced number of recorded repetitions of the stimuli in our study of this population phenomena implicated noticeable fluctuations in the pre-stimulus mean population activity for each train of stimuli.

In order to prevent our analysis from any possible bias produced by this slant, we normalized mean population activity vectors, using mean population activity of the previous second to the stimulus initiation for each ISI independently (Figure 4).

While the wide majority of studies of response to visual stimuli in different parts of the visual pathway focused on the response itself, addressing stimulus variables, we moved our scope on the population effect of interstimulus interval timing variations, hypothesizing that this may affect population time-persistent activity, thus population response. Our analysis suggested that ISI timing did affect population activity mediated by stimulus triggering. A more detailed and informative analysis should be performed by larger recordings in awake animals, as well as a more exhaustive analysis of this visual cortex neuronal populations property.

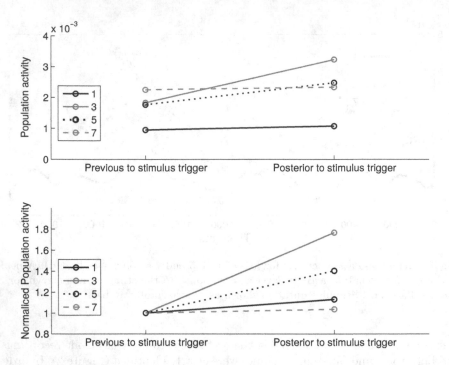

Fig. 4. Upper panel shows the mean population activity for the previous (left) and the following (right) second to the stimulus trigger for all de ISIs studied in example population 1. Bottom panel, data shown in the upper panel normalized to the mean population activity in the previous second to the stimulus trigger for each ISI.

4   Conclusion

Using multiunits recording in anesthetized Long Evans rats visual cortex, we measured the mean population response for a unique drifting square grating stimulus while modifying the time duration of its ISI. We found that mean neuronal population activity depends on ISI for longer times than those used to avoid SSA. We also found this effect to have an "optimal" ISI window, on which the population response to the stimulus is maximum. Future studies should extend these studies to awake animals and different stimuli and stimulus durations, recent papers [8] suggest that visual cortical responses in anesthetized and awake rodents have similar tuning properties; bearing this in mind, we expect to find similar results to what we have shown when recording in awake rats. Finally, our result should be considered in the design and development of visual brain-machine intracortical devices, as optimal population activation produced by ISI variations may improve the efficiently of cortical electrical stimulation with neurorehabilitation porpoises.

Acknowledgements. We would like to thank Lucas Jesús Morales Moya for helping in the creation of the code required for the performed analysis.

This work has been supported in part by grant MAT2012-39290-C02-01 from the Spanish Government, by the Bidons Egara Research Chair of the University Miguel Hernández and by a research grant of the Spanish Blind Organization (ONCE).

References

1. Ayaz, A., Saleem, A.B., Schölvinck, M.L., Carandini, M.: Locomotion controls spatial integration in mouse visual cortex. Curr. Biol. 23(10), 890–894 (2013)
2. Bathellier, B., Ushakova, L., Rumpel, S.: Discrete neocortical dynamics predict behavioral categorization of sounds. Neuron 76(2), 435–449 (2012)
3. Carandini, M.: Amplification of trial-to-trial response variability by neurons in visual cortex. PLoS Biol. 2(9), E264 (2004)
4. Druckmann, S., Chklovskii, D.B.: Neuronal circuits underlying persistent representations despite time varying activity. Curr. Biol. 22(22), 2095–2103 (2012)
5. Hubel, D.H., Wiesel, T.N.: Receptive fields, binocular interaction and functional architecture in the cat's visual cortex. J. Physiol. 160, 106–154 (1962)
6. Huberman, A.D., Niell, C.M.: What can mice tell us about how vision works? Trends Neurosci. 34(9), 464–473 (2011)
7. Martínez-Trujillo, J., Treue, S.: Attentional modulation strength in cortical area mt depends on stimulus contrast. Neuron 35(2), 365–370 (2002)
8. Niell, C.M., Stryker, M.P.: Modulation of visual responses by behavioral state in mouse visual cortex. Neuron 65(4), 472–479 (2010)
9. Rabinovich, M.I., Varona, P.: Robust transient dynamics and brain functions. Front. Comput. Neurosci. 5, 24 (2011)
10. Reynolds, J.H., Pasternak, T., Desimone, R.: Attention increases sensitivity of v4 neurons. Neuron 26(3), 703–714 (2000)
11. Shuler, M.G., Bear, M.F.: Reward timing in the primary visual cortex. Science 311(5767), 1606–1609 (2006)
12. Ulanovsky, N., Las, L., Nelken, I.: Processing of low-probability sounds by cortical neurons. Nat. Neurosci. 6(4), 391–398 (2003)

Towards the Reconstruction of Moving Images by Populations of Retinal Ganglion Cells

Ariadna Díaz-Tahoces[1(✉)], Antonio Martínez-Álvarez[2],
Alejandro García-Moll[1], Lawrence Humphreys[1], José Ángel Bolea[3],
and Eduardo Fernández[1]

[1] Institute of Bioengineering and CIBER BBN, University Miguel Hernández,
Alicante, Spain
adiaz@goumh.umh.es, e.fernandez@umh.es

[2] Department of Computer Technology, University of Alicante, Alicante, Spain
amartinez@dtic.ua.es

[3] Department of Materials Science and Engineering and Chemical Engineering.
University Carlos III. Madrid, Spain

Abstract. One of the many important functions the brain carries out is interpreting the external world. For this, one sense that most mammals rely on is vision. The first stage of the visual system is the image processing whose capture takes place in the retina. Here, photoreceptors cells transform light into electrical impulses that are then guided by amacrine, bipolar, horizontal and some glial cells up to the ganglion cells layer. Ganglion cells decode the visual information to be interpreted by the visual cortex. The understanding of the mechanism for decoding the visual information is a major task and challenge in neuroscience. This is especially true for images that change with time, for example during movement. For this purpose, extracellular recordings with a 100 multi-electrode-array (MEA) were carried out in the retinal ganglion cells layer of mice. Different moving patterns and actual images were used to stimulate the retina. Here, we present a new strategy for analysis over the spike trains recorded allowing the reconstruction of the actual stimuli with a reduced number of ganglion cell responses.

Keywords: Retina · Ganglion cells · Natural scene · Receptive fields

1 Introduction

Presently, how the nervous system interprets the outside world using neural messages through a sensory circuit remains a major challenge in neuroscience [1]. In vision, a light stimulus is transduced into an electrical impulse via the photoreceptors. This signal is then transmitted to the inner nuclear and ganglion cell layer which carry out the initial decoding of the stimulus [2].This information is then transmitted through the optic nerve to the visual cortex for further processing.

Normally, visual sensory neurons are characterized in a laboratory setting by their preference to light or dark, intensity, direction and receptive fields. However,

© Springer International Publishing Switzerland 2015
J.M. Ferrández Vicente et al. (Eds.): IWINAC 2015, Part I, LNCS 9107, pp. 220–227, 2015.
DOI: 10.1007/978-3-319-18914-7_23

this has usually been determined using simplified black and white stimuli as a representation of the real world. This is in contrast to the color and motion that we experience.

The problem is not in simulating a realistic visual signal; rather it lies in the interpretation of the various levels of decoding that is concurrently being carried out in the retinal layers [3][4]. This is in part due to the complexity of the signal as well as the many variables that modulate the electrical responses.

For this complicated task we have combined two methods, extracellular recordings from ganglion cell evoked responses under light stimuli using a MEA and a custom designed software to reconstruct complex visual stimuli.

Our software locates the receptive field of each cell based on the response to a moving bar crossing the visual field in 8 orthogonal directions. From this data we can identify their receptive field. Using this information we apply a visual stimuli composed from a natural scene and correlate electrical responses to this image. This allows us to accurately reconstruct our image using only the ganglion cell responses.

This novel method can be used as a tool to characterize electrical responses to complex visual stimuli. Here, we demonstrate with as little as 11 cells we can reconstruct natural images.

2   Material and Methods

Retina Preparation

Wild-type (C57BL/6J strain) mice were bred within a local colony established from purchased breeding pairs (Jackson Laboratories). Following anesthesia with 4% of isoflurane (IsoFlo®, Esteve Veterinaria) inhalational, cervical dislocation was performed. Then both eyes were removed. Animals were dark-adapted for one hour prior to sacrifice. All the experimental procedures were carried out in accordance with the ARVO and European Communities Council Directives (86/609/ECC) for the use of laboratory animals.

The cornea and lens were removed and discarded from the eyeball by a transverse cut along the *ora serrata* with a razor blade. Then, the retinas were removed from the remaining eyecup with the pigment epithelium and mounted on an agar plate with the ganglion cell side facing up. Finally, the tissue was covered with a piece of nitrocellulose paper in order to fix it and maintain the correct moisture. This paper had a small window cut into it to allow placement the electrode on the retinal ganglion cells layer.

This preparation was then mounted on a recording chamber, superfused with oxygenated Ringer medium (124mM NaCl, 2.5mM KCl, 2mM $CaCl_2$, 2mM $MgCl_2$, 1.25mM NaH_2PO_2, 26mM $NaCHO_3$ and 22mM Glucose) at physiological temperature.

These preparations was always performed under dim red illumination.

Multielectrode Recordings and Spike Sorting

Extracellular recordings were obtained from the retinal ganglion cell layer in the isolated mouse retina using an array of 100 electrodes with $400\mu m$ inter-electrode distances [5]. The electrical signals captured by the electrodes array were amplified with a 100-channel amplified (Bionic Technologies, Inc) with a gain of 5000 and a bandpass between 250 and 7500 Hz. The selected data from each channel as well as the state of the visual stimulus were digitized with a resolution of 16 bits at a sampling rate of 30 kHz and stored using a signal processor data acquisition system.

All neural spike events recorded exceeded at least 3.25 the standard deviation of noise level. When a supra-threshold event occurred, the waveform and time was stored together with the state of the visual stimulus for later offline analysis.

Each electrode can detect light evoked single- or multi-unit responses making the characterization and grouping of spikes necessary. The spike sorting was carried out by Nev2lkit program, free open source software based on principal component analysis (PCA) method and different clustering algorithms [6]. Time stamps for action potentials of each sorted unit were used to generate inter spike interval histograms (ISI), peristimulus time histograms (PSTH) and peristimulus spike raster analysis using NeuroExplorer®(Nex Technologies) as well as NeurALC software.

Visual Stimuli

All visual stimuli were programmed in Python and reproduced with Vision egg an open source library for real-time visual stimulus generation [7]. For this we used an area of 120×154 pixels of a 16-bit ACER TFT at 60 Hz refresh rate. The patterns displayed on this area were resized to a 4×4mm area with optical lenses and projected through a beam splitter focusing the stimulus onto the photoreceptor layer.

The retinas were then stimulated with three different types of stimuli. Several repetitions of a 700ms flash (196.25 cd$/m^2$) were displayed followed by darkness for 2300ms to classify the ganglion cells in ON, OFF, ON/OFF or spontaneous with no response to light (NRL) [8].

We then proceeded to stimulate the photoreceptors layer with $250\mu m$ wide white bars crossing a black screen at 0.5 & 1Hz. Four pairs of stimuli were used: 0°, 45°,90°,135°, 180°, 225°, 270° and 315°.

This was followed by the presentation of an animated panoramic natural scene projected on a virtual drum for 180 seconds at 0.7Hzfrequency. The image size was 1031×156 pixels grouped into squares of 15×15 pixels and was presented in black and white.

Delimit and Locate the Receptive Field

To determine the size and localization of the cells receptive field [9], the photoreceptor layer was stimulated with $250\mu m$ wide white bars crossing the black screen at 0.5Hz. To automatically map the response for each isolated ganglion cell to the corresponding squared-pixel from the stimulus image projection, an *ad-hoc* Python program was designed. Within this program, the responses to each pair of left-to-right and right-to-left moving-bars are processed separately to calculate their centroids.

To avoid measurement of unwanted firing responses such as noise signals, a custom weighting threshold was defined for filtering them. Then, both signals are set in phase to cancel inherent latency effects and locate the receptive field in every corresponding direction. Fig. 1 represents this automatic phasing process for a given cell response through each mentioned direction. Each phased contribution is added so that four $1 \times N$ matrices of responses are obtained: $M_{0,180}, M_{45,225}, M_{90,270}$ and $M_{135,315}$. Finally, the receptive field is calculated by multiplying M0,180, by Mt90,270 and M45,225 and Mt135,315 (their transposed orthogonal duals), and averaging the results.

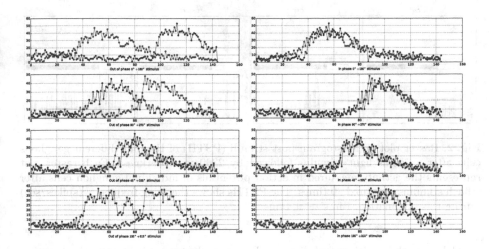

Fig. 1. Automatic phasing for $0°-180°$, $45°-225°$, $90°-270°$ and $135°-315°$ responses

To map the receptive field to the corresponding squared-pixel from the stimulus image, its centroid is calculated. The coordinates from this centroid reveals the actual image location from where the cell is integrating information.

Reconstruction of Natural Scenes

Once the process of determining and mapping every isolated ganglion cell is finished, a weighted set of ganglion cells responses (PSTH) is associated with every squared-pixel. As the original natural scene moves horizontally within a drum, every cell is mapped to a certain squared-pixel process as a $1 \times N$ row of image values in the range 0-255. For reconstructing the natural scene, the data provided by each different ganglion cells PSTH was normalized between the ranges of the corresponding row by means of a linear regression. In this way, the highest firing rate for a given row corresponds to the highest level of gray within the mentioned row for an ON-type cell. In addition, as some receptive field areas expand to more than one single row, a weighted sum of each rows adjacency is taken into account. In this analysis, OFF cells and NRL were rejected.

3   Results

We performed extracellular recordings in three wild-type adult mice retina. A minimum of 40 retinal ganglion cells were recorded in each animal during each experiment. All of these were classified according to their preference to light ON, OFF, ON/OFF and NRL after the flash stimulus (Fig. 2).

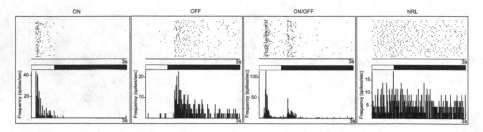

Fig. 2. Raster plot and histogram post stimuli (PSTH) of the four different ganglions cells after a 700ms light stimulus followed by 2300ms of darkness. Each raster plot represents the activity measured in action potentials of individual ganglion cells after a flash stimulus, repeated 30 times. The PSTHs shows the firing frequency (spikes/sec) during this stimulation. Bin size = 14ms.

Although the position and distance between electrodes is known, the location of each cell that they recorded from is not. Our goal was to identify the spatial position of each cell and outline the area of their receptive field. This was done using the responses recorded after light stimulation with bars (Fig. 3). The average value of the receptive field in our population was $0.201 \mu m \pm 0.026$ SE. From this information we established the actual stimulus area that each cell was able to decode.

The ganglion cells responses were recorded during the motion of the actual scene stimulus repeated 30 times within 3 minutes. After spike sorting, 11 cells were chosen depending on their location to cover the whole image size (Fig 4). The data provided by their PSTH were normalized by a linear regression between the gray values range of the corresponding rows. The image reconstructed with only 11 cells is visually similar to the actual one (Fig 5). Specifically, for lines

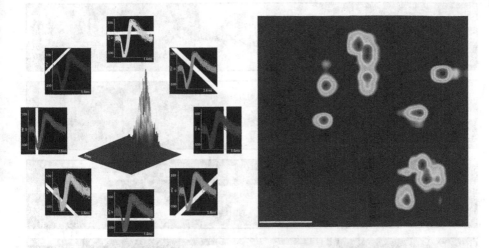

Fig. 3. Left) The action potential waveforms of a ganglion cell for each directional bar stimulus and a 3D plot of their receptive field. X axis = $\pm100\mu V$; Y axis = 1.6ms. Right) Graphical representation of 11 cell receptor fields and from their responses to the natural scene was reconstructed. Scale bar = 1mm.

Fig. 4. Two examples of ganglion cells receptive fields and how these cells are mapped to the actual scene location during drum rotation. Scale bar = 1mm.

1-6, 7, 10 and 11 ON-type cells responses were selected. The remaining image was reconstructed based on ON/OFF-type responses.

Bhattacharyya distance was calculated between the reconstructions and processed actual images in order to obtain quantifiable information about the similarity of both objects [10]. For this comparative measure the value 1 is assigned to the biggest difference and 0 to an equal distribution. In all experiments we obtained values below 0.35 for this index.

However, nonparametric and parametric statistical tests were also performed using U Mann-Whitney and T student, respectively. In both cases we observed that there are no significant differences between images, $p > 0.05$.

If the actual stimuli are unknown the reconstruction can be performed making the lineal regression between 0-255 gray scale and the ganglion cells responses, regardless of the receptive field location. In these cases Bhattacharyya distance value was within the range 0.4-0.5 (Fig 5D).

Fig. 5. A) The natural scene, 1031×159 pixels. B) Simplified image with pixels grouped into squares of 15 × 15. C) Image reconstructed based on ganglion cell responses of one animal, Bhattacharyya distance = 0.32. D) Image reconstructed without weighting between rows, Bhattacharyya distance = 0.42.

4   Conclusions

As a first approach, our preliminary results suggest that with the responses of only 11 ganglion cells we are able to reconstruct accurately a complicated natural image. This allows room for improvement for more accurate reconstruction if we were to incorporate more cells into our analysis. Moreover, using the proposed method it is not necessary to know what the natural image actually looks like for carrying out reconstructions as this can be done blind using the gray scale values to achieve reliable results. These experiments need to be repeated and analyzed using natural images presented at different frequencies of motion as well as in color to extend our knowledge for a more complete characterization of ganglion cell function. As this method is robust it can be adapted easily for characterization in other species.

Our ultimate goal is to apply the date acquired from this procedure and begin to compare ganglion cell visual responses in healthy retinas to those suffering from injury or neurodegenerative visual diseases. This could provide valuable information to the processes underlying the functional degradation of ganglion cells in visual impairments.

Acknowledgments. We are very grateful to Sonia Andreu for all her help and technical assistance. This work has been supported in part by grant MAT2012-39290-C02-01, by the Bidons Egara Research Chair of the University Miguel Hernández and by a research grant of the Spanish Blind Organization (ONCE).

References

1. Masland, R.H.: The Neuronal Organization of the Retina. Neuron 76(2), 266–280 (2012), http://www.cell.com/article/S0896627312008835/abstract, doi:10.1016/j.neuron.2012.10.002
2. Hoon, M., Okawa, H., Santina, L.D., Wong, R.O.: Functional architecture of the retina: Development and disease. Progress in Retinal and Eye Research 42, 44–84 (2014), http://www.sciencedirect.com/science/article/pii/S135094621400038X, doi:http://dx.doi.org/10.1016/j.preteyeres.2014.06.003
3. Gollisch, T., Meister, M.: Eye smarter than scientists believed: Neural computations in circuits of the retina. Neuron 65(2), 150–164 (2010), http://www.sciencedirect.com/science/article/pii/S0896627309009994, doi:http://dx.doi.org/10.1016/j.neuron.2009.12.00
4. Nirenberg, S., Pandarinath, C.: Retinal prosthetic strategy with the capacity to restore normal vision. Proceedings of the National Academy of Sciences 109(37), 15012–15017 (2012), arXiv:http://www.pnas.org/content/109/37/15012.full.pdf+html, doi:10.1073/pnas.1207035109
5. Fernández, E., Ferrández, J.-M., Ammermüller, J., Normann, R.A.: Population coding in spike trains of simultaneously recorded retinal ganglion cells1. Brain Research 887(1), 222–2229 (2000), http://www.sciencedirect.com/science/article/pii/S0006899300030729, doi: http://dx.doi.org/10.1016/S0006-89930003072-9
6. Bongard, M., Micol, D., Fernández, E.: NEV2lkit: A new open source tool for handling neural event files from multi-electrode recordings. International Journal of Neural Systems 24(04), 1450009, pMID: 24694167 (2014), arXiv:http://www.worldscientific.com/doi/pdf/10.1142/S0129065714500099, doi:10.1142/S0129065714500099
7. Straw, A.D.: Vision egg: An open-source library for realtime visual stimulus generation. Frontiers in Neuroinformatics 2, http://dx.doi.org/10.3389/neuro.11.004.2008, doi:10.3389/neuro.11.004.2008
8. Van Wyk, M., Wässle, H., Taylor, W.R.: Receptive field properties of on- and off-ganglion cells in the mouse retina. Visual Neuroscience 26, 297–308 (2009), http://journals.cambridge.org/article_S0952523809990137, doi:10.1017/S0952523809990137
9. Zhang, Y., Kim, I.-J., Sanes, J.R., Meister, M.: The most numerous ganglion cell type of the mouse retina is a selective feature detector. Proceedings of the National Academy of Sciences 109(36), E2391–E2398 (2012), arXiv:http://www.pnas.org/content/109/36/E2391.full.pdf+html, http://www.pnas.org/content/109/36/E2391.abstract, doi:10.1073/pnas.1211547109
10. Goudail, F., Réfrégier, P., Delyon, G.: Bhattacharyya distance as a contrast parameter for statistical processing of noisy optical images. J. Opt. Soc. Am. A 21(7), 1231–1240 (2004), http://josaa.osa.org/abstract.cfm?URI=josaa-21-7-1231, doi:10.1364/JOSAA.21.001231

FPGA Translation of Functional Hippocampal Cultures Structures Using Cellular Neural Networks

Victor Lorente[2], J. Javier Martínez-Álvarez[1(✉)], J.M. Ferrández-Vicente[1],
Javier Garrigós[1], Eduardo Fernández[2], and Javier Toledo[1]

[1] Dpto. Electrónica, Tecnología de Computadoras y Proyectos,
Universidad Politécnica de Cartagena, Cartagena, Spain
[2] Instituto de Bioingeniería,
Universidad Miguel Hernández, Alicante, Spain
jjavier.martinez@upct.es

Abstract. Electric stimulation in neural cultures in neural cultures may be used for creating adjacent physical or logical connections in the connectivity graph following Hebbs Law modifying the neural responses principal parameters. The created biological structure may be used for computing a certain function, however this achieved structure vanished with time as the stimulation stops. A DTCNN architecture, specifically designed for optimum parallel implementation over dedicated hardware, is proposed to emulate the behavior ans structure of the biological neuronal culture. The FPGA circuit can be used as a permanent model and is also intended to facilitate and speed up further experimentation.

Keywords: Cultured neural network · Hebbian Law · Induced plasticity · Learning · CNN · FPGA

1 Introduction

Biological brains use millions of biological processors, with dynamic structure, slow commutations compared with silicon circuits [1, 2, 22], with low power consumption and unsupervised learning. The use of dissociated cortical neurons cultured onto MEAs represents a useful experimental model to characterize both the spontaneous behavior of neuronal populations and their activity in response to electrical and pharmacological changes.

Learning is a natural process that needs the creation and modulation of sets of associations between stimuli and responses. Many different stimulation protocols have been used to induce changes in the electrophysiological activity of neural cultures looking for achieving learning [3–13] and low-frequency stimulation has brought good results to researchers enhancing bursting activity in cortical cultures [10, 11]. Hebbian learning describes a basic mechanism for synaptic plasticity wherein an increase in synaptic efficacy arises from the presynaptic cell's repeated and persistent stimulation of the postsynaptic cell continuously

© Springer International Publishing Switzerland 2015
J.M. Ferrández Vicente et al. (Eds.): IWINAC 2015, Part I, LNCS 9107, pp. 228–237, 2015.
DOI: 10.1007/978-3-319-18914-7_24

and repeatedly. The theory is commonly evoked to explain some types of associative learning in which simultaneous activation of cells leads to pronounced increases in synaptic strength. Basically the efficiency of a synaptic connection is increased when presynaptic activity is synchronous with post-synaptic activity. In this work, we use this kind of stimulation to create adjacent physical or logical connections in the connectivity graphs using Hebbs Law.

In previous papers, we used a specific low-frequency current stimulation on dissociated cultures of hippocampal cells to study how neuronal cultures could be trained with this kind of stimulation [14, 15]. We showed that persistent and synchronous stimulation of adjacent electrodes may be used for creating adjacent physical or logical connections in the connectivity graph following Hebbs Law. In later experiments, we have used different parameters for this stimulation to check if those connections can be created stimulating with different configurations. However this created biological structure vanishes with time as the stimulation stops. The stability of the achieved connectivity graph depends on the stimulation provided and biological culture characteristics. This means it is necessary to translate the biological culture structure and behavior to a more permanent substrate in order to study its computational capabilities. We propose to translate these parameters to a DTCNN over a FPGA element, as result of our previous implementation of visual systems for low vision devices [23].

The outline of the paper is as follows. Section 2 presents the methods for addressing Hebbian Learning throw electrical stimulation. Section 3 shows the results obtained using a specific stimulation with our experimental setup on hippocampal cultures to train them. Sections 4 and 5 detail a DTCNN architecture specifically designed for hardware projection on programmable devices, and an specific configuration for the modelling of the presented hippocampal cultures setup, respectively. Finally, Section 6 conclude by discussing some crucial aspects of the research and the remaining challenges.

2 Methods

2.1 Cell Culture Preparation

Dissociated cultures of hippocampal CA1-CA3 neurons were prepared from E17.5 sibling embryos (Figure 1). During the extraction of the hippocampus a small amount of cortical tissue will have inevitably also been included. Tissue was kept in 2 ml of HBSS. 10 mg/ml of trypsin was added to the medium and placed in a 37 °C water bath for 13 min for subsequent dissociation. The tissue was then transferred to a 15 ml falcon containing 4 ml of NB/FBS and triturated using combination of fine pore fire polished Pasteur pipettes (Volac). Cells were then transferred onto 12 well plates (Corning Incorporated) containing glass coverslips (Thermo Scientific).

The coverslips were pre-treated overnight with PDL (50 mg/ml), a synthetic molecule used as a coating to enhance cell attachment. The PDL was then aspirated away and the coverslips washed twice with PBS. This was then followed by a final coating of laminin (50 μg/ml), a protein found in the extracellular matrix, to

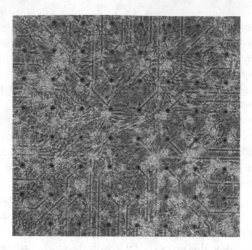

Fig. 1. Hippocampal CA1-CA3 culture (21 DIV) on a microelectrodes array

further help anchor the dissociated hippocampal cells. The cells were maintained in a mixture of 500 ml NB/B27 (promotes neural growth) and 500 ml NB/FBS (promotes glial growth), each supplemented with Glutamax and Pen/Strep (dilution 1/100). Glutamax improves cells viability and growth while preventing buildup of ammonia whereas Pen/Strep helps to prevent any infections. Cell density for each coverslip was roughly 200,000 cells. Cells were kept in an incubator at 37 °C in 6% CO_2.

2.2   Experimental Setup

Microelectrode arrays (Multichannel systems, MCS) consisted of 60 TiN/SiN planar round electrodes (200 μm electrode spacing, 30 μm electrode diameter) arranged in a 8x8 grid. The activity of all cultures was recorded using a MEA60 System (MCS). After 1200X amplification, signals were sampled at 10 kHz and acquired through the data acquisition card and MCRack software (MCS). Electrical stimuli were delivered through a two-channel stimulator (MCS STG1002) to each pair of electrodes.

2.3   Analysis Performed

We observed the spontaneous activity of the cultures before and after the stimulation experiments, as well as their evoked response to the applied stimulus. Extensive burst analysis, post-stimulus time histograms and functional connectivity were the main analysis performed to the registered data.

Functional connectivity [16, 17] captures patterns of deviations from statistical independence between distributed neurons units, measuring their correlation/covariance, spectral coherence or phase locking. Functional connectivity is often evaluated among all the elements of a system, regardless whether these elements are connected by direct structural links; moreover, it is highly time-dependent (hundreds of milliseconds) and model-free, and it measures statistical interdependence (e.g. mutual information) without explicit reference to causal effects.

Correlation and information theory-based methods are used to estimate the functional connectivity of in-vitro neural networks: Cross-correlation, Mutual Information, Transfer Entropy and Joint Entropy. Such methods need to be applied to each possible pair of electrodes, which shows spontaneous electrophysiological activity. For each pair of neurons, the connectivity method provides an estimation of the connection strength (one for each direction). The connection strength is supposed to be proportional to the value yielded by the method. Thus, each method is associated to a matrix, the Connectivity Matrix (CM), whose elements (X,Y) correspond to the estimated connection strength between neuron X and Y.

High and low values in the CM are expected to correspond to strong and weak connections, respectively. By using such approach, inhibitory connections could not be detected because they would be mixed with small connection values. However, non-zero CM values were also obtained when no apparent causal effects were evident, or no direct connections were present among the considered neurons.

In our experiments, connectivity maps offered a visualization of the connectivity changes that occur in the culture. Connectivity maps were generated using the connectivity matrix (CM) obtained after applying the analysis and Cross-Correlation or Mutual Information. By setting thresholds in the CM, it is possible to filter out some small values that may correspond to noise or very weak connections. In consequence, these maps show the strongest synaptic pathways, and can be used for visualizing the neural weights dynamics, and validate the achieved learning.

3 Results

Low-frequency current stimulation and tetanic stimulation had both an impact on the electrophysiological responses of the cultures, as previous studies had reported [13,6]. Raster plots showed that all of the stimulations provided induce changes in the firing frequency of the cultures. We can observe some kind of reorganization in the firing activity, from a nearly random continuous spiking activity to a more discrete bursting spiking activity. After the third week in vitro, this bursting activity becomes more frequent and robust and this effect is much more evident than during the first weeks (Figure 2). First experiments showed that this behavior takes effect initially with low-frequency stimulation, however from our last experiments it may be concluded that both stimulations have a frequency impact on the spiking activity of the culture. It may be concluded that this effect is more evident using low-frequency neural stimulation.

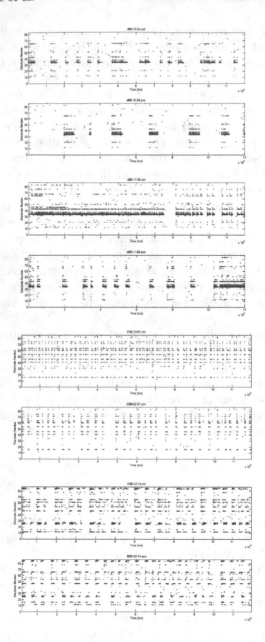

Fig. 2. Raster plots extracted from cultures of experiments E2 and E5. (a) (21DIV) and (b) (32DIV) belong to ID68 from E2, (c) (23DIV) and (d) (30DIV) belong to ID86 from E5. Each figure is divided in two graphs, which show the spiking activity of the culture before and after stimulation. Raster plots show a change in the spiking activity, changing from a uniform activity before stimulation to a more concentrated activity after stimulation. This result is emphasized after the third week in vitro due to maturing occurred in the cultures.

Connectivity diagrams based on cross-correlation between electrodes showed some kind of connections reorganization after stimulations, concentrating them in a few electrodes. Furthermore, adjacent physical or logical connections in the connectivity graph following Hebbs law appeared in some pairs of stimulated electrodes (31 and 42, in red in Figure 3). Electrodes with created connections between them can distinctly be detected with the instantaneous firing frequencies graphs. The two pairs of stimulated electrodes before and after the stimulation session follow exactly each other, whereas the firing periods of the first pair of electrodes do not match. Furthermore, the electrodes of the second pair change both the firing periods after stimulation. This feature indicates that there exists a strong connection between them.

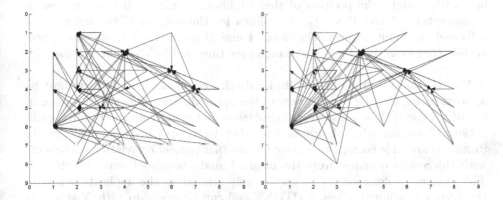

Fig. 3. Connectivity graphs based on cross-correlation between electrodes. The graph belong to the culture ID48 (E1) at 25 DIV. Pair of electrodes 31, 42 and 52, 53 were stimulated with low-frequency current stimulation with $50\,\mu A$ biphasic pulses. (a) No logical connections were observed before stimulation. (b) A connection (red arrow) between electrodes 31 and 42 has appeared.

Analyzing spike parameters such as peaks heights and widths and number of spikes have lead us to an important result. Both types of stimulation, low-frequency and tetanic stimulation, produced a reactivation of neurons over time which lead to the creation of adjacent physical or logical connections in the connectivity graph following Hebb's Law.

4 Non-linear Space-Variant DTCNN Model for FPGA Implementation

The DTCNN model used in this work has been derived from the original analogue model of CNN [18, 19] by means of the Euler method [20]. In comparison with other approaches [21], the Euler method offers an excellent behavior with minimum use of hardware resources.

With the Euler-based approach, the DTCNN model is defined through the recursive equations 1 and 2:

$$X_{ij}[n] = \sum_{k,l \in Nr(ij)} A_{kl}^{ij}[n-1]Y_{kl}[n-1] \; + \sum_{k,l \in Nr(ij)} B_{kl}^{ij}[n-1]U_{kl} \; + \; I_{ij} \;, \tag{1}$$

$$Y_{ij}[n] = \frac{1}{2}\left(|X_{ij}[n]+1| - |X_{ij}[n]-1|\right) \tag{2}$$

where I, U, Y and X denote input bias, input data, output data and state variable of each cell, respectively. $Nr(ij)$ gives the neighborhood distance r for cell (i,j), with i and j the coordinates of the position of the cell in the network, and with k and l the position of the neighboring cells relative to the cell in consideration. A and B are the templates for the outputs of the neighboring cells and the input, respectively. Both A and B are k×l non-linear and space-variant templates, i.e., they can change over time and be also different for each cell.

Eqs. (1) and (2) imply an infinite feedback loop in the model which must be approximated. Clearly, the accuracy of the approximation depends on the number of iterations to be considered. Simulations of typical video processing applications (Gaussian, blur, sharpness and edge detections) have revealed that 10 iterations are wide enough to achieve results that present correlation coefficients with the results obtained from the original analogue model around 9.997e−1. The number of iterations N is very closely related to the application and to the accuracy required. Thus, a DTCNN cell can be made up with N stages in pipeline, as shown in Fig. 4. The figure also shows the graphical model of each stage of the cell.

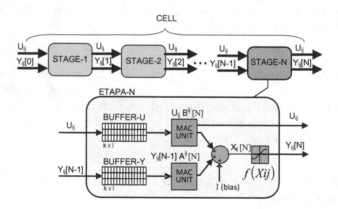

Fig. 4. The cell consists of N identical pipelined stages

The stage has two input ports: the input U_{ij} to the cell, which is weighted by template B, and the output $Y_{ij}[n-1]$ from previous stages, weighted by template

A. In each stage, the inputs can be adjusted by different templates $A^{ij}[n]$ and $B^{ij}[n]$. This allows to design powerful non-linear CNNs.

The stage has two output ports: the output data $Y_{ij}[n]$ and the input data U_{ij}. The presence of the input data in an output port makes it possible the pipelining of stages all working with the same input data. In order to ensure the data synchronization, input data is delayed as many cycles as the stage latency requires.

With this port structure, three different basic configurations can be built according to the values of the templates. Thus, the output will depend on both the data input and the neighbors if A and B are non-null; it will depend only on the input if A is null and B non-null; and the output will depend only on the neighbors if A is non-null and B is null. This flexibility together with the stage pipelining approach allow for the design of complex cells with non-linear and space-variant behavior. With the suitable connections and values for the templates, it is possible to design networks where the radius of influence of the input on the output is far more than the $k \times l$ template. It is also possible to design networks with more than one layer, each layer with its own configuration.

5 Hardware Implementation of a DTCNN Model for Hippocampal Cell Cultures

As can be observed in Fig. 4, a stage is made up of two convolution modules, a ternary adder and a linear activation function with symmetrical saturation. Both the convolution of the input U_{ij} with the template B and the convolution of the output $Y_{ij}[n-1]$ with the template A have been implemented using Multiply-And-Accumulate (MAC) structures. The number of MAC units for a given application would come from the tradeoff between circuit area and performance. MACs are common blocks in signal processing algorithms, so FPGAs come with dedicated resources to implement thousands of these blocks efficiently. The FIFOs manage the data streams between stages and have been designed to ensure that all the data required for the convolutions are available at the right moment. With this solution the system can work on data streaming, and so the problem of the data storage between stages is overcome. As with MACs, FIFOs have been implemented carefully to make efficient use of the internal BlockRAM resources available in the Xilinx FPGAs.

In order to implement the model depicted from the analysis in Section 2.3, that uses a 8×8 MEA, a DTCNN with a 8×8 bidimensional array of cells will be required. For such an small experiment, the connectivity maps in Figure 3 indicate that full connectivity between cells must be provided. Thus, in this case, templates $A[n]$ and $B[n]$ would be 16×16 matrices. Further experiments, using larger MEAs are planned, in which case the size of templates should be adjusted to meet the observed connectivity neighborhood.

Template $A[n]$, linked to the outputs of the neighboring cells, will then represent the functional connectivity map for cell n resulted after tetanic stimulation was applied as training method. On the other hand, template $B[n]$, coupled to the cell's inputs, represents the influence of the input (external) stimuli. Individual values weigh the influence of cell/input ij on current cell and will be normalized between 0 and 1.

6   Conclusions

Learning in biological neural cultures is a challenging task. Different authors have proposed different methods for inducing a desired and controlled plasticity over the biological neural structure. Low-frequency stimulation and tetanization has brought good results to researchers enhancing bursting activity in cortical cultures. In previous papers, we have shown that using these kind of stimulations it is possible to create adjacent physical or logical connections in the connectivity graph following Hebbs Law and such connections induce changes in the electrophysiological response of the cells in the culture, which can be observed in the different analysis performed. Furthermore, low-frequency stimulation induces permanent changes in most experiments using different values of current amplitude and stimulation patterns. Persistent and synchronous stimulation of relevant adjacent electrodes may be used for strengthen the efficiency of their connectivity graph. These processes may be used for imposing a desired behaviour over the network dynamics. In this work, we translate the biological neural network behavior and structure to a DTCNN over a FPGA device. The modularity of the solution and the reconfiguration capability of the device permits to adapt the system to other dimensions and functionalities. In future works, the working behavior of the translated system will be compared with the biological one in order to analyze the principal components and critical elements that constitute the whole system.

Acknowledgements. This work has been partially supported by the Fundación Séneca de la Región de Murcia through the research project 15419/PI/10.

References

1. Anderson, J.A., Rosenfeld, E.: Neurocomputing: Foundations of research. MIT Press, Cambridge (1988)
2. The Facets Project, http://facets.kip.uni-heidelberg.de/ (accessed March 2013)
3. Maeda, E., Kuroda, Y., Robinson, H.P., Kawana, A.: Modification of parallel activity elicited by propagating bursts in developing networks of rat cortical neurones. Eur. J. Neurosci. 10(2), 488–496 (1998)
4. Jimbo, Y., Robinson, H.P., Kawana, A.: Strengthening of synchronized activity by tetanic stimulation in cortical cultures: application of planar electrode arrays. IEEE Trans. Biomed. Eng. 45(11), 1297–1304 (1998)
5. Jimbo, Y., Tateno, T., Robinson, H.P.C.: Simultaneous induction of pathway specific potetiation and depression in networks of cortical neurons. Biophys. J. 76(2), 670–678 (1999)
6. Tateno, T., Jimbo, Y.: Activity-dependent enhancement in the reliability of correlated spike timings in cultured cortical neurons. Biol. Cybern. 80(1), 45–55 (1999)

7. Shahaf, G., Marom, S.: Learning in networks of cortical neurons. J. Neurosci. 21(22), 8782–8788 (2001)
8. Ruaro, M.E., Bonifazi, P., Torre, V.: Toward the neurocomputer: image processing and pattern recognition with neuronal cultures. IEEE Trans. Biomed. Eng. 52(3), 371–383 (2005)
9. Wagenaar, D.A., Pine, J., Potter, S.M.: Searching for plasticity in dissociated cortical cultures on multi-electrode arrays. Journal of Negative Results in Biomedicine 5(16) (2006)
10. Madhavan, R., Chao, Z.C., Potter, S.M.: Plasticity of recurring spatiotemporal activity patterns in cortical networks. Phys. Biol. 4(3), 181–193 (2007)
11. Chao, Z.C., Bakkum, D.J., Potter, S.M.: Region-specific network plasticity in simulated and living cortical networks: comparison of the center of activity trajectory (CAT) with other statistics. J. Neural. Eng. 4(3), 294–308 (2007)
12. Ide, A.N., Andruska, A., Boehler, M., Wheeler, B.C., Brewer, G.J.: Chronic network stimulation enhances evoked action potentials. J. Neural. Eng. 7(1), 16008 (2010)
13. Bologna, L.L., Nieus, T., Tedesco, M., Chiappalone, M., Benfenati, F., Martinoia, S.: Low-frequency stimulation enhances burst activity in cortical cultures during development. Neuroscience 165, 692–704 (2010)
14. Lorente, V., Ferrández, J.M., de la Paz, F., Fernández, E.: Training hippocampal cultures using low-frequency stimulation: Towards Hebbian Learning. In: 8th International Meeting on Substrate-Integrated Microelectrodes Arrays, pp. 90–91 (2012)
15. Ferrández, J.M., Lorente, V., de la Paz, F., Fernández, E.: Training Biological Neural Cultures: Towards Hebbian Learning. Neurocomputing 114, 3–8 (2013)
16. Sporns, O., Tononi, G.: Classes of network connetivity and dynamics. Complexity 7, 28–38 (2002)
17. Friston, K., Frith, C., Frackowiak, R.: Time-dependent changes in effective connectivity measured with PET. Human Brain Mapping 1, 69–79 (1993)
18. Chua, L.O., Yang, L.: Cellular Neural Networks: Theory. IEEE Trans. on Circuits and Systems CAS-35(10), 1257–1272 (1988)
19. Chua, L.O.: The CNN: A brain-like computer Neural Networks. In: IEEE Intern. Joint Conference, vol. 1, pp. 25–29 (2004)
20. Martínez-Álvarez, J.J., Garrigós-Guerrero, F.J., Toledo-Moreo, F.J., Ferrández-Vicente, J.M.: Using reconfigurable supercomputers and C-to-hardware synthesis for CNN emulation. In: Mira, J., Ferrández, J.M., Álvarez, J.R., de la Paz, F., Toledo, F.J. (eds.) IWINAC 2009, Part II. LNCS, vol. 5602, pp. 244–253. Springer, Heidelberg (2009)
21. Martínez, J.J., Garrigós, F.J., Villó, I., Toledo, F.J., Ferrández, J.M.: Hardware Acceleration on HPRC of a CNN-based Algorithm for Astronomical Images Reduction. In: Int. Work. on Cellular Nanoscale Networks ans their Applications, pp. 1–5 (2010)
22. Ferrández, J.M., Bolea, J.A., Ammermüller, J., Normann, R.A., Fernández, E.: A neural network approach for the analysis of multineural recordings in retinal ganglion cells. In: Mira, J. (ed.) IWANN 1999. LNCS, vol. 1607, pp. 289–298. Springer, Heidelberg (1999)
23. Toledo, F.J., Martínez, J.J., Garrigós, F.J., Ferrández, J.M.: FPGA implementation of an augmented reality application for visually impaired people. In: Int. Conf. on Field Programmable Logic and Applications, pp. 723–724 (2005)

Parkinson's Disease Monitoring
from Phonation Biomechanics

P. Gómez-Vilda[1], M.C. Vicente-Torcal[2], J.M. Ferrández-Vicente[3(✉)],
A. Álvarez-Marquina[1], V. Rodellar-Biarge[1], V. Nieto-Lluis[1],
and R. Martínez-Olalla[1]

[1] Neuromorphic Speech Processing Lab, Center for Biomedical Technology,
Universidad Politécnica de Madrid, Campus de Montegancedo,
28223 Pozuelo de Alarcón, Madrid, Spain
[2] Speech Therapist, Hospital Ruber Internacional, Madrid, Spain
[3] Universidad Politécnica de Cartagena, Campus Universitario Muralla del Mar,
Pza. Hospital 1, 30202 Cartagena, Spain
jm.ferrandez@upct.es
pedro@fi.upm.es

Abstract. Organic as well as neurologic diseases leave important correlates in phonation. Parkinson's Disease (PD) may leave marks in vocal
fold dystonia and tremor. Biomechanical parameters monitoring vocal
fold tension and unbalance, as well as tremor are defined in the study.
These correlates are known to be of help in tracing the neuromotor activity of both laryngeal and articulatory pathways. As the population
affected by PD is mainly above 60, the main problem found is how to
differentiate PD phonation correlates from aging voice (presbyphonia).
An important objective is to explore which correlates react differentially
to PD than to aging voice. As an example a study is conducted on a
set of male PD patients being monitored in short intervals by recording
their phonation. The results of these longitudinal studies are presented
and discussed.

Keywords: Neurologic disease · Parkinson's Disease · Speech neuromotor activity · Aging voice · Dysarthria

1 Introduction

Parkinson's Disease is an illness derived from deterioration of substantia nigra
in midbrain, mainly with age, which affects a larger population each year. It is
estimated that *substantia nigra* cell decay responsible for Parkinson's Disease
(PD) is about 5% per decade in a normal subject, and ten times larger in a
PD patient [1, 2]. The prevalence of PD is less than 0.4% among the population
under 40 years whereas it is around 1.5% in the population over 65. The population affected by PD will double in 2030 with respect to 2005 in the average.
PD affects voice and speech even at an early stage, when other symptoms are
not yet evident [3, 4]. As illness progresses the impairment of the patient's self-
care capabilities and well-being worsens, demanding more specialized assistance.

© Springer International Publishing Switzerland 2015
J.M. Ferrández Vicente et al. (Eds.): IWINAC 2015, Part I, LNCS 9107, pp. 238–248, 2015.
DOI: 10.1007/978-3-319-18914-7_25

PD affects mainly to neuromotor activity involving most biomechanical systems in the body, among these limbs, head and neck muscles. Malfunction of these systems affect walking, arm and hand fine movements, body balance, and speech and phonation [5]. Therefore gait, handwriting and drawing, as well as speech features have been routinely used to detect, grade and monitor PD by clinicians [6, 7]. In PD patient clinical evaluation a commonly used scale is that of Hoehn and Yahr (H&Y) [8], which classifies patients in five stages (1-5) according to unilateral or bilateral involvement, balance, reflexes, equilibrium, walk or stand unassisted, or handling an independent life activity. The purpose of these evaluations is to monitor illness evolution, but they are time-consuming and costly, and affected by the subjectivity with which evaluators and patients may quantify a specific feature. On the other hand, it is believed that PD effects on speech and phonation are good descriptors of general neuromotor deterioration, in the sense that there is "compelling evidence to suggest that speech can help quantify not only motor symptoms ... but generalized diverse symptoms in PD" [6]. Speech could be used as a handy signal to monitor PD evolution subsequent to chirurgical, pharmacological or rehabilitative treatment [9]. During the last decade, important advances in PD evaluation and monitoring have been produced [10]. These are mainly based in estimates of phonation quality, prosody or fluency. Phonation correlates are based in distortion measures as jitter, shimmer, harmonic-noise-ratios (HNR), dysphonia severity index (DSI) or mel-frequency cepstral coefficients (MFCC's), among others [6]. Prosody marks are based on the temporal evolution of fundamental frequency (f_0) and envelope energy (E_e), which may be parameterized and quantified using nonlinear techniques [11]. Fluency correlates include vocal-onset-time (VOT), pause and silence intervals, syllable rate, etc [12]. Massive pattern-matching and statistical machine learning techniques have been used in PD grading reproducing subjective evaluations in the Unified PD Rating Scale (UPDRS) [6, 13, 14]. The problems in using these massive data-driven methodologies are on one hand their relatively large computational cost, and on the other hand, the loss of semantics due to rather obscure parameterization methods (as is the case of MFCC's, for instance). A loss of semantics may imply a too large payoff severing further advancement in the generation of new hypotheses and experiments to carry on further research. The present approach is intended to advance in the study of PD monitoring keeping the semantics of the problem alive. For such, a set of biomechanical parameters derived from phonation have been proposed which are known to express their capability to monitor PD evolution in early work, preserving parameter semantics [15, 16]. These are related with vocal fold mechanical stress statistical and stochastic behaviour: average, dispersion, skew, kurtosis and cyclical variability (tremor). Tremor is a specific mark which may be perceived in around 60% of the PD cases. It is known to be distributed in bands around 2-4 Hz (physiological), 6-10 (neurological) or above 10 Hz (sometimes addressed as jitter or flutter). Empirical mode decomposition has been proposed for its estimation [17]. The present study is intended to explore the capability of vocal fold stress unbalance and tremor in monitoring the timely evolution of PD dysphonia in longitudinal studies. A brief description of how PD affects the biomechanics of phonation and a method for tremor estimation from vocal fold stress are provided in Section 2. Section 3 presents the methodology of the study. Results are shown and discussed in Section 4. Conclusions are presented in Section 5.

2   Phonation Model and Tremor Characterization

Speech production is planned and instantiated in the linguistic neuromotor cortex (Broadmann's areas 4, 6, 8, and 44-47 [18]). The neuromotor speech sequence activates the muscles of the pharynx, tongue, larynx, chest and diaphragm through sub-thalamic secondary pathways. Fine muscular control is provided by a neuro-sensory feedback regulation system, in which *substantia nigra* is involved. Degeneration of this substance reduces the production of the specific neurotransmitters responsible of the regulatory function and this eventually results in the appearing of the PD syndrome characterized by perturbations in the respiration, phonation and articulation giving place to specific dysphonias and dysarthrias which may characterize PD speech: hoarseness, breathiness, hypokinetic dysarthria, tremor, raised f0, jitter, low intensity (bradykinesia and rigidity), poor prosody (monotonous speech), poor VOT (especially when switching nasal to oral sounds), and deficient fluency (low syllable rate, longer inter-syllable pauses, etc). Tremor in voice is one of the symptoms associated to PD, although it may not be perceived in all cases. The reason is that either its intensity may be very low, or that it appears in bands where it may not be perceived by simple listening. Classically tremor in voice is separated into three bands (physiological, neurological and flutter). If tremor is predominant over 10 Hz (flutter) it will go unnoticed except for a trained expert. It is believed that tremor as a correlate to PD appears first in these higher bands (flutter, jitter) to intensify in amplitude and move to lower bands (neurological) as illness progresses. That is why automatic detection of tremor becomes essential in PD studies. Tremor in phonation may be characterized in different ways. The usual method is to derive it from sustained long vowels. Frequency Modulation techniques and Empirical Model Decomposition have been successfully used [19, 17]. In the present work a different approach has been proposed. It is based on the detection of vocal fold body mechanical stress, as this magnitude is directly related with neuromotor activity of laryngeal muscles and preserve the semantic value of the estimates. As tremor in voice may be the consequence of neuromotor instability in the respiration, phonation or articulation, this approach presents the advantage with respect to the others mentioned of ensuring that only phonation will be expressed in the results. For such, it is essential to obtain precise estimates of the biomechanical stress acting on *musculus vocalis*. The methods used in the estimation of this correlate are vocal tract inversion by a lattice adaptive filter [20], and biomechanical inversion of a 2-mass model of the vocal folds [21]. As a result, an estimate of the vocal fold body mechanical stiffness ξ_n is produced for each phonation cycle n. Afterwards, this parameter is modeled by an order-K autoregressive system

$$\xi_n = \sum_{i=1}^{K} a_i \xi_{n-i} + \epsilon_n \qquad (1)$$

where $\mathbf{a} = \{a_i\}$ are the model parameters and ϵ_n is the estimation error. This modeling is carried out by another adaptive lattice inverse filter. The coefficients of the equivalent transversal model $\mathbf{a}_{Kn}$ and the filter pivoting coefficients $\mathbf{c} = \{c_i\}$ may be used as tremor descriptors (see [15] for more details). Tremor is expressed in terms of frequency, relevance and amplitude from the inverse model in the domain of $z = e^{j\omega}$

$$H(z) = \frac{1}{1 - \sum_{1}^{K} a_i z^{-i}} = \prod_{i=1}^{K} \frac{z}{z - z_i}; \quad z_i = r_i e^{j\varphi_i} \qquad (2)$$

where $\mathbf{z} = \{z_i\}$ are the poles of the transfer function $H(z)$, with modulus and phase given by r_i and φ_i. These are used to estimate tremor frequency f_{ti} and relevance ρ_{ti} as

$$f_{ti} = \frac{\varphi_i}{2\pi} \langle f_0 \rangle; \quad \rho_{ti} = \frac{1}{1 - r_i} \qquad (3)$$

$\langle f_0 \rangle$ being the average phonation fundamental frequency. The relevance (relative amplitude) of the estimate will be given by the modulus of the pole (r_i). Another important parameter is the root mean square tremor amplitude (rMSA), given by

$$\eta_t = \frac{\frac{1}{N_k} \sum_{n \in W_k} \left[\xi_{Kn} - \bar{\xi}_K \right]^2}{\bar{\xi}_K^2} \qquad (4)$$

where N_k is the number of phonation cycles in the estimation window W_k. An example of these parameter estimates from a sustained vowel /a/ uttered by a PD patient (72 year old female) is shown in Fig. 1 modeled with a $K = 3$ order filter.

3   Materials and Methods

The present study has a marked exploratory nature, as 3-band tremor parameters estimated from the vocal fold body stiffness have not been used before in PD detection, grading or monitoring. The intention of the study is to show their performance in monitoring longitudinal studies of PD patients. Therefore a database of recordings from a set of 50 male and 50 female normative subjects free from organic or neurologic pathology selected by the ENT services of Hospital Gregorio Marañón of Madrid has been used. Long sustained vowels (/a, e, i, o, u/), a short sentence, and a text reading were recorded at a 44,000 Hz sampling frequency and 16 bits from each subject. Fragments of 500 ms long

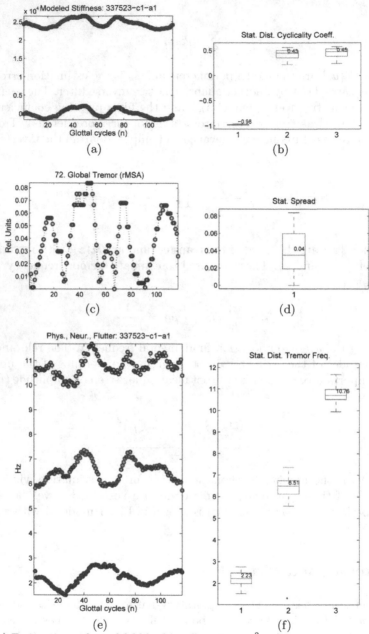

Fig. 1. a) Estimations of vocal fold body stiffness in 10^{-3} N/m per each glottal cycle n, corresponding to a fragment of vowel /a/ 500 ms long. The upper trace is the absolute value of the estimate ξ_n, the lower trace is the same estimate after de-biasing: $\xi_n - \langle \xi \rangle$. b) Values and distribution boxplots of the cyclical coefficients c_{1-3}. c) Global tremor root mean square amplitude η_t. d) Distribution boxplot of η_t. e) Estimations of the frequency components of tremor in physiological, neurological and flutter bands. f) Values and distribution boxplots of frequency components, showing well-conditioned distributions (small skew, no outliers).

Table 1. Parameters used in the study

P1. Fundamental frequency f0	P44. Cover mass unbalance
P2. Jitter	P46. Cover stiffness unbalance
P3. Shimmer	P60. Glottal gap due to contact defect
P5. Noise-harmonic ratio	P61. Adduction defect
P35. Body mass (dynamic)	P62. Closure permanent defect
P37. Body stiffness (ξ_n)	P63. First order pivoting coefficient (c_1)
P38. Body mass unbalance	P67. Physiological tremor amplitude
P40. Body stiffness unbalance	P69. Neurological tremor amplitude
P41. Cover mass	P71. Flutter amplitude
P43. Cover stiffness	P72. rMSA

of /a/ recordings were analyzed and the 20 means and std. deviations of the parameters in Table 1 were used in the study.

The estimates produced from the male set were used as a normative dataset to monitor four male PD patients from the Neurology Services of Hospital Ruber Internacional of Madrid. The description of the cases is given in Table 2. These patients were under pharmacologic treatment and speech therapy, which included voice recordings identical to the ones mentioned for the normative subjects, taken at their homes. Each patient's stage was scored according to the H&Y scale.

Table 2. Summary results for the four cases studied

Case	Gender	Age	Diagnose	Grade (S)	Syndrome
S1	M	70	PD	2 (H&Y)	Bilateral, mild walk and stability affection
S2	M	54	PD	2 (H&Y)	Bilateral, mild walk and stability affection
S3	M	74	PD	4 (H&Y)	Limited walk, rigidity, bradykinesia, dependent
S4	M	69	PD	1 (H&Y)	Unilateral, rest tremor, mild walk and gait affection

A simple way to present estimation results vs the normative database is given in Fig. 2. Each parameter mean i from subject j and phonation cycle n given as P_{ijn} is estimated on n as $P_{aij} = ave_n\{P_{ijn}\}$. This mean is compared to the sample population mean for the same parameter $P_{ai} = ave_k\{P_{ikn}\}$ for k the speakers in the normative sample to produce a normalized estimate $P_{rij} = P_{aij}/P_{ai}$. Each normalized estimate P_{rij} corresponding to parameter i from subject j is represented on the sundial plot.

The normalized sample population averages correspond to the unity circle (in melba). The dispersion values corresponding to 1.5 times the standard deviation $P_{si} = std_k\{P_{ikn}\}$ over the mean of each parameter are represented as the melba

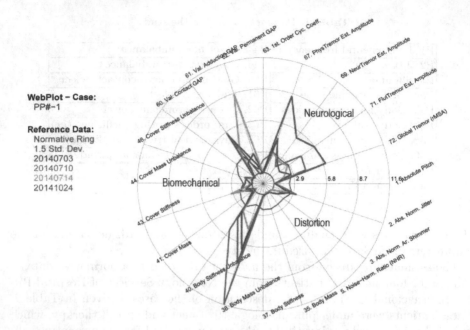

Fig. 2. Comparison of evaluation results for case S3. Red: 3.7.14. Yellow: 7.7.14. Green: 14.7.14. Blue: 24.10.14. Values are normalized to the normative male database. The upper right quadrant is associated to tremor parameters (P63, 67, 69, 71 and 72). The lower right quadrant is associated to distortion parameters (P1, 2, 3 and 5). The upper and lower left quadrants are associated to biomechanical parameters (P35, 37, 38, 40, 41 43, 44, 46, 60, 61 and 62). The inner melba circle marks the unity normative circle. The outer melba polygon marks the 1.5 standard deviation boundary over the mean.

outer polygon. This polygon ensures that at least 93% of the normative population is under this limit for a given parameter. This representation devotes the first quadrant to neurological sensitive parameters (tremor), the second quadrant to distortion parameters (jitter, shimmer, etc.) and the third and fourth quadrants to biomechanical parameters. Case S3 as depicted in Fig. 2 is a representative example. It may be seen that during the first evaluation this patient showed a strong alteration in the tremor indices (P67, 69, 71 y 72) as signaled by the red polygon pointing to P67. During the next three evaluations (yellow, green and blue) these indices became sensibly more reduced in response to treatment (although not normalizing completely). On the other side, the biomechanical unbalance parameters (P38 and 40) remained sensibly out of limits during the four evaluations (feather pointing to P38). Adduction defects (P60 and 61) were out of normative limits through all evaluations. This observation could be interpreted in the sense that unbalance and adduction defects (both biomechanical) do not seem to be sensitive to specific neurological treatment. Their deviation from normative values may be pointing to aging voice distortion (presbyphonia) or to other organic dysphonias. This is a common situation in the other PD cases studied (see specific results in Table 3).

Table 3. Estimation results for the most relevant parameters (%). Relative grade agreement is signaled in bold.

Case	Date D.M.Y	P2	P38	P40	P61	P67	P69	P71	P72	Grade(S)	Grade (O)
S1	26.6.14	1.61	0.84	9.86	2.41	3.28	13.59	3.95	9.58	2	4,19
	3.7.14	3.72	8.66	15.82	15.86	1.19	1.89	1.01	3.95	2	**2,15**
	14.7.14	3.10	9.41	15.69	19.47	2.24	0.58	1.61	3.22	2	**2,02**
	24.7.14	3.80	9.87	17.22	13.57	2.62	1.69	1.63	4.30	2	**2,39**
S2	2.7.14	1.33	0.78	3.41	2.74	0.77	0.43	0.62	1.54	2	0,66
	8.7.14	3.29	11.18	17.37	11.92	2.69	3.31	6.66	8.02	2	3,85
	10.7.14	1.69	3.20	6.53	9.60	1.88	2.60	2.50	3.76	2	**2,00**
	14.7.14	1.42	0.91	3.76	13.35	1.08	0.80	0.63	1.42	2	0,96
S3	3.7.14	2.63	3.62	8.81	6.53	0.78	0.40	0.46	2.05	4	1,00
	10.7.14	3.45	16.17	22.83	14.01	2.67	1.79	1.39	11.04	4	**3,84**
	14.7.14	1.83	1.65	5.38	8.59	1.77	0.65	0.80	2.10	4	1,10
	24.7.14	2.61	4.86	9.89	12.03	0.72	0.50	0.84	3.89	4	1,53
S4	8.1.15	1.08	1.01	3.08	21.05	1.62	0.62	0.46	2.95	1	**1,38**
	21.1.15	0.06	0.19	1.25	15.21	0.20	0.27	0.50	1.18	1	**0,67**
	28.1.15	1.87	1.76	5.54	9.01	0.41	0.26	0.28	1.37	1	**0,73**
	4.2.15	1.16	0.76	3.09	37.45	1.09	0.42	0.22	2.07	1	**1,43**

The confirmation of this hypothesis is being evaluated in other ongoing studies. These plots will be used to present information to clinicians and speech therapists for fast patient evaluation.

4 Results and Discussion

An important issue in monitoring pathology is that of grading, as short-term timely monitoring of PD may be highly relevant for patient treatment and rehabilitation [9]. One of the intentions of the study was to relate speech correlates (objective grading) with this evaluation scale (subjective grading), in timely evolution. For such, a metrics had to be defined based on feature fusion, accordingly to a weighting functional as:

$$r_o = f \left\{ \sum_{i \in S_N} w_i P_i \right\} \tag{5}$$

where P_i are the feature estimates considered in the grading, w_i are the weights of the fusion, f is the function (usually a sigmoid mapping from 0 to 5) and S_N is the normative set of features considered (in this case the parameters listed in Table 3). All patients were evaluated four times in a 3-4 week period except one of them who was evaluated during a shorter interval (S2). The distortion, unbalance and tremor estimates from each phonation produced by each patient at each evaluation date are compared. Patient S1 had a first evaluation (26.6.14) showing moderate unbalances (P38 and P40) and a strong neurological tremor (P69). Subsequent evolution showed large unbalances, but tremor experienced a strong reduction to more moderate values. The objective grade was very large at the first evaluation. In subsequent evaluations it showed good agreement with subjective grade. Patient S2 suffered four evaluations within two weeks. The second one (8.7.14) showed a very much dysphonic condition with large unbalances (P38 and P40) and a high neurological tremor (P.69). The total rMSA was also very large. The next two evaluations (10.7.14 and 14.7.14) showed a significant reduction in biomechanical unbalance (P38 and 40) and in all tremor bands. The agreement between subjective and objective grades was not stable, as only one of the evaluations agreed (10.7.14). Patient S3 was evaluated four times within four weeks. The results of the first evaluation were relatively normal (3.7.14). One week later the evaluation (10.7.14) showed a more complicate condition with large unbalances (P38 and 40) and relatively large tremor, both the objective and subjective grades being in agreement. The next two evaluations (14.7.14 and 24.7.14) showed a relative reduction in unbalance, and a larger reduction in tremor, therefore the objective grade was well below the subjective. Finally patient S4 showed good agreement between the objective and the subjective grade, as the neurological parameters (P67, 69, 71 and 72) gave moderate estimates, whereas the adduction defect showed to be quite large in most of the cases (P61). This parameter, as well as the biomechanical ones, seem to be more related with organic pathologic or presbyphonic voicing than with neurologic alterations, therefore the relative influence of these parameters should be minimized in grading neurological disease.

5   Conclusions

The timely monitoring of PD using voice is a strong challenge. Through the present study a first approach to this problem under a systematic methodology has been presented. Although the limited nature of the number of cases presented, the most important findings established in their timely evaluation are the following:

- An association of feature estimates under a given functional may help in producing an objective estimate of the patient's pathological severity, although this study is being conducted on a larger database.
- It seems that there may be sudden surges of instability and tremor (critical epochs) in the evaluation results from relatively short intervals in the two cases graded 2 and in the case graded 4.
- In general, the objective and the subjective grades agree better during the epochs than during the steady evaluations.
- The critical epochs in tremor and phonation stability do not seem to persist in consequent evaluations, possibly due to the effects of medication and rehabilitation.
- The sets of features monitoring biomechanics and tremor evolve differentially with treatment, possibly pointing to different causes (neurologic vs organic etiology).

An important part of the study which has to be resolved prior to continue further research is the differentiation in the sensitivity and specificity of monitoring response associated to the biomechanical and tremor parameters, as these seem to be influenced differently by organic than by neurologic pathologies. Another open question is the best separation interval between subsequent evaluations, either daily, weekly, or monthly, etc., and the adequate recording of medical and rehabilitation protocols to better associate drastic changes in feature patterns with treatment. Doubtless, these studies will help in the personalized care of patients and in the optimization of treatment protocols.

Acknowledgements. This work is being funded by grants TEC2012-38630-C04-01 and TEC2012-38630-C04-04 from Plan Nacional de I+D+i, Ministry of Economic Affairs and Competitiveness of Spain. Special thanks are due to Dr. M. Kurtis Urra, Head Neurologist of the Movement Disorders Services of Hospital Ruber Internacional of Madrid.

References

1. De Lau, L.M.L., Breteler, M.M.B.: Epidemiology of Parkinson. Lancet Neurol. 5, 525–535 (2006)
2. Dorsey, E.R., et al.: Projected number of people with Parkinson disease in the most populous nations, 2005 through 2030. Neurology 68, 384–386 (2007)
3. Falk, T., Chan, W., Shein, F.: Characterization of atypical vocal source excitation, temporal dynamics and prosody for objective measurement of dysarthric word intelligibility. Speech Communication 54(5), 622–631 (2012)
4. Yunusova, Y., Weismer, G., Westbury, J., Lindstrom, M.: Articulatory movements during vowels in speakers with dysarthria. Journal of Speech, Language, and Hearing Research 51, 596–611 (2008)
5. Chenausky, K., MacAuslan, J., Goldhor, R.: Acoustic Analysis of PD Speech. Parkinson's Disease (2011), doi:10.4061/2011/435232
6. Tsanas, A.: Accurate telemonitoring of Parkinson's disease symptom severity using nonlinear speech signal processing and statistical machine leaning. PhD. Thesis, U. of Oxford, U.K. (June 2012)
7. Tsanas, A., Little, M.A., McSharry, P.E., Spielman, J., Ramig, L.O.: Novel speech signal processing algorithms for high-accuracy classification of Parkinson's disease. IEEE Transactions on Biomedical Engineering 59, 1264–1271 (2010)
8. Hoehn, M.M., Yahr, M.D.: Parkinsonism: onset, progression, and mortality. Neurology 17(5), 427–442 (1967)
9. Gamboa, J., Jiménez, F.J., Nieto, A., Montojo, J., Ortí, M., Molina, J.A., et al.: Acoustic Voice Analysis in Patients with Parkinson's Disease Treated with Dopaminergic Drugs. J. Voice 11, 314–320 (1997)
10. Little, M.A., McSharry, P.E., Hunter, E.J., Spielman, J., Ramig, L.O.: Suitability of dysphonia measurements for telemonitoring of Parkinson's disease. IEEE Trans. on Biomed. Eng. 56(4), 1015–1022 (2009)
11. López-de-Ipiña, K., et al.: On the Selection of Non-Invasive Methods Based on Speech Analysis Oriented to Automatic Alzheimer Disease Diagnosis. Sensors 13, 6730–6745 (2013)

12. Sapir, S., Sprecher, E., Skodda, S.: Early motor signs of Parkinson's Disease detected by acoustic speech analysis and classification methods. In: Manfredi, C. (ed.) Proc. MAVEBA 2013, pp. 3–5. Florence University Press (2013)
13. Tsanas, A., Little, M.A., McSharry, P.E., Scanlon, B.K., Papapetropoulos, S.: Statistical analysis and mapping of the Unified Parkinson's Disease Rating Scale to Hoehn and Yahr staging. Parkinsonism & Related Disorders 18(5), 697–699 (2012)
14. Little, M., Wicks, P., Vaughan, T., Pentland, A.: Quantifying Short-Term Dynamics of Parkinson's Disease Using Self-Reported Symptom Data From an Internet Social Network. J. Med. Internet Res. 15(1), e20 (2013)
15. Gómez, P., et al.: Estimating Tremor in Vocal Fold Biomechanics for Neurological Disease Characterization. In: Prof. of the 18th Int. Conf. on Digital Signal Processing (DSP 2013), Santorini, Greece, M1C-2 (June 2013)
16. Gómez, P., et al.: Characterizing Neurological Disease from Voice Quality Biomechanical Analysis. Cognitive Computing 5, 399–425 (2013)
17. Mertens, C., Schoentgen, J., Grenez, F., Skodda, S.: Acoustical Analysis of Vocal Fold Tremor in Parkinson Speakers. In: Manfredi, C. (ed.) Proc. of MAVEBA 2013, pp. 19–22. Florence University Press (2013)
18. Demonet, J.F., Thierry, G., Cardebat, D.: Renewal of the Neurophysiology of Language: Functional Neuroimaging. Physiol. Rev. 85, 49–95 (2005)
19. Pantazis, Y., Koutsogiannaki, M., Stylianou, Y.: A novel method for the extraction of tremor. In: Manfredi, C. (ed.) Proc. of MAVEBA 2007, pp. 107–110. Florence University Press (2007)
20. Deller, J.R., Proakis, J.G., Hansen, J.H.L.: Discrete-Time Processing of Speech Signals. Macmillan, NewYork (1993)
21. Gómez, P., et al.: Glottal Source Biometrical Signature for Voice Pathology Detection. Speech Communication 51, 759–781 (2009)

Retinal DOG Filters:
Effects of the Discretization Process

Adrián Arias[1], Eduardo Sánchez[1(✉)], and Luis Martínez[2]

[1] Grupo de Sistemas Inteligentes (GSI)
Centro Singular de Investigación en Tecnologías de la Información (CITIUS)
Universidad de Santiago de Compostela, 15782,
Santiago de Compostela, Spain
{adrian.arias.abreu,eduardo.sanchez.vila}@usc.es
[2] Instituto de Neurociencias de Alicante
CSIC-Universidad Miguel Hernández, 03550
Alicante, Spain
l.martinez@umh.es

Abstract. This paper aims at analyzing the effects of the discretization process of the continuous Difference-of-Gaussians models obtained empirically by Enroth-Cugell and Robson (1966). The filter properties of the Discrete DoG kernels were analyzed in the frequency domain and their effects on input images were characterized by means of GLCM descriptors. The results demonstrate that the DoG Kernels behaviour range between true High-Frequency Enhancing filters and Band-pass filters depending on the discretization parameters. Moreover, the analysis of filtered images suggest that those kernels that enhance contrast come at a cost of higher entropy as well as lower spatial correlation.

Keywords: Retina · Difference of Gaussians · Discrete kernels · Contrast · Correlation · Entropy

1 Introduction

Kuffler found that retinal ganglion cells (RGCs) possesed Receptive Fields (RFs) arranged into a center-surround organization. They were excited when a bright stimulus was applied to the center region; and, conversely, were inhibited when the same stimulus was applied to its surround [1]. This center-surround receptive fields (RFs) were later mathematically described by Rodieck (1965) as the sum of two Gaussian functions: a positive one, representing the center, and a wider negative one representing the surround. This model was called the Difference-of-Gaussians (DoG) model and has been widely used as a model to represent the RFs of RGCs ever since. The parameters of the DoG model applied to RGCs were first estimated by Enroth-Cugell and Robson (1966, 1984). They fitted the model against contrast sensitivity curves that were obtained from recorded responses of ganglion cells. For each recorded RGC, the parameters of the DoG model, the radius and maximum amplitudes of both the center and surround Gaussians, were obtained.

© Springer International Publishing Switzerland 2015
J.M. Ferrández Vicente et al. (Eds.): IWINAC 2015, Part I, LNCS 9107, pp. 249–257, 2015.
DOI: 10.1007/978-3-319-18914-7_26

Later on, the responses of mammalian RGCs were found to depend on the difference between the stimulus luminance applied to the center of their receptive fields (RF) relative to the luminance applied to its surround, as well as the mean levels of retinal illumination [5,6]. As the ratio between these parameters allows to compute local contrast, the RGCs were naturally viewed as contrast detectors. Under this paradigm, the role of the retina would be to transform the coding of retinal signals from absolute levels of illumination to contrast values [7]. The purpose of the early visual processing was later addressed by Marr and Hildreth on its theory of edge detection [8]. They argued that intensity changes, i.e local contrasts, could be the building blocks to construct a primitive description of the image. As those changes can occur at different scales, the detection process would have to operate at different resolutions of the incoming input. In order to satisfy these requirements, a new operator, the second derivative, or the Laplacian, of a Gaussian was proposed. The gaussian would work as an smoothing filter to set the image scale, and the zero-crossings obtained from the Laplacian would signal the intensity changes at that scale. An edge could then be perceived when a set of intensity changes occur at any given location. Moreover, they showed that an accurate approximation of the Laplacian of a Gaussian could be developed by using a DoG function with an appropriate ratio $\sigma_{surround}/\sigma_{center}$ about 1.6.

A different view of the retina and the role of RGCs has its roots in the field of information theory. Barlow (1961) suggested that as the visual stimulus is highly redundant, i.e presents a high degree of spatial correlation, RGCs would have to remove that redundancy in order to encode the information in a more compact way. Following this line of thought, RGCs were proposed to transform the input into a decorrelated output by means of either predictive [10] or transform coding [11]. The efficient RGC filters estimated under these schemes showed similar center-surround structures to those found experimentally.

This paper aims at analyzing the properties of the discrete versions of the DoG models obtained experimentally by Enroth-Cugell and Robson. First, we wonder about the dependency between the kernel properties with the parameters of the discretization process. Second, we want to quantify the effect of the discrete kernel on input images in terms of appropriate descriptors, such as contrast, correlation and entropy. Finally, we discuss the values of these descriptors in light of the predictions made by the aforementioned retinal paradigms.

2 Methods

2.1 Parameters of the DoG Model

The Difference-of-Gaussians model is made up with two gaussians: the first one representing the excitatory center of the RF, the second one the inhibitory surround of the RF. Figure 1 shows an schematic diagram of the model as it was used by Enroth-Cugell and Robson [3,4], with their relevant parameters: the maximum amplitudes k_c and k_s, and the radius r_c and r_s. The function is formalized as follows:

$$DoG(r) = k_c e^{-(r/r_c)^2} - k_s e^{-(r/r_c)^2} \tag{1}$$

being the relevant parameters: the maximum amplitudes k_c and k_s, and the radius r_c and r_s. For each RGC, 17 cells reported in Enroth-Cugell and Robson (1966) and 6 in Enroth-Cugell and Robson (1984), the theoretical contrast sensitivity function derived from the DoG model was fitted to the empirical contrast sensitivity function measured at different spatial frequencies. As a result, a set of parameters (r_c, r_s, r_s/r_c, and $k_s r_s^2/k_c r_c^2$) were estimated for each cell. For brevity these parameters are not shown here but can be found in the original papers [3,4].

2.2   Discrete DoG Kernels

The continuous DoG functions fitted to experimental data has to be converted into discrete DoG kernels to operate with input images. In what follows, the four steps required in that procedure are described.

As the continuous function can take infinite values, the first step consisted on truncating it in order to set a finite range of values. Figure 2 describes the process in one dimension. A variable T was defined to determine the width of the truncation: $T = n_T * r_c$, where n_T is an integer and r_c the radius of the center.

The second step consists on sampling the continuous function. By setting the size SxS of the kernel, the number of both the elements of the kernel matrix as well as the sampling points of the continuous DoG function are set. Both variables T and S define the step of the sampling process:

$$Step = \frac{2T}{S-1} \tag{2}$$

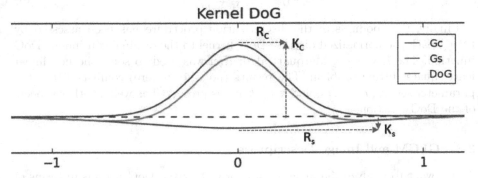

Fig. 1. The Difference-of-Gaussians model. The gaussian describing the excitatory center (upper blue colored gaussian) is combined with the inhibitory surround (lower blue colored gaussian). The result is the Difference-of-Gaussian function (pink). The relevant parameters, amplitudes (k_c, k_s) and radius (r_c, r_s), for each gaussian are also shown.

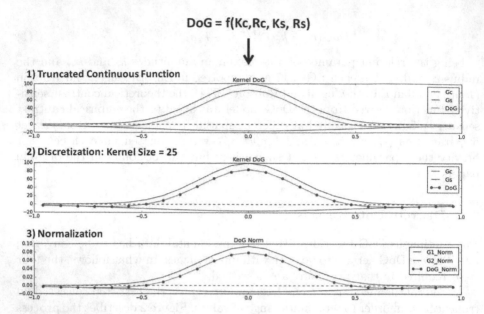

Fig. 2. Discretization of the continuous DoG model: truncation (upper inset), sampling (middle inset), and normalization (lower inset)

After the convolution of the input image with the discrete kernel, the output image has to preserve the intensity ranges of the input. The third step therefore involves the normalization of the elements of the DoG kernel matrix. Each element or weight w_{ij} of the matrix is normalized as follows:

$$w_{ij}^{norm} = \frac{w_{ij}}{\Sigma_i \Sigma_j w_{ij}} \tag{3}$$

Finally, the goodness of the discretization procedure has been assessed by fitting back the normalized discrete DoG kernel to the original continuous DoG function. The Levenberg-Marquart algorithm was used to solve the non-linear least squares fitting problem. The results (not shown here) confirmed that the parameter set (r_c, r_s, r_s/r_c, and $k_s r_s^2 / k_c r_c^2$) is preserved as well as other aspects of the DoG function.

2.3   GLCM and Image Descriptors

As we want to analyze the images transformed by the DoG kernels in terms of local contrasts as well as descriptors that quantify the efficiency of the transform, histogram information is not enough. We therefore require information regarding pixel intensities according to the spatial location of those pixels. There is a way to manage these considerations by means of the Gray-Level-Coocurrence-Matrix (GLCM) and the statistical descriptors derived from it. GLCM is a matrix where each element $p(i, j)$ represents the joint probability that a pixel with gray-level i occurs horizontally adjacent to a pixel with value j. Figure 3 shows an example of how to construct such a matrix using an image patch with $L = 5$ gray levels ranging from 0 to 4. The corresponding $5x5$ unnormalized GLCM matrix at an angle of 0 degrees, representing horizontal adjacency, is shown in the right inset. The final GLCM matrix is computed by normalizing this matrix so that the sum of its elements is equal to 1. The results shown in this paper were obtained with GLCM matrices of size $256x256$, to take into account the gray levels of an 8-bit image representation, and 4 possible angles, to consider horizontal, vertical and diagonal adjacencies.

A set of image descriptors can then be computed over the GLCM matrix. In this work we have focused in those descriptors that better describe the image in terms of contrast and efficient coding:

– Contrast. A measure of intensity changes between a pixel and its neighbor over the whole image. This measure will be 0 for a constant image.

$$\Sigma_i \Sigma_j (i - j)^2 p_{ij} \tag{4}$$

– Energy. A measure of uniformity of the image. It is 1 for a constant image.

$$\Sigma_i \Sigma_j p_{ij}^2 \tag{5}$$

– Correlation. A measure of how correlated a pixel is relative to its neighbor over the entire image. The range of values is between 1 to -1 corresponding to perfect positive and perfect negative correlations.

$$\Sigma_i \Sigma_j \frac{(i - \mu_i)(j - \mu_j)p_{ij}}{\sigma_i \sigma_j} \tag{6}$$

– Entropy. A measure of disorder or randomness of the elements of GLCM. The entropy is 0 when $p(i, j) = 1$ for any given i and j, and is maximum when all $p(i, j)$ are equal.

$$- \Sigma_i \Sigma_j p_{ij} log_2(p_{ij}) \tag{7}$$

2.4 Image Processing and Kernel Analysis

A comprehensive *python* simulation environment was developed to carry out the discretization procedure as well as the analysis of the kernels presented in the Results section. The environment takes advantage of some powerful *python* libraries, such as: *OpenCV*, to convolve an image with a kernel; *Numpy*, to generate 2D kernels and compute 2D discrete Fourier transforms; *Scipy*, to solve the non-linear least squares fitting problem described in section 2.2; and *Matplotlib* to plot graphs and view images.

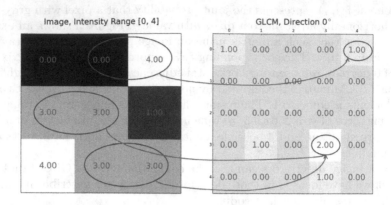

Fig. 3. GLCM matrix: Image patch of size 3x3 and gray-level range $[0, 4]$ (left) and absolute frequencies for horizontal adjacency (right).

3   Results

For each of the $23(= 17 + 6)$ cells reported by Enroth-Cugell and Robson (see section 2.1) a set of discrete DoG kernels were generated after varying the parameters of the discretization process being applied on the original continuous DoG function. Two discretization parameters, the integer n_T describing the width of the truncation T and the size S of the kernel, were modified in order to obtain each kernel of the set. The kernels were then analyzed in terms of its representation in both spatial and frequency domain, the Bode diagram, and the output after convolving the kernel with an standard test input (Lena image). It can be concluded that all set of kernels change with the discretization parameters in a quite similar way regardless of the analyzed cell. On the basis of this result, only cell number 1 of Enroth-Cugell and Robson (1966), was chosen to study the impact of the discretization procedure.

The first objective was to analyze the properties of the discrete DoG kernels corresponding to cell number 1. Figure 4 illustrates the Bode diagrams, which plots the filter gain for each spatial frequency, for truncation values $n_T = 3$ (first row) and $n_T = 19$ (second row) and kernels of increasing sizes. For $n_T = 3$, the shape of the filter gain changes with kernel size, starting as a High-Frequency Enhancing Filter (HFEF) at small sizes and becoming a Band-Pass Filter (BPF) at large sizes. The dependency is the same for $n_T = 19$, but bigger kernel sizes (more discretization points) are required to achieve the BPF filter. Furthermore, with this truncation value, the gain is almost negligible for small sizes. These results suggest that (1) the discretization or sampling of the continuous DoG function is crucial to determine the behaviour of the discrete DoG kernel, and (2) the response of the continuous DoG function in the frequency domain, which is of a band-pass type, is better approximated with more dense discretizations.

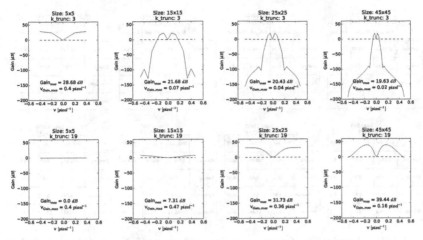

Fig. 4. Bode diagrams for different discretizations of the continous DoG kernel. The case corresponds to cell number 1 of Enroth-Cugell and Robson (1966) with continuous function parameterized as follows: $k_c = 100$, $k_s = 15.9$, $r_c = 0.32$, $r_s = 0.76$. Filter gain is shown for truncation value $n_T = 3$ (upper row) and $n_T = 19$ (lower row), for kernel sizes ranging from 5x5 to 45x45 (left to right columns).

Fig. 5. Output images for different discretizations of the continous DoG kernel. Results after filtering the test input (Lena image) through the discrete DoG kernels. Each output corresponds with the kernels shown in figure 4.

The output images (Fig. 6), obtained after convolving the input with the kernels, allow to analyze the filtering carried out by the different discrete versions of the DoG function. For $n_T = 3$, the first kernel ($n_T = 3$, S=5) amplifies high frequencies so much that the image output looks rather noisy. The output looks much better when the kernel behaves as a BPF at sizes S=15 and S=25. However, the last kernel (S=45) removes high frequencies outside the band, i.e suppresses

Size: 5x5 k_trunc: 3	Total_out	AVG_out	AVG_diff
Contrast	3178.28	794.57	717.03
Energy	0.05	0.01	-0.03
Correlation	3.34	0.84	-0.14
Entropy	54.54	13.64	2.59

Size: 15x15 k_trunc: 3	Total_out	AVG_out	AVG_diff
Contrast	577.99	144.5	66.96
Energy	0.09	0.02	-0.02
Correlation	3.89	0.97	-0.01
Entropy	49.53	12.38	1.33

Size: 25x25 k_trunc: 3	Total_out	AVG_out	AVG_diff
Contrast	262.23	65.56	-11.98
Energy	0.13	0.03	-0.01
Correlation	3.95	0.99	0.01
Entropy	46.61	11.65	0.6

Size: 45x45 k_trunc: 3	Total_out	AVG_out	AVG_diff
Contrast	106.68	26.67	-50.87
Energy	0.17	0.04	0.0
Correlation	3.98	1.0	0.02
Entropy	44.42	11.11	0.06

Size: 5x5 k_trunc: 19	Total_out	AVG_out	AVG_diff
Contrast	310.16	77.54	0.0
Energy	0.16	0.04	0.0
Correlation	3.92	0.98	0.0
Entropy	44.2	11.05	0.0

Size: 15x15 k_trunc: 19	Total_out	AVG_out	AVG_diff
Contrast	494.43	123.61	46.07
Energy	0.11	0.03	-0.01
Correlation	3.87	0.97	-0.01
Entropy	47.43	11.86	0.81

Size: 25x25 k_trunc: 19	Total_out	AVG_out	AVG_diff
Contrast	4029.39	1007.35	929.81
Energy	0.05	0.01	-0.03
Correlation	3.21	0.8	-0.18
Entropy	55.32	13.83	2.78

Size: 45x45 k_trunc: 19	Total_out	AVG_out	AVG_diff
Contrast	3648.53	912.13	834.59
Energy	0.14	0.04	-0.0
Correlation	3.48	0.87	-0.11
Entropy	55.14	13.78	2.73

Fig. 6. GLCM descriptors of output images. The set of 4 descriptors described in section 2.3 is computed for each output image of figure 6. Each descriptor is computed on the set of GLCM matrix, each one representing 4 adjacency types (see section 2.3). AVG_{out} represents the average of descriptor values along the 4 GLCM matrices, while AVG_{diff} indicates the difference between AVG_{out} and the average value of the image input.

a certain amount of detail, and generates a smooth version of the input image. For $n_T = 19$, the output is almost the same as the input at size S=5, it seems to improve a bit with the weak gain provided with the kernel at size S=15, and finally becomes noisy as the gain increases with the kernel showing a HFEF behavior at size S=25. Remarkably, the result is still noisy at size S=45 even though the kernel has became a BPF. It seems that the gain is indeed quite large at higher frequencies.

The effects of the DoG kernels are better quantified with the GLCM descriptors. Image contrast seems to depend on the gain of the filter at higher frequencies. For $n_T = 3$, the highest positive AVG_{diff}, the difference of average contrast between the output and the input, corresponds to the image filtered with kernel size S=5. This value becomes negative, i.e contrast is reduced when compared with the input, with kernel sizes S=25 and S=45. For $n_T = 19$, the AVG_{diff} peak is achieved with size S=25 and reduced when the BPF behavior appears at size S=45. Image entropy seems to possitive correlate with image contrast, meaning that contrast enhancements do correspond with entropy increases. On the other hand, image correlation seems to be at odds with contrast. Contrast enhancements appear to reduce the spatial correlation of the output image.

4   Discussion

The results indicate that the sampling of the continuous DoG function determine the behaviour of the corresponding discrete DoG kernel. The discretization process could yield filters with either an HFEF or a BPF behaviour. If the goal of the retina is to highlight or enhance contrast, it has to be done without amplifying high-frequency noise, thus restricting the filter gain to a band of frequencies. Moreover, enhancing average contrast comes at a cost of: (1) higher entropy, and (2) lower spatial correlation. In terms of the language of information theory, images with higher entropies require longer size codes, i.e the amount of data required to represent the image is increased rather than reduced. In terms of machine learning, removing correlations could impede the learning of visual patterns. In closing, it seems that the optimal discretization of the DoG models would have to satisfy a trade-off between contrast, entropy and spatial correlation.

Acknowledgements. This research was sponsored by the Ministry of Science and Innovation of Spain under grant TIN2011-22935.

References

1. Kuffler, S.W.: Discharge patterns and functional organization of mammalian retina. J. Neurophysiol. 16(1), 37–68 (1953)
2. Rodieck, R.W.: Quantitative analysis of cat retinal ganglion cell response to visual stimuli. Vision Res. 5, 583–601 (1965)
3. Enroth-Cugell, C., Robson, J.G.: The Contrast Sensitivity of Retinal Ganglion Cells of the Cat. J. Physiol. 187, 517–523 (1966)
4. Enroth-Cugell, C., Robson, J.G.: Functional characteristics and diversity of cat retinal ganglion cells. Basic characteristics and quantitative description. IOVS 25, 250–267 (1984)
5. Barlow, H.B.: Pattern recognition and the responses of sensory neurons. Annals of the New York Academy of Sciences 156(2), 872–881 (1969)
6. Shapley, R., Enroth-Cugell, C.: Visual adaptation and retinal gain controls. Progress in Retinal Research 3, 263–346 (1984)
7. Kandel, E.R., Schwartz, J.H., Jessell, T.M.: Essentials of neural science and behavior. Appleton-Lange, Norwalk (1995)
8. Marr, D., Hildreth, E.: Theory of Edge Detection. Procs. Royal Soc. of London, Series B, Biological Sciences 207, 187–217 (1980)
9. Barlow, H.B.: Possible principles underlying the transformations of sensory messages. In: Sensory Communications, pp. 217–234 (1961)
10. Srinivasan, M.V., Laughlin, S.B., Dubs, A.: Predictive coding: a fresh view of inhibition in the retina. Procs. Royal. Soc. of London B: Biol. Sciences 216(1205), 427–459 (1982)
11. Atick, J.J., Redlich, N.: What does the retina know about natural scenes? Neural Comp. 4(2), 196–210 (1992)

Computable Representation of Antimicrobial Recommendations Using Clinical Rules: A Clinical Information Systems Perspective

Natalia Iglesias[1], Jose M. Juarez[1(✉)], Manuel Campos[1],
and Francisco Palacios[2]

[1] Computer Science Faculty – Universidad de Murcia, Murcia, Spain
jmjuarez@um.es
[2] Intensive Care Unit – University Hospital of Getafe, Madrid, Spain

Abstract. The overuse of antimicrobials promotes the resistance of antibiotics, which is a great concern in hospitals. Clinical Guidelines are essential documents that provide useful recommendations to clinicians about the therapy. In order to obtain a Computerised Clinical Guideline, main efforts to represent this knowledge focus on ad-hoc data flow models. However, they have had a low impact in the industry since they generally neglect clinical standards or they are hard to maintain due to the model complexity. In this work, we propose to step backward to use rule-based approaches to obtain clinical rules, more simple to model and easier to manage. We also review and discuss main rule representation alternatives and we present a case study in the Ventilator Associated Pneumonia from a Clinical Guideline.

1 Introduction

Antibiotic administration is widely extended to all stages of healthcare to treat bacterial infections. Great efforts have been done in order to define policies to keep the effectiveness of antibiotics, palliating the occurrence of resistances due to an inappropriate and unnecessary use of them. In particular, hospital-acquired infections in Intensive Care Units are a major concern, due to the patients' conditions.

Clinical Guidelines (CGs) are a useful tool for intensivists to support their medical decisions. Essentially, a CG is a document that provides a set of recommendations based on the best medical evidence available regarding diagnosis or prescription among others. In this work, we draw our attention on CG for infection diagnosis and antibiotic prescription assessment.

CGs have been used as a source of knowledge to build clinical decision support system. According to [16] three key issues are identified: (1) modelling and computable representation, (2) acquisition and (3) verification and execution. Most efforts in the literature focus on the first issue in order to define a expressive and free-ambiguous model.

© Springer International Publishing Switzerland 2015
J.M. Ferrández Vicente et al. (Eds.): IWINAC 2015, Part I, LNCS 9107, pp. 258–268, 2015.
DOI: 10.1007/978-3-319-18914-7_27

There is a wealth of generic computerised CG languages such as ProForma, ASBRU, GLIF or GLARE among others [16]. Other approaches deal with specific medical problems such as ONCOCIN or T-HELPER, designed for oncology or HIV protocols. According to [6], key aspects to model computerised CG are the primitives of the language, the complexity of the CG type, the domain knowledge or the maintenance problems. Unlike the aforementioned models, Arden syntax [4] has some impact in the industry. Arden syntax is a rule-based language which simplicity allows modelling basic CGs using simple modules to share the knowledge.

From the implementing point of view, clinical rules-based systems benefit from solid formal models and mature technologies and there is a increasing interest on prescription supervision and pharmaceutical validation [2]. Moreover, semantic rule languages such as SWRL or production rule engines such as Drools provide excellent platforms for the development of CDSS.

From the clinical point of view, rules as declarative expressions, are generally easy to interpret by clinicians. Furthermore, in the infection and antibiotic management, there are evidences of effective knowledge representation using rules. In [12] the detection of hospital-acquired infections a production rule was proposed. Regarding the antibiotic selection issue, TREAT system uses a probabilistic causal networks [13]. The MoniICU system [1] provides an infection alert module for ICUs based on Arden modules. Recent works also propose an automatic translation of these modules to Drools framework [10].

This paper is structured as follows. In Section 2 we propose an implementation-oriented framework to analyse different rule-based approaches to model clinical guidelines. Section 3 describes a case of use in VAP, modelling a CG from a Ventilation-Associated Pneumonia guideline [5]. We discuss the results obtained in Section 4. Finally, we present our conclusions and further research in Section 5.

2 Rule Model Analysis

In this work, we analyse the use of rule models to represent pieces of a clinical guideline under the assumption that the results obtained must be adoptable by the industry according to available technology. We propose to use the following dimensions of study:

- Expressivity and capacity of the modeling language in terms of syntax flexibility, logic expressivity and time management.
- Suitability of the rule model tot be easily adopted in the clinical domain, considering the physician interpretability and the interoperability with Clinical Information Systems.
- Regarding the industry aspects, we consider the availability of reasoning engines, the industry support, standardization and the maintenance capacities.

There is a wealth of rule models proposed in the literature. We cluster the works reviewed in three main groups: specific clinical oriented languages to model clinical guidelines using rules, production rule languages and semantic rule languages. The rest of the section presents a description of the different languages reviewed in each group and we finally present our comparative analysis according to the abovementioned dimensions of study.

2.1   Clinical Rule Languages

Arden syntax [4] is a language for representing and sharing medical knowledge. The Medical Logic Modules (MLMs) are the key component of Arden syntax. MLMs are set of rules comprising the clinical knowledge to make a single medical decision. The language allows both the typical if-then constructs used in declarative production rules systems, as well as the use of classical procedural components . The Arden ML syntax is a HL7 language certified as standard. Although Arden ML has been available for more than 20 years now, it is not widely used mainly due to its complex syntax and the lack of compilers and an execution environment that allows validation.

2.2   Production Rule Languages

Drools [17] is a cross-platform Business Rule Management System (BRMS) written in Java and developed by the JBOSS community open source projects, that uses an enhanced implementation of the classical forward-chaining RETE algorithm. The Drools language to represent knowledge in form or rules is based on First Order Logic and with a Closed World Assumption (CWA). Drools reasoning capabilities are based on two main processes, the Authoring process and the Runtime process: (1) the authoring process: involves the creation of rules files (*.drl* files) which contain the representation of the rules knowledge base which is used for representing the domain knowledge in a formal way; (2) the runtime process: the rules knowledge base is also a runtime component of Drools, that can instantiate one or more Working Memories at any time. The Working Memory consists of a number of sub-components, including Event Support for Temporal Reasoning, Truth Maintenance System, Agenda and Agenda Event Support. Furthermore, Drools Fusion allows the use of different temporal operators, both for expressing discrete time or point-in-time or interval events. The temporal relationships that can be expressed cover all the 13 temporal constraints between events defined by Allen.

Finally, it is worth mentioning JESS [7], a production rule framework based on Java technology. Jess models is a CLIPS-like language, popular among researchers, and using an optimized version of the Rete algorithm with backwards-chaining support. Several systems combines Jess with semantic rules as SWRL.

2.3   Semantic Rules

The Semantic Web Rule Language (SWRL) [9] is a tightly coupled hybrid technology that combines ontological conceptual reasoning based on Description Logic with a Rule Language, extending the OWL expressiveness with Horn Logic, in an Open World Assumption (OWA) scenario. SWRL is a proposed language for the Semantic Web that can be used to express rules as well as logic, combining OWL DL with a subset of the Rule Markup Language (Rule ML , itself a subset of Datalog). SWRL is the most popular formalism to express knowledge in rule form within the web community since it is based on OWL-DL: all rules are expressed in terms of OWL concepts. SWRL has the full power of OWL DL and extends it with a specific form of Horn-like rules. Classical reasoning in OWL-DL allows only limited use cases such as consistency checks, class properties and relationships and instance classification.

The Rule Interchange Format (RIF) [11] was chartered by the World Wide Web Consortium in 2005 as a standard for exchanging rules among web rule engines that focuses on exchange rather than defining a single one-fits-all rule language. In contrast to other Semantic Web standards, such as RDF, OWL and SPARQL, it was immediately clear that a single language would not cover all rules for knowledge representation and business modeling. RIF in fact provides more than just a format: although RIF dialects were designed primarily for interchange, each dialect is a standard rule language and can be used even when portability and interchange are not required.

SparQL Inference Notation SPIN (SPIN)[18] is a W3C Member Submission that has become the de-facto industry standard to represent SPARQL rules and constraints on Semantic Web models. SPIN also provides meta-modeling capabilities that allow users to define their own SPARQL functions and query templates. Finally, SPIN includes a ready to use library of common functions. SPIN allows to represent a wide range of business rules in SPARQL as RDF Triples. In fact, SPIN is also referred to as SPARQL Rules. SPARQL is a well-established W3C standard implemented by many industrial-strength RDF APIs and all databases. This means that rules can run directly on RDF data without a need for materialization. SPIN provides a framework that helps users to leverage the fast performance and rich expressivity of SPARQL for various application purposes.

Finally, there is a query language based on SWRL called SQWRL, a concise, readable and semantically robust query language for OWL, as SPARQL understanding of OWL semantics is incomplete. SQWRL uses SWRL strong semantic foundation as its formal underpinning and provides a small but powerful array of operators. SPARQL cannot be considered to be a rule language, but a query language exclusively over RDF, allowing to make one step transformation from one RDF graph match to another, and iteratively apply SPARQL constructs to simulate the functionality of a rule language in a rather complex way, for example using Answer Set semantics or using the SPIN language, rule language expressed in SPARQL constructs.

In Table 1 we summarize our analysis of the use of rule models to represent pieces of a clinical guideline under the assumption that the results obtained must be adoptable by the industry.

Table 1. Rule language analysis. N/A:not available, FOL: first-order logic, CWA: Closed World Assumption; HLP: Horn Language Program; DL: Description Logics; DSL: Domain Specific Language.

	Language			Clinical Domain		Industry			
	Syntax	Logics	Time	Constructs	Interop	Engine	Support	Standards	Maintenance
Arden syntax	Arden MLM	N/A fuzzy logics	no	clinical oriented	interchange format	Arden Engine	(company) Medexer	ASTM HL7	no
Drools	Java/Mvel	supports FOL, CWA	yes	DSL available	via *drl* files	KIE platform	JBoss project	JRS-94	metadata
Jess	Java/CLIPS	supports FOL	no	no	via *clp* files	Jess Platform	(company) Sandia Labs	no	no
SWRL	OWL-DL	HLP∪DL	no	no	no	Bossam/Hoolet/Pellet (not full supported)	RuleML	W3C submission	yes
RIF	exchange rules dialects	HLP (dialects vary)	no	no	(generic) interchange format	N/A	N/A	W3C recomm.	yes
SPARQL	RDF query	N/A	no	no	no	Jena/Virtuoso	less support after RIF	W3C recomm.	–
SQWRL	OWL rules (over SWRL)	HLP∪DL undecidable	no	no	no	Protege OWL	–	N/A	–
SPIN	RDF syntax SPARQL rules	N/A	no	no	no	RDF/SPARQL platforms	–	W3C submission	–

3 Case Study

According to the Spanish Prevalence Study of Nosocomial Infections (EPINE) reported in 2012, 7.1% of hospital patients suffered an infection in which Pneumonia is the second most frequent hospital-acquired infection affecting 20.92% of all critically ill patients in Spain. Medical evidences highlight that about eighty percent of nosocomial lung infections are associated with mechanical ventilation [19]. In this case study, we focus on the Ventilator-Associated Pneumonia (VAP), a specific lung infection of inpatients on breathing machines.

The sources of medical knowledge used in this work are two international references of the medical literature: the VAP recommendations of the John Hopkins Antibiotics Guidelines [5] and the definitions of Ventilator-Associated Events [3].

3.1 VAP Knowledge

In essence, this piece of guideline referring VAP is composed by a natural language description of recommendations regarding:

1. The calculation of the Clinical Pulmonary Infection Score (CPIS) which is an indicator of the possible presence of an pulmonary infection, a surrogate tool to state a diagnosis.
2. Diagnosis recommendations based on CPIS value and other clinical factors.
3. Recommendations on antibiotic prescription according to the diagnosis hypothesis.
4. Antibiotic empiric treatment (including duration) recommendations for each diagnosis.
5. Antibiotic follow-up of treatment recommendations.

The above-mentioned recommendations refer to symptoms, signs and other information that can be obtain from the patient's medical record.

The CPIS description is an ambiguous free specification of the calculi of the score according to the following medical parameters: temperature, peripheral WBC, tracheal secretions, chest x-ray, progression of infiltrate from prior radiographs, culture of ET suction and oxygenation (PaO2/FiO2).

Time also plays a key role on this medical knowledge, expressed in the form of temporal information or thresholds on the duration of the temporal event. This is the case of the diagnosis recommendations that lay on the CPIS value and other conditions of the patient (e.g. hours of hospitalization and long-term facility origin), resulting in one of the following diagnoses: VAP unlikely, Early-onset VAP and Late-onset VAP. For instance: *the patient may have an early-onset VAP when CPIS is over 6 within first 72 hours hospitalization and the patient does not come from long-term facility.*

Similarly, in the recommendations on antibiotic prescription the therapy suggested includes the duration of the antimicrobial administered. For example: *"Administration of Moxifloxacin 400 mg IV Q24H during 7 days when early-onset VAP is suspected"*.

Some other recommendations combine quantitative temporal constraints as well as fuzzy expressions. The antibiotic follow-up of treatment recommendations suggest to revise the therapy according to patient's conditions. For example: *the doctor might consider to cancel te treatment or to alter the treatment duration to 3 days (by default 7) when VAP is unlikely and the current CPIS is bellow or equal to 6 after 3 days of treatment.*

3.2 Rule Modelling

According to the clinical guidelines requirements (described in Section 3.1) and the rule model analysis (presented in Section 2) we chose Drools as framework.

In particular, we have taken into account the following design considerations:

- Tracking: each rule must be traced back from the clinical guideline document, the knowledge source.
- Authoring: design and validation of each rule is an iterative process that requires version control.
- Modularity: rules must be grouped and activated (using the Drools *agenda* functionality) according to the current state of the medical process (e.g. diagnosing, therapy , following up, etc.).
- Temporal dimension: time constraints of rules can modeled using the Drools Fusion Complex Event Correlation capabilities, simulating real time scenarios that allow checking for example how a patient's response to treatment has evolved over time.

Figure 1 depicts an example of a rule modeling a recommendation of the VAP clinical guideline. The audit metadata block shows that the rule trace is *page* = 81 and *lines* = 34 − 35 of the clinical guidelines documents, describing

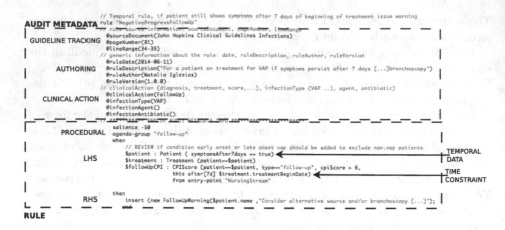

Fig. 1. Rule implemented

the author and version, as well as metadata of clinical actions for knowledge base maintenance issues.

The rule description block consists of: (1) procedural information including priority (*salience* = −50) and group execution block (*agenda − group* = "*follow − up*"); (2) the antecedent of the rule (left hand side, LHS) and (2) the consequence of the rule (right hand side, RHS). In the LHS the temporal constraint "*after 7 days of the beginning of treatment*" is expressed as *after*[7d] *treatment.treatmentBeginDate*. The fact *treatmentBeginDate* should be stated in the hospital Clinical Information System.

3.3 Knowledge Base Implementation

Rules were modelled from the clinical guideline under the supervision of experts. During this study 36 rules have been formalized that represent the VAP infection. The resulting rules have been grouped into execution blocks to allow for a correct rule execution flow, while rules remain declarative and therefore interpretable by experts.

Figure 2 depicts the general structure of the rules implemented and the RETE network using IDE Rule Workbench provided by Drools.

The resulting VAP rules can be classified in 6 types:

- CPI Score rules: given a set of patient's symptoms, that we assume to be part of the Patient Medical Record (PMR) in a Clinical Information System, these rules calculate the final value of the CPI Score for a specific patient.
- Diagnosis rules: the diagnosis rules are decoupled from the treatment rules, although the guidelines couple this knowledge. That means that first, a diagnosis is assessed for a given patient, and only when the diagnosis is known, treatment rules can be activated.

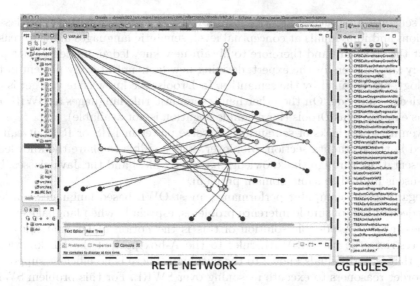

Fig. 2. Implementation of the knowledge base

- Treatment rules: these rules select the right treatment for a patient, using as input the assessed diagnosis and other patient information coming from the Patient Medical Record.
- Follow-up rules: these rules are in many cases temporal rules, as they check the patient CPI Score after some days of treatment to decide if treatment should be stopped, continued or complemented.
- Treatment notes rules: rules that show specific alerts regarding the treatment of, for example, certain bacterial agents.

The rules are executed in this specific order, grouping them in the so called Drools agenda-groups. Because when a fact is changed in the working memory with the *modify* operand, Drools does not know which data has been actually changed, all rules are activated again and therefore are candidates to fire again, possibly generating infinite loops. This situation is a well known problem in Rule Based Systems and Drools provide some special attributes and features to avoid many of these situations (but not all).

4   Discusion

As shown in the case study, modelling a piece of clinical guideline requires a large number of rules. Therefore, three key aspects should be discussed: language expressivity, rule maintenance and computer performance. In this work, we focus on production rule languages and semantic rule languages.

In general, production rule languages allow an easy maintenance of business logic. While semantic languages (as SWRL, RIF or SPARQL) represent semantic knowledge, Drools offers a pure syntactic knowledge representation.

In semantic approaches, knowledge is represented using axioms between classes, relations and individuals at conceptual level, semantic languages allow to reason about the structure and therefore to obtain new knowledge, hidden or not previously known even for the expert, deriving new facts not literally present in the ontology, but entailed by the semantics. In Drools, the knowledge you get is the knowledge you code. On the other hand, semantic rule languages as SWRL are not as expressive as Drools (for example negation is not available).

Even if Drools lacks of semantic expressive power in LHS or RHS, it can be placed in the metadata section. Moreover, it is possible to share the knowledge representation, through the knowledge base (.*drl* files) and the Java classes, but is re-use is limited to same domain problem.

Regarding the computer performance, most OWL-based languages can lead to undecidability of simple inference problems, especially when language expressiveness grows. A possible solution of this is the notion of DL-safe rules [14], that restricts the use of SWRL rules to the A-box part of the DL knowledge base, that is, individuals known to the ontology. There is no efficient support of first-order reasoners to execute reasoning over SWRL. For this problem SWRL rules are for example translated to existing rule systems as JESS [15].

The case study also highlights that time dimension plays a key role on the clinical guideline representation. In this sense, temporal rules can be easily added to the knowledge base by using the functionality provided by Drools Fusion. On the other hand, temporal reasoning with SWRL is still a research area in Semantic Web.

5   Conclusions

The aim of this work is to identify, in the antimicrobial prescription context, the most suitable model to represent CGs using a rule-based approach. To this end, we present a comparative analysis of current rule models and platforms considering expressivity, standardization and industry criteria. A second contribution of this work is to analyse the rule-based representation of a piece of CG front the VAP problem.

Previous studies on computerized CG deal with representing knowledge of most of the CG document. They mainly focus on the extraction of clinical rules, where the models proposed are highly expressive but complex. In this work, we highlight the advantages of representing knowledge using rules which simplicity allows an effective implementation and an easy transfer to the industry.

When medical knowledge is modeled, it is crucial to evaluate how the knowledge is modeled, where it is stored, the use of norms and standards, and scalability and performance. In [8], the authors review advantages and disadvantages of several rule languages for decision support on cardiology. They conclude that semantic rule technology provides reusable knowledge and the advanced reasoning capabilities are far better than those of Drools. The study also claims that Drools is more difficult to maintain as knowledge evolves and new or modified rules arise. Our conclusions differ from [8], since we also take into account the metadata elements of Drools. Drools however comes out as the champion regarding expressiveness, mainly due to the possibility to add Java code to rules.

The experiments carried out with Drools for modeling a CG for the VAP infection reveals the advantages of this rule approach. In particular, authoring and tracking the rules to the original CG text is essential in order to validate the rules by experts. Moreover, time management is essential and time constraint primitives must be included in the rule language.

Future works will be focused on including other parameters in our study (such as performance analysis and scalability) as well as the implementation of different CG for antimicrobial management.

Acknowledgment. This work was funded by the Spanish Ministry of Economy and Competitiveness under the WASPSS project (Ref: TIN2013-45491-R).

References

1. Adlassnig, K.P., Blacky, A., Koller, W.: Artificial-intelligence-based hospital-acquired infection control. Stud. Health Technol. Inform. 149, 103–110 (2009)
2. Boussadi, A., Bousquet, C., Sabatier, B., Caruba, T., Durieux, P., Degoulet, P.: A business rules design framework for a pharmaceutical validation and alert system. Methods Inf. Med. (1), 36 50 (2011)
3. Centers for Disease Control and Prevention. CDC NHSN patient safety component (2013)
4. Clayton, P.D., Pryor, T.A., Wigertz, O.B., Hripcsak, G.: Issues and structures for sharing knowledge among decision-making. approaches for creating computer-interpretable guidelines that facilitate decision support 25 systems: The 1989 arden homestead retreat. In: Proc. 13th Symp. on Computer Applications in Medical Care, pp. 116–121 (1989)
5. Cosgrove, S.E., Avdic, E.: John hopkins antibiotics guidelines 2013-2014: Treatment recommendations for adult inpatients (2013)
6. de Clercq, P.A., Blom, J.A., Korsten, H.H.M., Hasman, A.: Methodological review: Approaches for creating computer-interpretable guidelines that facilitate decision support. Artificial Intelligence in Medicine 31(1), 1–27 (2004)
7. Hill, E.F.: Jess in Action: Java Rule-Based Systems. Manning Publications Co. (2003)
8. Van Hille, P., Jacques, J., Taillard, J., Rosier, A., Delerue, D., Burgun, A., Dameron, O.: Comparing drools and ontology reasoning approaches for telecardiology decision support. In: Quality of Life through Quality of Information Proceedings of MIE 2012, The XXIVth International Congress of the European Federation for Medical Informatics, Pisa, Italy, August 26-29, pp. 300–304 (2012)
9. Horrocks, I., Patel-Schneider, P.F., Boley, H., Tabet, S., Grosof, B., Dean, M.: SWRL: A semantic web rule language combining OWL and RuleML. W3C Member Submission (May 21, 2004), http://www.w3.org/Submission/SWRL
10. Jung, C.Y., Sward, K.A., Haug, P.J.: Executing medical logic modules expressed in ardenml using drools. JAMIA 19(4), 533–536 (2012)
11. Kifer, M.: Rule interchange format: The framework. In: Calvanese, D., Lausen, G. (eds.) RR 2008. LNCS, vol. 5341, pp. 1–11. Springer, Heidelberg (2008)

12. Landers, T., Apte, M., Hyman, S., Furuya, Y., Glied, S., Larson, E.: A comparison of methods to detect urinary tract infections using electronic data. Jt. Comm. J. Qual. Patient Saf. (9), 411–417 (2010)
13. Leibovici, L., Paul, M., Nielsen, A.D., Tacconelli, E., Andreassen, S.: The treat project: decision support and prediction using causal probabilistic networks. Int. J. Antimicrob. Agents 30, 93–102 (2007)
14. Motik, B., Sattler, U., Studer, R.: Query answering for OWL-DL with rules. J. Web Sem. 3(1), 41–60 (2005)
15. O'Connor, M.F., Knublauch, H., Tu, S., Grosof, B.N., Dean, M., Grosso, W., Musen, M.A.: Supporting rule system interoperability on the semantic web with SWRL. In: Gil, Y., Motta, E., Benjamins, V.R., Musen, M.A. (eds.) ISWC 2005. LNCS, vol. 3729, pp. 974–986. Springer, Heidelberg (2005)
16. Peleg, M.: Computer-interpretable clinical guidelines: A methodological review. Journal of Biomedical Informatics 46(4), 744–763 (2013)
17. Proctor, M.: Drools: A rule engine for complex event processing. In: Schürr, A., Varró, D., Varró, G. (eds.) AGTIVE 2011. LNCS, vol. 7233, pp. 2–2. Springer, Heidelberg (2012)
18. Prud'hommeaux, E., Seaborne, A.: SPARQL Query Language for RDF. Technical report, W3C (January 2008)
19. Richards, M.J., Edwards, J.R., Culver, D.H., Gaynes, R.P.: Nosocomial infections in medical intensive care units in the united states. national nosocomial infections surveillance system. Crit. Care. Med. 27(5), 887–892 (1999)

Abstracting Classification Models Heterogeneity to Build Clinical Group Diagnosis Support Systems

Oscar Marin-Alonso[✉], Daniel Ruiz-Fernández, and Antonio Soriano-Paya

Dept. of Computer Technology. University of Alicante, Ctra. San Vicente del Raspeig s/n - 03690 San Vicente del Raspeig - Alicante - Spain
{omarin,druiz,asoriano}@dtic.ua.es

Abstract. Many diagnosis support systems (DSS) are focused on precise disorders, being not useful for differential diagnosis (DD) or facing comorbidities. Few DSSs offer a rich list of potential diagnoses and they do not reflect complex relations between diseases to be diagnosed. We present a model to allow collaboration of multiple heterogeneous diagnostic units (DU), which are actual DSSs, behaving as a whole system. The heterogeneity of the DUs refers to the disease they diagnose and the classification model they use to do so. This model offers a framework to build multi-purpose DSSs, assuring their operability and functioning despite the heterogeneity of the single diagnostic units.

Keywords: CDSS · Diagnosis support systems · Collaborative diagnosis · Group decision · Decision fusion

1 Introduction

The use of information and communication technologies (ICT) in the clinical field began in the early 60's of the last century, as computer science (CS) start its growing. CS areas of research like artificial intelligence (AI) are nowadays widely applied with this purpose. In fact many researches within AI, as other areas of CS, are inspired by biological systems studied in biomedical science, e.g., artificial neural networks, genetic algorithms, etc.

From the different ways CS technologies have been used to help practitioners in their tasks, supporting clinical decision-making was one of the first. This is not surprising since clinical practice is considered to be a continuous decision-making process that lasts from the first contact with a patient with a set of sings and symptoms, until the end of the cycle of clinical care, when the patient doesn't need any care. Clinical decision supports systems aim to help experts and other role users involved in healthcare (clinicians, staff, patients,...,) in any decision-making task that affects healthcare delivery. These systems provide knowledge and person-specific information intelligently filtered or presented at appropriate times, to enhance health and healthcare [1].

Diagnosis could be considered as the key stage during patients attention since its results will condition later stages like treatment assignment or prognosis.

© Springer International Publishing Switzerland 2015
J.M. Ferrández Vicente et al. (Eds.): IWINAC 2015, Part I, LNCS 9107, pp. 269–277, 2015.
DOI: 10.1007/978-3-319-18914-7_28

The process that leads to a final diagnosis is one of the most complex activities within clinical practice. This complexity is inherent to the decision-making task needed to perform diagnosis, and affects equally to artificial systems and human individuals or groups focused on diagnosing [2]. Besides, new ways of medical care are being developed, like evidence based medicine, which empowered the development of new laboratory tests leading to a new type of attention like personalized medicine. Nowadays diagnostic experts have at their disposal vast amounts of clinical data related to patients that they have to analyse and interpret at point of care in a very dynamic environment [3].

Is easy to imagine that all the previously shown difficulties can get worse in cases which are clinically difficult to deal with. Situations where diagnosis is a confusing and non obvious process, like comorbidities [4]. However, diagnosis support systems (DSS), have been traditionally focused on the diagnosis of one disease, not being helpful to face complex diagnostic situations which involve more than one disease and require the participation of multiple experts. There are few exceptions of systems which perform differential diagnosis processes (DDx)

Other common lack in current DSS due to its single-disease design, is that they miss to reflect complex relations between diseases. Sometimes the diagnosis of a sign is useful to be considered as an input to diagnose other diseases, e.g, hypertension and diabetes. The ability of dealing with diseases from multiple fields is a desired feature in a DSS. Furthermore, generalization capability improves the chances of a system to be accepted and used in daily practice. Diagnostic experts were more prone to use such systems that reflect more realistic diagnoses.

We have developed a model to build and configure group DSSs to overcome cited difficulties. Those group DSS are based on the collaboration of multiple heterogeneous diagnostic units (DU), which are actual DSSs, behaving as a whole system. The heterogeneity of the DUs refers to the disease they diagnose and the classification model they use to do so. This model offers a framework to build multi-purpose DSSs, assuring their operability and functioning despite the heterogeneity of the single diagnostic units. Besides DUs interaction can reflect relations between disorders are considered. The model is highly parametrized so experts can easily set preferences about the way final diagnoses should be emitted.

In the next sections we first give an overview of the fields related to our work (section 2), then a general description of the model we present (section 3), followed by a closer view on the output diagnose definition (section 4), and finally some conclusions reached throughout the work done (section 5).

2   Related Works

In the first times of CDSS's development they were mainly applied to diagnosis, following the model of the "Greek Oracle". Computers were supposed to replace human experts because they were expected to give the correct diagnose giving a rational explanation for it in all circumstances since they could store even more knowledge diagnostic expert. This view was proved to be false, and there was a shift of the views on these systems, to a more realistic scenario in which computers support clinicians tasks, being those the ones who, using the system, could offer better and more informed opinions. [5].

Nowadays there is a great consensus on the benefits that the use of CDSS offer in multiple ways to healthcare. Studies like [6] show the effect of these systems on healthcare actors (doctors, nurses, patients,....,), and the different sort of outcomes they produce.

Besides those benefits there still limitations and drawbacks in the use of CDSS. Recently, authors stated that the in the future DSS should cover the diagnosis of diseases from multiple clinical environments, since there are complex cases which involve many specialities. In [4], the authors show what they considered challenging topics in clinical decision support that should be faced in the next years. They highlight the suitability of combining recommendations for patients with comorbidities and the need of architectures for sharing executable CDS modules and services.

Traditionally, CDSSs were focused on a single disease. However, recently some works tried to fulfil the need for systems that consider a wider set of diseases, which could help to face comorbidities or cases in which a differential diagnose process is needed. The obvious way to do this is by combining single decisions related to different diseases. Some works tried this, like [7] and [8], but applied to treatment assignment and clinical guidelines merging respectively. Regarding diagnosis, authors of [9] present framework for decision fusion but having only mammographic masses as targeted diagnosis.

Talking about DDx, we should refer the review about DDx generators done by Bond et alt. [10]. In their review they compare recent and interesting DDx solutions scoring them using 14 criterion, resulting ISABEL[11] and DxPlain [12] with the highest scores.

These systems offer quite good diagnostic results but they are closed systems that don't take profit of other systems.Their knowledge comes from the accumulation of clinical knowledge related to diverse clinical field, nor from a collaboration between other expert systems. Besides they don't reflect complex relations between diseases, i.e., that one disease can be a sign of a second one.

3   Model's Description

In section 1 we introduced the convenience of having DSSs that support clinicians decision making task in complicated situations where diagnoses are unclear or many diseases are involved. In traditional clinical practice those situations are faced through processes that involve practitioners from different clinical specialities, e.g., DD. Besides, we established some ideas which drove the definition of a model to guide the building of group DSS to perform a sort of DD. This group DSS would be based on single specialized DUs that could interact reflecting the possible sign-of and symptom-of that could exist between the diseases they diagnose.

To build our model we follow a bottom-up reasoning strategy. We began focusing on the basis of a group DSS, the cited DUs. After that, we looked at them from the outside paying attention to their needs of interaction with other components of the system, i.e., other DUs, clinical databases (CDB) etc. Then, we saw the need to describe the ways all these elements will interchange data. Finally, when it came to data, we had to face the need for mechanisms to assure the understanding between heterogeneous systems that use that clinical data, and with the sources and destination of that data as well.

As a result, the model defines three basic components: diagnosis, control, and communication. They are characterized by defining their functionality and the sets of input and output data. In the following sections we go deeper into each one. We can also see a structural diagram of the described components in Fig.1.

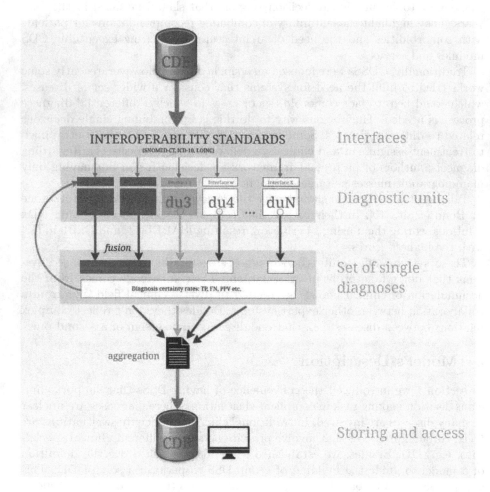

Fig. 1. Structural diagram of the main parts of the presented model and their connections

3.1   Diagnosis Component

This component mainly consists on the set of DUs that compound the group DSS at a certain time. From the model's point of view, each DU is a black box, that is, we only pay attention to the way they interact with other elements. Details about the classifying model the DU uses to diagnose and how this is performed are not considered.

Each DU is an actual DSS focused on the diagnose of a certain disease, being possible that more than one diagnoses the same disease. Every DU has a known set of input features that they need to perform this task. The description of these features is needed to feed the right clinical input data into the right way to the DU and to manage its functioning within the group DSS once it is part of it.

Regarding the DUs output, this would not only consist on a diagnostic value, but also on statistical values related to its diagnosing performance: accuracy rates, true positive and negative rates, false positive and negative rates, positive and negative predictive values etc. Those values will be used to conform the final diagnose as it is explained in section 4.

3.2   Control Component

The control component builds up the mechanisms to manage the global diagnosis task. This task entails having an updated and precise image of the DUs that form the group DSS at a time. Having a clear picture of each DU, we are able to reflect the existing relations between the set of targeted diseases by the group DSS. For instance, we can have a simple DU which diagnoses fever, whose output can be used as an input, with other patient-related data, to diagnose other diseases to which fever is a sign or a symptom.

Main functions of this component are:

- Management of DUs: discovering,information maintenance, interrelations, etc.
- Management of the updated set of diseases that the group DSS considers.
- DD process driving: DUs selection, clinical data handling (gathering and dissemination), global output emission (detailed in 3.1), etc.
- Interoperability maintenance, in a joint effort with communication component (section 3.3).

3.3   Communications Component

Communication between elements and components of a group that must interact to achieve their common goal is a key question. Moreover, in the clinical field such thing is not easy due to the lack of standard frameworks and communication protocols, being diverse the ways and languages used to make clinical systems work together.

Communication component of the model will assure the interoperability among its components and between them and CDBs or other systems that are the source and/or the destination of the clinical data. These origins of data could use different clinical standards to store and transmit patient-specific data, but mainly standards such as SNOMED-CT, ICD-10 and LOINC. Each of them is thought to codify or classify clinical terminology, diseases/symptoms/signs, and clinical tests.

The group DSS would have to take into account all of those standards using them all together. There could be cases in which an input, (e.g., a clinical test value), is received using a standard that differs from the one that a DU expects. In such cases, mapping and translation tasks would be needed in order to abstract model's components and users from this variety of communication protocols.

A way to do this is to define a simple new protocol to allow the communication among model's elements. For the sake of simplicity, this could be done using a language like XML, and defining messages which encapsulate meta-data related to the data that is to be transferred, e.g., standard used to codify the data, if a value is missing or not, etc. This helps the communication and control components to manage the sending and reception of data within the group DSS, and from it to the outside.

4 Diagnosis Selection

One of the difficult questions that arise during a group diagnosis process of a pathology and a patient is how to choose between divergent opinions when all of them come from experts and are based on the same evidences. Besides this, if we want to offer a list that includes diagnosis of multiple diseases, we have to fuse decisions coming from multiple experts regarding multiple diseases. To do so we will need on one hand to measure the goodness of each expert's decision and on the other hand, a procedure, which using such measurement's results, to choose between decisions through fusion and aggregation to build a final list of diagnoses. In the following sections we show how we face both needs in our proposal.

4.1 Measurement of Diagnosis Goodness

Analysing these situations we notice that this is a multi-objective problem where a certain set of requirements should be met before accepting a decision. if we have multiple options that do so, then we can use optimization procedures of prioritised variables to choose between the candidate decisions which met the set requirements.

For instance, when deciding about which treatment should be assigned to a patient, first, it has to be considered the minimum criteria established by the public health administration, regarding safety measures, possible secondary effects and treatment efficacy. After meeting those requirements, it is time to decide which variables optimization will be estimated. These variables could be treatment's associated cost (prescribe the cheapest drug) or efficacy (treatment with a higher rate of recovery after its use).

A problem happens when the chosen decision does not imply an optimization of the variables or neither meets some of the requirements. Then, subjective arguments like the expertise of the ones involved in the decision-making task should be used to minimize the risk of the decision.To quantify such subjective arguments is difficult. For instance, talking about expertise, we know that it depends on the years of practise and the amount of times the decision expert has faced the same or a similar situation. Moreover, subjective arguments don't assure the suitability of a decision. In group decision-making is common to avoid these situations by reaching an agreement between the experts involved in the decision group.

One of the main parts of the work we present is the mechanism used to reach a final set of diagnoses. This model's component contains a set of minimum needs, i.e., requirements that DUs must meet, and zero or more variables to optimize inspired by the way human experts groups come to a decision. In addition, DUs have some attributes that allow to rank them using weights which are associated with the diagnose they propose. We can group these attributes as follows:

- **DUs variables (Du_Var):** this tag includes all the variables related to DUs, which are used to score the diagnostic opinion a DU emits. This score is used to weight each DU diagnose to build a global final diagnostic list. Examples of these variables are: priorly estimated accuracy rate, sensitivity, specificity and other clinical statistics related to diagnosis.
- **Diagnostic variables (Di_Var):** within this group we include the variables associated with the diagnose decision whose values optimization will be estimated for each DU decision. For instance: decision's cost, prevalence, morbidity etc.
- **DUs description data (Du_Desc):** these data includes all th information useful to identify each DU of the group DSS: diagnosed disease, id value, etc.
- **Diagnostic description data (Di_Desc):** this tag includes data associated to the actual diagnose that a DU emits: description, diagnostic value etc.

All the attributes and data associated with DUs used in this process are included in the metadata that model's elements interchange using XML messages, as it was explained in section 3.3.

4.2   Group Diagnosis Algorithm

The group decision algorithm used won't be unique, on the contrary, different versions of it will be configured depending on the precise DUs involved in the diagnostic process. Diagnosis requirements and variables to optimize (section 4.1) will be considered also into the configuration process. All this information will be included as metadata in configuration XML-based messages, containing also the criteria for consensus or decision fusion. The criteria will be weighted using a weighting function applied to each DU.

An example could be an algorithm to diagnose a patient who shows certain set of signs and symptoms. The algorithm states that each DU should offer specificity and sensitivity rates over 50% diagnosing the targeted disease to consider their decision (minimum requirements). In addition, the set of diagnoses will be ordered by its complexity, giving priority to those whose treatment is more simple (variables optimization). Finally, decisions from DUs will be weighted using its sensitivity criteria.

5   Conclusions

In this paper we present a model that can be helpful to build group DSSs. What is new in this model is that we follow a collaborative strategy to reach a common diagnose. Multiple heterogeneous diagnostic units), which are actual DSSs specialised in the diagnosis of different, or coincident in some cases, diseases will collaborate to fulfil a differential diagnosis process, behaving as a whole system.

One of the main advantages of the model is that it defines ways to hide the heterogeneity of the DUs when it comes the classification model they use to perform their task. This idea clearly simplifies the management and growing of the global system, improving its modularity and scalability, which means that new DUs would be easily added to the system. This property is interesting because allows the system to easily adapt to changes in diagnostic procedures: new laboratory tests, clinical findings, new relations between signs and symptoms etc.

Heterogeneity associated with the disease targeted by each DUs is also hidden to users. This way the users have the perception of a global system able to go through the diagnosis of the combined set of diseases.

Another important issue considered in the definition of the model its the importance of interoperability. This issue is viewed from two sides, first from the inner side of the group DSS. Each of its component could used different clinical standards, and despite this, they are expected to work together. Secondly, the group DSS implemented using this model will have to fit in actual clinical workflows, operating with other systems, i.e., electronic health records, laboratory tests repositories etc. So to assure interoperability needs in both cases, we have defined a new protocol based on XML, which includes metadata, that ease this communication.

At the time to emit the diagnose, the model offers multiple ways to show output diagnostic data thanks that it is highly parametrized. Depending on optimization of certain values or the fulfilling of set requirements diagnostics results would vary.

For all the highlighted reasons, we consider that the model presented in this paper is useful for building group DSSs, which can offer new and suitable ways ,through collaboration, to perform complex diagnostic processes, e.g., comorbidities, or others that entail the need of DDx.

References

1. Osheroff, J.A., Teich, J.M., Middleton, B., Steen, E.B., Wright, A., Detmer, D.E.: A Roadmap for National Action on Clinical Decision Support. Journal of the American Medical Informatics Association: JAMIA 14(2), 141–145 (2007)

2. Miller, R.A., Geissbuhler, A.: Diagnostic Decision Support Systems. In: Berner, E.S. (ed.) Clinical Decision Support Systems: Theory and Practice, 2nd edn.,
   pp. 99–125. Springer, New York (2007)
3. Musen, M.A., Blackford, M., Greenes, R.A.: Clinical Decision-Support Systems. In: Shortliffe, E.H., Cimino, J.J. (eds.) Biomedical Informatics: Health Informatics, 3rd edn., pp. 643–674. Springer, London (2014)
4. Sittig, D.F., Wright, A., Osheroff, J.A., Middleton, B., Teich, J.M., Ash, J.S., Campbell, E., Bates, D.W.: Grand challenges in clinical decision support. Journal of Biomedical Informatics 41(2), 387–392 (2008)
5. Miller, R.A.: Medical diagnostic decision support systems–past, present, and future: a threaded bibliography and brief commentary. Journal of the American Medical Informatics Association: JAMIA 1(1), 8–27 (1991)
6. Bright, T.J., Wong, A., Dhurjati, R., Bristow, E., Bastian, L., Coeytaux, R.R., Samsa, G., Hasselblad, V., Williams, J.W., Musty, M.D., Wing, L., Kendrick, A.S., Sanders, G.D., Lobach, D.: Effect of clinical decision-support systems: a systematic review. Annals of Internal Medicine 157(1), 29–43 (2012)
7. Riaño, D., Collado, A.: Model-based combination of treatments for the management of chronic comorbid patients. In: Peek, N., Marín Morales, R., Peleg, M. (eds.) AIME 2013. LNCS(LNAI), vol. 7885, pp. 11–16. Springer, Heidelberg (2013)
8. Jafarpour, B., Abidi, S.S.R.: Merging disease-specific clinical guidelines to handle comorbidities in a clinical decision support setting. In: Peek, N., Marín Morales, R., Peleg, M. (eds.) AIME 2013. LNCS(LNAI), vol. 7885, pp. 28–32. Springer, Heidelberg (2013)
9. Prasad, S., Bruce, L.M., Ball, J.E.: A multi-classifier and decision fusion framework for robust classification of mammographic masses. In: 30th Annual International Conference of the IEEE Engineering in Medicine and Biology Society, EMBS 2008, vol. 2008, pp. 3048–3051. IEEE, Vancouver (2008)
10. Bond, W.F., Schwartz, L.M., Weaver, K.R., Levick, D., Giuliano, M., Graber, M.L.: Differential diagnosis generators: an evaluation of currently available computer programs. Journal of General Internal Medicine 27(2), 213–219 (2012)
11. Ramnarayan, P., Kulkarni, G., Tomlinson, A., Britto, J.: ISABEL: A novel Internet-delivered clinical decision support system. In: Bryant, J. (ed.) Current Perspectives in Healthcare Computing, Harrogate, pp. 245–256 (2004)
12. Barnett, G.O., Cimino, J.J., Hupp, J.A., Hoffer, E.P.: DXplain: An Evolving Diagnostic Decision-Support System. JAMA 258(1), 67 (1987)

Using EEG Signals to Detect the Intention of Walking Initiation and Stop

Enrique Hortal[1], Andrés Úbeda[1], Eduardo Iáñez[1], Eduardo Fernández[2], and Jose M. Azorín[1]

[1] Brain-Machine Interface Systems Lab, Miguel Hernández University, Av. de la Universidad s/n, 03202 Elche, Spain
{ehortal,aubeda,eianez,jm.azorin}@umh.es
[2] Biomedical Neuroengineering Group, Miguel Hernández University, Av. de la Universidad s/n, 03202 Elche, Spain
e.fernadez@umh.es

Abstract. The ability of walking brings us a great freedom in our daily life. However, there is a huge number of people who have this ability diminished or are not even able to walk due to motor disabilities. This paper presents a method to detect the voluntary initiation and stop of the gait cycle using the ERD phenomenon. The system developed obtains a good accuracy in the detection of the rest and walking state (70.5 % and 75.0 %, respectively). Moreover, the average detection of the onset and ending instants of the gait is detected with a 65.2 % of accuracy. Taking into account the number of intentions of initiation and stop of the gait, the system reaches a good True Positive Rate (around 65%) but obtaining a still improvable False Positive Rate (15.4 FP/min in average). By reducing this factor, this detection system can be used in future works to control a lower limb exoskeleton or a wearable robot. These devices are very useful for rehabilitation and assistance procedures in patients with motor problems affecting their lower limb.

Keywords: EEG Signals · Gait analysis · ERD · Movement detection

1 Introduction

Brain-Machine Interfaces (BMI) represent a great help for people with disabilities or motor damage. Due to spinal cord injury, stroke or other causes, these patients might not have the chance of performing movements as common as picking up a glass of water or walking. Different methods have been applied in order to use BMIs to solve, or at least reduce, this kind of impediments [1,2]. Electroencephalographic (EEG) systems allow the measurement of the brain activity over the motor cortex while subjects are performing motor tasks [3,4]. This information can be used to control lower limb exoskeletons or wearable robots, providing an alternative communication path between the brain (by detecting the patient's movement intention) and these devices [5,6].

© Springer International Publishing Switzerland 2015
J.M. Ferrández Vicente et al. (Eds.): IWINAC 2015, Part I, LNCS 9107, pp. 278–287, 2015.
DOI: 10.1007/978-3-319-18914-7_29

Current technology allows registering and processing EEG signals that occur just before performing an action and thus it is possible to know the movement intention [7,8,9]. This methodology allows assisting motor movements when it is necessary. The detection of these movement intentions can be very useful in motor rehabilitation processes. For instance, through this detection, an exoskeleton attached to the lower limb [10,11] could allow patients with disabilities to walk. The coordination between the desire to execute a movement and the performance of the action itself increases the likelihood of the brain to create new communication channels due to neuronal plasticity [12]. Using this phenomenon, the effects of rehabilitation could increase more efficiently.

There are two widely used neurophysiological phenomena that begin before a voluntary action: the Bereitschaftspotential (BP or readiness potential) and the Event-Related Desynchronization (ERD). The slow potential BP is generally described as a decrease in the component closest to the DC component in EEG signals [13]. On the other hand, ERD represents a decrease in the spectral power of the EEG signals in the *mu* and *beta* frequency bands [14]. In this paper, a methodology to detect the walking intention onset using a non-invasive system based on ERD is presented. The main goal is the development of a system which allows controlling an exoskeleton attached to the lower limb. These system could be applied for both functional rehabilitation and assistance of walking.

2   System Architecture

The designed system is able to detect four different human gait states: *Relax*, *Start*, *Walking* and *Stop*. To that end, a motion capture system is used to analyze lower limb kinematic data to obtain real indices of these states. Afterward, the brain activity is analyzed to detect the current state in each moment. The brain signals are acquired using a non-invasive BMI system which provides 32 EEG channels.

2.1   Brain-Machine Interface

EEG signals are recorded through 32 active Ag/AgCl electrodes (g.LADYbird model - g.tec Medical Engineering GmbH, Austria) distributed over the scalp. These electrodes are placed on the positions Fz, FC5, FC3, FC1, FCz, FC2, FC4, FC6, C5, C3, C1, Cz, C2, C4, C6, CP5, CP3, CP1, CPZ, CP2, CP4, CP6, P3, P1, Pz, P2, P4, PO7, PO3, POz, PO4 and PO8 according to the International 10/10 System and covering central and parietal regions. A monoauricular reference is placed on the earlobe and the ground is located on AFz. To ensure a better placement of the electrodes, a g.GAMMAcap (g.tec Medical Engineering GmbH) is used. This cap allows a quick placement of electrodes. Moreover, this system is able to reduce motion artifacts and electromagnetic interference. The EEG signals are amplified using two g.USBamp (g.tec Medical Engineering GmbH). The sample frequency used to acquire the signals is 1200 Hz. A computer software developed in MatLab (The MathWorks, Inc., Natick, MA, USA)

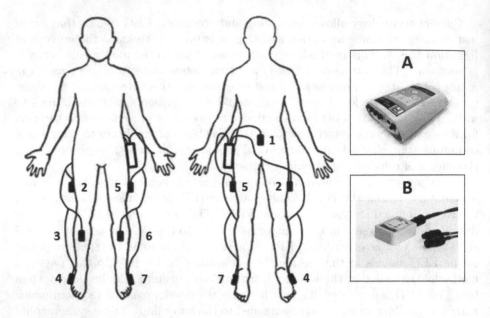

Fig. 1. INERTIAL MEASUREMENT SYSTEM. Position of the IMU sensors (1-Lumbar, 2-Right thigh, 3-Right leg, 4-Right foot, 5-Left thigh, 6-Left leg, 7-Left foot). Tech Hub that manages all the IMUs (A) and Inertial Measurement Unit (IMU) (B).

reads and processes the data acquired using the API (Application Programming Interface) provided by the manufacturer (gUSBamp MATLAB API).

2.2   Motion Capture System

The motion capture system used in this work is the Tech MCS (Technaid S.L., Spain). This product is a complete wireless motion analysis system. It manages seven Inertial Measurement Units (IMUs) which are used in the experiments (see Figure 1). Each Tech-IMU (Technaid S.L.) integrates three different types of sensors as an accelerometer, a gyroscope and a magnetometer.

In this work, the seven IMUs are distributed as follows: three sensors are placed on each lower limb (foot, thigh and leg) and the last one is placed on a lumbar position. Each IMU registers 19 variables corresponding to different parameters as rotation (nine parameters corresponding to the rotation matrix), acceleration (three parameters, m/s^2), angular velocity (three parameters, rad/s), magnetic field (three parameters) and temperature. Rotation parameters are used to detect gait initiations and stops. This data are acquired through a HUB that is connected to the PC USB port with a data acquisition frequency of 30 Hz.

3   Experimental Procedure

3.1   Test Protocol

To detect the different states of human gait, brain activity will be analyzed. For this purpose, a proper test protocol has been designed. The protocol consists of the execution, on a voluntary basis, of several changes in the gait state of the user. To that end the user performs, for each session, several initiations and stops of walking without using any external stimulus to indicate each of these changes. Voluntarily, users are asked to perform a total of 10 initializations and stops of the gait process, with a waiting period of more than two seconds between each of the changes. This requisite is applied to assure the minimum window time required to detect the onset and the end of movement using the ERD phenomenon (see Section 3.3).

Three male users aged between 22 and 29 years old (26.7±4.0) took part in the experiment. Each of them carried out a total of 8 runs per session with 10 complete cycles of motion (relax/start/walking/stop). All of them completed two sessions performed in two different days.

3.2   EEG Signals Processing

In order to enhance the EEG signals quality it is necessary to increase the signal-to-noise ratio. The amplifier includes several internal filters that can be applied to the input signals. Due to the fact that EEG signals are very noisy, two of these internal filters are applied. In the current work, a low pass filter with a cut off frequency of 100 Hz and a 50 Hz notch filter to eliminate the power line interference have been applied. Moreover, an 8th order Butterworth band pass filter programmed in MatLab from 5 Hz to 40 Hz is applied to remove artifacts and the DC component, preserving only the information of the frequencies of interest, which are *mu* and *beta* frequency bands (between 8 and 30 Hz).

Then, a spatial filter is applied to all EEG channels to reduce the contribution of the remaining electrodes in each channel and therefore to better isolate the information collected from each sensor. To do that, a Laplacian algorithm is applied to all the electrodes. This algorithm uses the information recorded from all the remaining electrodes and their distances to the sensor of interest. The visual result is a smoother time signal which should contain only the contribution coming from the particular position of the electrode. The Laplacian is computed according to the formula:

$$Vi^{LAP} = Vi^{CR} - \sum_{j \in Si} g_{ij} Vj^{CR} \tag{1}$$

where Vi^{LAP} is the result of applying this algorithm to the electrode i, Vi^{CR} is the electrode i signal before the transformation and,

$$g_{ij} = \frac{\frac{1}{d_{ij}}}{\sum_{j \in Si} \frac{1}{d_{ij}}} \tag{2}$$

where Si contains all the electrodes except from the electrode i and d_{ij} is the distance between electrodes i and j.

3.3   Data Selection

ERD is a phenomenon which refers to the decrease of the EEG signal power in the *mu* and *beta* bands related to the preparation and performance of voluntary motor tasks. This desynchronization starts about two seconds before the movement onset as it is stated in [14]. The study shows that ERD appears over the contralateral Rolandic region and becomes bilaterally symmetrical immediately before execution of a right hand movement. Although the movement performed in our experiment is not the same, we hypothesize that it may occur in the same time interval and also over the motor cortex. When the performance of the movement ends, the *mu* and *beta* bands recover the power and produce the event-related synchronization (ERS).

Typically, ERD-based research uses around two seconds width windows to analyze this phenomenon. As it is explained in Section 2.2, the kinematic data recorded are used to determine the current state of the walking cycle. The detection of the initialization and the stop of the gait allows the classification of the data in four different groups (Figure 2). Data between three seconds prior to gait onset and the onset itself is established as *Start* state. The same occurs with the *Stop* state, which is considered between three seconds prior to the end of the walking process and the end itself. Data between these states are considered as *Relax* state (between a *Stop* and a *Start*) and *Walking* state (between a *Start* and a *Stop*). With this procedure, a bigger amount of data is obtained for the *Walking* and *Relax* states (twice the size of *Start* and *Stop* data approximately). Using these data, the training model allows a better identification of these two states (*Walking* and *Relax*) avoiding false detection (False Positive or FP) of the *Start* and *Stop* states.

3.4   Feature Extraction

The four data groups obtained in Section 3.3 are segmented in windows of 1 second each 0.2 seconds (overlap of 0.8 seconds). Each window is processed separately to extract the features which represent the task. The selected EEG data are processed with a Fast Fourier Transform (FFT) to compute the spectral power. The features are the sums of three frequency bands, 8-12 Hz, 13-24 Hz and 25-30 Hz per each electrode which represent *mu* and *beta* bands, so 96 features define each class (32 electrodes, 3 features per electrode).

3.5   Classification

To determine the state of the walking cycle, a SVM-based (Support Vector Machine) classifier is used. The SVM classifier is a very useful technique for data classification [15]. To do the classification, SVM makes use of a hyperplane or

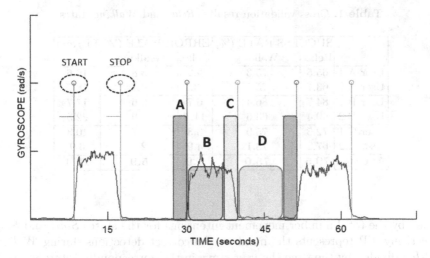

Fig. 2. DATA SELECTION. In this figure, the gyroscopes analysis is shown. *START* and *STOP* detections are marked. Moreover, the data used as *Start* (A), *Walking* (B), *Stop* (C) and *Relax* (D) are also represented.

groups of it in a very high (even infinite) dimensional space to distinguish the different classes to classify. The accuracy of the SVM-based classifier depends on the kernel used. In the case of a BMI system, generally a Gaussian kernel or a Radial Base Function (RBF) is applied [16]. In this case, a SVM-based system with a RBF kernel is used. A one-step multiclass strategy is used in the SVM system. In order to create the model and to detect the gait state, the data obtained following the procedure described in Section 3.4 are used.

After performing the classification of the four states (R:*Relax*, S:*Start*, T:*Stop* and W:*Walking*) the following confusion matrix is obtained:

$$
\begin{array}{c}
Detected \\
\begin{array}{cccc} R & S & T & W \end{array} \\
Real\ \begin{array}{c} R \\ S \\ T \\ W \end{array}
\begin{bmatrix}
c_{11} & c_{12} & c_{13} & c_{14} \\
c_{21} & c_{22} & c_{23} & c_{24} \\
c_{31} & c_{32} & c_{33} & c_{34} \\
c_{41} & c_{42} & c_{43} & c_{44}
\end{bmatrix}
\end{array}
$$

where c_{11} is the *Relax* Success Rate, c_{22} the *Start* Success Rate, c_{33} the *Stop* Success Rate and c_{44} the *Walking* Success Rate. The *Relax* Error Rate and the *Walking* Error Rate correspond to c_{12} and c_{43}, respectively. The remaining elements of the confusion matrix do not affect a correct performance of the system, as a consequence, they are not taken into account when calculating the accuracy of the system.

To validate this system in the control of real devices such as an exoskeleton, two more parameters are defined: True Positive Rate (TP) and False Positive Rate (FP). TP represents the number of valid commands sent to the exoskeleton

Table 1. Cross validation results. *Relax* and *Walking* states

	SUCCESS RATE (%)		ERROR RATE (%)		FP/min
	Relax	Walking	Relax	Walking	
User A.1	65.5	75.3	8.2	3.8	16.6
User A.2	63.7	75.7	13.1	6.7	18.7
User B.1	84.7	86.4	6.7	9.6	17.7
User B.2	69.4	63.6	11.1	8.9	22.3
User C.1	72.5	75.9	3.3	3.7	10.3
User C.2	67.2	73.1	4.9	2.4	6.9
Average	**70.5**	**75.0**	**7.9**	**5.9**	**15.4**

divided by the total number movement intentions for the states *Start* and *Stop*, respectively. FP represents the number of incorrect detections during *Walking* or *Relax* divided by the time the user stays in the corresponding state.

4   Results and Discussion

The results for the classification of the different gait states are shown in Tables 1 and 2. These results are calculated offline, by analyzing the data after an 8-fold cross validation (each session run is used as a fold). In Figure 3, an example of classification is shown (User A, one fold). In order to design a useful detection system for the control of rehabilitation or assisting devices, it is important to obtain a reliable behavior in the execution of control commands (*Start* and *Stop* states in this case). As it was mentioned in Section 3.3, the method followed allows a better detection of the rest periods (*Relax* state) and the continuous walking (*Walking* state). In Table 1, the behavior of the system in the detection of these states is shown.

Firstly, columns labeled as "SUCCESS RATE (%)" show the system accuracy in the detection of *Relax* and *Walking* states. In columns labeled as "ERROR RATE (%)" the error in the classification of these states is shown. An erroneous *Start* detections during a *Relax* state or wrong *Stop* detections during walking are considered as an error (or False Positive). Finally, the number of False Positives detected per minute (column "FP/min") is represented. This parameter is very important in the design of this kind of control systems. The error rates show a good classification index regarding this parameter (7.9% in *Relax* periods and 5.9% in *Walking* periods). However, the number of FP/min is too high (15.4 FP/min on average). This parameter should be reduced to be useful in a real time application. Taking into account these accuracies, all the users obtained similar results. However, User C achieved lower FP/min than the rest of the users. Furthermore, Table 1 shows a good success rate for the *Relax* and the *Walking* periods for all users, reaching and average of 70.5% and 75.0% respectively.

On the other hand, in Table 2, the success rate for the *Start* and the *Stop* states are shown (columns labeled as "SUCCESS RATE (%)"). The number of

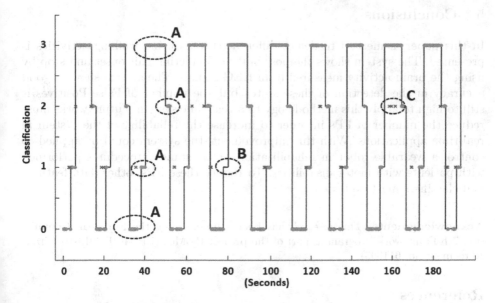

Fig. 3. DATA CLASSIFICATION. Y-axis represent the task performed (0-Relax, 1-Start, 2-Stop and 3-Walking). Data marked as A represent the correct detection of the four different states of the gait cycle. B is a not detected gait intention. C represents an error in the classification which provoke False Positives.

Table 2. Cross validation results. *Start* and *Stop* detection

| | SUCCESS RATE (%) | | TP rate (%) |
	Start	Stop	
User A.1	34.4	26.3	82.2
User A.2	49.5	36.5	94.1
User B.1	32.2	16.4	62.9
User B.2	42.0	13.0	66.2
User C.1	15.1	8.5	55.8
User C.2	10.3	1.1	30.2
Average	**30.6**	**17.0**	**65.2**

events properly classified is represented in the "TP rate (%)" column. Regarding success rates, these values are not very high. However, this accuracy does not represent the actual control commands of the system, but the number of trials that have been detected inside the movement window selected to analyze movement intention (*Start* or *Stop*, according to the method explained in Section 3.3). In relation to the "TP rate (%)", User A obtained clearly a better classification than User B and a remarkably higher rate than User C, who achieved the worst results.

5   Conclusions

In this paper, a method to detect different states during walking activities is presented. The system shows the possibility of detecting gait onset and stop by using the brain activity measured from EEG signals. The system shows a good accuracy in the detection of these states but the number of False Positives is still to high to apply this methodology to a real-time system. Future works must reduce the number of FPs in order to increase the reliability of the system in real-time applications. With this improvement, the system could be applied to control a wearable robot in rehabilitation or assistance procedures performed with patients with motor disabilities. To reduce these FPs, other data features and classifiers must be tested.

Acknowledgments. This research has been supported by the European Commission 7th Framework Program as part of the project BioMot (FP7-ICT-2013-10, Grant Agreement no. 611695).

References

1. Daly, J.J., Wolpaw, J.R.: Brain-computer interfaces in neurological rehabilitation. The Lancet Neurology 7(11), 1032–1043 (2008)
2. Wei, L., Yue, H., Jiang, X., He, J.: Brain Activity during Walking in Patient with Spinal Cord Injury. In: International Symposium on Bioelectronics and Bioinformatics (ISBB), pp. 96–99 (2011)
3. Hortal, E., Úbeda, A., Iáñez, E., Azorín, J.M.: Control of a 2 DoF Robot Using a Brain-Machine Interface. Computer Methods and Programs in Biomedicine 116(2), 169–176 (2014), New methods of human-robot interaction in medical practice,
4. Wolpaw, J.R., Birbaumerc, N., McFarland, D.J., Pfurtscheller, G., Vaughan, T.M.: Brain-computer interfaces for communication and control. Clinical Neurophysiology 113, 767–791 (2002)
5. Moreno, J.C., Collantes, I., Asin, G., Pons, J.L.: Design of better robotic tools adapted to stroke rehabilitation practice. In: World Congress on Medical Physics and Biomedical Engineering (2012)
6. Bortole, M., del Ama, A.J., Rocon, E., Moreno, J.C., Brunetti, F., Pons, J.L.: A Robotic Exoskeleton for Overground Gait Rehabilitation. In: IEEE International Conference on Robotics and Automation (ICRA), pp. 3356–3361 (2013)
7. Bai, O., et al.: Prediction of human voluntary movement before it occurs. Clinical Neurophysiology 122, 364–372 (2011)
8. Ibáñez, J., Serrano, J.I., del Castillo, M.D., Barrios, L., Gallego, J.Á., Rocon, E.: An EEG-Based Design for the Online Detection of Movement Intention. In: Cabestany, J., Rojas, I., Joya, G. (eds.) IWANN 2011, Part I. LNCS, vol. 6691, pp. 370–377. Springer, Heidelberg (2011)
9. Planelles, D., Hortal, E., Costa, A., Iáñez, E., Azorín, J.M.: First steps in the development of an EEG-based system to detect intention of gait initiation. In: 8th Annual IEEE International Systems Conference, Ottawa, Canada, pp. 167–171 (2014)
10. Dollar, A.M., Herr, H.: Lower Extremity Exoskeletons and Active Orthoses: Challenges and State-of-the-Art. IEEE Transactions on Robotics 24(1), 144–158 (2008)

11. Moreno, J.C., del Ama, A.J., de los Reyes-Guzmán, A., Gil-Agudo, A., Ceres, R., Pons, J.L.: Neurorobotic and hybrid management of lower limb motor disorders: a review. Medical & Biological Engineering & Computing 49(10), 1119–1130 (2011)
12. Koralek, A.C., Jin, X., Long, J.D., Costa, R.M., Carmena, J.M.: Corticostriatal plasticity is necessary for learning intentional neuroprosthetic skills. Nature 483, 331–335 (2012)
13. Shibasaki, H., Hallett, M.: What is the Bereitschaftspotential? Clinical Neurophysiology 117, 2341–2356 (2006)
14. Pfurtscheller, G., Lopes da Silva, F.H.: Event-related EEG/MEG synchronization and desynchronization: Basic principles. Clinical Neurophysiology 110(11), 1842–1857 (1999)
15. Hsu, C.W., Chang, C.C., Lin, C.J.: A practical guide to support vector classification (2003), http://www.csie.ntu.edu.tw/~cjlin/libsvm/
16. Flórez, F., Azorín, J.M., Iáñez, E., Úbeda, A., Fernández, E.: Development of a low-cost SVM-based spontaneous Brain-Computer Interface. In: International Conference on Neural Computation Theory and Applications, pp. 415–421 (2011)

Low-cost Remote Monitoring
of Biomedical Signals

J.M. Morales[1]([✉]), C. Díaz-Piedra[2], L.L. Di Stasi[2], P. Martínez-Cañada[1],
and S. Romero[1]

[1] Brain Computer Interface Lab, Department of Computer Architecture and
Technology, University of Granada, Granada, Spain
jm3661@correo.ugr.es, {pablomc,sromero}@ugr.es
[2] Mind, Brain, and Behavior Research Center (CIMCYC), University of Granada,
Granada, Spain
{dipie,distasi}@ugr.es

Abstract. The great usefulness of remote recording of biomedical signals in most aspects of daily life has generated an increasing interest in this field. Traditionally, monitoring devices from clinical enviroments are bulky, intrusive, and expensive. Thus, the development of wearable, mobile, and low-cost applications is desirable. Nevertheless, recent improvements in open-hardware allow developing low cost devices and portable designs for biosignal monitoring in out-of-lab applications, such as sports, leisure, e-Health, etc. This paper presents a low-cost wearable system able to simultaneously record electrical brain and heart activity (i.e. electroencephalography and electrocardiography). The system is able to send biomedical data to a platform for remote analyses. Both software and hardware are open-source. We assessed the system for its validity and reliability in a real road environment.

Keywords: Brain waves · eHealth · Electrocardiogram · Electroencephalogram · Wearable platform

1 Introduction

Nowadays, humans face increasing operational task demands, where cognitive skills are more important than physical ones [10, 13]. When operational safety is a prime concern, it's very relevant to know the actual operator's cognitive state (CS). Electroencephalographic (EEG) metrics are the most reliable current measures to assess operator's CS [9, 14, 15]. Electrocardiographic (ECG) data (heart rate and other indices) also seems to be useful providing information about operator's CS [19]. However, the ability to objectively and sensitively measure CS online in real scenarios remains a major challenge [12]. Both, EEG and ECG have failed to gain traction in some applied domains -as driving safety-, due to the technical and methodological difficulties of measuring these signals in everyday tasks, and the intrusiveness and bulkiness of the equipment. In recent years, user-friendly commercial mobile devices have overcome many of these

© Springer International Publishing Switzerland 2015
J.M. Ferrández Vicente et al. (Eds.): IWINAC 2015, Part I, LNCS 9107, pp. 288–295, 2015.
DOI: 10.1007/978-3-319-18914-7_30

1997　　　　　　　　　　2015

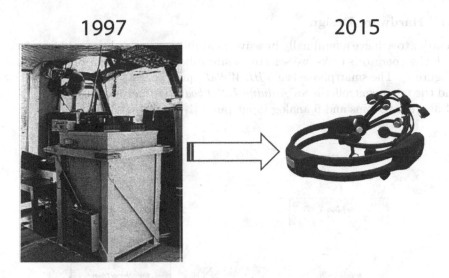

Fig. 1. The evolution of the EEG monitoring systems from the Airborne Spectrum 32 (left) to the contemporary Emotiv Epoc Headset EEG system (right) (adapted from [3, 11]).

barriers [14]. The miniaturization of systems and new electronic approaches (e.g. open-hardware platforms) provide solutions for mobile and wearable biosignals recording [16–18]. Thus, current studies can rely on off-the-shelf systems to assess operator's CS non-invasively, via unobtrusive devices. These new devices offer high recording quality, comparable to professional clinical systems, and permit a good trade-off between costs and performance.

Examples of low-cost available EEG recording devices are the *Emotiv EPOC* (Emotiv, San Francisco, CA, USA) - a 14 EEG-channels device, with wet electrodes, able to measure attention levels [3]-, the *Emotiv Insight* (Emotiv, San Francisco, CA, USA) -a 5 channels EEG device, able to detect facial expressions (smile, blinking) [4]-, or the most affordable *MindWave* ($80, *NeuroSky*, San Jose, CA, USA, see below), that provides EEG signals from a single dry-sensor [5].

Here, we developed a low-cost remote bio-monitoring platform for using in real life tasks. The main features of our system include: a) low-cost recording solutions, b) user-friendly interface for analyzing and visualizing biosignal data, and c) low-invasive, comfortable, and portable elements. Finally, the system allows continuous monitoring of operator, giving instant feedback about his/her CS and location.

2 Platform Design

In the following sections, we will describe the platform's hardware and software design and an example of information recorded during its use during driving tasks.

2.1   Hardware Design

In order to achieve a minimally invasive wearable system and to avoid interfering with the operator's tasks, we selected a smarphone/tablet and a microcontroller (Figure 2). The smartphone is a *THL W200* (quad-core 1,5 GHz, 1GB RAM) [8] and the microcontroller is an *Arduino UNO board* (Atmel ATMega328, 16 MHz, 14 digital I/O pins and 6 analog input pins) [1].

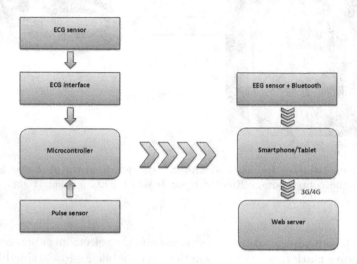

Fig. 2. Main hardware blocks of the recording system. The smartphone *THL W200* and the microcontroller *Arduino UNO* collect data from different sensors (here, ECG, EEG and pulse sensors) and send these data to the web server through a 3G/4G data connection.

The microcontroller unit (MCU) collects data from two sensors. Heart rate is detected by a *PulseSensor* [7] device, directly connected to one of the Arduino's analog inputs. Additionally, a set of three electrodes -connected to an *Arduino eHealth Shield* [2]- records ECG signal.

An *RN-XV* (Figure 3) Wi-Fi Shield is superposed to the previous shield to allow *Arduino* sending ECG data to the smartphone/tablet. EEG data is collected by the *NeuroSky MindWave Mobile* headset [6] (Figure 4). This product employs a single frontal dry electrode to record brain activity, and includes a Bluetooth link to send EEG data to the smartphone/tablet.

As the smartphone/tablet only allows for one active connection, we chose a combined Bluetooth/Wi- Fi scheme to connect both the EEG sensor and the ECG to the MCU. The connection from the smartphone/tablet to the web server uses a 3G/4G data connection. Once the connection type is programmed, the Arduino script selects the Wi-Fi network to which it will be connected. The connection of the PulseSensor consists of three terminals: GND (ground), V+ (power supply) and ANALOG (analog input). The sampling rate for PulseSensor is two milliseconds, signaled by an interrupt, measuring systole, and dyastole.

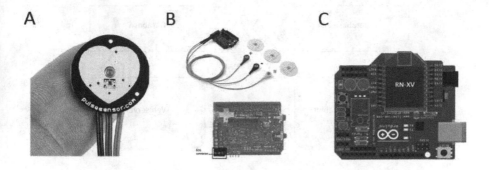

Fig. 3. Elements connected to the microcontroller unit: (A) *PulseSensor* (PulseSensor, NY, USA)(http://pulsesensor.com/), (B) electrocardiogram Shield (Libelium Comunicaciones Distribuidas S.L., Zaragoza, Spain), (C) Wi-Fi Shield (Microchip, AZ, USA). The microcontroller unit records data from two sensors (A, B) and sends these data through the Wi-Fi shield (C) to the smartphone/tablet.

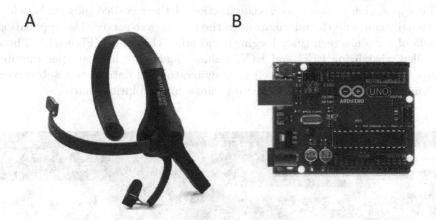

Fig. 4. (A) *Neurosky Mindwave Mobile* (NeuroSky, San Jose, CA, USA) used to record EEG activity (http://store.neurosky.com) (B) *Arduino UNO* microcontroller unit (adapted from http://en.wikipedia.org/wiki/Arduino).

The total cost of the components employed should cost less than $800 (excluding the web server).

2.2 Software Design

The software subsystem is composed of two main blocks: the smartphone/tablet application and the web server application. The first one is intended for wearable biosignals recording, while the server acts as a remote collector of the recorded data for later analysis.

The basic software architecture is exhibited in Figure 5.

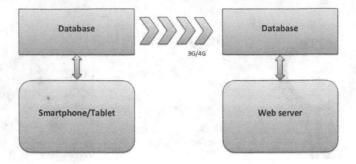

Fig. 5. Software elements used to build the platform

Smartphone. The *THL W200* smartphone terminal runs an application under *Android* operating system, designed to collect the data from the three sensors (EEG, ECG, and pulse sensor).

The application allows *in-situ* visualization of the recorded biosignals, while it optionally can send the information to the remote web server. The application consists of a main screen (user logging) and other three tabs (Figure 6). These tabs allow visualizing EEG and ECG values, heart rate in beats per minute, Bluetooth connection state, as well as a switch to send data to the web server. Biosignals can be shown both in instant values and evolution charts.

Fig. 6. BioTracker® screeshots for the smartphone/tablet. From left to right: *login* screen, *brain activity* screen, *real-time charts* screen and *ECG, heart rate* screen.

The application manages two packages: adapter and biotracker, containing a single class and eight classes respectively. The first package is responsible for managing the application tabs (sliding between tabs). The second package is responsible for managing the smartphone/tablet database to record biosignals. It also includes the managing of user's profiles, GPS parameters and real-time charts. The application uses Bluetooth, GPS, Wi-Fi and 3G/4G data connection.

Web Server. The web server receives all operator's profiles and biosignals data, and creates a list of routes. Each route (Figure 7) associates a sequence of GPS coordinates to the biosignals recorded at that location, so a geographical tracking of the evolution of the operator's CS can be done.

The website contains a database that stores the values sent by the smartphone application. Once the website receives the data, these are represented in different charts and maps, for later analysis.

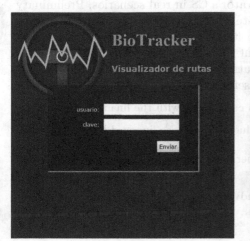 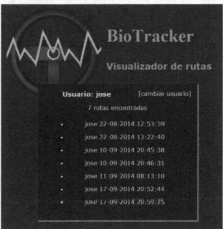

Fig. 7. BioTracker® screenshot, web application. Login page(Left panel) and routes recorded from a user (Right panel).

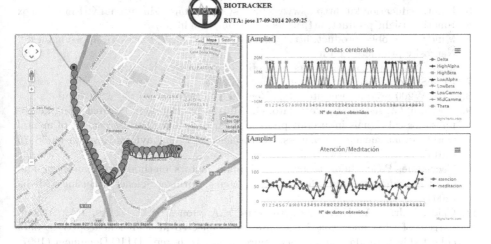

Fig. 8. Example of a route (Left panel), and biosignals for a selected point: brain activity (Upper right panel), attention/meditation levels (Lower right panel) as provided by NeuroSky's algorithms.

When the user logs in, the website shows a list with all routes recorded. Then, an individual route can be selected, showing the map and the charts associated to this route (Figure 8). Some additional parameters as altitude, latitude, and speed can be shown.

3   Conclusions

The platform we developed represents a low-cost, wearable and non-invasive solution able to on-line monitoring operator's CS in real scenarios. Preliminary tests have shown that our recording system works successfully, and it is able to provide real-time biosignals, even when it is used during complex tasks, as on road driving. In the next future, relevant sensors (e.g. optical sensors to record gaze behavior) will be added to the platform and minor modifications will be implemented to improve the system's usefulness.

Acknowledgements. This work has been carried out with the financial support of:

1. Project P11-TIC-7983, Junta of Andalucia (Spain), co-financed by the European Regional Development Fund (ERDF).
2. Project SPIP2014-1426, Dirección General de Tráfico (DGT) Spanish Ministry of Interior.

We also thank Prof. M.G. Arenas for her advice in LaTeX editing.

References

1. Arduino UNO product website, http://arduino.cc/en/Main/ArduinoBoardUno
2. eHealth product website,
   http://www.cooking-hacks.com/documentation/tutorials/ehealth-biometric-sensor-platform-arduino-raspberry-pi-medical
3. Emotiv education kit, http://www.cbinfosystems.com/EducationalKitEmotiv.aspx
4. Emotiv insight product, https://emotiv.com/insight.php
5. Mindwave mobile product, http://neurosky.com/career/hardware-apps-engineer/
6. Neurosky products website,
   http://neurosky.com/products-markets/eeg-biosensors/hardware/
7. PulseSensor website, http://pulsesensor.com/
8. THL W200 product website, http://en.thl.com.cn/product/thl-w200.html
9. Borghini, G., Astolfi, L., Vecchiato, G., Mattia, D., Babiloni, F.: Measuring neurophysiological signals in aircraft pilots and car drivers for the assessment of mental workload, fatigue and drowsiness. Neuroscience & Biobehavioral Reviews 44, 58–75 (2014)
10. Cacciabue, P.C.: Guide to applying human factors methods: Human error and accident management in safety-critical systems. Springer Science & Business Media (2004)
11. Caidwell Jr., J.A., Kelly, C.F., Roberts, K.A., Jones, H.D., Lewis, J.A.: A comparison of EEG and evoked response data collected in a uh-1 helicopter to data collected in a standard laboratory environment. Tech. rep., DTIC Document (1997)
12. Di Stasi, L.L., Catena, A., Canas, J.J., Macknik, S.L., Martinez-Conde, S.: Saccadic velocity as an arousal index in naturalistic tasks. Neuroscience & Biobehavioral Reviews 37(5), 968–975 (2013)

13. Di Stasi, L.L., Antolí, A., Cañas, J.J.: Main sequence: An index for detecting mental workload variation in complex tasks. Applied Ergonomics 42(6), 807–813 (2011)
14. Di Stasi, L.L.: Task complexity modulates pilot electroencephalographic activity during real flights. Psychophysiology (2015), doi: 10.1111/psyp.12419(-)
15. Jap, B.T., Lal, S., Fischer, P.: Comparing combinations of eeg activity in train drivers during monotonous driving. Expert Systems with Applications 38(1), 996–1003 (2011)
16. Patel, S., Park, H., Bonato, P., Chan, L., Rodgers, M.: A review of wearable sensors and systems with application in rehabilitation. Journal of Neuroengineering and Rehabilitation 9(1), 21 (2012)
17. Perego, P., Andreoni, G., Zanini, R., Bellu, R.: Wearable biosignal monitoring system for newborns. In: 2014 EAI 4th International Conference on Wireless Mobile Communication and Healthcare (Mobihealth), pp. 271–274. IEEE (2014)
18. Snäll, J.: Software development of biosignal pi: An affordable open source platform for monitoring ECG and respiration. DIVa (2014)
19. Thayer, J.F., Hansen, A.L., Saus-Rose, E., Johnsen, B.H.: Heart rate variability, prefrontal neural function, and cognitive performance: the neurovisceral integration perspective on self-regulation, adaptation, and health. Annals of Behavioral Medicine 37(2), 141–153 (2009)

Asynchronous EEG/ERP Acquisition
for EEG Teleservices

M.A. Lopez-Gordo[1,2,3(✉)], Pablo Padilla[3], F. Pelayo Valle[4],
and Eduardo Fernández[5]

[1] Dept. of Engineering in Automatic and Electronics, Electronic, Architecture of
Computers and Networks, University of Cadiz, Cadiz, Spain
miguel.lopez@uca.es
[2] Nicolo Association, Churriana de la Vega,
Granada, Spain
malg@nicolo.es
[3] Dept. of Signal Theory, Communications and Networking,
University of Granada, 18071, Granada, Spain,
pablopadilla@ugr.es
[4] Dept. of Computer Architecture and Technology,
University of Granada, 18071, Granada, Spain,
fpelayo@ugr.es
[5] Institute of Bio-engineering, University Miguel Hernndez
and CIBER BBN, 03202, Alicante, Spain,
e.fernandez@umh.es

Abstract. The aging issue threatens to collapse health public systems
in some regions of first world. Although telemedicine is one of the solutions to avoid people insti-tutionalization, it has severe limitations and
not all medical services can be of-fered. While few years ago the electrical
complexity and cost of EEG systems prevented execution of clinical EEG
tests out of hospital, now services such as home-based video-EEG are possible. Conversely, some important clinical tests such as event-related potentials cannot be executed remotely. The reason for that is the accurate
synchrony between local stimulus onset and remote starting of EEG acquisition. In hospital, synchrony is guaranteed by means of a wired connection between stimulus display that triggers EEG recording while in home-
based testing this link normally does not exist. In this study we show an
effective way to execute event-related potentials based on asynchronous
EEG data transmission. We executed a dichotic listening paradigm with
forced-attention modality. The user goal was to detect the attended audio sentence from the analysis of evoked auditory event-related potentials.
The rate of successful detection in both synchronous and asynchronous
modalities was compared and results revealed no significant difference.
Our asynchronous approach can be used in on-line acquisition of home-
based event-related potentials with remote processing.

Keywords: EEG Teleservices · Brain-computer interface · Brain area
networks

J.M. Ferrández Vicente et al. (Eds.): IWINAC 2015, Part I, LNCS 9107, pp. 296–304, 2015.
DOI: 10.1007/978-3-319-18914-7_31

1   Introduction

Event-related potentials (ERPs) are EEG signals elicited as response to a stimulus (exogenous ERP) or internal event (endogenous ERP). Analysis of exogenous ERPs constitutes a valuable technique in clinical assessment of the whole sensorial path-ways from the periphery up to the cortex. In an EEG trace, they appear as deflections that occur at a certain number of milliseconds after stimulus onset (e. g. N100, P300). They are extensively used in clinical practice because their latencies point out well-known cognitive or physiological impairment.

Clinical ERP acquisition is performed in hospitals. Traditional reasons are high cost of EEG devices, need of isolated chamber for experimentation and clinical staff with expertise in EEG set up and electrical montages. Currently this reason no longer exists. Modern wireless EEG headsets with dry electrodes are capable of cheap and ubiquitous EEG acquisition by users with application in Brain-computer Interfaces ([1],[2]). However, there is still an issue not resolved yet by BCIs technology, which is synchronization between the stimulus display and the EEG acquisition unit. ERPs acquisition requests accuracy in the range of few milliseconds for an accurate clinical diagnosis. For this reason, clinical EEG systems integrate both stimulation display and EEG acquisition unit and interconnect them at hardware level by means of a communication port. Conversely, low-cost wireless EEG headsets do not offer this, thus being not adequate to execute event-related paradigms. In summary, for both behavioral tests and ERPs analysis, clinical EEG systems do offer a level of syn-chrony that wireless EEG headsets cannot. This is the reason why most of the home-based EEG services are not meant for ERPs. Next paragraph shows some representa-tive examples of them.

The home-based and mobile EEG teleservice is gaining supporters. For instance, in [3], ninety-nine percent of patients expressed a high degree of satisfaction with re-mote video-EEG service. In [4] a home-based polysomnography system for obstruc-tive sleep apnea diagnosis was proposed. The system was equipped with wireless access for data and video communication via Skype. In [5] four mobile EEG systems acquired epileptiform episodes and compared their performances with satisfaction in users. Many other examples of teleservices exist with of mobile and home-based acquisition of EEG and other biosignals [6][7][8][9][10].These and other cases of mobile EEG tele-services have some clear advantages (e. g. access to rural population, cost savings, etc.) and also some downsides.The examples of mobile EEG tele-services mentioned before did not require stimulation. Epileptic seizures or apnea normally manifest during many seconds (or even minutes) and are not caused by exogenous events. Then, precise synchrony between the starting and detection times is not necessary. In summary, mobile EEG tele-services are not meant for the execution of paradigms that requires precise synchronization such as even-related paradigms.

Modern BCIs [1] are based on wireless, wearable and dry EEG headsets meant not only for clinic, but for other personal uses [11][12][13], including mobile EEG ser-vices. Mobile EEG headsets offer synchronization with the monitoring server via proprietary protocols and techniques based on sequencing and time stamps inserted in the EEG raw data. However, the synchronization between the EEG headset and a mobile media player (i.e. the stimulus display) does not exit. Out of the lab, users are free to use any mobile device as media player (e.g. iPads, smartphones, MP4 players, etc.) without guarantee of hardware compatibility with the EEG headset. The wired connection between the media player and the EEG headset is not only unfeasible in terms of hardware requirements, costs and compatibility, but also a questionable approach from the usability point of view in the context of Wireless Body Area Net-works. For this reason, most of the applications developed for mobile EEG systems are not meant for ERP paradigms. Conversely, they are intended for low frequency cerebral rhythms or steady-state EEG responses [8] (e. g. alpha band, steady-state visual evoked responses). In summary, ERPs measurement is not offered as a mobile EEG as a teleservice.

In this paper, we propose an audio-media preamble as a way for synchronization that virtually generates the wired connection between the media player and the EEG headset. This preamble is a header that encapsulates the multimedia data (i.e. the stimulation). It generates a synchronization preamble meant to evoke a quasi-deterministic brain response that can be detected by the monitoring unit with preci-sion of milliseconds.

2   Methodology

2.1   Subjects and Recordings

A total of two people (both males) participated in this study (31 and 41 years old). The experiment was conducted in a laboratory of the Institute of Bioengineering, University of Miguel Hernndez of Elche (Spain). The study was full auditory without any type of visual stimulation or feedback. The volume of the auditory stimulation was manually adjusted to the comfort level of each participant. It was presented by means of earphones. All participants were previously instructed with the procedure and methodology of this study and signed the informed consent.

The electrical montage consisted of just one EEG active channel located on the vertex (Cz, of the International 10-20 system) and referenced to the mean value of the ear lobes. These positions of the were chosen because they match reports of successful studies of auditory event-related potentials [14]. The ground electrode was placed between Fpz and Fz. The recordings were acquired on a Synamps 2, by Compumedics Neuroscan, were band-pass filtered between 1 and 100 Hz and were sampled at a rate of 1 KHz.

2.2   Auditory Message

The auditory message consisted of two parts, a preamble, which is intended for syn-chronization purposes, and the stimulus itself, which is intended to evoke a response for classification. Each auditory message was z-scored. Then, assuming independent and uncorrelated messages the total energy was constant across trials (see Fig. 1).

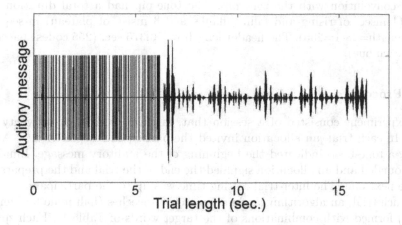

Fig. 1. Auditory message structure. The auditory message is composed of a preamble (left part of the auditory message) and a speech randomly chosen from CRM.

Speeches. Auditory messages were sentences from the Coordinate Response Measure speech corpus (CRM) [15]. The corpus is commonly used in selective attention experiments [16], The auditory messages consist of seven words containing three target words. They follow the structure Ready call-sign go to color number now. Table 1 shows all possible target words.

Table 1. Target words of Coordinate Response Measure

Call-sign	Color	Number
"arrow","baron", "charlie"	"blue", "green"	"one", "two", "three"
"hopper", "laker", "ringo"	"red", "white"	"four", "five", "six"
"eagle", "tiger"		"seven", "eight"

Each speech was tagged to elicit the BPSK constellation. It was achieved by modulating the amplitude (100% depth) of two carriers with a frequency of 5 Hz by the two respective speeches. The two carriers were counter-phased and with the same frequency, namely 5 Hz. This procedure gave rise to two auditory messages that were delivered one to left and the other to the right ear. There is numerous papers that used this principle to elicit a reliable constellation of BPSK signals [17]. Please refer to them for further details about the psycho physiologic principles that justifies this.

Preamble. The preamble was generated by means of a pseudo-random code convoluted with a tone-pip. The pseudo-random code consisted of a binary m-seq of 255 codes length (8 taps). Each binary value was spaced out 25 msec before convolution with the tone-pip. The tone-pip had a total duration of 5 msec (1 msec. of rising and falling flanks and 3 msec. of plateau) m-seq of 8 taps (length=28-1=255). The header length was 6.375 sec. (255 codes spaced 25 msec. each one).

2.3   Procedure

The experiment consisted of a session that, in turns, consisted of twenty one trials. In each trial, an allocution invited the participant to press a key. Afterwards, a tone-beep indicated the beginning of the auditory message. Finally a beep sounded and an allocution signaled the end of the trial and the preparation for the next one. The inter-trial resting time was up to the participant.

In each trial, an algorithm randomly selected speeches (half male half female voices) formed with combinations of the target words of Table 1). Each speech was encapsulated with the auditory preamble, thus forming the auditory message presented in each trial (see Fig. 1).

We formed two auditory messages per trial, one was the target and the other was considered a distracter. The auditory messages were concurrently delivered although the onset of the second auditory message was delayed half a second (see Fig. 2). The delivery order was first and second auditory message to left and right ear respectively. Participants were cued to pay selective attention to one of them and ignore the other. Cues to left and right were counterbalanced, as well as the male and female voices of speeches (always male and feminine voices to left and right ear respectively.

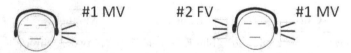

#1<P> Ready **Ringo** go to **blue one** now...   #1 <P>Ready **Ringo** go to **blue one** now...

#2 <P>Ready **Laker** go to **white four** now...

0.5 sec.

Fig. 2. Auditory messages delivery. The speech of the first auditory message was always a male voice delivered to the left ear. The second one was always a female voice delivered to the right ear with a delay of 0.5 sec. In the figure, target words are in bold. ¡P¿ stands for preamble. MV and FM stand for male and female voice respectively.

2.4   Feature Extraction and Classification

There are many approaches for the extraction and classification of EEG features in either clinical or BCI applications or for brain insight. In this study we used a simple approach to build a BPSK receiver. EEG acquisition of each trial was multiplied by a Tukey window (=0.20) and then, the DFT computed. Only the coefficient corre-sponding to 4 Hz (C4) was extracted for posterior classification.

Classification of the attended auditory message was performed by mapping the extracted feature (C4) into the BPSK constellation and by applying the principle of minimum Euclidean distance. This principle, under some general assumptions, is the optimal Bayesian classifier [36].

As BPSK constellation is composed of two counter-phased symbols. They repre-sent the DFT components at 4 Hz. That means that both symbols are 180 separated or the equivalent in time 125 msec. Then, small errors in detection of stimulus onset, for instance 14 msec, would give rise to a rotation of the classification boundary of 20, thus running the classification results (see Fig. 3).

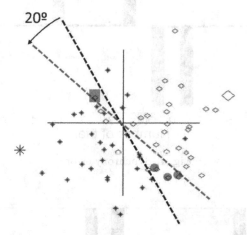

Fig. 3. Error in detection of classification boundaries. This example simulates extracted features (small stars and diamonds) in a constellation of BPSK signals (big star and diamond). The black dotted line represents the decision boundary based on the Euclidean distance. A simple error of 14 msec in the detection of the stimulus onset, at 4 Hz, would cause a phase error of 20 in the decision boundary (grey dotted line). In this example it could cause three additional errors in classification (grey circles) and one success (grey rectangle).

3   Results

We mentioned in the methodological section that two people participated in this study. After data analysis, we discover that EEG signals from one of the partic-ipants were contaminated with high energetic electrical artifacts, thus making his EEG rec-ord useless. Then, this section shows results of only one of them. Table 2 shows the performance of the binary classification. The output of the BPSK receiver was compared with the cue given participants in the 21 trials.

Bars of Fig. 4 represent, for each of the 21 trial, the time error in asynchronous de-tection of the preamble taking as reference the synchronous onset detection. Synchronous detection was considered the gold pattern.

Table 2. Classification performance

SUT	Synchronous		Asynchronous	
	acc(%)	C.I.(%)	acc(%)	C.I.(%)
S01	52	[32..71]	57	[36..75]

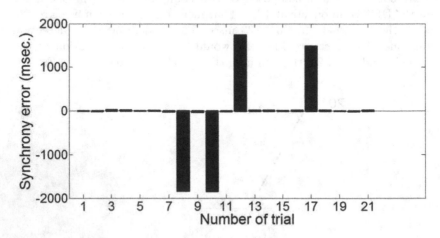

Fig. 4. Synchrony error

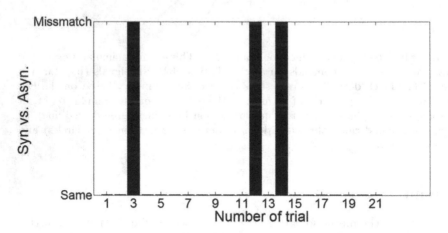

Fig. 5. Classification mismatch

Bars of Fig. 5 represent differences in classification of the attended stimulus. In this figure, the output of the BPSK receiver of trials number 3, 12 and 14, yielded different values when asynchronous and synchronous detection was used.

4   Discussion

Table 2 shows the classification performance with synchronous and asynchronous preamble detection. As we can see, the accuracy of the binary classification, taken into consideration the confidence intervals (C.I.), is no better than chance level. The methodology based on BPSK constellation of EEG signals for detection of attention in dichotic listening task was demonstrated in previous studies [17][18][19] and is not under discussion in the present study. However, in this experiment we could not prove that detection of attention was any better than chance level. We may have an explanation for that. Studies based on BPSK receiver, use a in the experiment a stimulation frequency at 5 Hz. The reason for that is because its half period, namely 100 msec, coincides with that of intervals of potentials N100 and P200. So the BPSK receiver is optimized for 5 Hz. We used 4 Hz instead of 5 Hz because the uncertainty of asynchronous detection of the stimulus onset, could lead to severe error in the classification boundary, and hence, in the final performance (see Fig. 3 for a detailed explanation) but at the cost of a suboptimal experiment design.

We must keep in mind that the objective of this study is not the validation of the approach based on the BPSK receiver for detection of attention, but to check if the asynchronous detection of preamble could yield similar results to that of the synchro-nous detection. In this regards, Fig. 4 shows that, a part of few trials, namely four, the asynchronous detection of the preamble had approximately zero error taken as refer-ence the synchronous one. Furthermore, only in three out of twenty-one trials, the BPSK receiver yielded different classification.

Acknowledgement. This study has been co-financed by Nicolo Association for the R+D in Neurotechnologies for disability, the regional project P11-TIC-7983, Junta of Andalucia (Spain), the National Grant TIN2012-32039 (Spain), co-financed by the European Regional Development Fund (ERDF) and the MAT2012-39290-C02-01.

References

1. Lee, S., Shin, Y., Woo, S., Kim, K., Lee, H.-N.: Review of Wireless Brain-Computer Interface Systems. In: Fazel-Rezai, R. (ed.) Brain-Computer Interface Systems - Recent Progress and Future Prospects. InTech (2013)
2. Lopez-Gordo, M., Morillo, D., Pelayo, F.: Dry EEG Electrodes. Sensors 14(7), 12847–12870 (2014)
3. Campos, C., Caudevilla, E., Alesanco, A., Lasierra, N., Martinez, O., Fernndez, J., Garca, J.: Setting up a telemedicine service for remote real-time video-EEG consultation in La Rioja (Spain). Int. J. Med. Inf. 81(6), 404–414 (2012)
4. Bruyneel, M., Van den Broecke, S., Libert, W., Ninane, V.: Real-time attended home-polysomnography with telematic data transmission. Int. J. Med. Inf. 82(8), 696–701 (2013)
5. Askamp, J., van Putten, M.J.A.M.: Mobile EEG in epilepsy. Int. J. Psychophysiol. 91(1), 30–35 (2014)

6. Brunnhuber, F., Amin, D., Nguyen, Y., Goyal, S., Richardson, M.P.: Development, evaluation and implementation of video-EEG telemetry at home. Seizure 23(5), 338–343 (2014)
7. Goodwin, E., Kandler, R.H., Alix, J.J.P.: The value of home video with ambulatory EEG: A prospective service review. Seizure 23(6), 480–482 (2014)
8. Alix, J.J.P., Kandler, R.H., Mordekar, S.R.: The value of long term EEG monitoring in children: A comparison of ambulatory EEG and video telemetry. Seizure 23(8), 662–665 (2014)
9. Patwari, N., Wilson, J., Ananthanarayanan, S., Kasera, S.K., Westenskow, D.R.: Moni-toring Breathing via Signal Strength in Wireless Networks. IEEE Trans. Mob. Comput. 13(8), 1774–1786 (2014)
10. Guo, L., Zhang, C., Sun, J., Fang, Y.: A Privacy-Preserving Attribute-Based Authenti-cation System for Mobile Health Networks. IEEE Trans. Mob. Comput. 13(9), 1927–1941 (2014)
11. van Gerven, M., Farquhar, J., Schaefer, R., Vlek, R., Geuze, J., Nijholt, A., Ramsey, N., Haselager, P., Vuurpijl, L., Gielen, S., Desain, P.: The brain computer interface cycle. J. Neural Eng. 6(4), 041001 (2009)
12. Liao, L.-D., Chen, C.-Y., Wang, I.-J., Chen, S.-F., Li, S.-Y., Chen, B.-W., Chang, J.-Y., Lin, C.-T.: Gaming control using a wearable and wireless EEG-based brain-computer in-terface device with novel dry foam-based sensors. J. NeuroEngineering Rehabil. 9(1), 5 (2012)
13. Matthews, R., Turner, P.J., McDonald, N.J., Ermolaev, K., Manus, T.M., Shelby, R.A., Steindorf, M.: Real time workload classification from an ambulatory wireless EEG system using hybrid EEG electrodes, pp. 5871–5875 (2008)
14. Hillyard, S.A., Hink, R.F., Schwent, V.L., Picton, T.W.: Electrical Signs of Selective Attention in the Human Brain. Science 182(4108), 177–180 (1973)
15. Bolia, R.S., Nelson, W.T., Ericson, M.A., Simpson, B.D.: A speech corpus for multi-talker communications research. J. Acoust. Soc. Am. 107(2), 1065 (2000)
16. Shinn-Cunningham, B., Ihlefeld, A.: Selective and divided attention: Extracting information from simultaneous soundsources. In: Proceedings of ICAD 2004-Tenth Meeting of the International Conference on Auditory Display, Sidney (2004)
17. Lopez-Gordo, M.A., Pelayo, F.: A Binary Phase-Shift Keying Receiver for the Detection of Attention to Human Speech. Int. J. Neural Syst., p. 130418190845004 (March 2013)
18. Lopez-Gordo, M.A., Fernandez, E., Romero, S., Pelayo, F., Prieto, A.: An auditory brain computer interface evoked by natural speech. J. Neural Eng. 9(3), 1–9 (2012)
19. Lopez-Gordo, M.A., Pelayo, F., Prieto, A., Fernandez, E.: An Auditory Brain-Computer Interface with Accuracy Prediction. Int. J. Neural Syst. 22(3), 1–14 (2012)

A Machine Learning Approach to Prediction of Exacerbations of Chronic Obstructive Pulmonary Disease

Miguel Angel Fernandez-Granero[1(✉)], Daniel Sanchez-Morillo[1],
Miguel Angel Lopez-Gordo[1], and Antonio Leon[2]

[1] Biomedical Engineering and Telemedicine Research Group, University of Cadiz.
Avda. de la Universidad, 10, 11519 Puerto Real, Cadiz, Spain
[2] Neumology and Allergy Unit, Puerta del Mar University Hospital, Cadiz, Spain
ma.fernandez@uca.es

Abstract. Chronic Obstructive Pulmonary Disease (COPD) places an enormous burden on the health care systems and causes diminished health related quality of life. The highest proportion of human and economic cost is associated to admissions for acute exacerbation of respiratory symptoms. The remote monitoring of COPD patients with the view of early detection of acute exacerbation of COPD (AECOPD) is one of the goals of the respiratory community. In this study, machine learning was used to develop predictive models. Models robustness to exacerbation definition was analyzed. A non-knowled-ge based approach was followed on data self-reported by patients using a multimodal tool during a remote monitoring 6 months trial. Comparison of different classifier algorithms operating with different AECOPD definitions was performed. Significant results were obtained for AECOPD prediction, regardless of the definition of exacerbation used. Best accuracy was achieved using a PNN classifier independently of the selected AECOPD definition. Our study suggests that the proposed data-driven methodology could help to design reliable predictive algorithms aimed to predict COPD exacerbations and therefore could provide support both to physicians and patients.

Keywords: COPD · Exacerbation · Telehealth · Symptoms · Questionnaire · Early detection · Data-Driven · Machine Learning

1 Introduction

COPD places an enormous burden on the health care systems and causes diminished health related quality of life. The highest proportion of human and economic cost is associated to the use of expensive urgent healthcare and to admissions for acute exacerbation of respiratory symptoms[1].

Telehealth enabled chronic care management services can effectively support people with long term conditions at home. Recent studies have analysed how home telemonitoring may affect to clinical outcomes, health relate quality of life

© Springer International Publishing Switzerland 2015
J.M. Ferrández Vicente et al. (Eds.): IWINAC 2015, Part I, LNCS 9107, pp. 305–311, 2015.
DOI: 10.1007/978-3-319-18914-7_32

(HRQOL) and cost of interventions[2]. Since early detection and treatment of exacerbations may improve outcomes, the remote monitoring of COPD patients with the view of early detection of acute exacerbation of Chronic Obstructive Pulmonary Disease (AECOPD) is one of the goals of the respiratory community. However, only a few studies have attempted to achieve early detection of AE-COPD and consequently, currently available tools for monitoring and managing COPD exacerbations are limited.

In clinical practice, telemonitoring of COPD is usually performed with patient diaries. Assessment in detecting exacerbations using traditional paper-based methods of collecting symptoms have been reported but with a moderate sensitivity and specificity pair [3]. However, studies focused on prediction of AECOPD on a day-to-day basis through telehealth approaches have been reported scarcely. The works published have been supported on using weekly or biweekly reported physiological data, clinical diaries or a combination of them [4,5]. Diary-keeping in COPD and ascertained items that best predicted emergency attendances for exacerbations has been also evaluated [6].

In recent reported results from the exacerbations of COPD Tool (EXACT), a daily diary for evaluating COPD exacerbation severity through monitoring of patients was evaluated. EXACT scores increased from baseline were analyzed at exacerbation onset. However, concerns remained about the ability of the EX-ACT to accurately detect exacerbations[7]. In addition, unsupervised physiological home telehealth measurements has been evaluated to predict the patient's condi-tion in advance[8,9]. Very recently, a Probabilistic Neural Network (PNN) classifier was reported to be able to predict COPD exacerbations early with 4.8 days as average prior to AECOPD onset[10]. In late 2014, authors in [11] tested a personalized thresholds algorithm using physiological measurements, self-reported symptom scores and medicines. More lately, researchers in [12] applied a Radial Basis Function (RBF) classifier to early detect AECOPD with a margin of 4.5 days as average prior to medical attention.

Despite of these reported efforts, the evidence to support the effectiveness of home telemonitoring interventions for patients with COPD is limited and further work is required[13]. The purpose of home telemonitoring of patients with COPD to early detect and address AECOPD may have not been reached because of poor patients' compliance[14] and for the lack of useful early predictors[15] and reliable predictive algorithms since conventional COPD monitoring systems apply simple thresholds to patient data. Therefore, developing predictive algorithms with clinical reliability is a priority for the future development of telemonitoring of COPD.

On the other hand, many definitions of COPD exacerbations have been reported. About 40 % of studies used events-based AECOPD definitions and 78% applied symptoms-based criteria[16]. Events-based exacerbation is defined as an attendance at a hospital emergency department or primary care unit with worsening clinical symptoms. In some studies, self-administration of antibiotics and/or corticosteroids in the case of patients with a self-management plan is also considered in the definition of event-based exacerbation. The alternative approach of symptom-based definitions of exacerbation uses increasing of specific and nonspecific symptoms. The choice for a definition determines the number of exacerbations observed and as a consequence the algorithm performance. Therefore, predictive algorithms should be robust independently of the applied AECOPD definition.

In this study, data driven predictive models and their robustness to AECOPD were analyzed. A non-knowledge-based approach using machine learning which allowed to learn from past experiences and find patterns in clinical data was followed. Machine learning and statistical techniques, by learning or training, were used on data self-reported by patients using a multimodal tool during a remote monitoring 6-months trial[17]. Comparison of different classifier algorithms operating with different AECOPD definitions was performed. This study aimed to provide further evidence in support of the hypothesis of the feasibility of early prediction of COPD exacerbations based on the daily patient's self-report of symptoms applying pattern recognition techniques.

2   Patients and Methods

A sample of 16 COPD patients were equipped with a home base station to daily respond to a questionnaire during 6 months.They were recruited in the Pneumology and Allergy Department of the University Hospital Puerta del Mar of Cadiz (Spain). Participants were all aged over 60 years and had a diagnosis of COPD confirmed by spirometry, classified in groups C and D according to GOLD guidelines [18]. Patients had cumulative tobacco consumption greater than 20 Packs-Year and at least two exacerbations treated with oral antibiotics or corticosteroids or one hospital admission for exacerbation in the past year. The patients were guided at home by a multimodal interface to record their symptoms through a daily questionnaire. Details about the interface and the questionnaire can be found in [17]. Likert item responses were assigned scores and forward and backward imputation was used to handle missing diaries. Twelve predictor variables were processed. Four additional input parameters were calculated: average scores for symptoms associated with minor, major and complementary symptoms and 3-days moving average applied to the total score. The target for the classifier was defined as a categorical binary variable. The day of the exacerbation onset depended on the AECOPD definition applied. The procedure for event-based exacerbations is detailed in [12]. For symptom-based episodes, the methodology can be found in [10]. Therefore, prediction of exacerbations was addressed as a classification problem.

Three different classifiers were trained, validated and compared: 1) a radial basis function neural network (RBF); 2) a k-means classifier and 3) a probabilistic neural network (PNN). RBF was introduced by Powell [19] in 1987 for the purpose of exact interpolation. A RBF network has three layers completely linked: input, hidden layer of radial units and an output layer of linear units with a feedforward. Neurons in the hidden layer have Gaussians functions[20]. Secondly, k-means non hierarchical cluster analysis was used. K-means is an unsupervised classifier that finds statistically similar groups in multifeatures space. K-means iterative algorithm assigns each observation to the cluster with the nearest mean. Within-groups the criterion used to minimize an objective function was the sum of squares [21]. Finally, a PNN classifier was designed. PNNs have a similar structure to back-propagation neural networks. The main difference is in that the sigmoid activation function is replaced by a statistically derived one. The decision boundary implemented by the PNN asymptotically approaches the Bayes optimal decision surface under certain easily met conditions [22].

For each of the aforementioned classifiers, the output was forwarded to a simple decision rule in order to reduce the false positive rate. Two consecutive days with a positive output in the classifier were needed to rise an alarm state. 10-Cross-validation was used to ensure stability of the results and to gauge the generalizability of the classifiers. Evaluating the performance of the monitoring system to early detect AECOPD is the primary objective of this study. Performance of the classifier was assessed ac-cording to accuracy, sensitivity, specificity, confusion matrix, positive predictive value (PPV) and negative predictive value (NPV). MathWorks MATLAB® was used for graphical representation, signal processing and statistical analysis.

3 Results

Along the 6-months pilot, 33 exacerbations were detected because of unscheduled attendance at a hospital emergency or primary care units or self-administration of antibiotics and/or corticosteroids. Table 1 illustrates the diagnostic performance of the predictive designed classifiers considering an event based definition of exacerbation episodes and before applying the 2-days decision rule.

Table 1. Diagnostic performance assessment of predictive models trained and validated using event-based definition of AECOPD. TP: true positives (n); TN: true negatives (n); FP: false positives (n); FN: false negatives (n); Se: sensitivity (%); Sp: specificity (%); Acc: accuracy (%).

Algorithm	Records	TP	FP	TN	FN	Acc	Se	Sp
RBF	789	186	70	467	66	82.8	73.8	87.0
K-means	789	197	54	471	67	84.7	74.6	89.7
PNN	789	260	36	444	49	89.3	84.1	92.5

Concerning symptom based episodes, 41 events of exacerbation were accounted in the group of remote monitored patients. 33 out of this 41 episodes corresponded to reported events while 8 events were non reported exacerbations detected by symptoms monitoring. Table 2 details the diagnostic performance of the classifiers designed according to a symptom based definition of exacerbation, before applying the 2-days decision rule.

Table 3 summarizes the comparison results of the different validated predictive algorithms depending on the selected AECOPD definition and the prodrome period (7 or 14 days) used to define alarm states.

Table 2. Diagnostic performance assessment of predictive models trained and validated using symptom-based definition of AECOPD. TP: true positives (n); TN: true negatives (n); FP: false positives (n); FN: false negatives (n); Se: sensitivity (%); Sp: specificity (%); Acc: accuracy (%).

Algorithm	Time Segments	TP	FP	TN	FN	Acc	Se	Sp
RBF	99	15	6	26	52	67.7	36.6	89.7
K-means	117	30	15	61	11	77.8	73.2	80.3
PNN	94	33	3	50	8	88.3	80.5	94.3

Table 3. Diagnostic performance of predictive models from each classification methodologies, exacerbation definition and prodrome period (days) used for alarms definition. Predictive range is expressed in days (mean ± std).

Algorithm	AECOPD Definition	Prodrome	Predicted	Predictive Range	False alarms
RBF	Event-based	7	31	5.3 ± 2.1	20
		14	31	7.1 ± 3.2	11
	Symptom-based	7	15	4.4 ± 1.8	10
		14	13	5.7 ± 3.4	6
K-means	Event-based[12]	7	31	4.5 ± 2.1	23
		14	32	6.3 ± 3.3	15
	Symptom-based	7	29	4.4 ± 2.0	24
		14	30	5.7 ± 3.3	15
PNN	Event-based	7	32	4.5 ± 2.0	18
		14	33	6.3 ± 3.3	8
	Symptom-based[10]	7	31	4.5 ± 1.5	5
		14	33	4.8 ± 1.8	3

4 Discussion

Results from this study provide evidence in support of the hypothesis of the feasibility of early prediction of COPD exacerbations based on the daily remote patient's self-report of signs and symptoms. Data driven predictive algorithms for predicting the occurrence of an exacerbation of COPD of a patient in the near future were trained, validated and compared. Data were daily acquired during a 6-months telemonitoring pilot.

Significant results were obtained for AECOPD prediction, regardless of the definition of exacerbation used. Best accuracy was achieved using a PNN classifier and an interval for seeking (i.e. predicting) events of two weeks (prodrome) independently of the selected AECOPD definition. For the events-based episodes, 100% of AECOPD were detected with a margin of 6.3 ± 3.3 days prior to onset. For symptoms-based episodes, 80.5% of AECOPD were predicted (33 out of 41) with 4.8 ± 1.8 days prior to onset. Noteworthy is the low number of false alarms generated by the system and high percentage of success in prediction of non-reported exacerbations.

The choice for a definition of COPD exacerbation determines the number of exacerbations observed. This could lead to the presentation of opportunistic results in predicting AECOPD[16]. In this sense, the proposed method is robust since achieves good prediction results with two different widely used AECOPD definitions.

Finally, a larger sample of patients would assist in determining whether the findings reported in this work are replicable and generalizable, using leave-one-subject-out cross validation (LOSO-CV) to avoid optimistic bias.

In conclusion, our study suggests that the proposed data-driven methodology could help to design reliable predictive algorithm aimed to early detect COPD exacerbations and therefore could provide support both to physicians and patients. Importantly, we have shown that predictive algorithms can operate robustly with independence of the adopted definition for COPD exacerbations, which is often a controversial. Further studies are required to extend the conclusions and to check the consistency of the results obtained.

Acknowledgments. This work was supported in part by the Ambient Assisted Living (AAL) E.U. Joint Programme, by grants from the Spanish Ministry of Education and Science (Ministerio de Educacion y Ciencia) and the Carlos III Health Institute (Instituto de Salud Carlos III) under projects PI08/90946 and PI08/90947.

References

1. Toy, E.L., Gallagher, K.F., Stanley, E.L., Swensen, A.R., Duh, M.S.: The economic impact of exacerbations of chronic obstructive pulmonary disease and exacerbation definition: a review. COPD 7, 214–228 (2010)
2. Lundell, S., Holmner, Å., Rehn, B., Nyberg, A., Wadell, K.: Telehealthcare in COPD: A systematic review and meta-analysis on physical outcomes and dyspnea. Respiratory Medicine 109(1), 11–26 (2015)
3. Trappenburg, J., Touwen, I., Oene, G., Bourbeau, J., Monninkhof, E., et al.: Detecting exacerbations using the Clinical COPD Questionnaire. Health Qual Life Outcomes 8, 102 (2010)
4. Jensen, M.H., Cichosz, S.L., Dinesen, B., Hejlesen, O.K.: Moving prediction of exacerbation in chronic obstructive pulmonary disease for patients in telecare. J. Telemed. Telecare 18, 99–103 (2012)
5. van der Heijden, M., Lijnse, B., Lucas, P.J.F., Heijdra, Y.F., Schermer, T.R.J.: Managing COPD exacerbations with telemedicine. In: Peleg, M., Lavrač, N., Combi, C. (eds.) AIME 2011. LNCS, vol. 6747, pp. 169–178. Springer, Heidelberg (2011)
6. Walters, E., Walters, J., Wills, K., Robinson, A., Wood-Baker, R.: Clinical diaries in COPD: compliance and utility in predicting acute exacerbations. International Journal of Chronic Obstructive Pulmonary Disease 7, 427–435 (2012)
7. Mackay, A.J., Donaldson, G.C., Patel, A.R., et al.: Detection and severity grading of COPD exacerbations using the exacerbations of Chronic Obstructive Pulmonary Disease Tool (EXACT). Eur. Respir. J. 43(3), 735–744 (2014)
8. Mohktar, M.S., Redmond, S.J., Antoniades, N.C., Rochford, P.D., Pretto, J.J., Basilakis, J., McDonald, C.F.: Predicting the risk of exacerbation in patients with chronic obstructive pulmonary disease using home telehealth measurement data. Artificial Intelligence in Medicine (2014), doi:10.1016/j.artmed.2014.12.003
9. Burton, C., Pinnock, H., McKinstry, B.: Changes in telemonitored physiological variables and symptoms prior to exacerbations of chronic obstructive pulmonary disease. Journal of Telemedicine and Telecare, 1357633X14562733 (2014)

10. Fernández-Granero, M.A., Sánchez-Morillo, D., León-Jiménez, A., Crespo, L.F.: Automatic prediction of chronic obstructive pulmonary disease exacerbations through home telemonitoring of symptoms. Bio-Medical Materials and Engineering 24(6), 3825–3832 (2014)

11. Rutter, H., Velardo, C., Toms, C., Williams, V., Tarassenko, L., Farmer, A.: Using a Mobile Health Application to Support Self-Management in COPD-Development of Alert Thresholds Derived from Variability in Self-Reported and Measured Clinical Variables. Am. J. Respir. Crit. Care Med 189, A1396 (2014)

12. Sanchez-Morillo, D., Fernandez-Granero, M., Leon, A.: Detecting COPD exacerbations early using daily telemonitoring of symptoms and k-means clustering: a pilot study. Med. Biol. Eng. & Comp. (2015), doi:10.1007/s11517-015-1252-4

13. McKinstry, B.: The use of remote monitoring technologies in managing chronic obstructive pulmonary disease. QJM 106, 883–885 (2013)

14. Sanders, C., Rogers, A., Bowen, R., Bower, P., Hirani, S., et al.: Exploring barriers to participation and adoption of telehealth and telecare within the whole system demonstrator trial: a qualitative study. BMC Health Serv. Res. 12, 220 (2012)

15. Hurst, J.R., Donaldson, G., Quint, J.K., Goldring, J.J.P., Patel, A.R.C., Wedzicha, J.A., et al.: Domiciliary pulse-oximetry at exacerbation of chronic obstructive pulmonary disease: prospective pilot study. BMC Pulm. Med. 10, 52–58 (2010)

16. Effing, T.W., Kerstjens, H.A., Monninkhof, E.M., van der Valk, P.D., Wouters, E.F., Postma, D.S., van der Palen, J.: Definitions of exacerbations: Does it really matter in clinical trials on COPD? CHEST Journal 136(3), 918–923 (2009)

17. Sanchez-Morillo, D., Crespo, M., Leon, A., Crespo, F.L.: A novel multimodal tool for telemonitoring patients with COPD. Inform Health Soc. Care 40, 1–22 (2013)

18. Global Strategy for the Diagnosis, Management and Prevention of COP (2014), Global Ini-tiative for Chronic Obstructive Lung Disease (GOLD), http://www.goldcopd.org/ (accessed November 24, 2014)

19. Powell, M.: Radial Basis Functions for Multivariable Interpolation: A Review. In: Mason, Cox (eds.) Algorithms for Approximation, pp. 143–167. Clarendon Press, Oxford (1987)

20. Begg, R., Kamruzzaman, J., Sarker, R.: Neural Networks in Healthcare. Potential and Challenges. Idea Group Publishing (2006)

21. Hartigan, J.A., Wong, M.A.: A K-means clustering algorithm. Applied Statistics 28, 100–108 (1979) [A8]

22. Specht, D.F.: Probabilistic neural networks. Neural Networks 3, 109–118 (1990)

Brain-Computer Interfacing to Heuristic Search: First Results

Marc Cavazza[⊠], Gabor Aranyi, and Fred Charles

School of Computing, Teesside University, Middlesbrough, United Kingdom
m.o.cavazza@tees.ac.uk

Abstract. We explore a novel approach in which BCI input is used to influence the behaviour of search algorithms which are at the heart of many Intelligent Systems. We describe how users can influence the behaviour of heuristic search algorithms using Neurofeedback (NF), establishing a connection between their mental disposition and the performance of the search process. More specifically, we used functional near-infrared spectroscopy (fNIRS) to measure frontal asymmetry as a marker of approach and risk acceptance under a NF paradigm, in which users increased their left asymmetry. Their input was mapped onto a dynamic weighting implementation of A* (termed WA*), modifying the behaviour of the algorithm during the resolution of an 8-puzzle problem by adjusting the performance-optimality tradeoff. We tested this approach with a proof-of-concept experiment involving 11 subjects who had been previously trained in NF. Subjects were able to positively influence the behaviour of the search process in over 58% of the NF epochs, resulting in faster solutions.

1 Introduction and Rationale

In this paper, we introduce a novel BCI approach which maps neural signals onto the behaviour of heuristic search algorithms. The theoretical inspiration for this work is to return to the cognitive roots of heuristic functions, as representing intuitive knowledge, meant to guide search behaviour in search spaces too large to be explored systematically.

Our experimental setting consists in letting users influence the behaviour of a heuristic search algorithm by expressing their eagerness to reach a solution for a computation of unknown, but potentially significant, duration. The objective of this research is to explore how a real-time neurophysiological measure of approach, which has been associated to eagerness [22] as well as risk acceptance [21], can be translated into potentially faster heuristic search, implemented through a principled deviation from admissibility of the heuristic [14]. We thus aim at reconciling a natural and a computational version of eagerness in search, one that trades optimality for speed.

In terms of neural correlates, there is extensive work relating approach to prefrontal cortex asymmetry [5,22], which can serve as a basis for EEG Neurofeedback BCI [4]. In the work reported here, the input signal consists in a

© Springer International Publishing Switzerland 2015
J.M. Ferrández Vicente et al. (Eds.): IWINAC 2015, Part I, LNCS 9107, pp. 312–321, 2015.
DOI: 10.1007/978-3-319-18914-7_33

measure of prefrontal asymmetry using fNIRS-based NF, which is translated into a dynamic modification of the heuristic function, rendering it less admissible, but potentially more efficient. Successful BCI input should then result in the faster computation of a near-optimal solution.

2   Previous and Related Work

Relationships between BCI and Intelligent Systems are complex [13]. Some of the most significant research in BCI interfacing to intelligent systems has taken place in the field of BCI-enhanced Information Retrieval. Gerson et al. [9] have demonstrated increased human performance in satellite image analysis when a BCI system was used to detect regions of interest despite very fast, almost subliminal, visual scanning by users. Kapoor et al. [12] have used EEG-based BCI in combination with computer vision to improve image categorisation. Eugster et al. [8] have described how BCI could assist to automatically detect term relevance during Information Retrieval tasks. In terms of BCI technology, frontal asymmetry has been shown to be amenable to NF as part of clinical applications [2], and more recently in affective BCI [4].

There has been significant previous research in the use of fNIRS for BCI [20], including the measurement of task difficulty in conjunction with computer gameplay [10], or as an additional input channel to interactive systems [19]. This research has, however, primarily investigated fNIRS for passive BCI. In addition, Doi et al. [6] have shown fNIRS to be well-suited to the study of emotional responses in the prefrontal cortex, which is our target area.

3   Heuristic Search Properties

The study of mathematical properties of heuristic search has been pioneered by Pearl [14], who has provided a solid foundation to the complexity and performance analysis of heuristic search. In particular, he has identified the conditions under which heuristic search could be made more computationally efficient by relaxing some optimality requirements, which is since known as the precision-complexity exchange [15]. This phenomenon serves as a basis for the implementation of our BCI-based influence on heuristic search.

We will be using the original dynamic weighting variant of A* proposed by Pohl [15], in which the evaluation function is implemented as a weighted formula of the type:

$$f(n) = (1 - w) * g(n) + w * h(n) \tag{1}$$

where w can be dynamically altered during the search process itself [7,11]. It can be shown to be ϵ-admissible [7].

The 8-puzzle is a traditional search problem whose formal properties are well described (Figure 1a), and on which the impact of dynamic heuristic weighting could be properly assessed. The entire solution set of the 8-puzzle has been studied by Reinefeld [16], who has generated all $9!/2$ solvable tile configurations, and

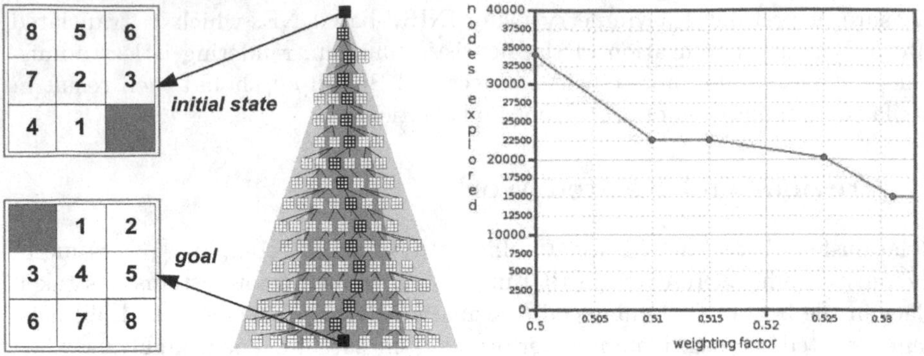

(a) Initial and goal configurations for the 8-puzzle problem considered

(b) Impact of dynamic weighting on the size of the 8-puzzle search space. In this sample, dynamic weighting is applied once 23% of default search space expansion (measured for $w = 0.5$) has been explored.

Fig. 1. (a) WA* search algorithm on 8-puzzle; (b) Impact of dynamic weighting.

computed all optimal solutions for all problem instances. These specific configurations provide an interesting test bed to explore how search can be influenced, by departing from admissibility of the heuristic. In particular, we have decided to use the two configurations with the greatest number of solutions in our experiments (64 solutions [16]) to allow interventions at various stages of the search without compromising the path to the solution.

In their study of weighted A*, Hansen and Zhou [11] have shown that significant performance improvements could be obtained with only marginally different weightings, sometimes differing by as little as 1%. Further, they report that the benefits of weighted A* applied to n-puzzle problems rest mostly in the reduced number of nodes expanded. We have thus explored the impact of various weightings on search speed and number of nodes explored (see Figure 1b). Our results confirm that the most dramatic improvements both in space complexity and time complexity take place for moderate deviations from standard A* evaluation function, namely in the $w = [0.5 - 0.53]$ range (standard A* corresponding to $w = 0.5$).

Interestingly, this departure from admissibility did not affect the quality of the solution itself, as evidenced by tests with a configuration pair admitting one single solution. Since the heuristic modification is meant to take place during the search itself, we have explored the impact of weight modification at various stages of the search, measured through the number of nodes visited. We found the impact of dynamic weighting to be most significant when applied at an early stage of the search progression (when less than 33% of nodes of a complete solution search have been expanded). We also note that a 55% reduction in search space can correspond to a much larger reduction in CPU time depending on the computer configuration (up to a seven-fold acceleration of computation), because of the exponential complexity of A*. This will be detailed in Figure 5b with results presentation.

Our A* implementation is a traditional one based on the version described in Pearl [14]: it accepts dynamic weighting at a pre-defined stage of search, defined in terms of % node expansions, which is passed as a parameter when launching the search on a configuration pair. Its only 8-puzzle specific enhancement is the use of a transposition table to avoid expanding inverse moves at a depth of two, which would simply cancel the previous one [16]. It is developed in Allegro Common Lisp from which a standalone executable has been generated for integration with the fNIRS software analysis and NF platform. We used a traditional, admissible, heuristic function for the 8-puzzle, defined as the sum of all Manhattan distances between individual tiles in their current position to the goal position.

4   Neurofeedback Experiment Design

We used fNIRS to operationalise BCI input, which measures changes in blood oxygenation associated with the functional activation of the cerebral cortex (see [3] for a general description of fNIRS).

Eleven adults (Age: M=37.18 years, SD=11.21, range−[20,52]; 3 female) provided written consent and participated in the experiment approved by a research ethics committee at the authors' institution. Subjects were seated in a dimly-lit room in a comfortable chair to minimise movements, with the fNIRS probe positioned over their forehead and covered with non-transparent fabric to prevent ambient light reaching the sensors. Subjects only provided input through the fNIRS probe. Each subject was compensated with an online retailer voucher equivalent to $30.

Data was collected with an fNIR Optical Brain Imaging System (fNIR400) by Biopac Systems, with 2Hz sampling rate. Raw fNIRS data and oxygenation values were acquired using software provided by the device manufacturer (COBI Studio and fNIRSoft v3.3). A 16-channel sensor with a fixed 2.5cm source-detector separation was placed on the subjects' forehead. We used measurements of changes in oxygenated hemoglobin (Oxy-Hb), as opposed to deoxygenated or total hemoglobin, because it is more commonly associated with neural activity [17]. Oxy-Hb values were averaged over four leftmost and four rightmost channels (located over the left and right dorsolateral prefrontal cortex, respectively). Average right Oxy-Hb was subtracted from average left Oxy-Hb to derive a simple, real-time prefrontal asymmetry score reflecting differential changes in oxygenation. An example of left-asymmetric prefrontal oxygenation is presented in Figure 2a.

We developed bespoke experimental software for generating real-time feedback and interfacing with the WA* algorithm. Response time is an important component of NF systems; however, Zotev et al. [23] reported successful fMRI-based NF despite the approximately 7s delay of the BOLD signal. Since delay

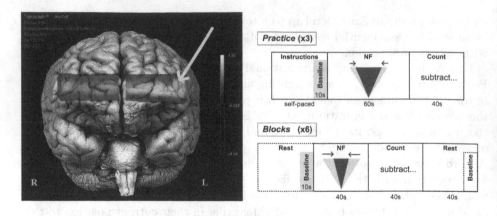

(a) Topographic image of asymmetric dorsolateral prefrontal oxygenation increase during a successful NF epoch, overlaid on a brain surface image.

(b) The NF protocol is composed of epochs of NF, Counting and Rest (see text).

Fig. 2. (a) Topographic image (fNIRSoft by Biopac Systems); (b) NF protocol.

using fNIRS is comparable, we sought inspiration from the experimental protocol of Zotev et al. [23].

In order to avoid the necessity of a lengthy NF training for the purpose of the current research (i.e. influencing the search process with NF), we gave specific directions to the subjects in terms of the cognitive strategies that are most likely to increase prefrontal asymmetry, without being prescriptive about the actual thought contents. Additionally, we recruited subjects from a pool of volunteers who have previously participated in EEG-based NF experiments and thereby were familiar with associated cognitive strategies. None of the subjects had previous experience with fNIRS.

Subjects received visual feedback in the form of a red cone symbolising the search space. Successful NF (corresponding to left prefrontal asymmetry) resulted in the cone shrinking, corresponding to a more focused search. At the beginning of the protocol, subjects completed three free-practice runs lasting one minute each to familiarise them with the system and allow them to explore cognitive strategies. The practice runs were separated by rest periods and further instructions (e.g. to expect a slight delay and some jitter in the feedback). Subjects were also introduced to a counting task, where they were instructed to mentally count backwards from 100 by subtracting a given integer. This task was included to distract subjects attention from the cognitive strategies used during NF (see [23]) and to promote prefrontal activation converging to baseline after NF blocks.

Following practice, subjects completed six blocks with the following structure of epochs: NF, Count, and Rest. Each epoch lasted 40 seconds. The last 10 seconds of each Rest epoch (20 observations sampled at 2Hz) was used to calculate the

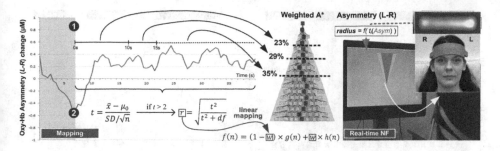

Fig. 3. System Overview. Subjects equipped with a fNIRS sensor engage in a Neurofeedback task whose visual display is a metaphor for the search space they are trying to reduce. Note that since each epoch (without the first 7s, shown in gray) contained at least 66 observations (33s with 2Hz sampling frequency), we applied as threshold criterion the t critical value for $p = .05$ (two-tailed) with 65 degrees of freedom (df), $t_{crit}(65) = 2.00$.

baseline for the next block (see [17]). Rest epochs contained no specified cognitive tasks and no feedback on prefrontal asymmetry, and they were used to calculate the baseline for successive blocks; therefore, Rest epochs were not analysed. Because the hemodynamic response measured by fNIRS occurs in approximately 7 seconds [3], the first 7 seconds of data in each NF and Count epoch were not analysed. An overview of the experimental protocol is presented in Figure 2b.

Success in each NF epoch was determined by performing a one-sample t-test on the asymmetry scores collected during the epoch against the test value of zero, upon the completion of the epoch. This tested whether there was a statistically significant increase in asymmetry against the baseline (measured during Rest), since the mean of prefrontal asymmetry during baseline was defined as zero [1]. If this test was significant, the effect-size measure r was calculated to characterise the magnitude of difference from the baseline (see Figure 3).

The system operates through mapping the outcome of fNIRS NF onto parameters of the 8-puzzle heuristic search mechanisms. The main mechanism consists in mapping the NF success score r onto the weighting coefficient w of WA*, at the end of each NF epoch (see Figure 3). This takes place via a linear mapping function, and ensures that departure from admissibility is proportional to the level of success of NF. However, simply launching a heuristic search at the end of an NF epoch would not benefit fully from the concept of dynamic weighting, which is also based on a choice of when w should be updated during the search process. The second mapping mechanism monitors the onset of fNIRS changes during NF to determine one of three possible stages at which w will be modified (23%, 29% or 35% of search, for a specific configuration pair). The entire NF sequence thus proceeds as follows: i) after the start of NF, the system detects the temporal pattern of variation and determines at which stage w is to be modified; ii) WA* is launched and will pause at the predefined stage of search to receive a new w value (Figure 3-1); iii) after completion of the NF block, the new w value

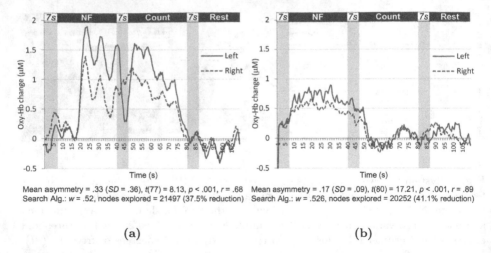

Mean asymmetry = .33 (SD = .36), t(77) = 8.13, p < .001, r = .68   Mean asymmetry = .17 (SD = .09), t(80) = 17.21, p < .001, r = .89
Search Alg.: w = .52, nodes explored = 21497 (37.5% reduction)   Search Alg.: w = .526, nodes explored = 20252 (41.1% reduction)

(a)   (b)

Fig. 4. Examples of average left and right oxygenation (Oxy-Hb) changes over time during two successful blocks. Areas in gray represent the first 7s of each epoch (i.e. the approximate delay of the hemodynamic response). Note that during the NF epoch, Oxy-Hb increases bilaterally, with asymmetry to the left; during the Count epoch following NF, Oxy-Hb decreases on both sides; during Rest, Oxy-Hb further decreases towards baseline.

is passed to WA* (Figure 3-2), which resumes operation until reaching a solution (which was guaranteed with the selected problem set). Thus, the outcome is the reduction in search space (nodes explored) [1] obtained from successful NF input.

5   Results and Discussion

Out of all 66 blocks completed by the 11 subjects, 38 (58%) contained an NF epoch with statistically significant left-side asymmetry; these blocks were considered successful during the experiment which triggered changes in the heuristic search process[2]. Each subject had at least one successful block, and 8 subjects (73%) had at least 3 successful blocks (i.e. half of blocks successful). No subject achieved NF success on all 6 blocks.

Since fNIRS signals are relative values, it can be difficult to compare them across subjects [18]; moreover, the magnitude of oxygenation changes can also

---

[1] Reduction in nodes explored was used as the outcome metric, since this is machine-independent (unlike, for example, reduction in solution time).

[2] Note that NF success was determined real-time by the experimental software by running t-tests on asymmetry scores after the completion of each epoch. The assumption of normality was not tested by the experimental software; however, post-hoc analyses using bootstrapping resampling method on a subset of epochs resulted in accepting the same epochs as successful. Additionally, post-hoc correction for family-wise error yielded the same result, with the exception of a single epoch.

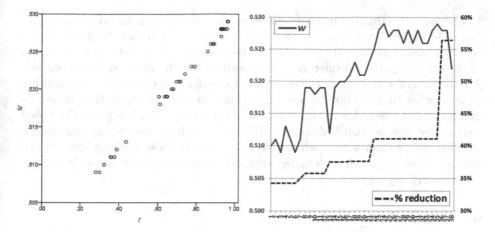

(a) Scatterplot showing the distribution of scores over the 38 successful NF epochs: r and w values are aligned on mapping function.

(b) Graph showing search space reduction across successful NF epochs (ranked low to high) and corresponding w values. Performance across subjects lead to significant difference in terms of search behaviour due to the exponential complexity of A*.

Fig. 5. (a) Mapping from r to w; (b) Search space reduction in successful epochs.

differ substantially across blocks within the same subject. Our mapping strategy (described in the previous section) was designed to mitigate the issue of comparability. We demonstrate this through two examples of a successful block from two different subjects (see Figure 4).

Figure 4a shows a larger mean asymmetry during the NF epoch than Figure 4b (ΔOxy-Hb = .33 and .17, respectively); however, the dispersion of asymmetry scores during the epoch was also larger compared to the mean (SD = .36 and .09, respectively). In other words, left-side oxygenation was more consistently above right-side throughout the NF epoch in Figure 4b; consequently, the t value was larger (t is calculated using the mean, standard deviation and number of data points), leading to a larger r value (r is based on t and the degrees of freedom), mapped to a larger w value, leading to greater reduction in search space (see Figure 3 for formulae).

The distribution of r scores mapped to w scores in successful NF epochs is presented in Figure 5a. We only calculated r values for epochs where left-side asymmetry was statistically significantly above 0 (i.e. the NF signal was used to influence the algorithm only when there was statistical evidence for left asymmetry). The lowest r value (.28) in Figure 5a shows that we could reliably detect medium (and large) effect sizes ($r \geq .30$) during 40 second-long NF epochs with 2Hz sampling frequency. Although several NF epochs approached the maximum r value of 1, which determined the maximum of the dynamic weighting of the search algorithm, the distribution of scores demonstrates that differential weighting was successfully applied based on the r effect-size measure.

The percent reduction in search space as a result of successful NF input is presented in Figure 5b. Average reduction in nodes visited across the 38 successful NF epochs was 39.5% (SD = 5.65). We used search space reduction compared to the default search parameters as a measure of search behaviour improvement. A reduction of search space of 50% can correspond to a reduction in computing time between 50% and 90% depending on the computer configuration, in particular its memory size, due to the exponential complexity of A*. It should be noted that in the vast majority of epochs, the system still returns an optimal length solution rather than a near-admissible one. Overall, these findings support the validity of our approach to defining NF success and mapping it to the behaviour of the search algorithm.

6   Conclusions

We have introduced a hybrid approach to BCI, in which cognitive attitudes could be used to guide computation. These preliminary results have established the feasibility of BCI-based control of algorithmic behaviour. In particular, the variations in BCI signal magnitude, across subjects and trials, can be mapped onto several bands of search improvements (e.g. space reduction, Figure 5b), which opens up the possibility of tuning of search algorithms' behaviour.

Our approach may also suggest research directions for a new sort of integration between cognitive processes and AI software, potentially applicable to a large set of problem-solving applications. The various neural signals that could support this process are still to be explored, and the approach could be refined to investigate more specific attitudes, such as risk acceptance.

References

1. Ayaz, H.: Functional near infrared spectroscopy based brain computer interface. PhD thesis, Drexel University (2010)
2. Baehr, E., Rosenfeld, J.P., Baehr, R.: Clinical use of an alpha asymmetry neurofeedback protocol in the treatment of mood disorders: Follow-up study one to five years post therapy. Journal of Neurotherapy 4(4), 11–18 (2001)
3. Bunce, S.C., Izzetoglu, M., Izzetoglu, K., Onaral, B., Pourrezaei, K.: Functional near-infrared spectroscopy. IEEE Engineering in Medicine and Biology Magazine 25(4), 54–62 (2006)
4. Cavazza, M., Charles, F., Aranyi, G., Porteous, J., Gilroy, S.W., Raz, G., Keynan, N.J., Cohen, A., Jackont, G., Jacob, Y., et al.: Towards emotional regulation through neurofeedback. In: Proceedings of the 5th Augmented Human International Conference, p. 42. ACM (2014)
5. Davidson, R.J., Ekman, P., Saron, C.D., Senulis, J.A., Friesen, W.V.: Approach-withdrawal and cerebral asymmetry: Emotional expression and brain physiology: I. Journal of Personality and Social Psychology 58(2), 330–341 (1990)
6. Doi, H., Nishitani, S., Shinohara, K.: NIRS as a tool for assaying emotional function in the prefrontal cortex. Frontiers in Human Neuroscience 7 (2013)
7. Ebendt, R., Drechsler, R.: Weighted A* search–unifying view and application. Artificial Intelligence 173(14), 1310–1342 (2009)

8. Eugster, M.J., Ruotsalo, T., Spapé, M.M., Kosunen, I., Barral, O., Ravaja, N., Jacucci, G., Kaski, S.: Predicting term-relevance from brain signals. In: Proceedings of the 37th International ACM SIGIR Conference on Research & Development in Information Retrieval, pp. 425–434. ACM (2014)

9. Gerson, A.D., Parra, L.C., Sajda, P.: Cortically coupled computer vision for rapid image search. IEEE Transactions on Neural Systems and Rehabilitation Engineering 14(2), 174–179 (2006)

10. Girouard, A., Solovey, E.T., Hirshfield, L.M., Chauncey, K., Sassaroli, A., Fantini, S., Jacob, R.J.K.: Distinguishing difficulty levels with non-invasive brain activity measurements. In: Gross, T., Gulliksen, J., Kotzé, P., Oestreicher, L., Palanque, P., Prates, R.O., Winckler, M. (eds.) INTERACT 2009. LNCS, vol. 5726, pp. 440–452. Springer, Heidelberg (2009)

11. Hansen, E.A., Zhou, R.: Anytime heuristic search. J. Artif. Intell. Res(JAIR) 28, 267–297 (2007)

12. Kapoor, A., Shenoy, P., Tan, D.: Combining brain computer interfaces with vision for object categorization. In: Conference on Computer Vision and Pattern Recognition, pp. 1–8. IEEE (2008)

13. Nijholt, A., Tan, D.: Brain-computer interfacing for intelligent systems. IEEE Intelligent Systems 23(3), 72–79 (2008)

14. Pearl, J.: Heuristics: intelligent search strategies for computer problem solving. Addison-Wesley Pub. Co., Inc., Reading (1984)

15. Pohl, I.: Heuristic search viewed as path finding in a graph. Artificial Intelligence 1(3), 193–204 (1970)

16. Reinefeld, A.: Complete solution of the eight-puzzle and the benefit of node-ordering in IDA*. In: Procs. Int. Joint Conf. on AI, Chambery, Savoie, France, pp. 248–253 (September 1993)

17. Ruocco, A.C., Rodrigo, A.H., Lam, J., Di Domenico, S.I., Graves, B., Ayaz, H.: A problem-solving task specialized for functional neuroimaging: validation of the Scarborough adaptation of the Tower of London (S-TOL) using near-infrared spectroscopy. Frontiers in Human Neuroscience 8 (2014)

18. Sakatani, K., Takemoto, N., Tsujii, T., Yanagisawa, K., Tsunashima, H.: NIRS-based neurofeedback learning systems for controlling activity of the prefrontal cortex. In: Oxygen Transport to Tissue XXXV, pp. 449–454. Springer (2013)

19. Solovey, E., Schermerhorn, P., Scheutz, M., Sassaroli, A., Fantini, S., Jacob, R.: Brainput: enhancing interactive systems with streaming fNIRS brain input. In: SIGCHI Conference on Human Factors in Computing Systems, pp. 2193–2202. ACM (2012)

20. Solovey, E.T., Girouard, A., Chauncey, K., Hirshfield, L.M., Sassaroli, A., Zheng, F., Fantini, S., Jacob, R.J.: Using fNIRS brain sensing in realistic HCI settings: experiments and guidelines. In: Proceedings of the 22nd Annual ACM Symposium on User Interface Software and Technology, pp. 157–166. ACM (2009)

21. Studer, B., Pedroni, A., Rieskamp, J.: Predicting risk-taking behavior from prefrontal resting-state activity and personality. PLoS One 8(10), e76861 (2013)

22. Sutton, S.K., Davidson, R.J.: Prefrontal brain asymmetry: A biological substrate of the behavioral approach and inhibition systems. Psychological Science 8(3), 204–210 (1997)

23. Zotev, V., Krueger, F., Phillips, R., Alvarez, R.P., Simmons, W.K., Bellgowan, P., Drevets, W.C., Bodurka, J.: Self-regulation of amygdala activation using real-time fMRI neurofeedback. PloS one 6(9), e24522 (2011)

English Phonetics: A Learning Approach Based on EEG Feedback Analysis

Luz García Martínez[1]([✉]), Alejandro Álvarez Pérez[2], Carmen Benítez Ortúzar[1],
Pedro Macizo Soria[2], and Teresa Bajo Molina[2]

[1] Department of Signal Theory, Telematics and Communications,
University of Granada, Granada, Spain
luzgm@ugr.es
[2] Department of Experimental Psychology,
University of Granada, Granada, Spain

Abstract. This work proposes a procedure to measure the human capability to discriminate couples of English vocalic phonemes embedded into words. Using the analysis of the EEG response to auditory contrasts in an oddball paradigm experiment, the Medium Mismatch Negativity potential (*MMN*) is evaluated. When the discrimination is achieved, MMN has a negative amplitude while positive or zero MMN amplitudes correspond to the confusion of the two vocalic phonemes heard by the subject performing the experiment. The procedure presented has many potential usages for phonetic learning tools given its capability to automatically analyze discrimination of sounds. This permits its usage in interactive and adaptive applications able to keep track of the improvements made by the users.

1 Introduction

Within the last years big amounts of economic and human resources have been invested for the learning of a second language (*L2*). From an educative perspective the objective is to develop programs that maximize the efficacy of the methods used for learning other languages. In many occasions it becomes difficult for the *L2* learners to achieve fluency, specially when acquisition of *L2* takes place at a late age [13]. One of its main problems associated is the difficulty for the oral comprehension [12]. Such problem, comes many times motivated by the inability to discriminate the oral phonemes of the second language.

This work presents a EEG monitoring experiment to analyze the phonetic discrimination process that takes place when hearing auditory stimuli. Native Spanish speakers with limited English knowledge perform a passive learning task with 10 phonological contrasts of English vowels embedded into words. For each phonological contrast frequent and deviant stimuli are presented, and the Medium Mismatch Negativity (MMN) [10] is proposed as phonological discrimination index.

The usage of signal processing permits the automation of the analysis, extracting information otherwise masked by the low SNRs characteristic of electrophisiological signals. It also offers the possibility to analyze on *quasi real-time* the

J.M. Ferrández Vicente et al. (Eds.): IWINAC 2015, Part I, LNCS 9107, pp. 322–329, 2015.
DOI: 10.1007/978-3-319-18914-7_34

auditory discrimination results of the subjects, becoming a potential feedback tool for language learning applications based on information and communication technologies, capable to adapt to individual user needs.

The rest of work is organized as follows. Section 2 describes the Medium Mismatch Negativity and its application to measure the phonological discrimination capability of the subjects under EEG monitoring. Section 3 details the procedure followed in the experiment proposed. Section 4 describes the signal processing done in the experiment to extract the discrimination analysis from the raw EEGs registers. Results and conclusions are presented in Section 5.

2 Medium Mismatch Negativity

2.1 The MMN Component: An Index for Phonologic Discrimination

Speech is a sound wave with acoustic and temporal properties. Our perceptive system extracts from it features that allow, finally, to identify speech phonemes. Nevertheless humans do not make use of speech parameters in a sharp manner. On the contrary, different sounds are grouped as examples or realizations of a same phoneme. In other words our phonologic perception is categorical. Speech signal is not perceived in a continuous way, but in a discrete manner [8]. Such behavior becomes adaptive, in a way that the speakers from a certain language categorize any sound they heard as an example of a certain phonetic category existing in such language. Up to the age of 6-12 months humans can discriminate the complete repertory of phonemes of any language. But after that age, the discriminative capacity is limited and sounds are grouped according to the language they are exposed to [3],[6]. Adults have categories of phonemes useful to understand the language they listen to.

A way to check the existence of phonetic categories for the sounds of a language is through the analysis of an electrophisiological component sensitive to the acoustic disparity among its phonemes named *Mismatch Negativity (MMN)*. The MMN component also called *auditory disparity index* when considering the cortical evoked potentials associated to events (ERPs), was isolated at the end of the 60's [9]. MMN is obtained through an oddball paradigm experiment characterized by the listening of auditory stimuli while the participants watch a movie without sound (that is, a passive listening task). The stimuli presented can be repetitive and frequent with a high probability of occurrence (standard stimuli) or, they can be infrequent or deviant stimuli. The variation of the deviant stimulus (consisting on its intensity, duration, frequency, acoustic complexity, etc) compared to the standard stimulus is enough to produce the MMN. The MMN component is characterized by a wave with a higher negative amplitude in the case of infrequent stimuli compared to that of frequent stimuli, around 100-250 ms. after the presentation of the auditory stimulus. The electrophisiological activity involved is located in the cortex frontal and central area. Several studies have confirmed the utility of MMN as an index of phonologic discrimination among phonemes of their language [10], proving that speakers have phonologic categories associated to the sounds of the language they listen to and speak.

2.2   Proposal: Phonological Discrimination of a Second Language L2 Based on MMN

The catalog of phonemes is very different among languages, that often do not present a fixed nor common set of them. For this reason, and given the categorization of sounds made by adult language users, learners of a second language *L2* will present difficulties to perceive phonemes that do not exist in their mother tongue *L1*.

Such difficulty increases with age and can vary for the different phonemes and learners. It will be high in the case of learners with disorders in phonological coding like, for example dyslexia, which deserves a special remark. The difficulties to integrate visual and auditory stimuli make the learning of *L2* very hard for dyslexic learnes[11]. The dual route model for reading [4] states two simultaneous paths to arrive to the meaning of a word during the reading process: the visual route (words are compared without dividing them into phonemes) and the phonological route (graphemes are converted into phonemes and gathered to obtain the correct pronunciation of the word). If the reader suffers from dyslexia, one or both routes do not work properly, and the auditory discrimination will become difficult. The problem becomes bigger in languages classified as *opaque* or *deep*, like English. Such languages present a complex phoneme-grapheme correspondence and irregularities opposite to transparent languages that present clear letter-sound correspondence and much regularity.

In such learning framework, this work proposes to use of the monitoring of the MMN component as a *feedback-index of phonological discrimination*. If the *L2* learning cognitive process is monitored in a personalized manner, the troublesome phonological categories could be identified. Individually adapted strategies to strengthen the learning process could then be proposed. The usage of the communication and information technologies plays a fundamental role for this purpose. The automatic monitoring and processing of the learner's electrophisiological answers permits the creation of adaptive, auto-configurative learning tools personalized for each particular individual.

3   Description of the Experiment

3.1   Participants

The oddball paradigm experiment has been carried out for a set of 10 participants with an average age of 25.30 years (standard deviation $(sd) = 4.42$). All of them had learned English as a second language (*L2*) starting at the average age of 7.52 years ($sd = 2.67$). 6 of them had lived in an English speaking country for an average time of 4.80 months ($sd = 5.55$). In order to know in detail their level of English, the participants answered a questionnaire about their oral and reading comprehension, and their oral and written communication skills. Table 1 shows the average results over a 1 to 10 scale [1].

Table 1. Average and standard deviant of English skills self-valuation of the participants on a 1 to 10 scale

Reading comprehension	6.80 ($sd = 2.09$)
Oral comprehension	6.30 ($sd = 1.88$)
Writing skills	5.20 ($sd = 2.29$)
Oral skills	5.50 ($sd = 2.27$)

3.2 Material and Procedure

Participants heard 10 auditory contrasts or *experimental conditions*. In addition a *control condition* was created in order to evaluate the acoustic perception skills of the participants. The control condition was made up of pure tones with a duration of 50 ms, at 2 different frequencies: 1000 Hz. for standard tones and 1050 Hz. for deviant tones. The rest of experimental conditions were made up of phonological contrasts of English vowels. Table 2 shows the set of vowels selected for the study, taken from [2].

Table 2. Phonemes of English vowels used in the experiment

/ɛ/	/ɔː/	/i/	/ə/	/ae/	/ʊ/	/I/	/ʌ/	/ɒ/	/e/

All the stimuli (standard and deviant) had equivalent acoustic characteristics. Their duration was 300 ms, being the difference among stimuli determined by the relative location of their *F1*, *F2* and *F3* formants. Table 3 shows the phonological contrasts used.

Each contrast (pure tones and English vowels) had a total of 400 standard stimuli and 100 deviant stimuli. The order of presentation in the experiment was pseudo-random assuring that each condition started with at least 5 standard stimuli and never having two consecutive deviant stimuli. The time gap between 2 stimuli was set to 500 ms, and the order of presentation of the 11 conditions was randomize for each participant. The subjects were instructed to the different conditions passively while watching a movie without sound. Stimuli were presented binaurally with an approximate intensity of 70 dB. The experimental session lasted around 1 hour 45 minutes, including 5 minutes rests after each contrast session.

4   Signal Processing

The continuous EEG of 9 Ag/Ag-Cl electrodes inserted on a Neuroscan Quick-cap elastic cap were registered during the experiment (500 Hz of sampling rate). Electrodes F3, Fz, F4, Fc3, Fc4, C3, Cz and C4 of the international 10-20 system [5] were chosen to detect the cortical endogenous answer generated with

Table 3. Numbering of the phonological contrasts used in the study. *Condition* stands for the type of stimulus -standard/deviant; *Vowel* specifies the vowels used. *Word* stands for the lexical context on which the vowel is inserted. *Spanish* -yes/no- stands for the existence/non-existence of the English vowel (*L2*) in Spanish (*L1*).

Contrast	Condition	Vowel	Word	Spanish
1	standard	ɒ	hod	no
1	deviant	ɔː	hawed	no
2	standard	ɒ	hod	no
2	deviant	ɛ	head	no
3	standard	ə	whod	yes
3	deviant	ʊ	hood	yes
4	standard	ʌ	hud	yes
4	deviant	ɒ	hod	no
5	standard	*I*	hid	no
5	deviant	ɛ	head	no
6	standard	ɛ	head	no
6	deviant	ae	had	no
7	standard	ae	had	no
7	deviant	ɒ	hod	no
8	standard	*e*	hayed	yes
8	deviant	ae	had	no
9	standard	*e*	hayed	yes
9	deviant	ɛ	head	no
10	standard	*i*	heed	yes
10	deviant	*I*	hid	no

contrasts described in the former section. Vertical ocular movements and blinking were registered with electrodes located on the external side of each eye. EEG potentials were referred to electrodes located on the left and right mastoids. Impedances were kept below 5 $K\Omega$. EEGs were amplified with Neuroscan Nuamps ^{TM}. The following steps were taken to process the EEG registers:

- EEGs were downsampled to 200 Hz and band-pass filtered in the band 0.5Hz-30Hz.
- The eye-blinking noise was removed from the continuous EEG registers, using Principal Component Analysis (PCA) by Singular Value Decomposition (SVD) [7]. Multichannel EEGs were separated into linearly independent (temporally and spatially noncorrelated) components. Eye movement artifacts appear as one of this independent components that can be eliminated before reconstructing the original EEG. This procedure can be implemented as a spatial filter in the form of matrix applicable to the channel EEG in real-time without redoing the time-consuming SVD.
- Continuous EEGs were epoched into temporal windows around the expected stimulus evoked response: EEG time windows starting from 100 ms before to 600 ms after the stimulus is presented, were selected. The first prior 100 ms

were used to eliminate artifacts or low frequency drifts through a baseline correction.

- In order to increase the SNR of the ERPs, averages were done for each experimental condition, and each type of stimulus. Averages of responses per subject have also calculated.
- Medium Mismatch Negativity components were calculated.

5   Results and Discussion

After checking the capability of the subjects to discriminate standard and deviant pure tones, experiments were conducted following the description done in Section 3. Figure 1 depicts five examples of contrast experiments. Average ERPs for the standard stimuli are plotted in gray, while average ERPs for deviant stimuli are plotted in black. Left column subfigures show examples of subjects that discriminated correctly the deviant vowel sound when heard, and produce a higher negative amplitude of the ERP (black line) compared to that of the standard stimulus (gray line). If no auditory differences are perceived both ERPs have similar amplitudes. Examples of this latter case are shown in the right column subfigures.

In order to evaluate the possible differences between standard and deviant stimuli on each vowel contrast, a T-Student analysis has been done with a 95% confidence interval comparing the amplitudes on each contrast separately [1]. Table 4 shows in columns 2 and 3 the average amplitude (M) obtained in all standard and deviant contrasts. Column 4 presents the Medium Mismatch Negativity amplitude (infrequent condition minus standard condition).

Table 4. Average amplitudes (microvolts) for standard and deviant contrasts. Medium Mismatch Negativity).

Contrasts	Standard M	Deviant M	MMN
/ɒ/-/ ɔː/	0.87	0.61	-0.26
/ɒ/-/ɛ/	0.54	0.58	0.04
/ə/-/ʊ/	1.35	1.27	-0.08
/ʌ/-/ɒ/	1.05	0.90	-0.15
/I/-/ɛ/	0.57	0.49	-0.08
/ɛ/-/ae/	1.36	0.98	-0.38
/ae/-/ɒ/	0.63	0.66	0.03
/e/-/ae/	0.70	0.40	-0.30
/e/-/ɛ/	0.58	0.36	-0.22
/i/-/I/	1.16	0.22	-0.93

Aside the average outcomes for the different contrasts, the individual results (per subject and per contrast) obtained in this work have a direct application

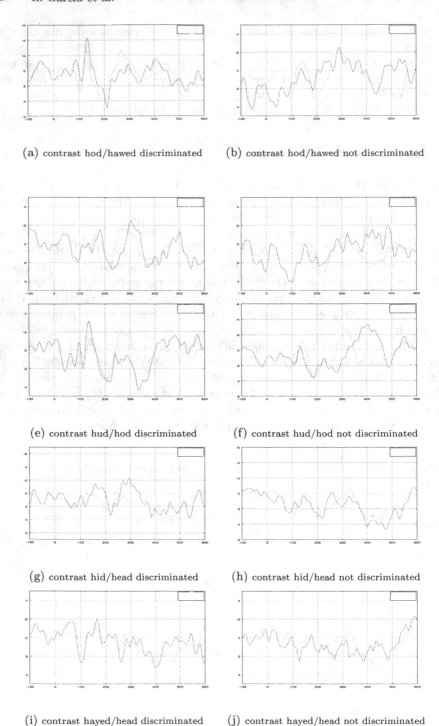

(a) contrast hod/hawed discriminated   (b) contrast hod/hawed not discriminated

(e) contrast hud/hod discriminated   (f) contrast hud/hod not discriminated

(g) contrast hid/head discriminated   (h) contrast hid/head not discriminated

(i) contrast hayed/head discriminated   (j) contrast hayed/head not discriminated

Fig. 1. Average ERPs for each vowel phoneme contrast presented to several subjects. Black/gray lines depict the ERP after the standard/deviant stimulus.

in the learning of a second language. Once they are pointed at, learners can focus on the phonological discriminations they have difficulties with. Given the advantages of the automatic signal processing proposed, the tool offers an on-line mechanism to individually adapt the learning process to the needs of the user with the possibility to keep track of the progresses achieved. The fact that the oddball paradigm experiment proposed does not need active attention makes the procedure optimal for children with difficulties of attention.

References

1. Álvarez Pérez, A.: Discriminacin fonológica en aprendices de una segunda lengua. Evidencia electrofisiológica. Trabajo fin de Mster. Departamento de Psicologa Experimental, Universidad de Granada. Dirigido por Pedro Macizo Soria (2014)
2. Assmann, P.F., Katz, W.F., Jenouri, K.M., Hamilton, P.R.: Identification of natural and synthesized vowels produced by children and adults: Effects of formant frequency variation. Journal of the Acoustical Society of America 98, 2964 (1995)
3. Cheour, M., Ceponiene, R., Lehtokoski, A., Luuk, A., Allik, J., Alhoi, K., Ntnen, R.: Development of language-specific phoneme representations in the brain. Nature Neuroscience 1(5), 351–353 (1998)
4. Coltheart, M., et al.: DRC: A dual route cascaded model of visual word recognition and reading aloud. Psychological Review 108, 204–256
5. Jasper, H.: The Ten Twenty Electrode System of the International Federation. Electroenceph. Clin. Neurophysiol. 10, 371–375 (1958)
6. Khul, P.K., Williams, K.A., Lacerda, F., Stevens, K.N., Lindblom, B.: Linguistic Experiende Alters Phonetic Perception in Infants by 6 Months of Age. Science 255(5044), 606–608 (1992)
7. Lagerlund, T.D., Sharbrough, F.W., Busacker, N.E.: Spatial Filtering of Multichannel Electroencephalographic Recordings Through Principal Component Analysis by Singular Value Decomposition. Journal of Clinical Neurophysiology 14(1), 73–82 (1997)
8. Liberman, A.M., Mattingly, I.G.: The motor theory of speech perception revised. Cognition 4(21), 1–36 (1985)
9. Ntnen, R., Gaillard, A.W.K., Mantysalo, S.: Early selective-attention effect on evoked potential reinterpreted. Acta Psychologica 42, 313–329 (1978)
10. Ntnen, R.: The perception of speech sounds by the human brain as reflected by mismatch negativity. Psychophysiology 38, 1–21 (2001)
11. Ramus, F., White, S., Frith, U.: Weighing the evidence between competing theories of dyslexia. Developmental Science 9, 265–269 (2006)
12. Shestakova, A., Huotilainen, M., Ceponiene, R., Cheour, M.: Event-related potentials associated with second language learning in children. Clinical Neurophysiology 8(114), 1507–1512 (2003)
13. Weber-Fox, C.M., Neville, H.J.: Maturational Contrains on Functional Specializations for Language Processing: ERP and Behavioral Evidence in Bilinguals Speakers. Journal of Cognitive Neuroscience 8(3), 231–256 (1996)
14. Ziegler, J., et al.: Developmental dyslexia in different languages: Language-specific or universal? J. Experimental Child Psychology 86, 169–193 (2003)

Dynamic Modelling of the Whole Heart
Based on a Frequency Formulation and Implementation
of Parametric Deformable Models

Rafael Berenguer-Vidal[1]([✉]), Rafael Verdú-Monedero[2],
and Álvar-Ginés Legaz-Aparicio[2]

[1] UCAM, Universidad Católica San Antonio de Murcia, 30107, Guadalupe, Spain
rberenguer@ucam.edu
[2] Dept. Tecnologías Información y Comunicaciones, Universidad Politécnica de Cartagena,
30202, Cartagena, Spain
rafael.verdu@upct.es, alvarlegaz@gmail.com

Abstract. In the past few years, numerous efforts have been devoted to the segmentation and characterization of the human heart from various medical image techniques. This paper addresses the first results of parametric deformable models defined in the Fourier domain as a tool to characterize the shape of the heart. The main advantage of these models is their high speed of adaptation to the dataset and their robustness against noise. In addition, due to their explicit parametric typology, different parameters of its dynamical behaviour can be derived from their mathematical expression. This article details the mathematical framework of deformable models defined in the frequency domain as well as the preprocessing and practical implementation of the model used in this application to model the cardiac cycle of the whole heart.

Keywords: Parametric deformable model · Fourier domain · Cardiac modelling · Motion tracking · B-*splines*

1 Introduction

The analysis of the motion and deformation of the heart is a topic with a considerable interest in the literature due to the great impact of the cardiovascular disease (CVD) as a cause of death. In recent years, several approaches have been proposed [1–3]. In [4–6], a detailed review of the most known methods for cardiac motion analysis are presented.

Due to the elastic nature of the heart and the robustness to noise of B-*splines*, deformable models based on B-*splines* are an interesting approach for cardiac characterization [2, 7–9]. This work proposes the use of frequency-based multidimensional parametric deformable models [10] instead of commonly used spatial-based implementations for the motion characterization of human heart, given that the frequency-based framework offers an efficient implementation [11].

The paper is organized as follows: In Section 2 we describe the formulation of a multidimensional parametric deformable model and we provide the equation of motion of the model defined in the frequency domain. Section 3 details the practical implementation of the algorithm, Section 4 shows the results of the segmentation of the whole heart in 4D CT data, and finally Section 5 closes the paper with the conclusions.

© Springer International Publishing Switzerland 2015
J.M. Ferrández Vicente et al. (Eds.): IWINAC 2015, Part I, LNCS 9107, pp. 330–339, 2015.
DOI: 10.1007/978-3-319-18914-7_35

2 Multidimensional Parametric Deformable Models

Multidimensional parametric deformable models stand for the generalization of the active contours described by Liang et al. [12]. These model are defined as an e-dimensional time-varying hypersurface evolving in the d-dimensional space $\mathbb{R}^d$ by means of a parametric function,

$$\mathbf{v} := \mathbf{v}(\mathbf{s},t) = [v_1(\mathbf{s},t), \ldots, v_d(\mathbf{s},t)]^\top, \tag{1}$$

where $\mathbf{s} := [s_1, \ldots, s_e]$ with $s_j \in [0, L_j]$, $e \leqslant d$ and $e, d \in \mathbb{N}$ is the vector of parametric variables in the spatial domain, t is time and $v_i(\mathbf{s},t)$ represents the coordinate function for each dimension i.

We consider the process of data modelling as a dynamic process. This means that the dataset under analysis changes over time. Under this assumption, the shape of the model is governed by the following energy functional,

$$\mathcal{E}(\mathbf{v},t) = \mathcal{S}(\mathbf{v},t) + \mathcal{P}(\mathbf{v},t). \tag{2}$$

The first term, $\mathcal{S}(\mathbf{v},t)$, is related to the internal deformation energy, which characterizes the elastic deformation of a flexible model [13]. This component smooths the shape of the model and rules its behaviour of elasticity and rigidity. Besides, the energies represented by the term $\mathcal{P}(\mathbf{v},t)$ commonly generates forces to attract the model to the edges of the dataset. Consequently, this term is obtained from the multidimensional data, although different restrictions such as non-lineal internal forces inside the model can also be included.

The shape of the model is obtained by minimizing the term of energies $\mathcal{E}(\mathbf{v},t)$. By applying the calculus of variations, the energy functional (1) reaches a minimum when the Euler-Lagrange equation is satisfied [12]. This can be generalized for the multidimensional case resulting in a system of d decoupled partial differential equations (PDEs)[10],

$$\mu(\mathbf{s})\partial_{tt} v_i(\mathbf{s},t) + \gamma(\mathbf{s})\partial_t v_i(\mathbf{s},t) - \partial_{s_1}\left(\alpha(\mathbf{s})\partial_{s_1} v_i(\mathbf{s},t)\right) - \cdots - \partial_{s_e}\left(\alpha(\mathbf{s})\partial_{s_e} v_i(\mathbf{s},t)\right) +$$
$$\left(\partial_{s_1 s_1} + \cdots + \partial_{s_e s_e}\right)\left(\beta(\mathbf{s})\partial_{s_1 s_1} v_i(\mathbf{s},t) + \cdots + \beta(\mathbf{s})\partial_{s_e s_e} v_i(\mathbf{s},t)\right) =$$
$$\mathbf{q}\left(\mathbf{v}(\mathbf{s},t)\right), \quad 1 \leqslant i \leqslant d, \quad i \in \mathbb{N}, \tag{3}$$

where ∂_t and ∂_{tt} denote, respectively, first and second partial derivative with respect to time, and ∂_{s_j} and $\partial_{s_j s_j}$ with respect to the parametric variables s_j. The parameters $\alpha(\mathbf{s})$ and $\beta(\mathbf{s})$ control the elasticity and rigidity of the model respectively, and $\mu(\mathbf{s})$ and $\gamma(\mathbf{s})$ provide its mass and damping densities. Finally $\mathbf{q}\left(\mathbf{v}(\mathbf{s},t)\right)$ represents the external forces derived from $\mathcal{P}(\mathbf{v},t)$. The equation (3) stands for the necessary condition for the model at equilibrium, enabling the calculation of each coordinate function separately.

In order to apply Eq. (3) by means of a discrete processing system, both spatial and temporal variables should be discretized. The discretization in the spatial domain is done by means of the finite element method (FEM) [14]. The domain $[0, L_j]$ of each parametric variable s_j is partitioned into N_j finite subdomains. Hence, the model $\mathbf{v}(\mathbf{s},t)$ can be expressed as the union of $N = N_1 N_2 \cdots N_e$ elements $v_i^{\overline{n}}(\mathbf{s},t)$, each one represented using shape functions $\mathbf{N}(\mathbf{s})$ and nodal variables $\mathbf{u}_i^{\overline{n}}(t)$, i.e.

$$v_i^{\bar{n}}(\mathbf{s},t) = \mathbf{N}(\mathbf{s})\mathbf{u}_i^{\bar{n}}(t), \tag{4}$$

where $\bar{\mathbf{n}} = [n_1, \ldots, n_e]$ are the indexes of the element in each coordinate function.

By applying the Galerkin's method [15] to Eq. (3) for each finite element and assembling both expressions and nodal variables for all elements, the motion equation of the whole system can be written in matrix form as [10],

$$\mathbf{M}d_{tt}\mathbf{u}_i(t) + \mathbf{C}d_t\mathbf{u}_i(t) + \mathbf{K}\mathbf{u}_i(t) = \mathbf{q}_i(t), \tag{5}$$

where $\mathbf{M}$, $\mathbf{C}$ and $\mathbf{K}$ represent the global matrices of mass, damping and stiffness respectively, and $\mathbf{q}_i(t)$ is the external forces vector.

The matrices $\mathbf{M}$, $\mathbf{C}$ and $\mathbf{K}$ depend on the shape function $\mathbf{N}(\mathbf{s})$ used in the spatial discretization and the parameters of the model α, β, μ and γ. These matrices show interesting properties, i.e. they are dispersed and nested circulant, which allows their simple calculation for models of any number of dimensions from the matrices of one-dimensional models. The system matrices for one and two-dimensional models are provided in [16], being computed for finite differences and B-*spline* shape functions. As stated in [16], through the use of different shape functions, the behaviour of the resulting model can be influenced to fit the desired application.

The time variable in Eq. (5) is discretized, $t = \xi \Delta t$, i.e. $\mathbf{u}_\xi = \mathbf{u}_i(\xi \Delta t)$, where Δt is the time step and $\xi \in \mathbb{N}$ is the iteration index. The time derivatives are also replaced by their discrete approximations, resulting in a second-order iterative system, which provides the equation of motion of the model.

Then, the discrete spatial domain is translated into the frequency domain by using the e-dimensional discrete Fourier transform ($e\mathcal{DFT}$). The frequency approach reduces the computational cost by several orders of magnitude in each of the dimensions of the model, as can be seen in [11]. The Fourier formulation also allows us to isolate the spectral components of the nodes of the model for the current iteration ξ,

$$\hat{\mathbf{u}}_\xi = \hat{\mathbf{h}}\left(a_1\hat{\mathbf{u}}_{\xi-1} + a_2\hat{\mathbf{u}}_{\xi-2} + (\eta\hat{\mathbf{f}})^{-1}\hat{\mathbf{q}}_{\xi-1}\right), \tag{6}$$

where $\hat{\mathbf{u}}$, $\hat{\mathbf{f}}$ and $\hat{\mathbf{q}}$ are the $e\mathcal{DFT}$'s of their respective spatial sequences, $\eta = m/\Delta t^2 + c/\Delta t$, $\gamma = \Delta t\, c/m$, $a_1 = 1 + (1+\gamma)^{-1}$ and $a_2 = -(1+\gamma)^{-1}$.

Internal forces are imposed by the e-dimensional low-pass filter $\hat{\mathbf{h}}$ and external forces are applied by means of $\hat{\mathbf{q}}$. The filters $\hat{\mathbf{h}}$ for one and two dimensional models are detailed in [17]. The spectral characteristics of filter $\hat{\mathbf{h}}$ depend on the parameters α, β and η, influencing in the behaviour of the model.

Given that the processing of each spectral component of the model is completely independent of the other components, the high degree of parallelization of the iterative system (6) is noteworthy. Thus, software implementations adapted to multicore hardware are able to exploit this feature increasing the speed of processing.

3   Practical Implementation of the Method

The aim of this work is to segment and characterize the shape and volume of a human heart and analyse their evolution over time. For this purpose, 4D medical data must be processed using the method described in Section 2. The four dimensions of the provided dataset involve three spatial variables and one temporal variable.

Therefore, the proposed formulation should be applied in $\mathbb{R}^3$, i.e., $d = 3$. In addition, since the region of interest is the outer edge of the heart, it is appropriate the use of a two-dimensional model, i.e. $e = 2$, which behaves as a balloon that inflates and deflates following the shape of the heart. The dynamic analysis is achieved by applying the whole process for all 3D cardiac frames n_f, throughout the whole cardiac cycle. The template used for the model is depicted in Fig. 1, where parametric variables s_1 and s_2 portray the edges of the data along the xy plane and z axis respectively.

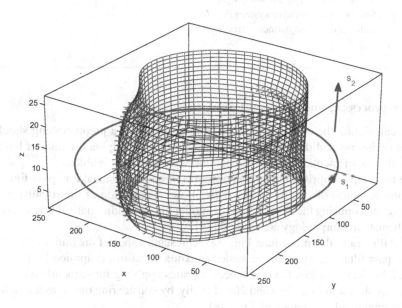

Fig. 1. Model used for the motion analysis of the heart

Algorithm 1 summarizes the procedure for the fitting of the model to the data. This method allow us to calculate the nodes of a parametric two-dimensional model from the three-dimensional scenario for each cardiac frame. Then, from the time-varying surface of the heart defined by these nodes, deformation and motion analysis can easily be performed.

Parameters $\alpha = \beta = 0.1$ and $\eta = 1$ are fixed based on the dynamics and assumed elasticity and rigidity for the deformable model, whereas parameter $\gamma = 0.48$ is calculated from the spectra of the dataset, in order to optimize the speed of convergence [18, 19]. The size of the model $\{N_1, N_2\} = \{64, 45\}$ is fixed according to the dimensions of the dataset. Finally, B-*spline* is chosen as shape function [20], since it enables the minimization of noise and data artefacts, specially when the spectrum of the data is mainly low pass. The performance of B-*spline* and finite difference deformable models to characterize 3D data in noisy environments is addressed in [16].

Algorithm 1. Iterative process

Data: 4D medical dataset, **I**.
Result: Nodes of 2D model for each frame n_f: $\mathbf{u}_{n_f}$.

1 Parameter setting
2 2D system filter computing, $\hat{\mathbf{h}}$
3 Data preprocessing, $\mathbf{I}'$
4 Gradient of data calculation, $F_{I'}$
5 **for** $n_f := 1$ **to** $MaxFrame$ **do**
6   │   Model initialization and freq. translation $\hat{\mathbf{u}}$
7   │   **while** ($\bar{e}_\xi >$ TOL **and** $\xi \leq maxiter$) **do**
8   │   │   Computing of $\hat{\mathbf{q}}_{\xi-1}$ for each $\hat{\mathbf{u}}_{\xi-1}$
9   │   │   Second order iterative system Eq. (6)
10   │   │   Calculation of adjustment error, $\bar{e}_\xi$
11   │   │ $\xi \leftarrow \xi + 1$

3.1 Preprocessing and Gradient Calculation

Before calculating the external forces used in the method, a preprocessing should be applied to the medical data under analysis. Note that although a CT dataset has been used in this experiment, the use of the method with MRI data is alike.

The first step comprises the application of a three dimensional low pass filter with a impulse response of size $3 \times 3 \times 3$. This allows a reduction of the impulsive noise without compromising the accuracy of the model fitting. Additionally, basic operations of mathematical morphology are applied [21].

Fig. 2 illustrates this procedure. Fig. 2(a) represents a slice of the initial volume after the low-pass filtering. Then, the extended-maxima transform is applied [22], depicted in Fig. 2(b). Next, we join the disconnected points applying subsequently a closing to the data, as shown in Figs. 2(c) and 2(d). Finally, by considering the dataset as a binary array internal pixels are removed, Fig. 2(e).

A two-dimensional filter based on B-*splines* is then applied to each slice. The remaining outline, depicted in Fig. 2(f), defines the edge of the heart and is used for the external forces calculation. Since the coordinate z of the nodes is held constant allowing motion only in x and y directions, gradients of the data are calculated for both directions, $F_{I;x}$ and $F_{I;y}$. Figs. 2(g) and 2(h) show the slice 32 of these three-dimensional gradient arrays.

3.2 Filtering Process

Once the spectrum of the system filter $\hat{\mathbf{h}}$ has been calculated [17], the fitting process can be applied by the iterative process described by Eq. (6). As detailed in Algorithm 1, the entire process is applied to each cardiac frame. For the first frame, $n_f = 1$, the model is initialized as a cylinder around the heart, as shown in Fig. 3(a). In subsequent frames instead, initialization is performed from the position of the model in the previous frame.

At each iteration ξ, the array of external forces, $\mathbf{q}$, is calculated from the gradient of the data F_I as well as the position of the nodes $\mathbf{u}$. Note that the overall iterative

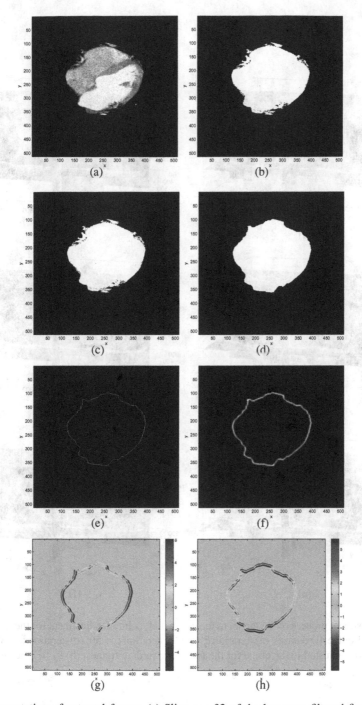

Fig. 2. Computation of external forces: (a) Slice $z = 32$ of the low-pass filtered first frame of the 4D dataset; (b-e) Preprocessing steps; (f) B-*splines* based two-dimensional filtering, (g-h) Gradients in x and y directions, $F_{I;x}$ and $F_{I;y}$.

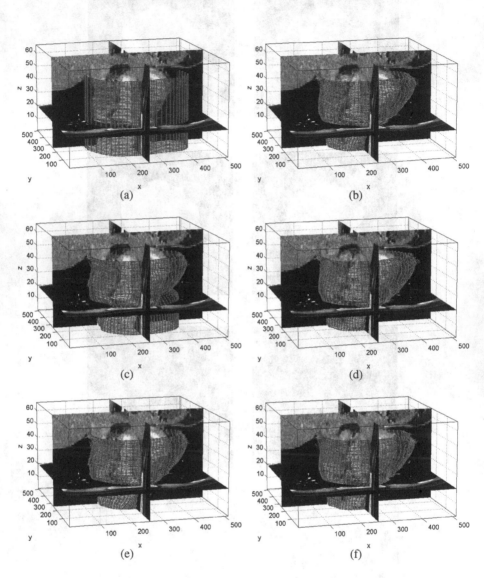

Fig. 3. Adjustment process of the model to the target. Left column, adjustment for the first cardiac frame $n_f = 1$ of the 4D dataset: (a) Iteration $\xi = 0$; (c) iteration $\xi = 50$; (e) iteration $\xi = 300$. Right column, model adjusted to the target for the following cardiac frames: (b) $n_f = 2$; (d) $n_f = 3$; (f) $n_f = 4$.

process is performed in the frequency domain. For this reason, external forces need to be translated into the Fourier domain, $\hat{\mathbf{q}}$. This process is executed until model is close enough to the outline of data.

4   Results and Discussion

Figure 3 depict the results of the experiment. The left column of Fig. 3 illustrates the process of adaptation of the model for the first cardiac frame $n_f = 1$. Likewise, the right column of Fig. 3 shows the final adjustment to the dataset for three cardiac frames, $n_f = \{2,3,4\}$. A similar process has been applied to the 20 frames of the dataset. As can be seen, the deformable model is able to follow the shape of the heart, allowing its characterization over time.

   Since the boundaries of the heart are now defined by a parametric function, the deformation and motion of the data can be easily analyzed. Fig. 4 shows an application of this study. Here, the model $\mathbf{v}$ has been used to estimate the volume of the heart along the cardiac cycle. Other parameters such as velocity or acceleration at any point of the data can easily be calculated from the model $\mathbf{v}$.

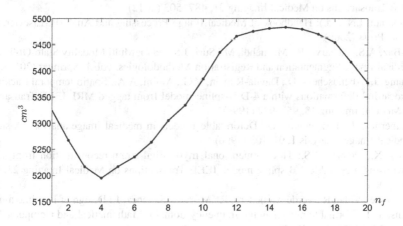

Fig. 4. Volume of the heart throughout the cardiac cycle

5   Conclusions

In this paper a parametric deformable model implemented in the frequency domain has been proposed to characterize the shape of a heart over time. The iterative process used for the adaptation of the model to the data is completely formulated in the frequency domain, providing a high computational efficiency.

   This paper has shown the preliminary results provided by the frequency-based deformable model to characterize the shape of the whole heart over time. The practical implementation and preprocessing steps have been described and applied to a 4D CT, where the parametric formulation of the model allows to derive dynamical parameters such as the volume over time.

As a future work, a comparative evaluation of existing methods to characterize the shape of the whole heart will be performed. Additionally, the proposed method will be applied to the segmentation and tracking of the left ventricle, due to the interest of this part of the heart in the literature [4, 6].

References

1. Zhong, J., Liu, W., Yu, X.: Characterization of three-dimensional myocardial deformation in the mouse heart: An MR tagging study. Journal of Magnetic Resonance Imaging 27, 1263–1270 (2008)
2. Schaerer, J., Casta, C., Pousin, J., Clarysse, P.: A dynamic elastic model for segmentation and tracking of the heart in MR image sequences. Medical Image Analysis 14, 738–749 (2010)
3. Constantinides, C., Aristokleous, N., Johnson, G.A., Perperides, D.: Static and dynamic cardiac modelling: Initial strides and results towards a quantitatively accurate mechanical heart model. In: 2010 IEEE International Symposium on Biomedical Imaging: From Nano to Macro, pp. 496–499. IEEE (2010)
4. Wang, H., Amini, A.A.: Cardiac motion and deformation recovery from MRI: A review. IEEE Transactions on Medical Imaging 31, 487–503 (2012)
5. Bankman, I.N. (ed.): Handbook of Medical Image Processing and Analysis. Elsevier, Academic Press (2008)
6. El-Baz, A.S., Acharya, R., Mirmehdi, M., Suri, J.S. (eds.): Multi Modality State-Of-The-Art Medical Image Segmentation and Registration Methodologies, vol. 1. Springer (2011)
7. Huang, J., Abendschein, D., Davila-Roman, V.G., Amini, A.A.: Spatio-temporal tracking of myocardial deformations with a 4-D B-spline model from tagged MRI. IEEE Transactions on Medical Imaging 18, 957–972 (1999)
8. McInerney, T., Terzopoulos, D.: Deformable models in medical image analysis: A survey. Medical Image Analysis 1, 91–108 (1996)
9. Deng, X., Denney, T.S.: Three-dimensional myocardial strain reconstruction from tagged MRI using a cylindrical B-spline model. IEEE Transactions on Medical Imaging 23, 861–867 (2004)
10. Berenguer-Vidal, R., Verdú-Monedero, R., Morales-Sánchez, J.: Design of B-spline multidimensional deformable models in the frequency domain. Mathematical and Computer Modelling 57, 1942–1949 (2012)
11. Weruaga, L., Verdú, R., Morales, J.: Frequency domain formulation of active parametric deformable models. IEEE Trans. Pattern Analysis Machine Intelligence 26, 1568–1578 (2004)
12. Liang, J., McInerney, T., Terzopoulos, D.: United snakes. Medical Image Analysis 10, 215–233 (2006)
13. Kass, M., Witkin, A., Terzopoulos, D.: Snakes: Active contour models. Int. Journal of Computer Vision 1, 321–331 (1988)
14. Zienkiewicz, O.C., Taylor, R.L. (eds.): The Finite Element Method: Its Basis and Fundamentals. Elsevier: Butterworth-Heinemann (2005)
15. Ern, A., Guermond, J. (eds.): Theory and practice of finite elements. Springer (2004)
16. Berenguer-Vidal, R., Verdú-Monedero, R., Menchón-Lara, R.-M., Legaz-Aparicio, Á.: Comparison of finite difference and B-spline deformable models in characterization of 3D data. In: Ferrández Vicente, J.M., Álvarez Sánchez, J.R., de la Paz López, F., Toledo Moreo, F. J. (eds.) IWINAC 2013, Part II. LNCS, vol. 7931, pp. 230–240. Springer, Heidelberg (2013)

17. Berenguer-Vidal, R., Verdú-Monedero, R., Morales-Sánchez, J.: Convergence analysis of multidimensional parametric deformable models. Computer Vision and Image Understanding (2015)
18. Verdú-Monedero, R., Morales-Sánchez, J., Weruaga, L.: Convergence analysis of active contours. Image and Vision Computing 26, 1118–1128 (2008)
19. Berenguer-Vidal, R.: Formulación de modelos deformables paramétricos multidimensionales en el dominio de la frecuencia. PhD thesis, Escuela Técnica Superior de Ingenieros de Telecomunicación, Universidad Politécnica de Cartagena (2013)
20. Unser, M., Aldroubi, A., Eden, M.: B-spline signal processing: Part I-theory; Part II-efficient design and applications. IEEE Trans. Signal Processing 41, 821–848 (1993)
21. Serra, J.: Image Analysis and Mathematical Morphology: Theoretical Advances, vol. II. Academic Press, London (1988)
22. Soille, P.: Morphological Image Analysis. Springer (1999)

Multimodal 3D Registration of Anatomic (MRI) and Functional (fMRI and PET) Intra-patient Images of the Brain

Álvar-Ginés Legaz-Aparicio[1]([⊠]), Rafael Verdú-Monedero[1], Jorge Larrey-Ruiz[1], Fernando López-Mir[2], Valery Naranjo[2], and Ángela Bernabéu[3]

[1] Universidad Politécnica de Cartagena, Cartagena, Spain
alvarlegaz@gmail.com, {rafael.verdu,jorge.larrey}@upct.es
[2] Universidad Politécnica de Valencia, I3BH LabHuman, Valencia, Spain
{ferlomir,vnaranjo}@labhuman.i3bh.es
[3] Inscanner S.L, Unidad de Resonancia Magnética, Alicante, Spain
angela.bernabeu@gmail.com

Abstract. This paper describes an application of variational image registration. The method is based on an efficient implementation of the diffusion registration formulated in the frequency domain. The goal is to register anatomical and functional brain images of the same patient to facilitate the process of functional localization. This non-rigid image registration of different modalities makes possible to obtain a geometric correspondence which allows for localizing the functional processes that occur in the brain. In order to evaluate the performance of the proposed method, visual and numeric results of registration are shown. The quality of the registration results is measured by considering the peak signal to noise ratio (PSNR), the mutual information (MI) and the correlation ratio (CR).

Keywords: Variational image registration · Non-rigid deformation · Medical image

1 Introduction

Image registration is the process of finding the optimum geometrical transformation which relates corresponding points of two dataset (images or volume images) of the same scene taken at different times, from different viewpoints, and/or by different sensors [1], [2], [3]. Geometrically, the image registration consist of aligning one of the datasets, known as the template set (T) with the other set, known as the reference set (R).

Applied to medical imaging, image registration tries to find the correspondence between datasets at different times or with different acquisition devices. The registration process helps to improve the diagnosis and tracking of a wide group of pathologies, as well as assist to plan the most appropriate treatment.

Among the applications of medical image registration, the registration of functional and anatomic images (intra-patient) of the brain of the same patient is

© Springer International Publishing Switzerland 2015
J.M. Ferrández Vicente et al. (Eds.): IWINAC 2015, Part I, LNCS 9107, pp. 340–347, 2015.
DOI: 10.1007/978-3-319-18914-7_36

an essential part of the functional localization process [4]. The functional images of the brain, for example, functional magnetic resonance images (fMRI) or positron emission topographies (PET) do not have detailed structural information and do not provide a specific anatomical location of the functional information, hence it is necessary to analyse them along with anatomical images as, for example, magnetic resonance images (MRI). This type of registration is known as multimodal registration due to the different contrast and intensity of the datasets. The most important part of the multimodal registration process is to find the univocal correspondence between the functional and anatomic images of a patient when usually the correspondences are not visible in both imaging modalities. The functional images have a spatial resolution, signal to noise ratio and contrast lower than anatomical images. PET and fMRI images provide very vague and imprecise brain structures; on the other hand, MR images usually have a high resolution and definition of the brain structures.

For many years, researchers have developed and implemented registration algorithms for medical imaging applications. In [5], non-rigid registration techniques are evaluated on thoracic CT images. Particularly, the liver segmentation is an open challenge [6], which provides an interesting setting for comparing image registration methods. A novel collection of medical image registration algorithms in C++ based on ITK [8] can be found in [7]. However, this collection restricts non-rigid transformations to B-spline models [9] or physical model-based splines [10], not taking into account non parametric registration methods (i.e. the approach proposed in this paper or, *i.e.* [11] and [12]).

In this work, we address the registration of anatomical images (MRI) and functional images (fMRI and PET) in order to obtain a correspondence between images of both modalities. The method is based on an efficient implementation of variational image registration, which is at least two times faster than other approaches in the spatial domain [13].

2   Variational Image Registration

In this work the datasets are volumes of images obtained from medical studies (MRI, fMRI, PET), $R, T : \mathbb{R}^3 \to \mathbb{R}$, and the registration will produce a non-rigid displacement field $\mathbf{u} : \mathbb{R}^3 \to \mathbb{R}^3$ that will make the transformed template dataset be similar to the reference dataset, $T(\mathbf{x} - \mathbf{u}(\mathbf{x})) \approx R(\mathbf{x})$, where $\mathbf{u}(\mathbf{x}) = (u_1(\mathbf{x}), u_2(\mathbf{x}), u_3(\mathbf{x}))^\top$ and $\mathbf{x}$ is the spatial position $\mathbf{x} = (x_1, x_2, x_3) \in \mathbb{R}^3$.

The non-parametric registration can be approached in terms of the variational calculus, by defining the joint energy functional to be minimized:

$$\mathcal{J}[\mathbf{u}] = \mathcal{D}[R, T; \mathbf{u}] + \alpha \mathcal{S}[\mathbf{u}]. \tag{1}$$

The energy term $\mathcal{D}$ measures the distance between the deformed template dataset and the reference dataset; $\mathcal{S}$ is a penalty term which acts as a regularizer and determines the smoothness of the displacement field; and $\alpha > 0$ weights the influence of the regularization.

The distance measure $\mathcal{D}$ is chosen depending on the datasets to be registered. When dealing with datasets from different sources or modalities (multimodal registration), statistical-based measures are more appropriate. In this application the correlation ratio [14] has been used.

The regularization term $\mathcal{S}$ gives the smoothness characteristics to the displacement field [11]. In this work we use the diffusion term, which is given by the energy of first order derivatives of $\mathbf{u}$:

$$\mathcal{S}^{\text{diff}}[\mathbf{u}] = \frac{1}{2} \sum_{l=1}^{3} \int_{\mathbb{R}^3} \|\nabla u_l\|^2 \, d\mathbf{x}. \tag{2}$$

As described in [12], the joint energy functional (1) can be translated into the frequency domain by means of Parseval's theorem, then $\mathcal{J}[\mathbf{u}] = \tilde{\mathcal{J}}[\tilde{\mathbf{u}}]$, where

$$\tilde{\mathcal{J}}[\tilde{\mathbf{u}}] = \tilde{\mathcal{D}}[\tilde{R}, \tilde{T}; \tilde{\mathbf{u}}] + \alpha \, \tilde{\mathcal{S}}[\tilde{\mathbf{u}}], \tag{3}$$

with $\tilde{\mathbf{u}}(\boldsymbol{\omega}) = (\tilde{u}_1(\boldsymbol{\omega}), \tilde{u}_2(\boldsymbol{\omega}), \tilde{u}_3(\boldsymbol{\omega}))^{\top}$ being the frequency counterpart of the displacement field, $\boldsymbol{\omega} = (\omega_1, \omega_2, \omega_3)$ is the three dimensional variable in the frequency domain, and where the distance measure $\tilde{\mathcal{D}}$ and the regularization term $\tilde{\mathcal{S}}$ are now defined in the frequency domain.

According to the variational calculus, a necessary condition for a minimizer $\tilde{\mathbf{u}}$ of the joint energy functional (3) is that the first variation of $\tilde{\mathcal{J}}[\tilde{\mathbf{u}}]$ in any direction (also known as the *Gâteaux* derivative) vanishes for all suitable perturbations. This leads to the Euler-Lagrange equation in the frequency domain:

$$\tilde{\mathbf{f}}(\boldsymbol{\omega}) + \alpha \, \tilde{\mathcal{A}}(\boldsymbol{\omega}) \, \tilde{\mathbf{u}}(\boldsymbol{\omega}) = \mathbf{0}, \tag{4}$$

where $\tilde{\mathbf{f}}$ is the 3D Fourier transform of the external forces, $\mathcal{FT}\{\nabla\mathcal{D}[R, T; \mathbf{u}]\}$, and $\tilde{\mathcal{A}}$ is a diagonal 3×3 matrix whose elements are scalar functions which implement the spatial derivatives in the frequency domain [13], allowing for their computation by means of products:

$$\tilde{\mathcal{A}}_{ii}(\boldsymbol{\omega}) = 2 \sum_{m=1}^{3} (1 - \cos \omega_m), i = 1, ..., 3. \tag{5}$$

The Euler-Lagrange equations (4) in the frequency domain provide a stable implementation for the computation of a numerical solution for the displacement field, and in a more efficient way than existing approaches if the three-dimensional fast Fourier transform is used [13]. To solve (4), formulated in the frequency domain, a time-marching scheme can be employed, yielding the following equation:

$$\partial_t \tilde{\mathbf{u}}(\boldsymbol{\omega}, t) + \tilde{\mathbf{f}}(\boldsymbol{\omega}, t) + \alpha \, \tilde{\mathcal{A}}(\boldsymbol{\omega}) \, \tilde{\mathbf{u}}(\boldsymbol{\omega}, t) = \mathbf{0}, \tag{6}$$

where $\partial_t \tilde{\mathbf{u}}(\boldsymbol{\omega}, t) = (\partial_t \tilde{u}_1(\boldsymbol{\omega}, t), \partial_t \tilde{u}_2(\boldsymbol{\omega}, t), \partial_t \tilde{u}_3(\boldsymbol{\omega}, t))^{\top}$ (in the steady-state $\partial_t \tilde{\mathbf{u}}(\boldsymbol{\omega}, t) = \mathbf{0}$ and (6) holds (4)). Equation (6) is solved by discretizing the time, $t = \xi\tau$, $\tau > 0$ being the time-step and $\xi \in \mathbb{N}$ being the iteration index, and the time derivative of $\tilde{u}(\boldsymbol{\omega}, t)$ is replaced by the first backward difference. Using the notation $\tilde{\mathbf{u}}^{(\xi)}(\boldsymbol{\omega}) = \tilde{u}(\boldsymbol{\omega}, \xi\tau)$, the following semi-implicit iterative scheme comes out:

$$\tilde{u}_l^{(\xi)}(\boldsymbol{\omega}) = H(\boldsymbol{\omega})\big(\tilde{u}_l^{(\xi-1)}(\boldsymbol{\omega}) - \eta^{-1}\,\tilde{f}_l^{(\xi-1)}(\boldsymbol{\omega})\big), \tag{7}$$

where $l = \{1, 2, 3\}, \eta = 1/\tau$ and $H(\boldsymbol{\omega})$ is the following 3D low pass filter $H(\boldsymbol{\omega}) = (1 + \eta^{-1}\alpha\tilde{A}_{ii}(\boldsymbol{\omega}))^{-1}$. An implementation based on the 3D FFT is, in terms of efficiency, two times faster than the fastest implementation available in the spatial domain [13], which is the DCT-based algorithm included in the FLIRT toolbox [15] for the diffusion and curvature registration methods [16].

3   Results

This section shows the results of the registration process between anatomic images (MRI) and functional images (fMRI and PET) of the brain. To evaluate the performance of the registration method, two experiments have been proposed. In Experiment 1 the MRI dataset and the fMRI dataset have been registered whereas in Experiment 2 the MRI dataset and the PET dataset have been registered. The sizes of the original datasets are $336 \times 336 \times 200$ voxels (MRI), $64 \times 64 \times 30$ voxels (fMRI) and $128 \times 128 \times 46$ voxels (PET). Due to the fact that registration method needs that the datasets have the same dimensions, initially the spatial range shared by the datasets is identified. Subsequently the datasets are re-sampled by a non-integer factor by means of a decimation and interpolation steps, achieving datasets with the same dimensions, $128 \times 128 \times 64$.

The registration parameters used in Experiment 1 are $\alpha = 20, \tau = 1, \epsilon = 100$. In Experiment 2, the registration parameters are $\alpha = 50, \tau = 1, \epsilon = 150$. The similarity measure minimized in both experiment has been the correlation ratio, given that the datasets are multimodal.

Figure 1 shows the registration process between the fMRI volume as the reference dataset (R) and the MRI volume as the template dataset (T). The first column shows three slices of the reference dataset, the third column shows three slices of the template dataset, and the second column shows three slices of the registered dataset. Numerical measurements of the registration process have been gathered in Table 1. This results have also been compared with the results provided by the *Elastix* toolbox [7]. In order to evaluate the results of the registration the following similarity measurements have been done: peak signal to noise ratio (PSNR), mutual information (MI) and correlation ratio (CR). As can be seen, the proposed registration method gets better results in the registration process than *Elastix*.

Figure 2 shows the results of Experiment 2. In this case, the reference dataset (R) is the MRI volume and the template dataset (T) is the PET volume. The first column shows three slices of the reference dataset, the third column shows three slices of the template dataset and the second column shows three slices of the registered dataset. Numerical measurements of the registration process have been gathered in Table 2, where the results provided by the proposed method have been compared with the results provided by *Elastix*. Once again

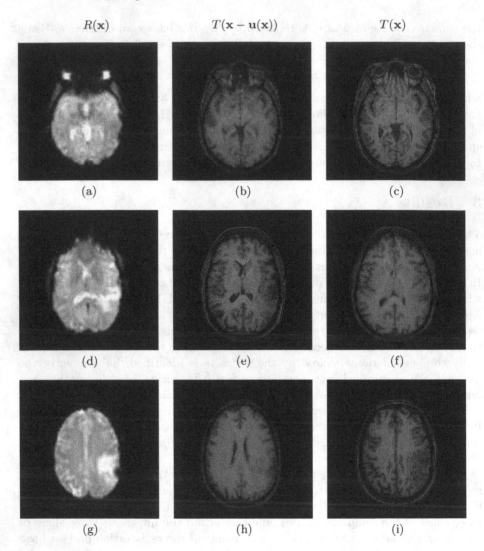

Fig. 1. Experiment 1: registration of fMRI and MRI ($128 \times 128 \times 64$). First column: reference dataset (fMRI). Second column: registered template. Third column: template dataset (MRI). First row: slice #20. Second row: slice #30. Third row: slice #40.

the proposed registration method achieves better results in the registration than *Elastix*.

Regarding the computational time, the proposed registration method, implemented in C++, needs 0.72 seconds per iteration to register datasets of $128 \times 128 \times 64$. On the other hand, *Elastix* needs 1.28 seconds per iteration to register the same datasets.

$R(\mathbf{x})$ $T(\mathbf{x} - \mathbf{u}(\mathbf{x}))$ $T(\mathbf{x})$

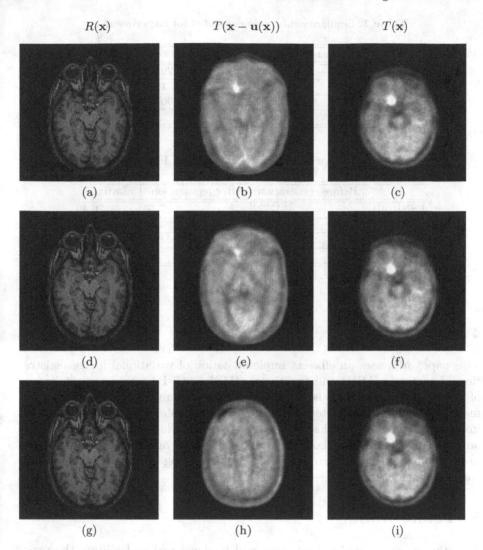

Fig. 2. Experiment 2: registration of MRI and PET (128 × 128 × 64). First column: reference dataset (MRI). Second column: registered template. Third column: template dataset (PET). First row: slice #20. Second row: slice #30. Third row: slice #40.

Finally, It is worth stressing that the registration process shows excellent results applied to multimodal datasets. However, in both experiments, the results of the registration are not perfect. Nevertheless, it should be also taken into consideration other factors that hinder the registration process (e.g. little shared information between datasets, different scales or the shape relation of the voxels).

Table 1. Similarity measures computed for Experiment 1

	Before registration	After registration	Elastix [7]
PSNR(dB)	21.64	29.33	28.70
MI(bits)	1.01	1.24	1.17
CR(%)	64.54	90.55	88.53
Time (s)	——	71	126

Table 2. Similarity measures computed for Experiment 2

	Before registration	After registration	Elastix [7]
PSNR(dB)	17.29	27.56	26.40
MI(bits)	0.82	1.17	1.12
CR(%)	62.85	88.27	85.53
Time (s)	——	108	192

4 Conclusion

This paper addresses an efficient implementation of variational image registration of anatomic (MRI) and functional (fMRI and PET) intra-patient images of the brain. The method is based on an efficient implementation of the diffusion registration formulated in the frequency domain. This method has been implemented in C++ and has been compared with *Elastix* toolbox [7]. Results on different experiments show the ability, efficiency and high accuracy of the proposed method to estimate the deformation existing in datasets with a lower computational cost.

5 Future Work

As future work, a graphical interface will be developed to facilitate the registration tasks between the brain images of different modalities. This interface will use the library developed in C++ that implements the proposed algorithm. The fact of using the programming language C++ along with the implementation of the registration algorithm in the frequency domain will provide high computational efficiency and minimum execution times.

References

1. Besl, P.J., McKay, N.D.: A method for registration of 3 D shapes. IEEE Trans. Pattern Anal. Mach. Intell. 14(2), 239–256 (1992)
2. Sotiras, A., Davatzikos, C., Paragios, N.: Deformable Medical Image Registration: A Survey. IEEE Transactions on Medical Imaging 32(7), 1153–1190 (2013)

3. Zitová, B., Flusser, J.: Imagen registration methods: A survey. Image and Vision Computing 21 (2003)
4. Gholipour, A., Kehtarnavaz, N., Briggs, R., Devous, M., Gopinath, K.: Brain Functional Localization: A Survey of Image Registration Techniques. IEEE Trans. Image Processing 26, 427–451 (2007)
5. Murphy, K., et al.: Evaluation of registration methods on thoracic CT: The EMPIRE10 challenge. IEEE Trans. Medical Imaging 30(11), 1901–1920 (2011)
6. Heimann, T., et al.: Comparison and evaluation of methods for liver segmentation from CT datasets. IEEE Trans. Medical Imaging 28, 1251–1265 (2009)
7. Klein, S., et al.: Elastix: A toolbox for intensity-based medical image registration. IEEE Trans. Medical Imaging 29(1), 196–205 (2010)
8. Ibáñez, L., Schroeder, W., Ng, L., Cates, J.: The ITK Software Guide. Kitware, Clifton Park (2005)
9. Rueckert, D., et al.: Nonrigid registration using free-form deformations: application to breast MR images. IEEE Trans. Med. Imaging 18, 712–721 (1999)
10. Davis, M.H., et al.: A physics-based coordinate transformation for 3D image matching. IEEE Trans. Medical Imaging 16(3), 317–328 (1997)
11. Fischer, B., Modersitzki, J.: A unified approach to fast image registration and a new curvature based registration technique. Linear Algebra & its Applications 308 (2004)
12. Larrey-Ruiz, J., Verdú-Monedero, R., Morales-Sánchez, J.: A Fourier domain framework for variational image registration. J. Math. Imaging Vis. 32(1), 57–72 (2008)
13. Verdú-Monedero, R., Larrey-Ruiz, J., Morales-Sánchez, J.: Frequency implementation of the Euler-Lagrange equations for variational image registration. IEEE Signal Processing Letters 15, 321–324 (2008)
14. Roche, A., Malandain, G., Pennec, X., Ayache, N.: The correlation ratio as a new similarity measure for multimodal image registration. In: Wells, W.M., Colchester, A., Delp, S.L. (eds.) MICCAI 1998. LNCS, vol. 1496, pp. 1115–1124. Springer, Heidelberg (1998)
15. Fischer, B., Modersitzki, J.: Flirt: A flexible image registration toolbox. In: Gee, J.C., Maintz, J.B.A., Vannier, M.W. (eds.) WBIR 2003. LNCS, vol. 2717, pp. 261–270. Springer, Heidelberg (2003)
16. Modersitzki, J.: Numerical Methods for Image Registration. Oxford University Press, USA (2004)

Localisation of Pollen Grains in Digitised Real Daily Airborne Samples

Estela Díaz-López[1](✉), M. Rincón[1], J. Rojo[2], C. Vaquero[2], A. Rapp[2], S. Salmeron-Majadas[1], and R. Pérez-Badia[2]

[1] Department of Artificial Intelligence. National University of Distance Education, Madrid, Spain
{ediazlopez,mrincon}@dia.uned.es
[2] Institute of Environmental Sciences, University of Castilla-La Mancha, Toledo, Spain
rosa.perez@uclm.es

Abstract Content analysis of pollen grains in the atmosphere is an important task for preventing allergy symptoms, studying crop production or detecting environmental changes. In the last decades, a lot of palynological labs have been created to collect, prepare and analyse airborne samples. Nowadays, this task is done manually with optical microscopes, requires trained experts and is time-consuming. The development of new computer vision systems and the low price of storage systems have improved the solutions towards an automated palynology. Some recognition problems have been solved with better quality images and other with 3D images, but localisation in real airborne samples, with debris, clumped and grouped pollen grains needs to be improved in order to achieve an automatic system useful for biological labs. In this manuscript, we analyse the advances achieved in the last years and explain a new low-cost methodology, that imitates the human expert labour using computational algorithms based on image characteristics and domain knowledge to detect pollen grains. The current results are promising (81.92% of recall and 18.5% of precision) but not enough to develop an automated palynology system.

Keywords: Automatic localisation · Pollen grain · Debris · Airborne · Light microscope · Digitised image · Data sharing

1 Introduction

The analysis of bio-images is fundamental to paleoenvironmental reconstruction, climate change, forensic science or medical studies. It is done by qualified experts that perform complex, tedious and time consuming tasks, detecting and counting biological particles like pollen grains. To improve the expert labor and the biological studies, in the last two decades a lot of research groups have been working in the development of automatic systems. The advances in hardware technology have provided new acquisition systems able to get better data, quicker and with higher resolution. Furthermore, thanks to the reduction of storage costs,

© Springer International Publishing Switzerland 2015
J.M. Ferrández Vicente et al. (Eds.): IWINAC 2015, Part I, LNCS 9107, pp. 348–357, 2015.
DOI: 10.1007/978-3-319-18914-7_37

it is possible to save much more additional data than before. So, nowadays, we have to deal with huge amounts of data that require to be processed, linked and explained.

In the last years, the pollen grain localisation and recognition problem has been studied by different research groups. Table 1 summarises the results of different studies using light microscope, the most common and economic system to analyse samples. Automatic localisation in pure samples with isolated objects shows very good results, with high recall and precision [2,3,15,1,14,5,10,13] but when the sample is real, with grouped or clumped pollen grains and debris, the results get worse [11,14]. Automatic recognition shows good performance when the images have good focus, isolated pollen grains and few number of taxa [10,12,15].

Table 1. Overview of published attempts about automatic pollen grain localisation

		Sample properties				Image properties			Detection		Identification	
	N. taxa	N. samples	N. pollen grains	Type	Stain	Size	Resolution	Appearance	Recall	Precision	Recall	Precision
France et al. (2000)	3	3	722	Real	-	748 x 576	x 250	Grey level	28,25% (isolated & no deformed)	-	86,8%	-
Bonton et al. (2001); Boucher et al. (2002)	30	-	350	Pure	Fuchsin	-	x 400 (x 600) Zstep 0.5 um	RGB	> 90% *estimated	-	77%	-
Hodgson et al. (2005)	4	25	120	Real (single object)	Fuchsin	-	x 200	Grey level	97%	-	96%	-
Rodriguez-Damian et al. (2006)	3 (Urtica family)	-	291	Pure	-	2048 x 3072	0,122 um/px	Grey level	82%	High	89%	-
Allen et all. (2006)	15 6	-	248	Pure	No	640 x 480 1024 x 768	3,8 um/px 0.414 um/px	Grey level	71,77%	-	89% 95%	-
Ranzato et al. (2007)	8	1429	3686	Real	Si	1024 x 1280	0,5 um/px	RGB	93.9% (isolated) 83% (ROI)	8.60%	77%	-
Landsmeer et al. (2009)	>2	61 stack images of 9 samples	65	Real	Safranin	2048 x 1536	0.18 um/px Zstep ~ 4 um	RGB	86%	61%	-	-
Holt et al. (2011)	6	4	810	Pure	No	640 x 480 1028 x 1024	3,8 um/px 0.138um/px	Grey level	-	-	85.31%	86.80%
Nguyen et al. (2013)	9	1	768 (isolated)	Pure	-	-	0.23 um/px	RGB	93.80%	89.50%	96,4%	-

The current methodology for airborne pollen quantification is widely adopted around the world, with volumetric Hirst sampler type [9] in Europe or Rotorod type [8] in North America. Pollen grains and other particles in the air impact on an adhesive-coated transparent tape that progress over time. Each daily adhesive tape is stained to improve the visibility of the pollen grains and later, a palynology expert analyses manually the sample, using an optical microscope according to the protocol. Other methods to quantify the particles of airborne have been studied recently, such as the Coriolis air sampler and Hirst sampler, which performance is compared in [4]. The results in quantity are quite similar but the recording methodology changes, causing problems in temporal studies and when comparing with old stored samples.

The use of bio-images with contextual features helps experts knowing more information to improve their decisions. All image analysis results need to be checked according to the domain knowledge of the task, but this knowledge is difficult to elicit. Therefore, we use the spiral methodology based on the comparison between automatic results with experts' ground truth to solve complex problems, in this case, the automatic detection and identification of pollen grains. Thanks to the reduction of storage costs, it is possible to save all samples in digital format, allowing data sharing and visual evaluation of results. The advances in image processing improve the palynology research minimising human errors, allowing reanalysis of data and increasing the information extracted from images. This advances together with a good organisation of all factors that take part in the biological process offer huge opportunities of new research projects, like prediction or reconstruction of palynological environments.

The novelty of this manuscript resides on using a semiautomatic system that combines scan and analysis of daily samples, offering a solution when the images are from real airborne, with imperfect focus and higher taxa number (around 40 types as recommend in [6]). The method for detecting and identifying pollen grains, explained in section 2, is composed of two steps: a low resolution analysis to find the objects and a high resolution analysis to identify them. In section 3 we explain the materials and methods to find pollen grains in real samples. In section 4 we show the results obtained, focusing our attention in the detection problem when the sample shows real context imperfections (bad focus, debris,...) and comparing the results with a ground truth done manually by palynological experts. Finally, in section 5 we discuss about the results and present conclusions.

2   Semiautomatic Pollen Grains Counting Methodology

The Spanish pollen counting protocol has beed defined by the Spanish Aerobiology Network (REA) [7] and fulfils the requisites of the European Aeroallergen Network (EAN), that provides a large data base when all members share their data. The protocol in Europe to analyse the content of pollen grains and spores in the atmosphere is based on air-suction to stick the suspension particles on a transparent tape. The tape is cut in daily fragments, mounted in a glass slide and stained to colour fuchsia the pollen grains and other vegetal particles. Then, an expert explores the sample manually with transmitted light microscope, counting and classifying different pollen grains.

To automate this process, the sample is scanned with digital and motorised transmitted light microscope and then, the real daily sample is analysed with a computer vision system. The schema of the semiautomatic pollen grains counting system is presented in figure 1. It is a low cost system, that can be easily installed in a biology lab.

The system is composed of two modules: i) a low resolution module where 2D images are used to detect pollen grains and ii) a high resolution module that uses 3D images (Z stack) to identify different pollen types.

As adhesive tape thickness is greater than pollen grain dimension, their position in Z axis vary as is shown in figure 2a. In the low resolution phase, the

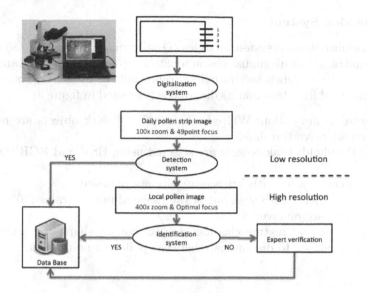

Fig. 1. Overview of the semiautomatic system developed

minimum digitised area contains several particles and each one could have different Z positions over the tape. In consequence, the image may present areas with imperfect focus. This problem has less influence in the high resolution phase because the digitised area is reduced and centred over a pollen grain.

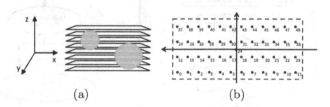

Fig. 2. (a) Pollen grain situation in Z axis over adhesive tape (b) Focus mesh of 49 equidistant dots

2.1 Digitalisation System

In low resolution (x 100 magnification), the scan system first analyses the focus in equally spaced grid inside the sample (figure 2b). Then, the daily sample is scanned using the best focus mesh. The focus values are used to adjust the Z level along the sample, improving the image quality.

In high resolution (x 400 magnification), the system acquires a stack of images centred in a specific point where it is supposed to exist a pollen grain. We use the focus mesh and pollen grain diameter to calculate Z position, number of images (Z stack) and thickness (Z step).

2.2   Detection System

The pollen grain detection system analyses 2D daily images (around 36480 x 14400 pixels) to find objects with similar characteristics to pollen grain. Some areas may have imperfect focus, debris and the pollen grains could be broken or grouped. The system consists of five steps and its diagram is presented in figure 3:

1. Background elimination: White background and dark objects are removed using an adaptive threshold
2. Colour threshold: Pink objects are selected using HSV and RGB threshold filter.
3. Shape filter: Objects with low eccentricity are removed.
4. Size filter: Objects with sizes out of the range of pollen grains (10 - 300 um of diameter) are removed.
5. Colour, shape and texture classifier: Object colour, shape and texture features are analysed to discriminate between pollen or non-pollen.

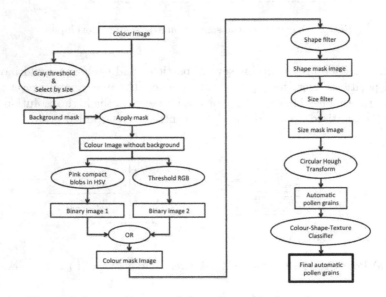

Fig. 3. Inference diagram of the pollen grain detection system

2.3   Identification System

The identification system utilises 3D images centred on each pollen grain to extract multiple features. The number of images in Z stack and their Z step depends on the object diameter detected in the previous phases. We use shape, size and texture features to identify the pollen grain types. Also, we use contextual variables like the probability of each pollen type to appear in a specific date to improve the results.

2.4   Verification System

An user-friendly program interface has been developed to show the image to the palynological expert to help deciding the pollen type in case the automatic identification system fails. If the expert, viewing the digital image, can not decide the type, the verification system will connect with the microscope, allowing the expert to adjust the focus to decide the type. Also, the verification system allows experts to evaluate the results of the automatic system. They can modify types or add/remove pollen grains.

3   Materials and Methods

In this study, we use 12 daily samples of airborne pollen grains. Each daily sample starts at 12:00 pm and corresponds to the 15th day of each month of 2012. The samples were collected using a Hirst-type volumetric spore trap [9] located in the city of Toledo (39°51'54.6"N 4°02'30.4"W, 529 m). The adhesive tape is stained with fuchsine and mounted in a glass slide.

The sample is scanned with an optic / digital microscope "NiKON 80i" with micro-positioning device "Prior H101 ProScann II" and NIS Elements Basic Research software (version 4.11) of NIKON. Daily sample size is 48 x 19 mm and is formed with a mosaic of RGB images of 640 x 480 pixels with 1.34 um/pixel. The focus mesh was manually done because the autofocus function stopped in debris instead of pollen grain.

Image analysis was done with Matlab software (version 8.1.0.604). In the first phase of analysis, the low resolution system detects possible pollen grains. In each candidate pollen grain, the high resolution system acquires a stack of RGB images of 640x480 pixel with 0.34 um/pixel. In the second phase of analysis, the system identifies the type of pollen grains with this image stack.

According to REA protocol, the experts analyse 4 horizontal lines for each sample of 48 mm x the diameter of the field of vision of the microscope (in our case 0.5 mm). To compare the results of three methods: 1) optical manual (analog method), 2) digital manual and 3) digital automatic, we analyse similar areas with the new methodology.

4   Results

The number of pollen grains detected manually by one expert in the four horizontal lines of each sample (REA protocol) is shown in table 2. It shows the number of objects detected with the analog method and with the digital method using both low resolution and high resolution images. The greatest number of pollen grains appears in spring, with levels around 1000-2000 pollen grains. In the rest of months the presence of pollen grains varies between 0 to 50. In the digital method the number of detected pollen grains is higher than in analog method because the localisation of each line has a variation of 2 mm (thickness of line mark) however the magnitude order is similar.

Table 2. Comparison of the number of pollen grains per month with different methods: analog (optic) and digital with low and high resolution

Location	Date	Detected			Identified			Not identified		
		Analog	Digital Manual		Analog	Digital Manual		Analog	Digital Manual	
		Optic	100x	400x	Optic	100x	400x	Optic	100x	400x
Toledo	15/01/12	6	13	13	6	4	4	0	9	9
Toledo	15/02/12	10	26	20	10	10	12	0	16	8
Toledo	15/03/12	1706	1792	1766	1705	809	1741	1	983	25
Toledo	15/04/12	120	161	152	119	98	138	1	63	14
Toledo	15/05/12	1157	1608	1608	1153	1190	1190	4	418	418
Toledo	15/06/12	168	202	202	166	23	139	2	179	63
Toledo	15/07/12	32	44	44	32	5	36	0	39	8
Toledo	15/08/12	7	37	35	7	11	22	0	26	13
Toledo	15/09/12	12	44	31	12	10	20	0	34	11
Toledo	15/10/12	1	51	51	0	4	4	1	47	47
Toledo	15/11/12	0	27	27	0	9	17	0	18	10
Toledo	15/12/12	45	56	50	43	36	40	2	20	10
Total		3264	4061	3999	3253	2209	3363	11	1852	636

It is worthy to note that, in real samples, there are a large variation in the visual characteristics of pollen grains. In the analysed samples exist 24 pollen types of the approximately 38 pollen types characteristic of this region, with different values of colour, size and date. The pollen type variation is shown in figure 4, on the left we show the pollen type distribution per sample and on the right different sub-images, on top a sample full of debris and on bottom some grouped pollen grains. Size variation is between 37 to 2564 pixels. In figure 5 left is shown the size variation of some pollen types. The colour pixel distribution per sample is shown in 5 right. Often, the identification of all pollen grains is not possible . In this study, the 0.33% of pollen grains were not identify in the analog method and the 15.9% in the digital manual method. It could be due a bad stack of images, without enough features to discriminate between various grain types or an incorrect focus.

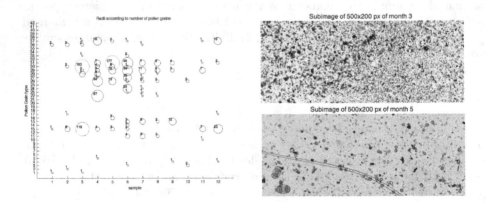

Fig. 4. Pollen type distribution (left) and images of some samples (right)

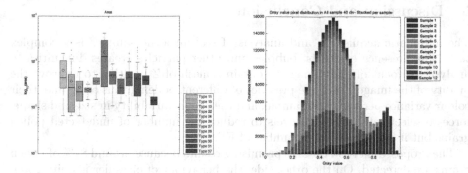

Fig. 5. Pollen size variability (left) and RGB colour distribution (right)

The results of our pollen grain detection method is shown in table 3. In this table, we show the number of pollen grains detected by experts (#expert), the number of pollen grains detected by automatic method (#detected) and the number of undetected pollen grains (#undetected). The mean true positive rate per sample is $89.4 \mp 16.9\%$.

The number of objects with similar colour and shape features than pollen grains is very high. To reduce this number we have used a decision tree classifier that uses colour, shape and texture blob features to recognise real pollen grains. Using the circular Hough transform and the classifier with local and regional features we reduce 86.5% the number of false negatives. After the classifier we obtain a TP rate of $81.92 \mp 13.24\%$ and a still elevated number of FP ($18.85 \mp 21.24\%$ of precision), that will be removed in the next phases.

Table 3. Automatic method results per month

Month	N. expert	N. detected	N. undetected	After classifier	
				TP	FP
1	13	13	0	12	150
2	20	21	1	19	357
3	1766	1697	200	1440	6398
4	152	142	6	120	164
5	1608	1554	15	1389	401
6	202	198	2	192	1020
7	44	42	0	39	584
8	35	33	1	27	186
9	31	30	1	24	201
10	51	39	11	23	216
11	27	61	16	22	1158
12	50	40	9	42	184
Total	**3999**	**3870**	**262**	**3349**	**11019**

5   Discussion and Conclusion

The automatic acquisition and analysis of real airborne samples is a complex task. The presence of debris, bubbles and other particles creates difficulties to analyse the focus of daily image to obtain a mesh of Z position to improve the quality of the image. Also, the presence of different pollen grains over time, their colour variance per type and sample, and their high variability in size and shape, force to select soft threshold levels to reduce the number of undetected pollen grains but it produces a high number of FP.

The proposed method obtains promising results because around 82% of pollen grains are detected. On the other side, the behaviour of classifier is quite good, removing 86% of false positives, however the number of non pollen grains that enter the system is also too high. Therefore, he initial detection method needs to be improved in order to reduce the number of false positives.

Our results are in line with other authors who work with real samples and a high number of pollen grains, but it is necessary to improve theses results to fulfil an automated palynology system.

Future works include the development of a focus function to stop in pollen grains and the reconfiguration of the colour detection method to reduce the number of initial objects.

Acknowledgments. Research described in this manuscript has been possible thanks to different support sources. Trainee research staff grants awarded by UNED (FPI-2012) and the Ministry of Education (FPU-2012), and the financial support provided through out the project POIC10-0302-2695 of the Consejería de Educación y Ciencia de la Junta de Comunidades de Castilla-La Mancha.

References

1. Allen, G.: An Automated Pollen Recognition System. PhD thesis, Massey University (2006)
2. Bonton, P., Boucher, A., Thonnat, M., Tomczak, R., Hidalgo, P.J., Belmonte, J., Galán, C.: Colour image in 2d and 3d microscopy for the automation of pollen rate measurement. Image Anal. Stereol. 1, 332–527 (2001)
3. Boucher, A., Hidalgo, P.J., Thonnat, M., Belmonte, J., Galan, C., Bonton, P., Tomczak, R.: Development of a semi-automatic system for pollen recognition. Aerobiologia 18, 195–201 (2002)
4. Carvalho, E., Sindt, C., Verdier, A., Galan, C., O'Donoghue, L., Parks, S., Thibaudon, M.: Performance of the coriolis air sampler, a high-volume aerosol-collection system for quantification of airborne spores and pollen grains. Springer Science+Business Media (2008)
5. Costa, C.M., Yang, S.: Counting pollen grains using readily available, free image processing and analysis software. Annals of Botany (2009)
6. Flenley, J.: The problem of pollen recognition. In: Problems in Picture Interpretation, pp. 141–145 (1968)
7. Galán, C., García, H., Cariñanos, P., Alcazar, P., Domínguez, E.: Manual de calidad y gestion de la Red Española de Aerobiología, REA (2007)

8. Giesecke, T., Fontana, S.L., Knaap, W.O., Pardoe, H.S., Pidek, I.A.: From early pollen trapping experiments to the pollen monitoring programme. Veget. Hist. Archaeobot. 19, 247–258 (2010)

9. Hirst, J.M.: An automatic volumetric spore trap. Ann. Appl. Biol. 39, 257–265 (1952)

10. Holt, K., Allen, G., Hodgson, R., Marsland, S., Flenley, J.: Progress towards an automated trainable pollen location and classifier system for use in the palynology laboratory. Review of Peleobotany and Palynology 167, 175–183 (2011)

11. Landsmeer, S.H., Hendriks, E.A., Weger, L.A., Reiber, J.C., Stoel, B.C.: Detection of pollen grains in multifocal optical microscopy images of air samples. Microscopy Research and Technique 72, 424–430 (2009)

12. Muradil, M., Okamoto, Y., Yonekura, S., Chazono, H., Hisamitsu, M., Horiguchi, S., Hanazawa, T., Takahashi, Y., Yokota, K., Okumura, S.: Reevaluation of pollen quantitation by an automatic pollen counter. Allergy and Asthma Proceedings 31(5), 422–427(6) (2010)

13. Nguyen, N.R., Donalson-Matasci, M., Shin, M.C.: Improving pollen classification with less training effort. In: IEEE Workshop on Applications of Computer Vision, pp. 421–426 (2013)

14. Ranzato, M., Taylor, P.E., House, J.M., Flagan, R.C., LeCun, Y., Perona, P.: Automatic recognition of biological particles in microscopic images. Pattern Recognition Letters 28(1), 31–39 (2007)

15. Rodriguez-Damian, M., Cernadas, E., Formella, A., Fernandez-Delgado, M., De Sa-Otero, P.: Automatic detection and classification of grains of pollen based on shape and texture. IEEE Transactions on Systems, Man, and Cybernetics, Part C: Applications and Reviews 36(4), 531–542 (2006)

Estimation of the Arterial Diameter in Ultrasound Images of the Common Carotid Artery

Rosa-María Menchón-Lara[1(✉)], Andrés Bueno-Crespo[2],
and José Luis Sancho-Gómez[1]

Dpto. Tecnologías de la Información y las Comunicaciones. Universidad Politécnica
de Cartagena, Murcia, Spain
Dpto. Informática de Sistemas, Universidad Católica San Antonio, Murcia, Spain
rmml@alu.upct.es

Abstract. This paper addresses a fully automatic segmentation method
for ultrasound images of the common carotid artery. The goal of this procedure is the detection of the arterial walls to assist in the evaluation of
the arterial diameter. In other words, the main objective is the segmentation of the region corresponding to the lumen of the vessel, where the
blood flows. The evaluation of the Lumen Diameter (LD) provides useful information for the diagnosis of arterial diseases. The monitoring of
LD and Intima-Media Thickness (IMT) is crucial in the early detection
of atherosclerosis and in the assessment of the cardiovascular risk. The
proposed methodology is completely based on Machine Learning and it
applies Auto-Encoders and Deep Learning to obtain abstract and efficient data representations. Thus, the segmentation task is posed as a
pattern recognition problem. The different architectures designed have
shown a good classification performance. In addition, the results obtained
for some ultrasound images of the common carotid artery can be visually validated in this work. The final automatic segmentation is quite
accurate, and it is possible to conclude that it will lead to a precise and
reliable measurement of the lumen diameter.

1 Introduction

An early medical diagnosis of arterial diseases, is crucial for prevention and
treatment of cardiovascular diseases (CVD), which remain the major cause of
death in the world [18]. In this sense, the study of anatomical features of the
human arteries by means of non-invasive ultrasound imaging is widely used as
form of diagnosis [12].

Among the arterial diseases, atherosclerosis stands out as the leading underlying pathological process that causes a large proportion of CVD, such as myocardial infarction and stroke [18]. Atherosclerosis involves a progressive thickening
of the arterial walls by fat accumulation, which hinders blood flow and reduces
the elasticity of the affected vessels. The Intima-Media Thickness (IMT) of the
Common Carotid Artery (CCA) is considered as an early and reliable indicator

© Springer International Publishing Switzerland 2015
J.M. Ferrández Vicente et al. (Eds.): IWINAC 2015, Part I, LNCS 9107, pp. 358–367, 2015.
DOI: 10.1007/978-3-319-18914-7_38

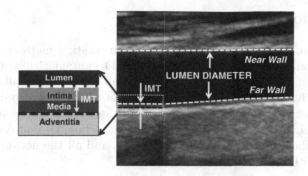

Fig. 1. Longitudinal view of the CCA in an ultrasound image

of atherosclerosis [16]. Besides, the evaluation of the artery diameter provides useful additional information for the diagnosis [13].

As can be seen in Fig. 1, the walls of blood vessels present three different layers, from innermost to outermost, intima, media and adventitia. The IMT is defined as the distance from the lumen-intima interface (LII) to the media-adventitia interface (MAI) and it should be measured preferably on the Far Wall (FW) of the CCA within a region free of atherosclerotic lesions (plaques) [16], where the double-line pattern corresponding to the intima-media-adventitia layers can be clearly observed (see Fig 1). This double-line pattern is not always visible in the Near Wall (NW) of the vessel. The internal diameter of the artery, or Lumen Diameter (LD), is defined as the distance from the LII in the NW, if it is visible (MAI, otherwise), to the LII in the FW.

In the last two decades, several solutions have been proposed for the segmentation of the carotid artery in ultrasound images [11,7]. Attending to the methodology applied, it can be found algorithms based on edge detection and gradient-based techniques, dynamic programming, active contours, statistical modelling, Hough transform or proposals based on combinations of the aforementioned techniques. Most of the proposed methods are not completely automatic and they require user interaction. Moreover, only a few of methods process both the NW and the FW in order to measure the LD [17,15,14,8,1].

This work addresses a fully automatic segmentation technique completely based on Machine Learning to extract the NW and FW boundaries from ultrasound CCA images and, therefore, to assist in the evaluation of the LD. The purpose of the proposed method is to complete the function of previous works by the same authors [9,10], which are focussed on the IMT measurement using related techniques.

The remainder of this paper is structured as follows. After this introduction, Sect. 2 describes the proposed methodology. In particular, Sect. 2.1 introduces the machine learning concepts used in this work, while in Sect. 2.2 and 2.3, the strategies for the arterial walls detection and the lumen boundaries extraction are explained in detail. Some of the results obtained are shown in 3. Finally, the main extracted conclusions close the paper.

2 Methodology

Fig. 2 shows an overview of the proposed segmentation methodology. Firstly, a given ultrasound CCA image is pre-processed to automatically detect the regions of interest (ROIs), which are the near wall and the far wall of the blood vessel. Then, those pixels belonging to the ROIs are classified to detect the lumen boundaries. As commented in Sect. 1, our method is completely based on Machine Learning. In particular, Deep Learning techniques and Auto-Encoders have been applied in the ROIs detection stage, and all the networks employed are founded on Extreme Learning Machine.

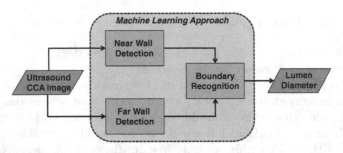

Fig. 2. Flow chart of the proposed method for the CCA segmentation

2.1 Machine Learning Techniques

In the last decade, Extreme Learning Machine (ELM) has emerged as a powerful tool in the learning process of Single-Layer Feed-Forward Networks (SLFN) by providing good generalization capability at fast learning speed [5]. Given N arbitrary distinct samples $(\mathbf{x}_n, \mathbf{t}_n)$, where $\mathbf{x}_n \in \mathbb{R}^d$ is an input vector and $\mathbf{t}_n \in \mathbb{R}^m$ its corresponding target vector, the output of a SLFN with M hidden neurons and activation function $f(\cdot)$ is given by

$$\mathbf{y}_n = \sum_{j=1}^{M} \beta_j f(\mathbf{w}_j \mathbf{x}_n + b_j), n = 1, ..., N; \tag{1}$$

where $\mathbf{w}_j = [w_{j1}, w_{j2}, ..., w_{jd}]$ is the input weight vector connecting the input units and the j-th hidden neuron, $\beta_j = [\beta_{j1}, \beta_{j2}, ..., \beta_{jm}]$ is the output weight vector connecting the j-th hidden neuron and the output units, and b_j is the bias of the j-th hidden neuron. If it is assumed that SLFN can approximate these N samples with zero error, then, there exist β_j, $\mathbf{w}_j$ and b_j such that

$$\sum_{i=j}^{M} \beta_j f(\mathbf{w}_j \mathbf{x}_i + b_j) = \mathbf{t}_i, i = 1, ..., N. \tag{2}$$

ELM is based on the randomly initialization of the input weights and biases of SLFN. Thus, the network can be considered as a linear system and the N equations in the expression (2) can be written compactly in the following form:

$$\mathbf{HB} = \mathbf{T};$$ (3)

where $\mathbf{T} \in \mathbb{R}^{N \times m}$ is the targets matrix, $\mathbf{B} \in \mathbb{R}^{M \times m}$ is the output weights matrix and $\mathbf{H} \in \mathbb{R}^{N \times M}$ is the hidden layer output matrix. Thereby, the training is reduced to solve the linear system in Eq. (3), whose smallest norm least-squares solution is given by:

$$\hat{\mathbf{B}} = \mathbf{H}^{\dagger}\mathbf{T};$$ (4)

where $\mathbf{H}^{\dagger}$ is the Moore-Penrose generalized inverse matrix of $\mathbf{H}$. Moreover, in order to improve the robustness and generalization performance, a regularization term (C) can be added to the solution [4]:

$$\hat{\mathbf{B}} = \left(\frac{\mathbf{I}}{C} + \mathbf{H}^{T}\mathbf{H}\right)^{-1} \mathbf{H}^{T}\mathbf{T}$$ (5)

Although ELM provides an efficient training for SLFN, the performance of machine learning methods and applications highly depends on the selected features for the representation of the problem. Thus, to make progress towards the Artificial Intelligence (AI), the new perspectives in Machine Learning are necessary based on learning data representations that make more accurate classifiers/predictors [2]. In this sense, Deep Learning has emerged as set of algorithms that attempt to model more abstract and useful representation of the data by means of architectures with multiple non-linear transformations [3].

Among the various deep learning architectures, this work focuses on deep networks based on Auto-Encoders (AE). In particular, the ELM Auto-Encoders (ELM-AE) introduced in [6] have been used in the detection of the arterial walls (NW and FW). Auto-encoders are SLFN performing unsupervised learning in the sense that an AE is trained to reconstruct its own inputs, i.e. $\mathbf{t}_n = \mathbf{x}_n$ (see Fig. 3). Therefore, in the hidden layer of the AE takes place a feature mapping: if $M < d$ (number of hidden neurons < input data dimension), a compressed data coding is obtained as hidden layer output; while if $M > d$, the result is a sparse representation of data.

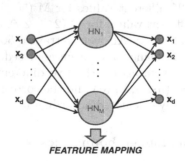

FEATRURE MAPPING

Fig. 3. Structure of a generic Auto-Encoder

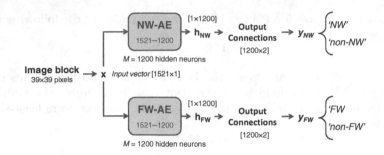

Fig. 4. Strategy for the detection of arterial walls in CCA ultrasounds

2.2 Detection of the Arterial Walls

This section describes the first stage of the proposed methodology, in which the carotid walls (NW and FW) are located in a completely automatic way by means of a deep architecture for Pattern Recognition (see Fig. 4).

Two different ELM-AE have been designed to obtain useful and efficient representations of image blocks. Then, the coding obtained from each AE is classified as 'NW', if the pattern of the near wall is recognized ('non-NW', otherwise) for the near wall auto-encoder (NW-AE), or as 'FW', if the pattern of the far wall is recognized ('non-FW', otherwise) for the FW-AE.

The size of the image blocks to process is 39×39 pixels, i.e. the ELM-AE has an input data dimension of 1,521 features. Two datasets have been used in the learning process of these systems. On one hand, the dataset employed to design the architecture for the near wall detection consists of 15451 samples (50% from each class: 'NW' and 'non-NW', respectively). On the other hand, the far wall dataset consists of 13776 observations equally distributed between 'FW' class and 'non-FW' class. Table 1 specifies the distribution of samples for each dataset. In both cases, two thirds of samples are intended for training and the remaining for testing.

For the configuration of the NW-AE and the FW-AE, an exhaustive search of the number of hidden neurons and the regularization parameter (M and C, respectively) by means of a cross-validation procedure has been performed in each case. In particular, 25 different values for M (50, 100, 150 ..., 1000, 1100, 1200, ..., 1500) and 20 different values for C (2^{-9}, 2^{-8}, ..., 2^{10}) have been considered. The NW-AE and the FW-AE were retrained 50 times for every pair of values and their mean performance have been analysed. The optimal coding is obtained with 1200 hidden neurons for the two different AE. The connections between these new features ($\mathbf{h_{NW}}$, $\mathbf{h_{FW}}$) and the corresponding system outputs ($\mathbf{y_{NW}}$, $\mathbf{y_{FW}}$) are analytically calculated according to Eq. (5).

2.3 Recognition of the Lumen boundaries

The segmentation of the carotid lumen in the ultrasound images is carried out by means of a classification of pixels belonging to the ROIs (NW and FW) previ-

Table 1. Specification of samples used in the design of the architectures for the detection of the arterial walls

	Near Wall Detection			**Far Wall Detection**		
	'NW'	'non-NW'	$\sum$	'FW'	'non-FW'	$\sum$
Training Data	5150	5150	10300	4592	4594	9186
Testing Data	2575	2576	5151	2296	2294	4590
$\sum$	7725	7726	**15451**	6888	6888	**13776**

ously detected. In particular, the intensity values from a certain neighbourhood centred on the pixel to classify provide the necessary contextual information to an ELM classifier for the recognition of the lumen boundaries. In the present study, the neighbourhoods consist of 45×3 pixels, i.e. input patterns with 135 features. Thus, two ELM classifiers have been implemented: one for processing the NW pixels (detected according to the strategy in Fig. 4) in order to extract the inner boundary of the near wall (*'NW-boundary'*), and other which classifies the FW pixels for detecting the LII in the far wall (*'FW-LII'*). Therefore, both classifiers perform a binary classification.

The labelled dataset used in the design and supervised training process of the NW-classifier consists of 8000 patterns, which are distributed equally among the two classes: *'NW-boundary'* and *'non-NW-boundary'*. In a similar manner, 8000 samples (50% from *'FW-LII'* class and 50% from *'non-FW-LII'* class) constitute the dataset used in the FW-classifier design. During the learning process, each dataset was randomly divided into two subsets: two thirds for training and the remaining for testing.

Once again, a cross-validation procedure over the number of hidden neurons and the regularization term is needed in each case to find the optimal design parameters of these architectures. In this case, M was varied in the range from 10 to 2000 hidden neurons and 38 different values for C (2^{-18}, ..., 2^{19}) were considered. The NW-classifier and the FW-classifier were retrained 50 times for every pair of values and their mean performance have been analysed. From this analysis, it is deduced that the optimal architecture for the NW-classifier consists of 1400 hidden neurons, whereas for the FW-classifier the optimal value of M is 1000 neurons.

3 Results

This section shows the results extracted from the present study. First of all, the performance of the architectures designed for the arterial walls detection and for the lumen segmentation is analysed in terms of classification accuracy (ACC), sensitivity (SEN) and specificity (SPEC). The last two describe the ability of the system to identify positive results and negative results, respectively. Table 2 shows this analysis for the proposed systems over the corresponding testing datasets. As can be seen, the different architectures have shown a good classification performance.

Table 2. Performance of the proposed architectures over the corresponding test sets. Mean ± standard deviation from 50 trials

	Near Wall Detection	Far Wall Detection	Lumen Segmentation NW-classifier	FW-classifier
ACC (%)	99.11±0.10	98.69±0.12	98.39±0.11	99.66±0.05
SEN (%)	100.00±0.01	99.50±0.12	99.81±0.09	99.95±0.04
SPEC (%)	98.23±0.21	97.88±0.21	96.98±0.21	99.36±0.10

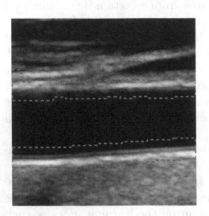 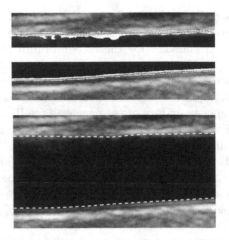

Fig. 5. Example of results: ROIs detection (left); NW with output of the NW-classifier and manual segmentations superimposed (right-top); FW with output of the FW-classifier and manual segmentations superimposed (right-central); final lumen contours (right-bottom)

In addition, the results obtained for some ultrasound images of the CCA are visually validated. In order to validate the precision of the proposed segmentation technique, the automatic results are compared with four manual tracings for each contour performed by two different experts. Thus, Figs. 5-7 show examples of processed ultrasound images. Left images depict the result of the stage for the NW and FW detection. Right-top images show the NW detected with the NW-boundary pixels recognized by the NW-classifier highlighted in white and the corresponding manual segmentations superimposed (dotted lines). In a similar manner, right-central images show the FW with the FW-LII pixels recognized by the FW-classifier and manual segmentations superimposed. As can be seen, the lumen boundaries are properly identified and the classification results cover the variability of the manual segmentations. However, it is necessary to define the final contours from the output of the classifiers. In this case, the final lumen boundaries are defined as the mean points of the positive outputs of the classifiers along the longitudinal axis. Finally, right-bottom images depict the final automatic segmentations for the carotid lumen (dashed lines). In view of these results, it is possible to appreciate that our automatic segmentation is quite accurate and, therefore, it will lead to a precise measurement of the LD.

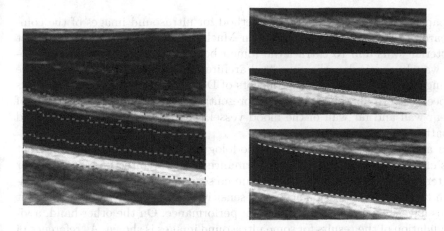

Fig. 6. Example of results: ROIs detection (left); NW with output of the NW-classifier and manual segmentations superimposed (right-top); FW with output of the FW-classifier and manual segmentations superimposed (right-central); final lumen contours (right-bottom)

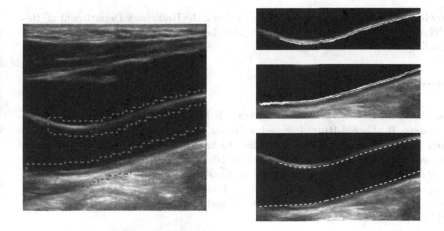

Fig. 7. Example of results: ROIs detection (left); NW with output of the NW-classifier and manual segmentations superimposed (right-top); FW with output of the FW-classifier and manual segmentations superimposed (right-central); final lumen contours (right-bottom)

4   Conclusions

This paper proposes a segmentation method for ultrasound images of the common carotid artery completely based on Machine Learning, in order to detect the arterial walls and to extract the lumen boundaries in a reliable and automatic way. In particular, the suggested architectures are based on the Extreme Learning Machine. Furthermore, concepts of Deep Learning and Auto-Encoders have been used to obtain useful data representations for solving the detection of the near wall and far wall of the blood vessel. The segmentation task is posed as a pattern recognition problem.

The accuracy of the proposed methodology is characterized from different points of view. On one hand, the performance of the architectures designed for the arterial walls detection and for the lumen segmentation is exhaustively analysed in terms of classification accuracy, sensitivity and specificity. The different systems have shown a good classification performance. On the other hand, a visual validation of the results for some ultrasound images is shown. As reference of target results, four manual tracings for each contour performed by two different experts are taken into account. In all the examples analysed, the lumen boundaries are properly identified and the classification results cover the variability of the manual segmentations. Moreover, the final automatic segmentation is quite accurate. Thus, it is possible to conclude that it will lead to a reproducible and reliable measurement of the lumen diameter. Future works must be focussed on a better characterization of these preliminary results on a larger image database.

Acknowledgements. Authors would like to thank the Radiology Department of 'Hospital Universitario Virgen de la Arrixaca' (Murcia, Spain) for their collaboration and for providing the ultrasound images used.

References

1. Bastida-Jumilla, M.C., Menchón-Lara, R.M., Morales-Sánchez, J., Verdú-Monedero, R., Larrey-Ruiz, J., Sancho-Gómez, J.L.: Frequency-domain active contours solution to evaluate intimamedia thickness of the common carotid artery. Biomedical Signal Processing and Control 16, 68–79 (2015)
2. Bengio, Y., Courville, A., Vincent, P.: Representation learning: A review and new perspectives. IEEE Trans. Pattern Anal. Mach. Intell. 35(8), 1798–1828 (2013)
3. Deng, L., Yu, D.: Deep learning: Methods and applications. Tech. Rep. MSR-TR-2014-21 (January 2014),
   http://research.microsoft.com/apps/pubs/default.aspx?id=209355
4. Huang, G.-B., Zhou, H., Ding, X., Zhang, R.: Extreme learning machine for regression and multiclass classification. IEEE Trans. Syst. Man Cybern. Part B-Cybern. 42(2), 513–529 (2012)
5. Huang, G.B., Zhu, Q.Y., Siew, C.K.: Extreme learning machine: Theory and applications. Neurocomputing 70(1-3), 489–501 (2006)
6. Kasun, L.L.C., Zhou, H., Huang, G.-B., Vong, C.M.: Representational learning with extreme learning machine for big data. IEEE Intelligent Systems 28(6), 31–34 (2013)

7. Loizou, C.: A review of ultrasound common carotid artery image and video segmentation techniques. Medical & Biological Engineering & Computing 52(12), 1073–1093 (2014)
8. Loizou, C., Kasparis, T., Spyrou, C., Pantziaris, M.: Integrated system for the complete segmentation of the common carotid artery bifurcation in ultrasound images. In: Papadopoulos, H., Andreou, A., Iliadis, L., Maglogiannis, I. (eds.) Artificial Intelligence Applications and Innovations. IFIP (AICT), vol. 412, pp. 292–301. Springer, Heidelberg (2013)
9. Menchón-Lara, R.M., Bastida-Jumilla, M.C., Morales-Sánchez, J., Sancho-Gómez, J.L.: Automatic detection of the intima-media thickness in ultrasound images of the common carotid artery using neural networks. Med. Biol. Eng. Comput. 52(2), 169–181 (2014)
10. Menchón-Lara, R.M., Sancho-Gómez, J.L.: Fully automatic segmentation of ultrasound common carotid artery images based on machine learning. Neurocomputing 151(pt. 1), 161–167 (2015)
11. Molinari, F., Zeng, G., Suri, J.S.: Review: A state of the art review on intima-media thickness (imt) measurement and wall segmentation techniques for carotid ultrasound. Comput. Methods Prog. Biomed. 100(3), 201–221 (2010)
12. Nikita, K.S.: Atherosclerosis: The evolving role of vascular image analysis. Comput. Med. Imaging Graph. 37(1), 1–3 (2013)
13. Reneman, R.S., Meinders, J.M., Hoeks, A.P.G.: Non-invasive ultrasound in arterial wall dynamics in humans: what have we learned and what remains to be solved. European Heart Journal 26(10), 960–966 (2005)
14. Rocha, R., Campilho, A., Silva, J., Azevedo, E., Santos, R.. Segmentation of ultrasound images of the carotid using ransac and cubic splines. Computer Methods and Programs in Biomedicine 101(1), 94–106 (2011)
15. Rossi, A.C., Brands, P.J., Hoeks, A.P.G.: Automatic localization of intimal and adventitial carotid artery layers with noninvasive ultrasound: A novel algorithm providing scan quality control. Ultrasound in Medicine and Biology 36(3), 467–479 (2010)
16. Touboul, P.J., et al.: Mannheim carotid intima-media thickness and plaque consensus (2004-2006-2011). Cerebrovasc. Dis. 34, 290–296 (2012)
17. Wendelhag, I., Liang, Q., Gustavsson, T., Wikstrand, J.: A new automated computerized analyzing system simplifies readings and reduces the variability in ultrasound measurement of intima-media thickness. Stroke 28(11), 2195–2200 (1997)
18. WHO: Global atlas on cardiovascular disease prevention and control. online, www.who.int/cardiovascular_diseases/en/

Comparison of Free Distribution Software for EEG Focal Epileptic Source Localization

Alexander Ossa[1], Camilo Borrego[2], Mario Trujillo[2], and Jose D. Lopez[1(✉)]

[1] SISTEMIC, Facultad de Ingenieria,
Universidad de Antioquia UdeA, Calle 70 No. 52-21, Medellín, Colombia
[2] Centro de investigaciones médicas de Antioquia CIMA,
Calle 63 No. 41-27, Medellín, Colombia
josedavid@udea.edu.co

Abstract. The effects of epilepsy in a patient can be significantly reduced with medical treatment. However, in some patients or after some time the anti-epileptic drugs do not take effect, being candidates to surgery. Preliminary studies of the patient are usually limited to EEG and MRI, and the epileptic focus is located using brain imaging algorithms that do not provide enough certainty to the specialist. In this work four of the most widely used free distribution neuroimaging software are tested with real epileptic data (EEGLab, SPM, LORETA, and Cartool), with the objective of illustrating their capabilities for locating the epileptic focus. As a result, a novel methodology for robust estimation that includes the advantages of the four software is proposed.

1 Introduction

Epilepsy is a neurological disorder which affects around 50 million people worldwide. Socially, epilepsy patients are target of discrimination and prejudgment that regularly lead to social stigma. The most common treatment for epilepsy is the use of anti-epileptic drugs (AEDs). When the treatment with AEDs fails, patients become candidates to surgical removal of the affected region. It is maybe the most underused treatment proved as effective, especially because of the potential side effects of the surgery due to wrong localization of the focus. Also, surgical procedure is commonly developed when the social and psychological effects are irreversible [5].

Surgical injury removal based on Positron Emission Tomography (PET) and Magnetic Resonance Images (MRI) gives free seizure results between 60-90 % of the patients after surgery. However, due to the possibility of epilepsy not related to cerebral injury, electroencephalography (EEG) appears as a complementary tool to localize the source of focal epilepsy [5,2].

EEG brain imaging aims to estimate the location and strength of current sources in the brain based on EEG data. Several studies have shown that is possible to locate the source of epileptic activity with certain accuracy using dipolar localization methods [9]. Also, with the implementation of more advanced techniques, it is possible to locate the epileptic source in patients with magnetic

J.M. Ferrández Vicente et al. (Eds.): IWINAC 2015, Part I, LNCS 9107, pp. 368–376, 2015.
DOI: 10.1007/978-3-319-18914-7_39

resonance negative focal epilepsy (MRN-E), i.e., no visible injury in MRI [2]. However, to the knowledge of the authors there is not a study comparing the performance of the software tools used to process the EEG data and locate the epileptic focus.

Several free distribution software are currently used to process EEG data, but not necessarily for localization of epileptic foci. Some of the most widely used are: EEGLab, SPM, LORETA, and Cartool. EEGLab is a toolbox for MATLAB that processes EEG, MEG and other electro-physiological data, it provides Independent Component Analysis (ICA) for artifact removal and neural activity reconstruction [4]. ICA consists in minimizing the statistical dependence among resulting components [8]. SPM is a MATLAB toolbox that includes the Multiple Sparse Priors algorithm for EEG brain imaging [10]. LORETA includes the inverse solution algorithms proposed by Dr. Pascual Marqui [11]. Finally, Cartool is a standalone software that provides the Lehman micro-states methodology that has been used for localizing epileptic foci [3]. The idea underlying this method is to select time periods of the EEG data for source analysis, on the basis that different scalp maps must have been originated by different configurations of neural sources [9].

In this paper real epileptic data are used to compare the mentioned free distribution software (EEGLab, SPM, LORETA, and Cartool), emphasizing in their advantages and disadvantages for processing EEG data and locating the epileptic focus. In the discussion section a novel methodology combining the advantages of the four software is proposed for increasing the robustness of the solution. Supplementary material with the dataset used and implementation details is available in the web-site: click here.

2   Materials and Methods

2.1   Data Description

The data consist of 100 spikes from an EEG study of a patient diagnosed with epilepsy. EEG signals were acquired using a model 15 Neurodata amplifier system with 64 channels from Grass Telefactor. The amplified signal was digitized using a 64 channels National Instruments A/D converter model 6071 with 256 Hz sampling rate and 12 bits of resolution. Spike selection was performed by a neurologist with experience in reading EEG data (Dr. Camilo Borrego). The spikes were visually identified over a single electrode (AFz), 500 ms before and after the peak of each spike were marked as trials. Then, all selected spikes were concatenated in a single file for processing. Fig. 1 shows 32 channels of five extracted trials.

2.2   Data Processing

The implementation details of the methodology used for processing the EEG data and obtaining the epileptic focus location, consist in data epoching and

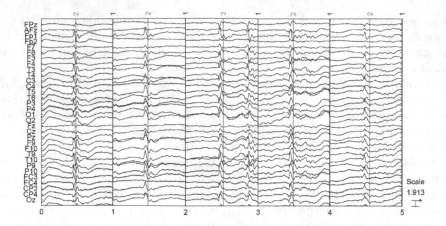

Fig. 1. 32 channels of 5 trials, each trial of 1 s long. Marks labeled as 1 and 2 are placed at the start of each trial and peak of each spike, respectively. This figure was obtained with EEGLab 10.2.5.8b.

baseline removal before extracting trials (500 ms around spikes peaks, see Fig. 1). Once the trials are concatenated into a single file, the EEG data is prepared for filtering. Frequencies within the range 1–30 Hz are of special clinical and psychological interest [1]; thus, a bandpass filter within these frequencies is performed.

An ICA analysis is performed over epoched data using INFOMAX algorithm (Called RunICA in EEGLab), in order to remove artifacts and not event related EEG activity. Those components with negligible contribution to data ERP ($<$ 2 %, unless suspects of deep low voltage epileptic foci) between -100 and 0 ms time range are discarded (i.e., out of the raising of the spike), and pruned EEG data is re-generated with the remained independent component ERPs. A second ICA is executed over the pruned EEG data, but this time only independent component ERPs with high contribution to the epileptic spike between -65 and 0 ms time range are selected. The objective of these two ICA runs is to remove not only external artifacts, but also non-related brain activity. Remaining independent components are expected to have high contribution during the spike raising phase, where epileptic foci are located [9]. Fig. 2 shows an example of the top 7 independent component ERPs with more contribution between -100 and 0 ms, obtained with EEGLab.

2.3   Software Implementation

EEGLab has been used as basis of this work because the other software used for comparison do not include an ICA implementation, and EEGLab processed data can be exported to the other software. Hence, source localizations with SPM and LORETA software were performed with both software specific and EEGLab processed data.

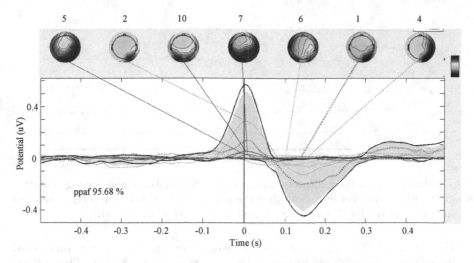

Fig. 2. Top 7 independent component ERPs with higher contribution between −100 and 0 ms. Note that they contribute with approximately 96 % to ERP. This figure was obtained using EEGLab 10.2.5.8b.

Specifically for SPM, data epoching and baseline removal stages are performed in order to extract data trials, and the bandpass filter (between 1 30 Hz) is applied. Subsequently, a single ERP is acquired using robust averaging with an offset about 3 for weighting function over data trials [10]. As robust averaging may re-introduce high-frequencies into the data [10], a second bandpass filter within the same frequencies (1 and 30 Hz) should be applied. SPM does not allow comparing results among time stamps of the spike as LORETA and Cartool do, but the robust averaging tool helps neglecting those solutions that do not correspond to the epileptic spike.

Cartool requires separation of trials in different files. Baseline correction and bandpass filtering (1–30 Hz) are applied over separated data trials. Once the data is averaged, micro-state segmentation using T-AAHC analysis is applied to ERP data as final stage. Fig. 3 shows the indicators that measure the reliability of each segmentation and the ERP GFP segmentation in micro-states.

2.4   Source Localization

Nowadays, several EEG source reconstruction algorithms are accepted as robust and reliable; however, they may not be as precise as expected for epileptic foci localization, or they may fail and the expert cannot notice immediately. In this sense, the fact that the four software compared in this work propose a different inverse solution, may provide enough confidence about the location of the epileptic foci. Also, not always the specialist has a structural MRI of the patient in preliminary studies. Hence, in this work source localization is computed with MRI templates provided by each software.

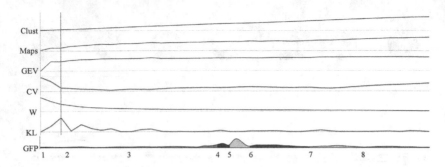

Fig. 3. Illustration of the indicators that measure the reliability and the segmentation selected with the marker line. Note that the micro-state labeled with 5 represents the raising phase of the GFP. This figure was obtained with Cartool 5.51.

The inversion algorithms used in this work (because they are implemented in the software) are: Dipole fitting, LORETA, MSP, and LAURA. They all are based on the general linear model:

$$Y = KJ + \epsilon \tag{1}$$

where the EEG data $Y \in \Re^{N_c}$ of N_c electrodes is generated from the linear relation between N_d dipolar sources of neural activity $J \in \Re^{N_d}$ distributed through the cortical surface (this may be a problem with some deep epileptic sources), and a forward model represented by the lead field matrix $K \in \Re^{N_c \times N_d}$. The data is affected by zero mean additive Gaussian noise $p(\epsilon) \sim \mathcal{N}(0, R)$, with $R \in \Re^{N_c \times N_c}$ the noise variance.

All solutions used in this work are based on the same covariance-weighted Tikhonov regularization (see [7] for a comprehensive review):

$$\widehat{J} = QK^T(KQK^T + R)^{-1}Y \tag{2}$$

with $(\cdot)^T$ the transpose operator. This solution is based on Bayesian assumptions for the prior neural activity: $p(J) \sim \mathcal{N}(0, Q)$, with $Q \in \Re^{N_d \times N_d}$; in order to estimate the neural distribution $\widehat{J} \approx J$. The difference between the inverse solutions relays in the way of computing Q.

EEGLab uses non-linear dipole fitting –DIPFIT– estimation. Dipole fitting is a non-linear single dipole estimation that assumes $N_d = 1$ with non-fixed position, and moves the dipole until it reaches the minimum norm error. This algorithm is known for having local minima that may affect the solution, and for having low noise rejection. The use of only those ICA components related with the phenomenon of interest is imperative.

LORETA software contains different implementations of the LORETA algorithm. It consists in defining Q as a weighted Laplacian that smooths the solution. This algorithm is robust and its variations sLORETA and eLORETA [11] are effective for single source problems such as the epileptic focus. However, the smoothed solution may affect the estimation of the epileptic focus extension.

SPM provides its own LORETA implementation, but its recommended algorithm is the Greedy Search implementation of the Multiple Sparse Priors algorithm. This algorithm divides the prior variance Q as the weighted sum of possible source locations (similar to LORETA), and it optimizes the estimation using the negative variational free energy as cost function [6].

Finally, Cartool is based on the LAURA algorithm [12]. This algorithm includes electromagnetic models in Q, in order to better explain the electric current propagation between the neural source and the electrodes. It is worth to mention that the main improvement of Cartool is not the inversion algorithm, it is the micro-state strategy that focuses the epilepsy specific problem.

3   Results

In this section, real epileptic data of a single selected subject is used for illustrating how the compared software can be used for increasing the robustness of the source localization. The data was processed with EEGLab and with the respective processing tools of each software. Where possible, the EEGLab processed ERP (using ICA) was compared with the software specific tools in order to determine the advantages and disadvantages of each software.

In EEGLab, DIPFIT was used for source localization. DIPFIT is an EEGLab plug in that can perform independent components source localization by fitting an equivalent current dipole model using a 4-shell spherical model. Independent components with activation within the time range of the raising phase of the spike were selected for inverse solution; however, only residual variance values of up to 45 % were achieved, making this method unfeasible without a subject specific head model. Also, the user experience selecting the correct independent components is highly related to good results.

In LORETA software, LORETA algorithm was applied using a 3-shell spherical model. Two EEG datasets were generated for reconstruction, the original one and the exported from EEGLab. Fig. 4 shows the inverse solution with LORETA algorithm to ERP processed with LORETA, and data ERP processed with ICA exported from EEGLab as mentioned above. Note that the right temporal lobe is active in both source reconstructions; However, source reconstruction with EEGLab processed data indicates a lower activation in left temporal lobe, and a more concentrated active area in the right temporal lobe. It may be associated to less noise levels and better artifact rejection due to the ICA processing.

For SPM source localization, the MSP algorithm was performed using the EEG BEM model template. A second EEG pruned data processed with ICA was exported from EEGLab. After uploaded, a robust averaging and a second bandpass filter stage were applied as well as mention above. A time window (-23 to -3 ms) focused on the raising phase was used for the inverse solution of each EEG data mentioned in previous section. Fig. 5 shows the inverse solution for both EEG datasets plotted at -16 ms. The data processed with SPM presented strong active areas in temporal and frontal lobes, and several blurred regions that can be associated with ghost sources introduced by noise and artifacts.

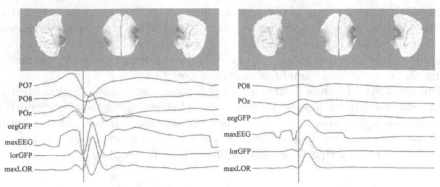

(a) Source localization for LORETA (b) Source localization for data pro-
processed data   cessed with ICA

Fig. 4. Source localization at 30 % of the GFP raising phase of software specific processed data, and data processed with EEGLab ICA, performed with LORETA algorithm. Note how data processed with ICA shows a smaller active region over the left
temporal lobe. This figure was obtained with LORETA 2014.06.25.

PPM at -16 ms (61 percent confidence)   PPM at -16 ms (60 percent confidence)
              512 dipoles   512 dipoles
Percent variance explained 64.24 (64.24)   Percent variance explained 85.02 (85.02)
           log-evidence = -215.6   log-evidence = -199.8

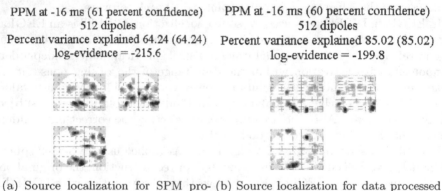

(a) Source localization for SPM pro- (b) Source localization for data processed
cessed data   with ICA

Fig. 5. Source localization computed with SPM, performed with MSP algorithm over
−23 to −3 ms time range. Note how data processed with ICA shows less phantom
regions with lower activity. In addition, note that explained variance is higher for ICA
processed data. This figure was obtained with SPM8.

Note that ICA processed EEG data presented a well defined inverse solution
with a strong active region in the right temporal lobe, and blurred regions with
less power.

For Cartool, LAURA algorithm was applied using L-SMAC model [3] with
5000 dipoles in the gray matter. Inverse solution was applied over the micro-
state marked as 5 (see Fig. 3). Fig. 6 shows the source localization performed
with LAURA algorithm. There is activity in cerebellar area and right temporal
lobe. In order to make the source localization visible, it was necessary to set

up the amplitude threshold such that the active area is displayed. However, the threshold setting was arbitrary and the extend area did not provide information about source strength.

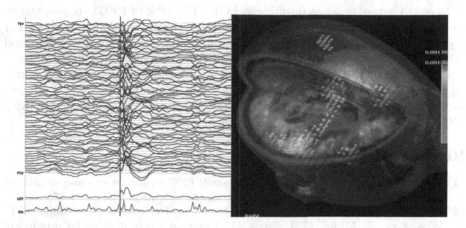

Fig. 6. Illustration of source localization performed with LAURA algorithm over GFP raising phase, computed with Cartool 3.51. There is a large active region in the right temporal lobe and cerebellar area.

4   Discussion and Conclusion

in this work four popular free toolbox software for neuroimaging (EEGLab, LORETA, SPM, and Cartool) were analyzed and compared, by processing and reconstructing real epileptic activity. The objective of this work was to propose a methodology for robust data processing and epileptic source localization, accounting the advantages and features of each software in order to improve the source estimation, and allow the user to compare and validate results.

In EEGLab, source localization is related to ICA component selection. Usually high amplitude components are related to spike peak and consequently with a propagated discharge of epileptic activity; however, components with lower amplitude present before the spike peak are suitable candidates to explain the epileptic source. Thus, the main limitation in EEGLab is that a good localization depends on the user experience.

SPM data processing was somewhat similar to EEGLab. However, each software provided different features for artifact rejection. EEGLab offers ICA, a well known tool for eliminating artifacts, but also for neglecting non-interest neural activity. Despite SPM is not designed to treat with spontaneous potential problems like epileptic discharges, it is possible to locate the source of the epileptic foci by limiting the source reconstruction to a specific time frame (spike raising phase).

LORETA software allows using previous processed ERP data from EEGLab. This software was easy to use with a intuitive user interface. Inverse solution results were faster than EEGLab and SPM, and its graphical interface was more suitable. Its inverse solution was performed over time (dynamic) allowing the user to acquire the source reconstruction at a mark time, a suitable characteristic to the selected criteria.

Despite Cartool is not as intuitive as LORETA, SPM or EEGLab, graphically Cartool is superior and faster than MATLAB based software. Cartool offers other processing tools such as micro-state segmentation, potential maps correlation, and other useful statistical approaches specific to epileptic focus source localization.

As a conclusion, Cartool can be used as a standalone software for source localization, and its results can be validated with a mixture of EEGLab data processing with ICA, followed by a source localization stage performed with SPM and LORETA.

References

1. Adeli, H., Zhou, Z., Dadmehr, N.: Analysis of EEG records in an epileptic patient using wavelet transform. Journal of Neuroscience Methods 123(1), 69–87 (2003)
2. Brodbeck, V., Spinelli, L., Lascano, A.M., Pollo, C., Schaller, K., Vargas, M.I., Wissmeyer, M., Michel, C.M., Seeck, M.: Electrical source imaging for presurgical focus localization in epilepsy patients with normal mri. Epilepsia 51(4), 583–591 (2010)
3. Brunet, D., Murray, M.M., Michel, C.M.: Spatiotemporal analysis of multichannel EEG: CARTOOL. Intell. Neuroscience 2011, 2:1–2:15 (2011)
4. Delorme, A., Makeig, S.: EEGLAB: an open source toolbox for analysis of single-trial EEG dynamics. Journal of Neuroscience Methods 134, 9–21 (2004)
5. Engel, J.: A greater role for surgical treatment of epilepsy: Why and when? Epilepsy Currents 3(2), 37–40 (2003)
6. Friston, K., Harrison, L., Daunizeau, J., Kiebel, S., Phillips, C., Trujillo-Barreto, N., Henson, R., Flandin, G., Mattout, J.: Multiple sparse priors for the M/EEG inverse problem. NeuroImage 39, 1104–1120 (2008)
7. Grech, R., Cassar, T., Muscat, J., Camilleri, K., Fabri, S., Zervakis, M., Xanthopoulos, P., Sakkalis, V., Vanrumste, B.: Review on solving the inverse problem in EEG source analysis. Journal of Neuro Engineering and Rehabilitation 5(1), 25 (2008)
8. Hyvarinen, A., Oja, E.: Independent component analysis: algorithms and applications. Neural Networks 13, 411–430 (2000)
9. Lantz, G., Spinelli, L., Seeck, M., de Peralta Menendez, R.G., Sottas, C.C., Michel, C.M.: Propagation of interictal epileptiform activity can lead to erroneous source localizations: a 128-channel eeg mapping study. Journal of Clinical Neurophysiology 20(5), 311–319 (2003)
10. Litvak, V., Mattout, J., Kiebel, S., Phillips, C., Henson, R., Kilner, J., Barnes, G., Oostenveld, R., Daunizeau, J., Flandin, G., et al.: EEG and MEG data analysis in SPM8. Computational Intelligence and Neuroscience (2011)
11. Pascual-Marqui, R.: Discrete, 3D distributed linear imaging methods of electric neuronal activity. Tech. rep., The KEY Institute for Brain-Mind Research. University Hospital of Psychiatry (2007)
12. Grave de Peralta, R., Hauk, O., Gonzalez, S.L.: The neuroelectromagnetic inverse problem and the zero dipole localization error. Intell. Neuroscience 2009, 1–11 (2009)

Weighted Filtering for Neural Activity Reconstruction Under Time Varying Constraints

J.I. Padilla-Buritica[1(✉)], E. Giraldo-Suárez[2], and G. Castellanos-Dominguez[1]

[1] Universidad Nacional de Colombia, sede Manizales, Colombia
[2] Universidad Tecnológica de Pereira, Pereira, Colombia,
egiraldos@utp.edu.co

Abstract. A novel Weighted Unscented Kalman Filtering method is introduced for neural activity estimation from electroencephalographic signals. The introduction of a weighting stage improves the solution by extracting relevant information directly from the measured data. Besides, a discrete nonlinear state space model representing the brain neural activity is used as a physiological constraint in order to improve the estimation. Moreover, time-varying parameters are considered which allow describing adequately healthy and pathological activity even for localized epilepsy events. Performance of the new method is evaluated in terms of introduced error measurements by application to simulated EEG data over several noise conditions. As a result, a considerable improvement over linear estimation approaches is found.

Keywords: Inverse problem · Unscented kalman filtering · Relevance analysis

1 Introduction

Neural activity estimation is the main task performed by brain mapping applications. The main challenge is to estimate the location, magnitude and time evolution of current sources that produce the measured electroencephalographic (EEG) signals [1]. This problem is considered as a mathematically ill-conditioned ill-posed inverse problem [8]. To make the problem tractable and to increase the reliability of the solution, a priori mathematical, anatomical and physiological information are included into the solution [7].

It is possible to use nonlinear dynamics models for neural activity as proposed by [1]. Those nonlinear models are more realistic and make possible better approximation of the real dynamic model of neural activity. However, at the same time, they increase the computational load and require more elaborate techniques for estimation of parameters. Nevertheless, nonlinear models could perform a better estimation for real EEG signals improving the analysis of the resultant estimated neural activity [7]. By considering time-varying parameters into the neural activity model a better estimation of neural activity can be performed since the variable model takes into account the variability of the process.

© Springer International Publishing Switzerland 2015
J.M. Ferrández Vicente et al. (Eds.): IWINAC 2015, Part I, LNCS 9107, pp. 377–387, 2015.
DOI: 10.1007/978-3-319-18914-7_40

The variability of the process should also be considered in the covariance matrices used for describing the neural activity variability from normal to pathological states [2].

We present a neural activity estimation method solving the dynamic inverse problem within a weighted Kalman filter framework and using a nonlinear physiology based constraints that consider parameters evolution. The introduction of a weighting stage improves the solution by extracting relevant information directly from the measured data. The analysis is made up from simulated EEG signals for several signal-noise ratios over a realistic head model calculated with the boundary elements method. The solution of the weighted inverse problem is achieved by using a Kalman filtering method where current densities and parameters of the physiological model are estimated simultaneously.

2 Materials and Methods

2.1 Inverse Problem Framework

The general dynamic state space system of the EEG inverse problem is as follows [4]:

$$y_k = Mx_k + \varepsilon_k, \tag{1a}$$

$$x_k = f\left(x_{k-1}, \cdots, x_{k-m}, w_k\right) + \eta_k, \tag{1b}$$

where vector $y_k \in \mathbb{R}^{d \times 1}$ holds the EEG electrical activity on the scalp measured by the d electrodes at the time instant k; vector $x_k \in \mathbb{R}^{n \times 1}$ is the current density of each source (i.e., neural activity), being n the number of distributed sources throughout the brain; and $M \in \mathbb{R}^{d \times n}$ is the lead field matrix that can be computed as a quasi-static approximation of the Maxwell and Poisson's equations for specific head models [5]. The lead field matrix associates the current density inside the brain, x_k, to the scalp measures, y_k.

Besides, the vector $\varepsilon_k \in \mathbb{R}^{d \times 1}$ represents the observation noise, $f : \mathbb{R}^n \mapsto \mathbb{R}^n$ is a non-linear vector function of order m that models the dynamics of the neural activity, the vector $w_k \in \mathbb{R}^{p \times 1}$ holds the p parameters of the function f, and $\eta_k \in \mathbb{R}^{n \times 1}$ is a random vector representing the additive process noise.

For describing the parameter evolution that is inherent to physiological models, we assume the dynamic behaviour of the forward problem ruled in the form:

$$w_k = g\left(w_{k-1}\right) + \epsilon_k, \tag{2}$$

where $g : \mathbb{R}^p \mapsto \mathbb{R}^p$ is a linear vector function of first order that models the dynamics of w_k, and $\epsilon_k \in \mathbb{R}^{p \times 1}$ is the noise affecting the p time-varying parameters. Here, the noise variables are assumed in the form: $\varepsilon_k \sim \mathcal{N}(0, C_\varepsilon)$, $\eta_k \sim \mathcal{N}(0, C_\eta)$, $\epsilon_k \sim \mathcal{N}(0, C_\epsilon)$, being $C_\varepsilon = \sigma_\varepsilon^2 I_d$, $C_\eta = \sigma_\eta^2 \left(L^\top L\right)^{-1}$ and $C_\epsilon = \sigma_\epsilon^2 I_p$ the covariance matrices of the measurement noise (see Eq. (1a)), the process noise (Eq. (1b)), and the vector parameter noise (Eq. (2)), respectively. The real-valued scalars σ_ε^2, σ_η^2 and σ_ϵ^2 are their corresponding variances.

Consequently, we get a discrete state space model representing the brain neural activity that assumes each source evolving independently from others so that each source dynamical behavior becomes time-varying [1]. Thus, the model-derived features allow describing adequately normal and pathological activity even for localized epilepsy events. Then, the time-varying model of the EEG inverse problem is defined as follows:

$$f\left(\boldsymbol{x}_{k-1}, \boldsymbol{x}_{k-2}, \boldsymbol{x}_{k-\tau}, \boldsymbol{w}_k\right) = \left(a_1 \boldsymbol{I}_n + b_1 \boldsymbol{L}\right) \boldsymbol{x}_{k-1}$$
$$+ a_2 \boldsymbol{x}_{k-1}^2 + a_3 \boldsymbol{x}_{k-1}^3 + a_4 \boldsymbol{x}_{k-2} + a_5 \boldsymbol{x}_{k-\tau}, \tag{3}$$

where $\boldsymbol{A}_1{=}a_1 \boldsymbol{I}_n{+}b_1 \boldsymbol{L}$, $\boldsymbol{A}_2{=}a_2 \boldsymbol{I}_n$, $\boldsymbol{A}_3{=}a_3 \boldsymbol{I}_n$, $\boldsymbol{A}_4{=}a_4 \boldsymbol{I}_n$, and $\boldsymbol{A}_5{=}a_5 \boldsymbol{I}_n$ with $\boldsymbol{I}_n$ the identity matrix sizing $n{\times}n$; $\boldsymbol{L}{\in}\mathbb{R}^{n \times n}$ is a spatial Laplacian matrix containing the spatial interactions among sources; f is a time varying vector function since $\boldsymbol{w}_k$ can change from one sample to another; and the set of parameters associated with the dynamics of Eq. (3) is $\boldsymbol{w}_k$ with $p{=}6$ defined as $\boldsymbol{w}_k^\top{=}\begin{bmatrix} a_1 & b_1 & a_2 & a_3 & a_4 & a_5 \end{bmatrix}$.

Therefore, the resulting nonlinear discrete representation in Eq. (3) can be assumed as the neural activity model that is grounded on physiological knowledge. Further, the model is reformulated in the form of a first order dynamic system, yielding:

$$\underbrace{\begin{bmatrix} \boldsymbol{x}_k \\ \boldsymbol{x}_{k-1} \end{bmatrix}}_{\boldsymbol{z}_k} = \underbrace{\begin{bmatrix} f\left(\boldsymbol{x}_{k-1}, \boldsymbol{x}_{k-2}\right) \\ \boldsymbol{x}_{k-1} \end{bmatrix}}_{f(\boldsymbol{z}_{k-1})} + \underbrace{\begin{bmatrix} \boldsymbol{I} \\ 0 \end{bmatrix}}_{\boldsymbol{B}} \eta_k \tag{4a}$$

$$\boldsymbol{y}_k = \underbrace{\begin{bmatrix} \boldsymbol{M} & 0 \end{bmatrix}}_{\boldsymbol{C}} \underbrace{\begin{bmatrix} \boldsymbol{x}_k \\ \boldsymbol{x}_{k-1} \end{bmatrix}}_{\boldsymbol{z}_k} + \varepsilon_k \tag{4b}$$

where $\boldsymbol{C}{\in}\mathbb{R}^{d \times 2n}$ is the augmented representation of the output matrix, $\boldsymbol{B}{\in}\mathbb{R}^{2n \times n}$ is the noise matrix, and vector $\boldsymbol{z}_k{\in}\mathbb{R}^{2n \times 1}$ includes the brain current densities reconstructing the brain neural activity.

2.2 Dynamic Inverse Problem using Weighted Unscented Kalman Filtering

The weighted dynamic inverse problem, based on the above-described time-varying dynamical model, is now stated within the following optimization framework:

$$\underset{\boldsymbol{x}_k, \boldsymbol{w}_k}{\text{minimize}} \qquad \left\| \boldsymbol{P}_k \left(\boldsymbol{y}_k - \boldsymbol{M} \boldsymbol{x}_k\right) \right\|^2 \tag{5}$$

$$\text{subject to} \qquad \left\| \boldsymbol{x}_k - f\left(\widehat{\boldsymbol{x}}_{k-1}, \cdots, \widehat{\boldsymbol{x}}_{k-m}, \boldsymbol{w}_k\right) \right\|^2 = 0,$$
$$\left\| \boldsymbol{w}_k - g\left(\widehat{\boldsymbol{w}}_{k-1}\right) \right\|^2 = 0.$$

where notation $\| \cdot \|$ stands for the Euclidean norm, $\widehat{x}_{k-i}$ is the state estimated at the $k{-}i$ instant, $\widehat{w}_{k-1}$ is the parameter estimated at the $k{-}1$ sample, whereas $P_k \in \mathbb{R}^{d \times d}$ is the weighting matrix directly associated to the noise measurement covariance.

Although the weighting matrix allows a non-uniform minimization of the terms of Eq. (5), this devised optimization task can be solved iteratively just for one variable at a time while the other variables remain fixed. As suggested in [9], calculation of the needed posterior mean and covariance of the process can be carried out trough the Unscented Kalman Filter that introduces a sigma matrix $\mathcal{X}_{k-1} \in \mathbb{R}^{2n \times (2n+1)}$ as follows:

$$\mathcal{X}_{k-1} = \begin{bmatrix} \xi_{k-1}^{(0)} & \xi_{k-1}^{(1)} & \cdots & \xi_{k-1}^{(2n)} \end{bmatrix}$$

being $\xi_{k-1}^{(i)} \in \mathbb{R}^{2n \times 1}$ the samples neighboring $\widehat{z}_{k-1}$ defined by:

$$\xi_{k-1}^{(i)} = \begin{cases} \widehat{z}_{k-1} + \mathrm{sgn}(i-n)\delta \overline{\Sigma}_{k-1}^{(n-i\,\mathrm{sgn}(i-n))}, & i = 0,\dots,2n, i \neq n \\ \widehat{z}_{k-1}, & i = n \end{cases}$$

where $\overline{\Sigma}_{k-1}^{(i)}$ is the i-th column of the matrix square root of Σ_{k-1}, and $\delta \in \mathbb{R}^+$ (usually set to a small positive value) determines the spread of the sigma points over the neighborhood of $\widehat{z}_{k-1}$.

The sigma matrix $\mathcal{X}_{k-1}$ is propagated through the nonlinear system as $\mathcal{X}_k = f(\mathcal{X}_{k-1})$, where $\mathcal{X}_k$ is the prior sigma matrix having the corresponding prior mean and covariance defined as follows:

$$\widetilde{z_k} \approx \sum_{i=0}^{2n} \omega_i^m \xi_k^{(i)}$$

$$\Sigma_k \approx \sum_{i=0}^{2n} \omega_i^c \left(\xi_k^{(i)} - \widehat{z_k} \right) \left(\xi_k^{(i)} - \widehat{z_k} \right)^\top + B\Sigma_\eta B^\top$$

where ω_i^m and ω_i^c are the weights associated to the mean an variance calculation, respectively.

For estimating the neural activity through the filtering task, however, the solution accomplishes two sequential UKFs (one for estimating states and another for parameters) that involve the weighting matrix P_k, as described in Algorithm 1.

3   Results AND Discussion

3.1   Experimental Set-up

The most common approach to assessing the performance of the inverse solution is to make use of simulated data taking advantage that the underlying source activity is known, and thus the methods can be objectively validated. With

---

Algorithm 1. Weighted Dual Kalman filtering under Non linear Constraints

---

Initialize with $\hat{z}_0$, $\hat{w}_0$, Σ_0, Σ_{w_0} ;

for $k = 1 \to T$ **do**

    Estimate P_k

    Time update equations for parameter filter:

    $\hat{w}_k^- = \hat{w}_{k-1}$

    $\Sigma_{w_k}^- = \alpha^{-1} \Sigma_{w_{k-1}}$

    Sigma points:

    $\mathcal{X}_{k-1} = \left[\xi_{k-1}^{(0)} \ \xi_{k-1}^{(1)} \ \cdots \ \xi_{k-1}^{(2n)} \right]$

    Time update equations:

    $\mathcal{X}_k = f_k (\mathcal{X}_{k-1})$

    $\hat{z}_k^- = \sum\limits_{i=0}^{2n} \omega_i^m \xi_k^{(i)}$

    $\Sigma_k^- = \sum\limits_{i=0}^{2n} \omega_i^c \left(\xi_k^{(i)} - \hat{z}_k^- \right) \left(\xi_k^{(i)} - \hat{z}_k^- \right)^\top + B \Sigma_\eta B^T$

    Measurement update equations:

    $G_k = \Sigma_k^- C^\top P_k^\top \left(P_k C \Sigma_k^- C^\top P_k^\top + \Sigma_\varepsilon \right)^{-1}$

    $\hat{z}_k = \hat{z}_k^- + G_k P_k \left(y_k - C \hat{z}_k^- \right)$

    $\Sigma_k = (I - G_k P_k C) \Sigma_k^-$

    Measurement update equations for the parameter filter:

    $G_k^w = \Sigma_{\hat{w}_k^-} (C_k^w)^\top \left(C_k^w \Sigma_{\hat{w}_k^-} (C_k^w)^\top + \Sigma_e \right)^{-1}$

    $\hat{w}_k = \hat{w}_k^- + G_k^w P_k \left(y_k - C \hat{z}_k^- \right)$

    $\Sigma_{w_k} = (I - G_k^w C_k^w) \Sigma_{w_k}^-$

end

---

this aim, a time-varying EEG simulation lasting $1s$ and sampled at $250Hz$ is carried out. Assuming that all sources have the same propagation model, the neural activity simulations that include normal and pathological behavior are accomplished using the discrete nonlinear model described in Eq. (3), where the following parameter values are fixed: $\tau=20$, $a_1=1.0628$, $b_1=-0.12$, $a_2=0.000143$, $a_3=-0.000286$, $a_4=-0.42857$, $a_5=0.008$, and $|\eta_k| \leq 0.05$. For simulation of the normal and pathological states, the parameter a_1 varies ranging from 1.0628 till 1.3 while a_4 ranges from -0.428 to -1, at the fixed time sample $k=125$ (i.e., $t=0.5s$).

Provided the simulated activity x_k, the resulting EEG measures y_k are the multiplication of the source current density by the lead field matrix, that is, $y_k = M x_k + \varepsilon_k$, where ε_k is set to get the following testing values of Signal-to-Noise Ratio (SNR)[dB]: 30, 20, and 10. The head structure used in the solution of the inverse problem is shown in Fig. 1, where the number of electrodes is fixed as $d=34$, and $n=5000$ sources are involved in the solution. All sources are perpendicularly located on the tessellated surface of the cortex. This assumption takes place since the main EEG generators are the pyramidal cortical neurons, whose dendrite trunks are locally oriented in parallel and pointing perpendicularly to the cortical surface [3]. Moreover, the lead field matrix M is calculated using a head model that considers the effects of the skin, skull, and cortex in the propagation of the electric fields. The computation of the matrix M is carried out using the Boundary Element Method approach that is explained in detail in [5].

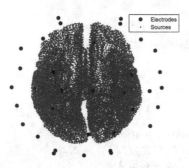

Fig. 1. Positions of the sources and electrodes used in the simulations and reconstructions

As regards the evaluation of the considered estimation methods of neural activity, their performance is computed in terms of the following error measure:

$$\kappa = \|\widehat{X} - X\|_F^2, \ \kappa \in \mathbb{R}^+$$

where the matrix $X \in \mathbb{R}^{d \times T}$ holds the dipole activity and the measuring sensors of electrical activity through all time instants, respectively. It must be noted that the projected error describes the inaccuracy relating to the recoverable solution, i.e., the part of the solution that is in the space rank of M, which is assumed, mathematically, to be the best reconstruction possible. Notation $\| \cdot \|_F^2$ stands for the Frobenious norm of the argument defined as $\|J\|_F^2 = \sqrt{\sum_{i=1}^{m_1} \sum_{j=1}^{m_2} j_{ij}^2}$, where j_{ij} is the ij element of $J \in \mathbb{R}^{m_1 \times m_2}$.

3.2   Validation of the Weighting Kalman Filtering

In order to carry out the proposed estimation method of neural activity in Eq. (3), we propose to compute the weighting matrix P_k involved in the cost function Eq. (5) through an introduced relevance analysis stage. To this end, the weighted Unscented Kalman filtering (WUKF) approach is developed for solving the optimization task in Eq. (5). Nonetheless, the concept of weighting matrix may be used in other approaches for accomplishing the Inverse Problem Solution.

For evaluation sake, the following meaningful versions of the weighting matrix are considered [2]:

a) $P = I$, being $I \in \mathbb{R}^{d \times d}$ the identity matrix, that is, the unweighted matrix.
b) $P = \text{diag}(\sqrt{\rho})^{-1}$, where $\rho \in \mathbb{R}^{d \times 1}$ is the variance of each j-th measuring channel, for $j=1, \ldots, d$, considering all samples $k = 1, \ldots, T$. (termed ρ method),
c) $P = \text{diag}(\sqrt{\alpha})^{-1}$, where $\alpha \in \mathbb{R}^{d \times 1}$ represents a relevance measure that is
   computed for each j- channel throughout the whole sample set (Q-α method).

The effect of the noise level on the estimation of P by calculating the corresponding matrix weights for several values of SNR (except for the case a)
assuming the unweighted matrix version). Figs. 2b and 2a show the estimated ρ
and α weight values, respectively, that turn out to be stable enough within the
testing range of noise.

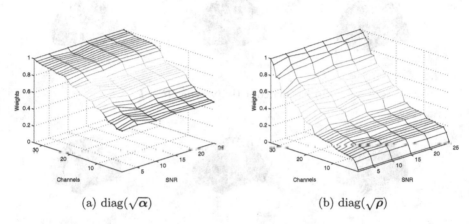

(a) $\text{diag}(\sqrt{\alpha})$ (b) $\text{diag}(\sqrt{\rho})$

Fig. 2. Computation of the preprocessing relevance weights for different values of SNR

In practice, the computational burden due to the weighting matrix calculation may be intractable because of the involved matrix sizes. To cope with this
drawback, estimation of the P is carried out in a moving frame version. Thus, a
downsized time-varying matrix $P_k \in \mathbb{R}^{d \times d}$ is now computed for each k-th frame.

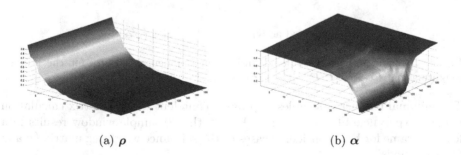

(a) ρ (b) α

Fig. 3. Weight estimation of P_k through the time framed approach using a 30-samples
frame for ρ and Q-α methods

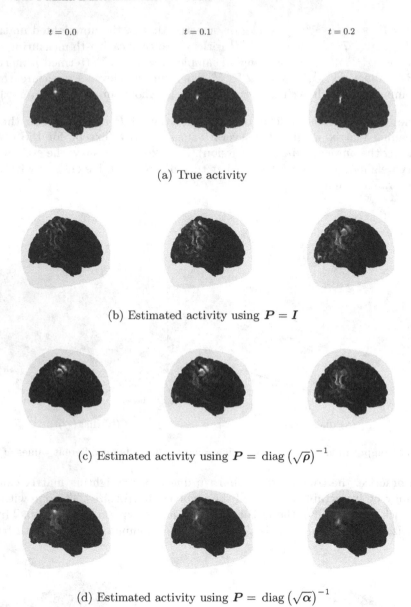

(a) True activity

(b) Estimated activity using $P = I$

(c) Estimated activity using $P = \mathrm{diag}\left(\sqrt{\rho}\right)^{-1}$

(d) Estimated activity using $P = \mathrm{diag}\left(\sqrt{\alpha}\right)^{-1}$

Fig. 4. Brain mapping of simulated and estimated neural activity for the studied weighting matrices: $P=I$, $P=\mathrm{diag}\left(\sqrt{\rho}\right)^{-1}$ and $P=\mathrm{diag}\left(\sqrt{\alpha}\right)^{-1}$

The minimal number of samples to provide confident framed matrix calculation is fixed experimentally. As shown in Fig. 3, the 30-sample window results in a feasible frame for both considered cases the relevance weighting matrix (ρ and Q–α methods).

As shown in Fig. 4 displaying the simulated and estimated neural activity for several weighting matrices at $t=0, 0.1, 0.2$ s, the use of WUKF leads to an im-

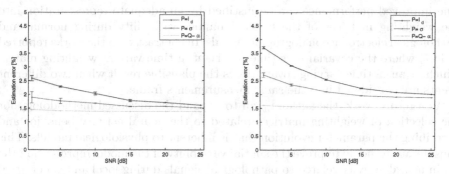

(a) Estimation using the nonlinear time-varying model

(b) Estimation using the linear time varying model

Fig. 5. Achieved κ error and dispersion values of the compared weighting matrices using linear or nonlinear dynamic models of the brain neural activity

provement of the estimation of the weighting matrix of either relevance method (ρ and $Q-\alpha$)

Fig. 5 shows the κ error of neural activity reconstruction computed for all compared weighting matrix. As seen, WUKF employing the $Q-\alpha$ method reaches better performance using either linear or nonlinear dynamic models representing the brain neural activity. The improved performance of $Q-\alpha$ method for source localization may be explained by the properties of vector α extracting a relevance measure from each measurement channel. In contrast, the ρ method only considers the variance of each channel [2]. In other words, the relevance value assessed by the $Q-\alpha$ method is more sensitive to significant changes of the electrical signal going through the channels. In contrast, the variance estimated by the ρ method only computes the dispersion just close the mean value.

4   Conclusions

We address the weighted dynamical inverse problem of EEG neural activity estimation, using a new method to improve the model through a weighted approach. The obtained results demonstrate that the models with including weighting matrices perform better than the assumed model. This improvement is because the model with weighting matrices corrects the initial assumptions of a uniformly distributed variance, from observations. These results are confirmed for simulated signals over several SNR values where the weighted model using the $Q - \alpha$ method provides the best performance.

In case of the Weighted Dynamic Inverse Problem solution based on a Kalman Filtering Framework, it is clear that the weighted method improves the model through the selection weighting matrices. Attained results demonstrate that the models with time varying weighting matrices perform better than models with fixed ones. These results are confirmed for simulated signals over several SNR values where the time varying weighted model reached the best performance. The improved performance can be justified by an adequate representation into the weighting matrices of the brain dynamics variability during normal and pathological electroencephalographic signals. In contrast with the works reported in [4,7] where the covariance is time invariant, a time variant weighting matrix which means a time varying covariance is the plausible result when two different events are considered into the same measurements frame.

As a future work, the authors plan to expand the discussed methodology for the selection of weighting matrices related to the neural activity behavior and describing the parameter evolution that is inherent to physiological models. This consideration should improve the spatial variability of the model representing the brain neural activity related to pathological signals during local and generalized epilepsy events.

Acknowledgments. This research was supported by COLCIENCIAS project *Evaluación asistida de potenciales evocados cognitivos como marcador del transtorno por déficit de atención e hiperactividad (TDAH)* and Programa Nacional de Formacion de Investigadores "GENERACION DEL BICENTENARIO", 2011.

References

1. Giraldo, E., den Dekker, A.J., Castellanos-Dominguez, G.: Estimation of dynamic neural activity using a Kalman filter approach based on physiological models. In: 32nd Annual International Conference of the IEEE Engineering in Medicine and Biology Society, Buenos Aires, Argentina, September 1-5 (2010)
2. Giraldo, E., Peluffo-Ordoñez, D., Castellanos, G.: Weighted time series analysis for electroencephalographic source localization. Journal of the Faculty of Mines Dyna Universidad Nacional de Colombia - Sede Medellın 79(176), 64–70 (2012)
3. Kim, J.W., Shin, H.B., Robinson, P.A.: Compact continuum brain model for human electroencephalogram. In: Society of Photo-Optical Instrumentation Engineers (SPIE) Conference Series. Society of Photo-Optical Instrumentation Engineers (SPIE) Conference Series, vol. 6802 (2007), doi:10.1117/12.759005
4. Giraldo, E., Castaño-Candamil, J.S., Castellanos-Dominguez, C.G.: A Weighted Dynamic Inverse Problem for Electroencephalographic Current Density Reconstruction. In: 6th International IEEE EMBS Conference on Neural Engineering, San Diego, California, November 6-8 (2013)
5. Hallez, H., Vanrumste, B., Grech, R., Muscat, J., Clercq, W., Velgut, A., D'Asseler, Y., Camilleri, K., Fabri, S., Van Huffel, S., Lemahieu, I.: Review on solving the forward problem in eeg source analysis. Journal of NeuroEngineering and Rehabilitation 4(46), 101–113 (2007)
6. Connors, W., Trappenberg, T.: Improved path integration using a modified weight combination method. Cognitive Computation 5(3), 295–306 (2013), doi:10.1007/s12559-013-9209-0
7. Barton, M., Robinson, P., Kumar, S., Galka, A., Durrant-White, H., Guivant, J., Ozaki, T.: Evaluating the performance of kalman-filter-based eeg source localization. IEEE Transactions on Biomedical Engineering 56(1), 435–453 (2009)

8. Grech, R., Tracey, C., Muscat, J., Camilleri, K., Fabri, S., Zervakis, M., Xanthoupoulos, P., Sakkalis, V., Vanrumste, B.: Review on solving the inverse problem in eeg source analysis. Journal of NeuroEngineering and Rehabilitation 5(25), 792–800 (2008)
9. Haykin, S.: Kalman Filtering and Neural Networks. Wiley (2001)

Neural Activity Estimation from EEG Using an Iterative Dynamic Inverse Problem Solution

E. Giraldo-Suárez[2(✉)] and G. Castellanos-Dominguez[1]

[1] Universidad Nacional de Colombia, sede Manizales, Colombia
[2] Universidad Tecnológica de Pereira, Pereira, Colombia,
egiraldos@utp.edu.co

Abstract. Estimation of neural activity using Electroencephalography (EEG) signals allows identifying with high temporal resolution those brain structures related to pathological states. This work aims to improve spatial resolution of estimated neural activity employing time-varying dynamic constraints within the iterative inverse problem framework. Particularly, we introduce the use of Dynamic Neural Fields (DNF) to represent neural activity directly related to epileptic foci localization adequately. So, we develop a DNF-based time variant estimation model in the form of an Iterative Regularization Algorithm (IRA) that carries out neural activity estimation at every time EEG sample. The IRA model performance that is evaluated on simulated and real cases is compared with the baseline static and dynamic methods under several noise conditions. To this end, we use different error measures showing that the IRA estimation model can be more accurate and robust than the other compared methods.

Keywords: EEG inverse problem · Neuroimaging · Time varying model · Dynamic regularization · Dynamic Neural Field

1 Introduction

Neuroimaging carries out mapping information analysis within the brain by non-invasively estimating activity of distributed neural networks. Among brain mapping techniques, neuroimaging based on Electroencephalography (EEG) signals is widely used due to its high temporal resolution and low implementation cost [1]. However, spatial resolution of the EEG-based neuroimaging is very poor since it reflects activity estimation triggered by a few thousands of neural generators, having just a very reduced set of scalp measures (typically, a couple of tens). The baseline linear solution of this estimation task (EEG inverse problem) is the static model (LOw REsolution TomogrAphy –LORETA) that does not account for temporal information, making statistically indistinguishable two sources next to each other. Meanwhile, to improve brain mapping quality, spatial and temporal dynamics inherently of neural activity should be considered within the solution framework [2].

© Springer International Publishing Switzerland 2015
J.M. Ferrández Vicente et al. (Eds.): IWINAC 2015, Part I, LNCS 9107, pp. 388–397, 2015.
DOI: 10.1007/978-3-319-18914-7_41

So as to include both dynamics, a dynamical estimation model can be used constraining the solution to some predefined geometric or physiological restrictions. Thus, Dynamic Neural Fields (DNF) approximate non-linear dynamic models improving brain mapping quality since these non-linear models are more realistic and allow better-representing neural activity dynamics [3]. However, their use increases the computational burden and requires more complex parameter estimation techniques. In practice, there are two main issues be considered for the dynamic inverse problem solution: A proper choice of the dynamical model for neural activity and the reduction of computational load preserving the spatial and temporal dynamics of measured EEG data [4].

Here, we introduce the use of Dynamic Neural Fields to estimate neural activity adequately during either normal or pathological states directly related to epileptic foci localization. So, we develop a DNF-based time variant estimation model in the form of an iterative regularization algorithm that carries out neural activity estimation at every time sample. At the same time, the iterative dynamic model changes over time without significantly increasing the computational cost.

2   Methods

2.1   Forward Problem for EEG Generation

We will consider dynamic systems described by the following equations:

$$y_k = M x_k + \varepsilon_k, \tag{1a}$$
$$x_k = f(x_{k-1}, \ldots, x_{k-m}, w_k) + \eta_k, \tag{1b}$$
$$w_k = g(w_{k-1}) + \epsilon_k, \tag{1c}$$

where vector $y_k \in \mathbb{R}^{d \times 1}$ holds measured EEG data on d scalp electrodes at the time instant k, vector $x_k \in \mathbb{R}^{n \times 1}$ is the current density of each source (neural activity), being n the number of distributed sources inside the brain. Vector $\varepsilon_k \in \mathbb{R}^{d \times 1}$ is the observation noise, $f \in \mathbb{R}^{n \times 1}$ is a non-linear vectorial function of order m modeling the current density dynamics, vector $w_k \in \mathbb{R}^{p \times 1}$ holds the p parameters of the function f, $\eta_k \in \mathbb{R}^{n \times 1}$ is a random variable describing additive process noise, $g \in \mathbb{R}^{p \times 1}$ is a linear vectorial function of first order modeling dynamics of w_k, $\epsilon_k \in \mathbb{R}^{p \times 1}$ is the noise of the $p \in \mathbb{N}$ time varying parameters, and $M \in \mathbb{R}^{d \times n}$ is the Lead Field Matrix that relates current density inside the brain x_k to the measure set y_k. Lead field matrix can be computed as a quasi-static approximation of the Maxwell and Poisson's equations head models [5].

Thus, Eq. 1a is a discrete time measure, Eq.1c is the parameter evolution, and Eq. 1b is the state evolution of current density dynamics to be modeled by DNF, incorporating corticothalamic connectivity and thalamic nonlinearity [3,6]:

$$\left(\frac{1}{\gamma_a^2} \frac{\partial^2}{\partial t^2} + \frac{2}{\gamma_a} \frac{\partial}{\partial t} + 1 - r_a^2 \nabla^2 \right) x(r, t) = q(\rho) \tag{2}$$

where $\gamma_a \in \mathbb{R}$ is the mean decay rate, $r_a \in \mathbb{R}$ is the mean range of axons a, and $q_{(r)} = q_{max} / (1 + \exp((-\rho_{(r)} - \theta)/\sigma))$, being $q_{(r)}$ the r-th element of vector $q \in \mathbb{R}^{n \times 1}$, that is the mean firing rate of excitatory and inhibitory neurons. Vector q is non-linear related to mean potentials $\rho \in \mathbb{R}^{n \times 1}$, being $\theta \in \mathbb{R}^+$ the mean firing threshold, $\sigma \in \mathbb{R}^+$ its standard deviation, and $q_{max} \in \mathbb{R}^+$ the maximum firing rate.

For simulation of EEG data, we use a discretized version of continuous time model in Eq. 2 as follows:

$$f(x_{k-1}, \ldots, x_{k-2}) = A_1 x_{k-1} + A_2 x_{k-1}^2 + A_3 x_{k-1}^3 + A_4 x_{k-2} + A_5 x_{k-\tau}, \quad (3)$$

where $A_1 = a_1 I_n + b_1 L$, $A_2 = a_2 I_n$, $A_3 = a_3 I_n$, $A_4 = a_4 I_n$, and $A_5 = a_5$, $I_n \in \mathbb{R}^{n \times n}$ is the identity matrix, $L \in \mathbb{R}^{n \times n}$ is the spatial Laplacian matrix holding all spatial interactions among sources, and $\tau \in \mathbb{R}^+$ is a delayed feedback.

2.2   Neural Activity Estimation within Inverse Problem Framework

As a concrete solution of the dynamic inverse problem, estimation of both neural activity, x_k, and discrete non-linear parameters, w_k, can be formulated from Eqs. (1a) to (1c) in the form of the following optimization task:

$$\underset{x_k, w_k}{\text{minimize}} \quad \| P(y_k - M x_k) \|^2 \quad (4)$$

$$\text{subject to} \quad \| Q(x_k - f(\hat{x}_{k-1}, \cdots, \hat{x}_{k-m}, w_k)) \|^2 = 0,$$

$$\| R(w_k - g(\hat{w}_{k-1})) \|^2 = 0.$$

where the state estimation $\hat{x}_{k-i}$ is done at k-i step, $\hat{w}_{k-1}$ is the parameter also estimated at the k-1 step, and $P \in \mathbb{R}^{d \times d}$, $Q \in \mathbb{R}^{n \times n}$, and $R \in \mathbb{R}^{p \times p}$ represent weighting matrices that are related to the noise covariance matrices of the measure set y_k, state, and parameter equations, respectively. Also, the following equivalent representation of the norm $\| \cdot \|$ is considered: $\| P(a-b) \|^2 = (a-b)^\top P^\top P(a-b)$.

Providing all weighting matrices allow non-uniform minimization, the regularized functional is obtained from Eq. 4 in terms of the following functional:

$$\Phi(x_k, w_k, \lambda, \gamma) = \| P(y_k - M x_k) \|^2$$
$$+ \lambda \| Q(x_k - f(\hat{x}_{k-1}, \cdots, \hat{x}_{k-m}, w_k)) \|^2$$
$$+ \gamma \| R(w_k - g(\hat{w}_{k-1})) \|^2, \quad (5)$$

where $\lambda \in \mathbb{R}^+$ and $\gamma \in \mathbb{R}^+$ are the regularization parameters ruling minimization of each functional term. For the simplification sake, the aforementioned relationship between weighting and the covariance matrices are redefined as follows: The covariance matrix the of scalp measures, $\Sigma = (P^\top P)^{-1}$, with $\Sigma \in \mathbb{R}^{d \times d}$, the state covariance, $\Lambda = (Q^\top Q)^{-1} / \lambda$, with $\Lambda \in \mathbb{R}^{n \times n}$, and the parameter noise covariance $\Gamma = (R^\top R)^{-1} / \gamma$, with $\Gamma \in \mathbb{R}^{p \times p}$.

The above multivariate optimization task is solved by iteratively optimizing one variable at a time, while remaining variables are kept fixed [4]. Consequently, given a known state estimation $\hat{x}_k$, estimation $\hat{w}_k$ is in the form: $\hat{w}_k = (G_k^\top \Lambda^{-1} G_k + \Gamma^{-1})^{-1} (G_k^\top \Lambda^{-1} \hat{x}_k + \Gamma^{-1} g(\hat{w}_{k-1}))$. Likewise, provided the known parameter set, $\widehat{w}_k$, estimation of the current density vector $\widehat{x}_k$ is also performed by minimizing functional in Eq. 5 with respect to x_k. In the end, we get the following estimation $\widehat{x}_k$:

$$\widehat{x}_k = \left(I_n - \Lambda M^\top \left(M \Lambda M^\top + \Sigma\right)^{-1} M\right) \left(\Lambda M^\top \Sigma^{-1} y_k + f\left(\widehat{x}_{k-1}, \ldots, \widehat{x}_{k-m}, \widehat{w}_k\right)\right)$$

Overall, the dual iterative estimation of $\widehat{w}_k$ and $\widehat{x}_k$ requires only some inverse calculations sizing $p \times p$ and $d \times d$, respectively, making this method suitable for tractable computational implementation of neural activity.

2.3   Inverse Problem with Dynamic Constraints

As dynamic constraints in the inverse problem solution, we consider the following two approaches based on the DNF model:

– *Linear model* that does not consider the delayed state as consequence of the introduced by the extra-cortical loop nor non-linear terms, i.e., $a_2 = a_3 = 0$:

$$f\left(x_{k-1}, x_{k-2} w\right) = A_1 x_{k-1} + A_4 x_{k-2}, \qquad (6)$$

  where the parameter set w ($p=3$) is assumed, i.e., $w^\top = \begin{bmatrix} a_1 & b_1 & a_4 \end{bmatrix}$.
– *Non-linear model* that includes the non linear terms of Eq. 3, but it ignores the extra-cortical neural feedback (higher order lag):

$$f\left(x_{k-1}, x_{k-2}, w\right) = A_1 x_{k-1} + A_2 x_{k-1}^2 + A_3 x_{k-1}^3 + A_4 x_{k-2}, \qquad (7)$$

  where $A_1 = a_1 I_n + b_1 L$, $A_2 = a_2 I_n$, $A_3 = a_3 I_n$, and $A_4 = a_4 I_n \in \mathbb{R}^{n \times n}$, being I_n the identity matrix and $L \in \mathbb{R}^{n \times n}$ a spatial Laplacian matrix holding all spatial interactions among sources [4]. In this case, the set of parameters w with $p=5$ is determined, namely, $w^\top = \begin{bmatrix} a_1 & b_1 & a_2 & a_3 & a_4 \end{bmatrix}$.

3   Experimental Set-Up

A common approach to assess the EEG inverse solution is the use of simulated EEG recordings where brain activity is known, so that estimation quality can be objectively validated. To this end, we simulate a set of EEG recordings lasting $1\,s$ at sample rate of $250\,Hz$. So as to to represent the simulated brain activity time series, the following discrete version of the DNF model is used:

$$x_k = A_1 x_{k-1} + A_2 x_{k-1}^2 + A_3 x_{k-1}^3 + A_4 x_{k-2} + A_5 x_{k-\tau} + \eta_k, \qquad (8)$$

where the parameter values are fixed to be as $\tau=20$, $a_1 = 1.0628$, $b_1 = -0.12$, $a_2 = 0.000143$, $a_3 = -0.000286$, $a_4 = -0.42857$, $a_5 = 0.008$, and $\|\eta_k\| \le 0.05$. Besides, to simulate normal and pathological states, value a_1 ranges from 1.0628 to 1.3, while a_4 from -0.428 to -1, at sample $k=125$ ($t=0.5\,s$). On the other hand, to get unbiased evaluation of IRA, different activity is simulated avoiding the generation model in Eq. 8. Instead, activity is simulated as a damped sinusoidal signal with $10\,Hz$ central frequency. Additionally, to obtain the simulated EEG $\boldsymbol{y}_k$ for a known activity $\boldsymbol{x}_k$, the current density of sources is multiplied by the lead field matrix in the form: $\boldsymbol{y}_k = \boldsymbol{M}\boldsymbol{x}_k + \boldsymbol{\epsilon}_k$, where $\boldsymbol{\epsilon}_k$ is set to achieve the following values of Signal-to-Noise Ratio (SNR): 30, 20, and 10 dB.

Here, IRA performance is assessed for several noise conditions under either linear or non-linear dynamic constraints. Obtained results are compared with both the LORETA and the baseline Kalman filter-based version given in [7] in terms of the following error measures:

$$\epsilon_r = \|\boldsymbol{M}\widehat{\boldsymbol{X}} - \boldsymbol{Y}\|_F^2; \ \epsilon_s = \|\widehat{\boldsymbol{X}} - \boldsymbol{X}\|_F^2; \ \epsilon_p = \|\widehat{\boldsymbol{X}} - \sum_{i \in \mathrm{rank}(\boldsymbol{M})} \langle \boldsymbol{X}, \boldsymbol{v}_i \rangle \boldsymbol{v}_i\|_F^2$$

where $\|\cdot\|_F$ is the Frobenious norm, $\boldsymbol{v}_i \in \mathbb{R}^{n \times 1}$ is the i-th right singular vector of matrix $\boldsymbol{M}$, matrices $\boldsymbol{X} \in \mathbb{R}^{n \times K}$ and $\boldsymbol{Y} \in \mathbb{R}^{n \times K}$ hold measured activity on dipoles and the sensor measures for all time instants, respectively. It must be quoted that the value $\varepsilon_p \in \mathbb{R}^+$ describes the error relating to the recoverable solution, i.e., to that solution part embedded on the row space spanned by matrix $\boldsymbol{M}$ and that is assumed mathematically the best possible reconstruction.

The testing head structure assumes $d=34$ and $n=5000$, where all sources are placed on the tessellated cortex surface and are perpendicular to it. Lead field matrix computation is carried out using the Boundary Element Method approach [5]. In practice, accuracy achieved by reconstruction methods is directly influenced by the used number of scalp electrodes. However, accuracy increases until 128-256 electrodes. From this point onwards, the use of additional electrodes should not increase reconstruction accuracy since supplied SNR gets worse. We use a very low dimensional model consisting in $d=34$ electrodes. So, reducing further the electrode number will worsen accuracy.

In the sequel, we make the following assumptions: during calculation of observation noise covariance, the measured noise is independent among sensors, i.e., $\boldsymbol{P} = \boldsymbol{I}_d$. In matrix $\boldsymbol{\Lambda}$, measured noise in each source is related to the noise coming from its respective neighbors. Thus, the weighting matrix $\boldsymbol{Q}$ is fixed to be as the Laplacian matrix containing all spatial relationships among sources, as suggested in [7]. Lastly, the weighting matrix $\boldsymbol{R}$ is set to represent independence among the p model parameters, i.e., $\boldsymbol{R} = \boldsymbol{I}_p$. Another aspect to be considered is the appropriate choice of the hyper-parameter values, λ and γ, that highly influences estimation quality. Specifically, estimation of covariance matrices $\boldsymbol{\Lambda}$ and $\boldsymbol{\Gamma}$ strongly depends on the fixed regularization parameters. To this end, several methods can be used, including L-curve, General Cross-Validation (GCV), and the Akaike's Bayesian Information Criterion, as described in [8]. Here, we calculate hyper-parameter values using the following minimization function:

$$\underset{\lambda, \gamma}{\arg\min} \frac{\|\boldsymbol{M}\boldsymbol{x}_k(\lambda, \gamma) - \boldsymbol{y}_x\|^2}{\mathrm{tr}(\boldsymbol{I}_n - \boldsymbol{M}\boldsymbol{T}(\lambda))^2}$$

where $\boldsymbol{T}(\lambda)$ is the inverse operator defined as $\boldsymbol{T}(\lambda) = \boldsymbol{\Lambda}\boldsymbol{M}^\top(\boldsymbol{M}\boldsymbol{\Lambda}\boldsymbol{M}^\top + \boldsymbol{\Sigma})^{-1}$.

4   Results

Fig. 1 displays the activity of current dipoles placed around the sensorimotor area and the achieved simulated activity reconstruction using each considered method for several noise conditions. The activity is generated using the DNF-based model shown in Eq. 8.

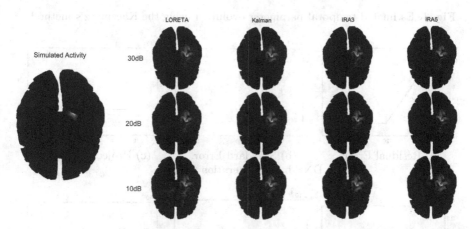

Fig. 1. Comparison of neural activity estimation for several noise conditions around the sensorimotor area, simulated by DNF-based generation model

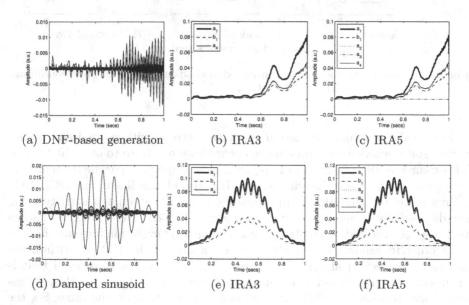

Fig. 2. Activity on the scalp EEG and temporal evolution of the DNF-based model parameters estimated using IRA

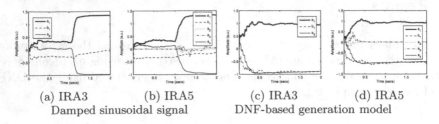

(a) IRA3   (b) IRA5   (c) IRA3   (d) IRA5
Damped sinusoidal signal   DNF-based generation model

Fig. 3. Estimated temporal parameter evolution using the Kaczmarz's method

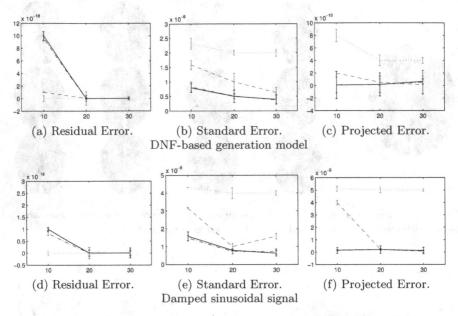

(a) Residual Error.   (b) Standard Error.   (c) Projected Error.
DNF-based generation model

(d) Residual Error.   (e) Standard Error.   (f) Projected Error.
Damped sinusoidal signal

Fig. 4. Performance for each of the considered methods under several noise conditions.
− · −IRA5 —IRA3 − −KAL · · · LORETA

Fig. 2a shows a concrete case of simulated activity with normal and pathological states where an abrupt change occurs from one state to other at $t=0.5\ s$. This change is also observed in Figs. 2b and 2c displaying parameter temporal evolution for the linear (herein, termed IRA3) and non linear models (IRA5) estimated by the IRA algorithm. Likewise, in case of simulated activity using the damped sine, simulated activity (see Fig. 2d) holds several smooth changes that are also reflected by similar parameter temporal evolution for both models.

In order to validate model identification ability of the IRA algorithm, the Kaczmarz projection algorithm of parameter estimation is considered that is an iterative approximated solution of linear equation systems. To this, we assume that dynamics of EEG data and neural generators are the same. So, the Kaczmarz algorithm is applied to the channel with the biggest variance estimated by either estimation linear or non-linear model. It must be quoted that

Kaczmarz-based parameter estimation is carried out apart from neural activity reconstruction using DNF-based (see Figs. 3a and 3b) or Damped sinusoidal (Figs. 3c and 3d) simulated activities.

As a result, for both cases of simulated activity, Fig. 4 shows obtained error measures varying noise condition. The results obtained for the standard error are statistically significant for every SNR value. On the other hand, for residual and projected errors, the results are statistically significant only for an SNR of 10 dB.

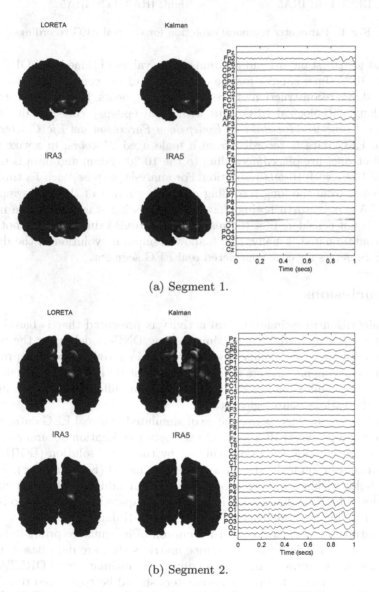

(a) Segment 1.

(b) Segment 2.

Fig. 5. Reconstruction of the brain activity for two real EEG recordings

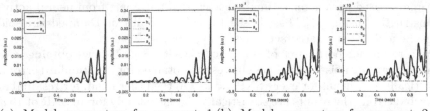

(a) Model parameters for segment 1. (b) Model parameters for segment 2.
Right: IRA3. Left: IRA5.   Right: IRA3. Left: IRA5.

Fig. 6. Parameter temporal evolution for the real EEG recordings

Note that for the standard and residual errors, values obtained by LORETA are shown 1×10^{-3} times smaller for the clarity sake of the comparison.

Fig. 5 shows reconstruction of two real EEG segments. Recordings are clipped from a long pathological EEG data with focal epilepsy collected during routine clinical practice (*Instituto de Epilepsia y Parkinson del Eje Cafetero from Pereira*). EEG data is recorded from a male aged 24 years, in awake resting state. Electrodes are placed according to the 10-20 system and data is sampled at rate of 1 *kHz* with 16-*bits*-resolution. For analysis purpose, each 1 *s* time series is segmented from the long recording at the beginning of the ictal event, that is at $t=0.5\,s$. It is worth noting that a preprocessing stage to remove noise or artifacts is not considered for the real EEG recordings since they are not significantly contaminated. Finally, Fig. 6 shows temporal evolution of the dynamic model parameters for each considered real EEG segment.

5   Conclusions

A new algorithm to estimate neural activity is presented that is based on the assumption that by introducing a time-varying DNF-model under the concrete regularization approach quality of estimated brain activity should improve. Besides, time-varying parameters needed to represent DNF-models allow detecting those brain dynamics changes clearly indicating on different transition intervals between the various neurological states.

Based on the obtained performance of simulated and real EEG data, we conclude that the proposed IRA method gets better estimation accuracy of neural activity reconstruction than one achieved by the static solution (LORETA) or the dynamic solution including a time-invariant model (Kalman filter). This advantage holds within a wide span of SNR values. In addition, the computational cost achieved by IRA algorithm is lower when compared to other methods where the model parameters are calculated off-line, i.e., Kalman filter.

As a future work, we will consider inclusion of informative priors in the estimation model. That is, using covariance matrices that are data-based, but not on pre-structured approaches as minimum norm estimates or LORETA. Also, a more complex model to update parameters should be considered to allow the method accurately identifying systems generating neural activity.

Acknowledgments. This research was supported by COLCIENCIAS project Evaluación asistida de potenciales evocados cognitivos como marcador del transtorno por déficit de atención e hiperactividad (TDAH) and Programa Nacional de Formacion de Investigadores "GENERACION DEL BICENTENARIO", 2011.

References

1. Giraldo, E., den Dekker, A.J., Castellanos-Dominguez, G.: Estimation of dynamic neural activity using a Kalman filter approach based on physiological models. In: 32nd Annual International Conference of the IEEE Engineering in Medicine and Biology Society, EMBC 2010, Buenos Aires, Argentina, September 1-5 (2010)
2. Plummer, C., Simon-Harvey, A., Cook, M.: Eeg source localization in focal epilepsy: Where are we now? Epilepsy 49(2), 201–218 (2008)
3. Kim, J.W., Shin, H.B., Robinson, P.A.: Compact continuum brain model for human electroencephalogram. In: Society of Photo-Optical Instrumentation Engineers (SPIE) Conference Series, Society of Photo-Optical Instrumentation Engineers (SPIE) Conference Series, vol. 6802 (2007), doi:10.1117/12.759005
4. Giraldo, E., Castaño-Candamil, J.S., Castellanos-Dominguez, C.G.: A Weighted Dynamic Inverse Problem for Electroencephalographic Current Density Reconstruction. In: 6th International IEEE EMBS Conference on Neural Engineering, San Diego, California, November 6-8 (2013)
5. Hallez, H., Vanrumste, B., Grech, R., Muscat, J., Clercq, W., Velgut, A., D'Asseler, Y., Camilleri, K., Fabri, S., Van Huffel, S., Lemahieu, I.: Review on solving the forward problem in eeg source analysis. Journal of NeuroEngineering and Rehabilitation 4(46), 101–113 (2007)
6. Connors, W., Trappenberg, T.: Improved path integration using a modified weight combination method. Cognitive Computation 5(3), 295–306 (2013), doi:10.1007/s12559-013-9209-0
7. Barton, M., Robinson, P., Kumar, S., Galka, A., Durrant-White, H., Guivant, J., Ozaki, T.: Evaluating the performance of kalman-filter-based eeg source localization. IEEE Transactions on Biomedical Engineering 56(1), 435–453 (2009)
8. Grech, R., Tracey, C., Muscat, J., Camilleri, K., Fabri, S., Zervakis, M., Xanthoupoulos, P., Sakkalis, V., Vanrumste, B.: Review on solving the inverse problem in eeg source analysis. Journal of NeuroEngineering and Rehabilitation 5(25), 792–800 (2008)

Supervised Brain Tissue Segmentation Using a Spatially Enhanced Similarity Metric

D. Cárdenas-Peña[✉], M. Orbes-Arteaga, and G. Castellanos-Dominguez[1]

Signal Processing and Recognition Group, Universidad Nacional de Colombia,
Manizales, Colombia
{dcardenasp,morbesa,cgcastellanosd}@unal.edu.co

Abstract. Many medical applications commonly make use of brain magnetic resonance images (MRI) as an information source since they provide a non-invasive view of the head morphology and functionality. Such information is given by the properties of head structures, which are extracted using segmentation techniques. Among them, multi-atlas-based methodologies are the most popular, allowing to consider prior spatial information about the distribution of brain structures. These approaches rely on a non-linear mapping of the information of the most relevant atlases to a query image. Nevertheless, methodology effectiveness is highly dependent on the mapping function and the atlas relevance criterion, being both of them based on the selection of an MRI similarity metric. Here, a new spatially weighting measure is proposed to enhance the multi-atlas-based segmentation results. The proposal is tested in an MRI segmentation database for state-of-the-art image metrics as means squares, histogram correlation coefficient, normalized mutual information, and neighborhood cross-correlation and compared against other spatial combination approaches. Achieved results show that our proposal outperforms baseline methods, providing a more suitable atlas selection.

Keywords: Magnetic resonance imaging · Image similarity metric · Multi-atlas segmentation · Template selection

1 Introduction

Segmentation of brain structures using magnetic resonance images (MRI) have been used in several medical applications, as the pathology progression analysis [1] and brain mapping [2]. Thus, many automatic approaches have been proposed, being the atlas-based ones the most employed. In these approaches, *a priori* knowledge about the structures of interest, e.g., shape and intensity distribution, can be better propagated from *atlases* to a query subject. To this end, the atlases are usually non-linearly mapped to the query image space so that they can serve as segmentation guiding references.

Recently, the use of multiple atlases has been proved to outperform single or averaged atlases, when each labeled image is correctly aligned with the target image independently, and their contributions are further combined [3]. The

© Springer International Publishing Switzerland 2015
J.M. Ferrández Vicente et al. (Eds.): IWINAC 2015, Part I, LNCS 9107, pp. 398–407, 2015.
DOI: 10.1007/978-3-319-18914-7_42

most widely used combination approach is the majority voting (also known as label voting or decision fusion) since it leads to fast implementation and accurate performance whenever the atlases are suitable selected. The majority voting strategy assigns to each voxel the label that most atlases agree. For example, the iterative STAPLE algorithm that linearly weighs each atlas according to its performance using the expectation maximization [4]. In [5], a similar approach is discussed that allows combining atlases at fine scales by weighting all atlases locally. Although presented results show that their proposal outperforms the global atlas weighting in segmentation accuracy, however, the mapping or registration of all atlases to a single query subject may become impractical for large image sets. Mostly, because the computational cost increases linearly with the number of atlases. Also, if the training set holds heterogeneous population, the achieved results can be biased because of anatomically unrepresentative images [6].

In order to overcome the above issues, atlas selection approaches are also included in the pipeline so that only the most appropriate candidate segmentations, for a given subject, are propagated and combined to provide the outcome. Moreover, the selection criterion is usually based on an introduced image similarity metric, being one the most popular the mutual information between query and atlas images. However, since the measure is assessed globally, it is biased towards large regions (as the background) instead of the small relevant structures, e.g. basal ganglia. In an attempt to cope with this restriction, structure-wise atlas selection is suggested in [7] to segment the brain MRIs based on the highest local mutual information. Also, an adaptive method for a local combination is proposed so that a subset of templates and their weighting are estimated independently at image localities [8]. The main drawback of these approaches lies in the requirement of a deformable registration stage measuring the image similarity for all atlases. That procedure is computationally much more expensive than linearly mapping all the images into a common reference space.

Bearing in mind all described above constraints, we propose a new spatially weighting procedure of the well-known image metrics for supporting the atlas selection within a multi-atlas-based segmentation scheme. Specifically, we study the mean squares (MS), histogram correlation coefficient (HCC), normalized mutual information (NMI), and neighborhood cross-correlation (NCC). Our approach computes independently the metric at regular image partitions; then all partition similarity values are linearly combined to get a single similarity outcome. The combination parameters are properly tuned to match the optimal atlas selection from a pre-labeled image set, being extracted by an offline exhaustive search. Our new linear combination scheme is compared with the global metric assessment, and with two other state-of-the-art combination approaches, for the subcortical brain MRI segmentation task. Carried out experiments show that our proposal outperforms the selection results for all considered metrics. In this paper, image similarity metrics for atlas selection and linear combination criteria are initially introduced. Then, all carried out experiments to evaluate the effectiveness of the metrics for atlas selection are described and discussed. Finally, some concluding remarks and future work are provided.

2   Materials and Methods

Let $\mathcal{X}=\{\boldsymbol{X}^n, \boldsymbol{L}^n:n=1,\ldots,N\}$ be a labeled MRI dataset holding N pairs of segmented images, where $\boldsymbol{X}^n=\{x_r^n\in\mathbb{R}:r\in\Omega\}$ is the n-th MR image, the value r indexes the spatial elements, and the matrix $\boldsymbol{L}^n=\{l_r^n\in[1,C]:r\in\Omega\}$ is the provided image segmentation into $C\in\mathbb{N}$ classes, which for 3D volumes holds dimension $\Omega=\mathbb{R}^{T_a\times T_s\times T_c}$, with $\{T_a, T_s, T_c\}$ as the Axial, Sagittal, and Coronal real-valued sizes, respectively.

2.1   Image Similarity Metrics

The similarity between a given image pair, $\{\boldsymbol{X}^n, \boldsymbol{X}^m\}$, can be assessed by using one of the following widely employed metrics:

Mean Squares (MS): This metric that is based on the average square difference along the space is embedded into a Gaussian kernel function, yielding the following bounded similarity measure:

$$s\{\boldsymbol{X}^n, \boldsymbol{X}^m\} = \exp\left\{-\frac{1}{2\sigma^2}\mathbb{E}\left\{(x_r^n - x_r^m)^2 : \forall r \in \Omega\right\}\right\} \in [0,1]; \tag{1}$$

where $\sigma\in\mathbb{R}^+$ is the kernel bandwidth. Notation $\mathbb{E}\{\cdot\}$ stands for the expectation operator.

Histogram Correlation Coefficient (HCC): This metric calculates similarity between image histograms as follows:

$$s\{\boldsymbol{X}^n, \boldsymbol{X}^m\} = \frac{\mathbb{E}\{h(x_r^n, x_s^m)(x_r^n x_s^m - \bar{x}^n\bar{x}^m) : \forall r, s \in \Omega\}}{\mathbb{E}\{h(x_r^n)(x_r^n - \bar{x}^n)^2\}\mathbb{E}\{h(x_r^m)(x_r^m - \bar{x}^m)^2\}} \in [0,1] \tag{2}$$

where $h(x^n, x^m)\in\mathbb{R}^+$ is the joint histogram between both input images, and $\bar{x}^v\in\mathbb{R}$, with $v\in\{m, n\}$, is the average intensity of the respective input image $\boldsymbol{X}^v$.

Normalized Mutual Information (NMI): This similarity value measures the normalized mutual information of a couple of images as:

$$s\{\boldsymbol{X}^n, \boldsymbol{X}^m\} = \frac{\mathbb{H}\{\boldsymbol{X}^n\} + \mathbb{H}\{\boldsymbol{X}^m\}}{\mathbb{H}\{\boldsymbol{X}^n, \boldsymbol{X}^m\}} - 1 \in [0,1] \tag{3}$$

where notation $\mathbb{H}\{\boldsymbol{X}^n, \boldsymbol{X}^m\}$ stands for the joint entropy between $\boldsymbol{X}^n$ and $\boldsymbol{X}^m$.

Neighborhood Cross-Correlation (NCC): This metric is widely used within the Advanced Normalization Tools (ANTs) framework, and computes the normalized cross correlation of voxel neighborhoods between two images [9]:

$$s\{\boldsymbol{X}^n, \boldsymbol{X}^n\} = \frac{\mathbb{E}\left\{(x_s^n - \bar{x}_s^n)^2 (x_s^m - \bar{x}_s^m)^2 : \forall s \in \Omega\right\}}{\mathbb{E}\left\{(x_s^n - \bar{x}_s^n)^2\right\}\mathbb{E}\left\{(x_s^m - \bar{x}_s^m)^2\right\}} \in [0,1] \tag{4}$$

where $x_s^n \in \mathbb{R}^{q \times q \times q}$ is the set of intensity levels in a q-sized neighborhood around the s-th voxel of the image X^n, and $\bar{x}_s^n = \mathbb{E}\{x_s^n : \forall s \in \Omega\}$.

It is worth noting that MS and NMI measures are re-written from their original definition so that all of the above similarity metrics share the same interpretation. Namely, $s=0$ implies the complete mismatch between images, while $s=1$ – an absolute match achieved only if $X_n = X_m$.

2.2 Spatial Enhancement of Image Metrics

Since all studied metrics are computed over the whole image, they do not account for local content similarities. Therefore, these measures are biased towards the large similar regions, as the background, masking the relationship between common image structures. Besides, those similarities lack in robustness against artifacts. For instance, the intensity inhomogeneity (being a low-frequency artifact) changes the image intensity distribution along the space. The most common approach to overcome this issue is to compute the metrics at all local regions, which should be further combined adequately into a single metric value.

To this end, the image X is split into P different regular blocks, $\{\Omega_p : p \in P\}$. Hence, each image is seen as a set of non-overlapped blocks $X = \{\Xi_p \in \mathbb{R}^{\rho_a \times \rho_s \times \rho_c}\}$, with $P = \prod_v P_v$, $\rho_v = T_v/P_v$, and P_v the number of partitions along the axis v ($v \in \{a, s, c\}$). Consequently, the following P-dimensional vector of metrics holding each the block-wise similarity value is obtained:

$$s\{X^n, X^m\} = \{s_p^{n,m} = s\{\Xi_p^n, \Xi_p^m\}; \forall p \in [1, P]\}$$

With the aim of building a new bounded scalar similarity metric, ζ, we make use of a linear combination of the elemental block-wise measures, namely,

$$\zeta^{n,m} = w^\top s\{X^n, X^m\}, \ \zeta^{n,m} \in \mathbb{R}[0, 1]$$

where $w = \{w_p\} \in \mathbb{R}^P$ is the combination vector that is subject to $\sum_{p \in P} w_p = 1$.

Since each weight has to account mostly for the influence of the corresponding region on the resulting metric $\zeta^{n,m}$, we assume the vector w to be dependent on the partition size. In the case of the equally-sized blocks, the combination vector is computed as follows [10]:

$$w_p = \frac{\rho_a \rho_s \rho_c}{T_a T_s T_c} = \frac{1}{P}, \tag{5}$$

As a result, the combination vector estimated in Eq. 5 leads to the plain averaged block distance, that is, $\zeta^{n,m} = \mathbb{E}\{s\{\Xi_p^n, \Xi_p^m : \forall p \in [1, P]\}\}$. In practice, each block holds a different amount of information depending on its content or its relevance to the task at hand. Regarding this, the contribution of each block can be achieved as its average intensity variance:

$$w_p = \frac{1}{\omega}\mathbb{E}\{\text{var}\{\Xi_p^n\}; \forall n \in [1, N]\} \tag{6}$$

where $\omega \in \mathbb{R}^+$ is the normalization factor.

2.3 Supervised Image Metric Learning

Basically, we are looking for a metric supporting the Atlas selection within the multi-atlas-based segmentation task. Therefore, we propose to use the provided set of segmented images of the dataset $\mathcal{X}$ to learn the corresponding combination weights for improving the segmentation accuracy. For this purpose, the following similarity matrix $\boldsymbol{Z_w} = \{z_w^{n,m} \in \mathbb{R}^+ : m, n = 1 \ldots N\} \in \mathbb{R}^{N \times N}$ holding all pair-wise metric values is built as a function of the estimated combination weights:

$$z_w^{n,m} = \zeta\{\boldsymbol{X}^n, \boldsymbol{X}^m\} = \sum_{p=1}^{P} w_p s_p^{n,m}; \forall n, m \in [1, N], \tag{7}$$

where $\boldsymbol{S}_p = \{s_p^{n,m} : m, n = 1 \ldots N\}$ is the similarity matrix attained at the p-th block. Since all considered metrics are equivalent to bounded similarity measures, each of the $\boldsymbol{S}_p$ matrices becomes a positive definite symmetric (PDS) kernel matrix, as well as their linear combination $\boldsymbol{Z_w}$.

On the other hand, for the n-th image in the dataset $\mathcal{X}$, the vector of optimal atlas selection order $\boldsymbol{o}^n$ can be found by an exhaustive search as

$$\boldsymbol{o}^n = \{o_m^n = i \in \mathbb{N} : \|\boldsymbol{L}^n - \boldsymbol{L}^i\|_\Omega \geq \|\boldsymbol{L}^n - \boldsymbol{L}^j\|_\Omega : \forall i < j\} \tag{8}$$

where $\| \cdot \|_\Omega$ is a norm in the image domain. Then, a supervised PDS kernel matrix $\boldsymbol{K} = \{k^{n,m} = k^{m,n}\} \in \mathbb{R}^{N \times N}$ holding the best possible symmetric selection order is computed as

$$k^{n,m} = (o_m^n + o_n^m)/2N \tag{9}$$

Thus, the similarity metric ζ can be learned by finding the weights w_p maximizing the correlation between $\boldsymbol{Z_w}$ and the objective kernel matrix $\boldsymbol{K}$ as follows:

$$\max_{w} \frac{\langle \boldsymbol{Z}_w', \boldsymbol{K}' \rangle_F}{\|\boldsymbol{Z}_w'\|_F \|\boldsymbol{K}'\|_F}, \tag{10}$$

where $\langle \cdot, \cdot \rangle_F$ denotes the inner product and $| \cdot |$ the Frobenius norm, $\boldsymbol{Z}_w'$ and $\boldsymbol{K}'$ are the centered kernel matrices of $\boldsymbol{Z_w}$ and $\boldsymbol{K}$, given by $\boldsymbol{K}' = \boldsymbol{HKH}$ with $\boldsymbol{H} = [\boldsymbol{I} - \boldsymbol{1}\boldsymbol{1}^\top/N]$, and $\boldsymbol{1} \in \mathbb{R}^{N \times 1}$ is the all-ones vector.

Consequently, the solution within the optimization problem in Eq. 10 (known as the kernel centered alignment –KCA) for calculating $\boldsymbol{w}$ is given by [11]:

$$\boldsymbol{w} = \frac{\boldsymbol{A}^{-1}\boldsymbol{b}}{\|\boldsymbol{A}^{-1}\boldsymbol{b}\|} \tag{11}$$

$$\boldsymbol{A} = \{a_{pq} = \langle \boldsymbol{S}_p', \boldsymbol{S}_q' \rangle_F; \forall p, q \in [1, P]\} \in \mathbb{R}^{P \times P}$$

$$\boldsymbol{b} = \{b_p = \langle \boldsymbol{S}_p', \boldsymbol{K}' \rangle_F\} \in \mathbb{R}^{P \times 1}$$

3 Experimental Set-Up

In order to evaluate all studied metrics within the multi-atlas-based segmentation task, they are performed to select the most similar labeled images for estimating the segmentation of a query image. To this end, the majority voting scheme is considered for labeling each voxel since the segmentation quality is mostly dependant on the selection strategy. Additionally, the well-known Dice Index similarity is measured to evaluate the segmentation performance.

3.1 Database

Here, the dataset tested is the one used in the MICCAI 2012 *Multi-Atlas Labeling and Statistical Fusion* challenge[1] that is a subset of the Open Access Series of Imaging Studies (OASIS) database. This data collection holds T1-weighted structural MRI scans from 35 subjects (13 males and 22 females), aging from 18 to 90 years. Each 256×256×287-sized MRI volume has a voxel size of 1×1× 1 *mm*. All images were expertly labeled with 26 structures. Due to our research interest lies in Parkinson surgery, only the following structures are considered: hypothalamus (HYPO), amygdala (AMYG), putamen (PUT), caudate nucleus (CAUD), thalamus (THAL) and pallidum (PAL)Fig. 1 shows a sample of an image subject as well as its provided segmentation.

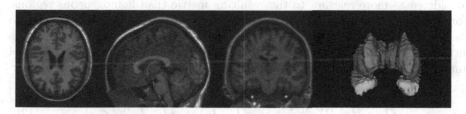

Fig. 1. Left to Right: Axial, Sagittal, and Coronal views as well as the ground-truth segmented structures

3.2 Image Preprocessing

For the sake of comparison within a single common space, all images are spatially normalized into the Talairach space. Thus, each image is rigidly aligned to the ICBM atlas (MNI305-template) allowing to extract the morphological feature set accurately from each considered image. To this end, the Advanced Normalization Tool (ANTS) is employed using a quaternion based mapping and the MI metric as parameters.

In order to perform the label propagation, every pre-labelled image is also spatially mapped into the query image spatial coordinates (*target space*) with a non-linear transformation so that query and atlas images match the best. Further, the registration procedure is performed using the ANTS tool having the following default parameters: elastic deformation as the mapping function (**Elast**), MI as the similarity metric, and 32-bins histograms for estimating the probability density functions. Lastly, to get a finer alignment, the registration

---

[1] https://masi.vuse.vanderbilt.edu/workshop2012

is performed at three sequential resolution levels: *i)* the coarsest alignment with a resolution of $1/8 \times$ *Original space*, and 100 iterations, *ii)* the middle resolution $1/4 \times$ *Original space* and 50 iterations, and *iii)* the finest deformation with a resolution of $1/2 \times$ *Original space* parameter and 25 iterations, the Gaussian regularization method is employed ($\sigma = 3$).

3.3   Metric Parameter Learning

For including spatial information, each similarity metric is assessed separately at different locations of the image. To this end, the MRI volumes are regularly partitioned into $P = 27$ blocks (3 partitions along each dimension). Then, the outcome metric is computed for all image pairs as a weighted linear combination of all local measures. Here, three different parameter tuning approaches are considered. The first one assumes that the contribution of each block to the similarity metric is equal for all of them. Therefore, the similarity metric corresponds to the averaging of the local measures. For the second approach, the weights are computed as the average intensity variance in each block as in Eq. 6. In this way, the shape differences on the brain structure intensity changes are considered as more relevant to the resulting metric than homogeneous regions. For the third approach, the weights are computed based on the contribution to the kernel centered alignment with respect to the objective kernel matrix K as introduced in Eq. 11. In this paper, $K \in [0, 1]^{N \times N}$ is built from the ordered label image similarity in the training set:

$$k_{nm} = (\mathcal{O}_{nm} + \mathcal{O}_{mn})/2N \tag{12a}$$
$$\mathcal{O}_n = \{i \in \{[1, N] - n\} : \varrho(\boldsymbol{L}_n, \boldsymbol{L}_i) \geq \varrho(\boldsymbol{L}_n, \boldsymbol{L}_j) \forall i < j\}, \tag{12b}$$

where $\varrho \in \mathbb{R}+$ is the known Dice Index similarity, defined as:

$$\varrho(\boldsymbol{L}, \boldsymbol{L}') = \mathbb{E}\left\{2\langle \boldsymbol{L}_c, \boldsymbol{L}'_c \rangle / (\|\boldsymbol{L}_c\|_1 + \|\boldsymbol{L}'_c\|_1) : \forall c \in [1, C]\right\} \tag{13}$$

Fig. 2 shows the 3D scatter plotting all resulting weights, where the coordinates of each element are the spatial location of the image partition while the color and size are directly proportional to the value. As shown in Fig 2a, the central image region has the highest variance. Anatomically, this partition corresponds to basal ganglia location having tissue structures with high variant shape and intensity. However, the scatter plot also shows a substantial amount of variance on the corners. This dispersion should be related to the partial presence of the scanned head and background, but not necessarily to the intensity dispersion. According with Fig. 2b to 2e, the weight distribution for all metrics exhibit the similar behavior. The corresponding weight to the central region is higher than other ones; i.e. the similarity metric assessed in this region is more correlated to the supervised kernel matrix than in boundary regions.

3.4   Evaluation of Similarity Metrics

We consider the leave-one-out validation scheme to evaluate the performance of the resulting metrics. In this case, all metrics are used to carry out an atlas

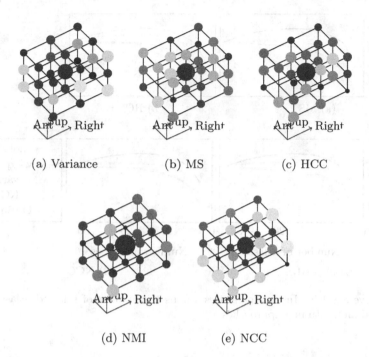

(a) Variance   (b) MS   (c) HCC

(d) NMI   (e) NCC

Fig. 2. Resulting weight distribution for the variance criterion and all considered metrics. Markers are located at the center of each partition. Color and size are directly proportional to weight parameter value.

selection task for the atlas-voting segmentation approach in the target image space. Finally, the metric performance is assessed with the Dice Index similarity.

Fig. 3 shows obtained results of multi-atlas segmentation obtained by each tested metric and using all templates selected within the common space. As seen, the achieved accuracy gets close to the global computation when all weights become equal. Although the accuracy obtained by the MS metric should tend to the one of equally-weighted case, in practice both outcomes differ because of computational accuracy issues, but their resulting selection curves are statistically similar. Moreover, the noise artifact produces a high-variability of intensity, leading to unsuitable weights and a biased metric. About the weights computation, the similarity metric slightly improves the accuracy in comparison to the global and equally-weighted metrics if the variance intensity is taken into account. On the other hand, our proposed approach achieves the highest accuracy with an optimal atlas selection for all tested metrics. The approach outperforms not only all other benchmark combination methods, but also the accuracy obtained with the whole data-set.

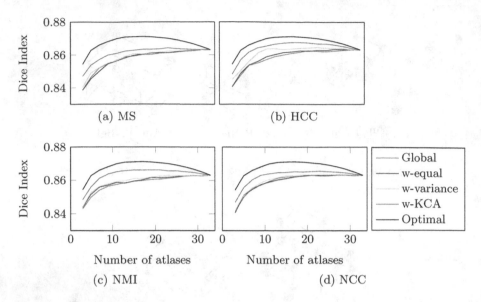

Fig. 3. Average Dice Index similarities versus the number of selected atlases for all considered metric tuning approaches

4   Discussion

We propose a new spatially weighting criterion to improve similarity metrics aiming to measure image correspondences to support atlas selection on a multi-atlas segmentation scheme. Our proposed measure outperforms the widely used equal and variance-based weighting on the tested similarity metrics.

In accordance with the obtained results, it is clear that the spatially weighted metrics outperform the global ones. However, the computation of weights becomes an important task. The equal weighting provides spatial information in terms of partition size, but the content inside each locality is not taken into account. Hence, it performs the worst among all weighting approaches.

With the purpose of capturing shape differences on brain structures, the variance criterion weighs each block according to the intensity variance on it. However, the background on the images biases the weights computation towards image edges, which contain just partially the scanned head. Moreover, the inherent noise reduces the performance of a variability-based criterion.

Meanwhile, we propose to use a supervised kernel matrix aiming to learn the combination weights for the similarity metric. In this sense, the weights are computed according to the contribution to the kernel centered alignment with respect to the supervised kernel matrix. An advantage of our proposal is that the closed form to the kernel alignment solution provides an easy implementation for weights computation, while the construction of supervised kernel matrix is carried out only once off-line. As a result, we assess a suitable image similarity metric as an atlas selection criterion. Therefore, the similarity-based atlas ranking correlates correctly with the segmentation accuracy for subcortical structures. Actually, the learned metric can provide a subset of atlases achieving a higher accuracy for the label propagation than the whole dataset provides.

Acknowledgments. This work was supported by *Programa Nacional de Formación de Investigadores "Generación del Bicentenario"*, 2011/2012, and the research project number 111056934461, both funded by COLCIENCIAS.

References

1. Bron, E.E., Cardenas-Pena, D., et al.: Standardized evaluation of algorithms for computer-aided diagnosis of dementia based on structural mri: The CADDementia challenge. NeuroImage (2015)
2. Joshi, S., Davis, B., Jomier, M., Gerig, G.: Unbiased diffeomorphic atlas construction for computational anatomy. NeuroImage 23(suppl. 1), S151–S60 (2004)
3. Rohlfing, T., Brandt, R., Menzel, R., Russakoff, D., Jr., M.C.: Quo vadis, atlas-based segmentation? In: Suri, J., Wilson, D., Laxminarayan, S. (eds.) Handbook of Biomedical Image Analysis. Topics in Biomedical Engineering International Book Series, pp. 435–486. Springer US (2005)
4. Warfield, S.K., Zou, K.H., Wells, W.M.: IEEE Transactions on Medical Imaging
5. Artaechevarria, X., Munoz-Barrutia, A., Ortiz do Solorzano, C. IEEE transactions on medical imaging
6. Aljabar, P., Heckemann, R.A., Hammers, A., Hajnal, J.V., Rueckert, D.: NeuroImage
7. Wu, M., Rosano, C., Lopez-Garcia, P., Carter, C.S., Aizenstein, H.J.: NeuroImage
8. van Rikxoort, E.M., Isgum, I., Arzhaeva, Y., Staring, M., Klein, S., Viergever, M.A., Pluim, J.P.W., van Ginneken, B.: Medical image analysis
9. Avants, B., Tustison, N., Song, G., Cook, P., Klein, A., Gee, J.: A reproducible evaluation of ants similarity metric performance in brain image registration. NeuroImage 54, 2033–2044 (2011) cited By 139
10. Studholme, C., Drapaca, C., Iordanova, B., Cardenas, V.: Deformation-Based Mapping of Volume Change From Serial Brain MRI in the Presence of Local Tissue Contrast Change, 626–639 (2006)
11. Cortes, C., Mohri, M., Rostamizadeh, A.: Algorithms for learning kernels based on centered alignment. The Journal of Machine Learning Research 13(1), 795–828 (2012)

iLU Preconditioning
of the Anisotropic-Finite-Difference Based
Solution for the EEG Forward Problem

E. Cuartas-Morales$^{(\boxtimes)}$, C. Daniel-Acosta, and G. Castellanos-Dominguez

Signal Processing and Recognition Group, Universidad Nacional de Colombia,
Bogptá, Colombia
ecuartasmo@unal.edu.co

Abstract. We investigate the use of the iLU preconditioning within the framework of the Anisotropic-Finite-Difference based Solution for the EEG Forward Problem. Provided the minimal error of representation, comparison of the convergence rate and computational cost is carried out for several competitive numerical solver combinations. From the testing on real data, we obtain that combination of the biconjugate gradient solver and incomplete LU factorization results in a numerical solution that outperforms the other considered approaches in terms of accuracy and computational cost. We validate this numerical solution combination against analytical spherical mode. Also, testing on realistic head models (with high anisotropic areas and heterogeneous tissue conductivities) shows high accuracy and low computational cost.

1 Introduction

There are several techniques for monitoring and extracting from the human brain more refined information allowing better results in clinical applications like medical treatment, surgery planning, or more generalized brain research tasks. Among those non-invasive techniques, the Magnetic Resonance imaging (MRI) or Computed Tomography (CT) have widely shown that its synergy with functional analysis techniques, particularly ElectroEncephaloGraphy (EEG), can overcome weakness of single modality analysis. Extracted multi-modal information is useful in diagnosis and preoperative stages of brain surgery, being usually the only suitable analysis tool due to the high risk of alternative surgical interventions [1]. The different tissues are segmented from neuroimages such MRI or CT. The MRI, CT scans conform a volume with a large number of slices in a series of two-dimensional images. Every slice must be registered in the same coordinate system in order to obtain a coherent three-dimensional volume. After the registration stage, the data set contains a gray scale volume of the head with the different tissues in it [2]. The general areas to be segmented from the volume are the scalp, where the EEG electrodes are placed, the skull, the cerebrum spinal fluid, the gray matter and the white matter, but, nowadays, several tissues are considered for the segmentation stage in order to obtain more realistic/acurate

© Springer International Publishing Switzerland 2015
J.M. Ferrández Vicente et al. (Eds.): IWINAC 2015, Part I, LNCS 9107, pp. 408–418, 2015.
DOI: 10.1007/978-3-319-18914-7_43

forward models [1]. Nowadays there are several toolboxes that handles the image processing stage with good results (SPM, FSL, FreeSurfer, Lony) [2]. As far as EEG is a meaningful signal, the electrical activity of a local group of neurons can be modeled as a current dipole [3]. Moreover, dipoles can be localized within the brain using a conductivity model of the human head volume (source localization problem). Therefore, accurate and realistic forward solver must be develop in order to improve the EEG source localization. There are several methodologies to solve the forward problem, each one having its own advantages and weaknesses depending on the necessity of the user, regarding to this, the most used solutions are: spherical models [4], Boundary Finite Elements (BEM), Finite Elements Method (FEM), Finite Difference Method (FDM) and Finite Volumetric Method [3]. Since image data (MRI,CT) are usually acquired in regular formats of digitalization (mostly, 1×1×1), we used FDM that can be straightforwardly adapted to the grid. The solution of the Poisson equation in a volume conductor medium (forward problem) using FDM numerical approximations derives in a large linear system with a ill-posed coefficient matrix. The numerical stability and computational burden of this type of problem highly depends on the conditional number of the coefficient matrix, therefore, in order to improve the numeric properties of the coefficient matrix, suitable preconditioners must be applied. In this sense, [5] introduce an anisotropic FDM formulation with a Successive Over Relaxation (SOR) solver, but, due to the stationary nature of the algorithm, the technique takes to much time to solve realistic head models, in contrast, [6] propose a fictitious domain data ordering allowing Fourier type preconditioning and obtaining fast and accurate realistic forward model calculations. In this work, we analyse several numerical solutions for the anisotropic FDM problem in the fictitious domain, finding a suitable preconditioning solver combination, allowing accurate and fast forward calculations in highly anisotropic and heterogeneous realistic head models.

2 Methods

2.1 Anisotropic-Finite-Difference Based Solution of the EEG Forward Problem

The forward problem consists in the calculation of the electrical potential field on the scalp surface (provided the geometry and electrical conductivity of the head volume) for a given position, orientation, and magnitude of the dipole current sources. Due to measured EEG/MEG frequencies are usually below $100\,Hz$ and electric and magnetic field time derivatives are typically much smaller than ohmic currents, we can neglect the nonstationary electromagnetic field terms [3]. Therefore, in order to determine the electrode potential field, $\boldsymbol{V} \in \mathbb{R}^m$ (being m the number of electrodes) generated by a brain current dipole with volumetric conductivity $\boldsymbol{\Sigma} \in \mathbb{R}^{3 \times 3}$, we make use of the quasi-static approximation of the Maxwell's and Poisson equations as follows:

$$\nabla \left(\boldsymbol{\Sigma}(\boldsymbol{r}) \nabla \boldsymbol{V}(\boldsymbol{r}) \right) = \nabla \boldsymbol{J}(\boldsymbol{r}), \tag{1}$$

where $J \in \mathbb{R}^m$ is the electric current density and $r \in \mathbb{R}^{3 \triangleleft}$ is the point of detection holding the direction and position dipole. $\Sigma(r)$ is the conductivity tensor with the direction-dependent conductivity that for isotropic conductivity becomes a diagonal matrix. For the anisotropic case, however, $\Sigma(r)$ varies according to the position in the anisotropic compartment. Namely, at the interface between two different tissue compartments (medium change), two boundary conditions take place. Particularly, if assuming all current leaving one compartment with conductivity $\sigma_1 \in \mathbb{R}^{3 \triangleleft 3}$ through the interface enters the neighboring compartment with conductivity $\sigma_2 \in \mathbb{R}^{3 \triangleleft 3}$, the current density at the head surface reads (termed *Neumann boundary condition*) [7]:

$$\begin{cases} J_1 \cdot e_n & = J_2 \cdot e_n \\ (\sigma_1 \nabla V_1) \cdot e_n & = (\sigma_2 \nabla V_2) \cdot e_n \end{cases} \tag{2}$$

where e_n is the normal component on the interface. Likewise, another boundary condition also holds for interfaces not connected with air, stating that by crossing the interface the potential cannot have discontinuities (*Dirichlet boundary condition*), i.e.: $V_1 = V_2$.

For numerical solving, Eq. 1 can be transformed into a set of linear equations modeling the volume conduction. For this end, we use a cubic grid in which each cube (or element) has a conductivity tensor that can vary between neighbouring elements, so that the directions of the anisotropy are changed along the coordinate system axes of the head model. In [5], an approach (termed *finite difference method in anisotropic media – aFDM*) is presented to handle anisotropic properties of tissues, where a finite difference formulation for the Laplace's equation is extended to the Poisson's equation (see Eq. 1) that is valid everywhere in a piecewise inhomogeneous anisotropic medium. Since image data (MRI,CT) are usually acquired in regular formats of digitalization (mostly, 1×1×1), aFDM can be straightforwardly adapted to the grid.

We make use of the *fictitous domain* S_j described in [8] to enclose the irregular volume conductor data in a rectangular grid with zero redundance with N (*discretization number*) partitions in each direction, allowing not only a direct data ordering, but also, a soft diagonal predominant, symmetric coefficient matrix. Finally, we use a set of finite-difference approximations of the spatial derivatives on the rectangular grid with a 19-point stencil with 8 voxels sharing the same vertex V^j, building a finite difference linear equation for the vertex V_i^j in the stencil S_j around V^j.

$$\sum_{i \in S_j}^{18} \sigma_i^j V_i^j - \left(\sum_{i \in S_j}^{18} \sigma_i^j \right) V^j = I \tag{3}$$

where $V_i^j \in \mathbb{R}$ is the scalar-valued potential at the i-th neighbor vertex of the j-th node in the stencil S_j. $I \in \mathbb{R}$ is the dipole current, and $\sigma_i^j \in \mathbb{R}$ are the coefficients depending on the conductivity tensor and the internode distance, which are fixed to ensure the Neumann and Dirichlet boundary conditions [9].

2.2   Anisotropic-Finite-Difference Linear System Solution

The finite difference formulation derived in Eq. 3 form a linear equation system (LES), $Ax=b$, with $A \in \mathbb{C}^{n \times n}$ nonsingular and $b \in \mathbb{C}^n$, with the following properties on the coefficient matrix A:

- The coefficient matrix is square and sparse with only 19 non zero entries per row. Matrix dimension becomes $N \times N \times N$, where N is the discretization number.
- The coefficients connecting the same pair of neighbouring voxels are identical, resulting in a symmetric matrix A having weak diagonal dominance.
- The LSE $Ax=b$ possesses infinite solutions differing only in an additive constant.

To date, several iterative solvers have been developed for a regular LES that can also be extended to the system matrix A holding the above-described restrictions. Yet, the choice of the solver depends to a large extent on two considerations: the convergence speed to achieve a given accuracy and the computational complexity of each iteration. The baseline stationary solver is the Successive Over-Relaxation (SOR) introducing a relaxation factor that reduces errors of succeeding approximations until all errors are within specified limit. This factor depends strongly upon the properties of the matrix, A, and its choice is not necessarily easy, resulting in algorithms that are not quite efficient in terms of their computational cost.

Regardless the solver, however, matrix convergence in finite-precision arithmetic remains a crucial issue. Namely, the more ill-conditioned the system matrix A (that is, the larger its condition number defined as $\|A\| \|A^{-1}\|$), the slower the convergence of the steepest descent method. To cope with this drawback, preconditioning techniques are introduced to improve the condition number of a nonsingular matrix, M, making the resulting condition number $M^{-1}A$ much smaller than in the original matrix A.

In practice, the combination of solvers with preconditioning procedure (*numerical solution combination* – NSC) based on incomplete LU (iLU) factorizations constitutes an effective class of methods for solving the sparse linear systems arising from the numerical approximation of partial differential equations. In this work, we used two different solvers (GMRES and BiCG-Stabilized) and different preconditioners (including Choletsky, LU, iLU and also the adapt isotropic circulant Fourier-Jacobi deffined by [6]) to compare different NSC's in terms of computational time and accuracy.

3   Experimental Set-Up

In order to test the effectiveness of the iLU preconditioning within the Anisotropic-Finite-Difference based Solution for the EEG Forward Problem, we compare its convergence rate and numerical stability to other competitive numerical solution combinations, for a given accuracy. Then, we validate the best iLU-based numerical solution against a concrete analytic method to demonstrate the efficiency of the proposed numerical approach in the highly heterogeneous anisotropic case. All tests are carried out using a 7-shell spherical head model with anisotropic skull and white matter compartments. Lastly, we also validate the iLU numerical solution in a realistic head model white anisotropic skull and white matter derived from high-resolution MRI data.

Multi Shell Anisotropic Spherical Model: The validating spherical head model is a 7-shell anisotropic skull and white matter layers [4]. The shells represent the scalp, the skull, the cerebral spinal fluid (CSF), gray matter (external and thalamic inner sphere) and white matter. We use the following external radius [m]: 0.092 (scalp), 0.084 (skull), 0.076 (CSF), 0.068 (GM), 0.050 (WM), 0.020 (GM); all tissues having conductivity values [S/m]: 0.33 (scalp), 0.018 (anisotropic skull with 1:10 radial/tangential ratio), 1.79 (CSF), 0.33 (brain), and 0.14 (anisotropic white matter with 9:1 radial/tangential ratio). For the anisotropic skull and white matter, we apply rotational transformation to the local coordinate system for reorienting the eigenvectors in a normal direction from the concentric spheres. Additionally, we carry out testing for 10 different values of image resolution in order to analyze the conditional number of the system matrix.

3.1 Testing of Convergence Rate and Computational Cost

We compare a set of Anisotropic-Finite-Difference based solutions that include combination of four widely-known preconditioners (Cholesky, LU, iLU, Fourier-Jacobi) together with two baseline non-stationary solvers (GMRES, BiCG-Stabilized) and the baseline stationary SOR solver (without any preconditioning). Besides, since the Cholesky preconditioner is just devoted to the GMRES solver and the Fourier Jacoby for BiCG solver, only these concrete combinations are considered, respectively. As a result, we get the following set of numerical solution combinations: *SOR, GMRES, GMRES–Cholesky, GMRES–LU, GMRES–iLU, BiCG, BiCG–LU, BiCG–iLU, BiCG–Fourier-Jacoby.* It must be noted that the used version of the BiCG is the stabilized one.

The convergence rate is calculated as the lowest number of iterations that each NSC at hand requires for reaching its minimal error of representation. Specifically, We calculate this amount as the residual error value assuming the more complex image resolution, that is, 6×6×6, mm. Another important aspect of comparison is the minimal amount of the residual set that each NSC must reach to make it workable in practical applications. Here, we fix this value as 10^{-13}. Numerical computation shows that the plain SOR gets the worse convergence since it needs 350 iterations to reach its lowest error of representation close to 1.8×10^{-7}; this amount remains far from the fixed minimal residual set. As seen in Fig. 1 showing the achieved outcomes of residual error as dependence of the testing iteration number, the GMRES algorithm performs a bit better than SOR, and BiCG improves even better. Besides, the use of either preconditioning, iLU or LU, leads to a less number of iterations, improving the NSC convergence. Moreover, the latter preconditioner improves the convergence

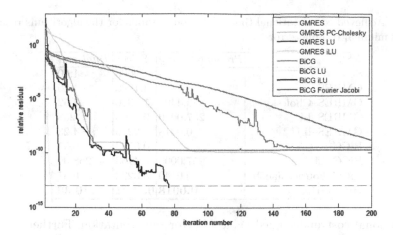

Fig. 1. Convergence rate of the compared NSC algorithms. Dashed line remarked in red stands for the fixed minimal residual set.

rate remarkably. In fact, the best NSC is the BiCG-Stabilized LU that reaches the needed residual value just after five iterations. Likewise, the Cholesky and Fourier Jacoby preconditioners improve the convergence rate of their corresponding solvers. Nevertheless, the use of preconditioning does not mean that the NSC algorithm should reach the fixed minimal amount of residual set error (dashed line remarked in red). Thus, the GMRES, BiCG, and BiCG–Fourier-Jacoby algorithms hang up on some intermediate values close to~10^{-10}. In turn, the GMRES–iLU, GMRES–Cholesky, and GMRES–LU converge towards a lower residual value (~10^{-12}); still, that amount is not enough. Certainly, the BICG–LU and BICG–iLU are the only algorithms to reach the fixed minimal error of representation.

On the other hand, the computational cost of each iteration turns out to be different in dependence on the used preconditioning approach. For having a common base time to compare computational cost units, we make the time unit as the one spent by the baseline stationary SOR solver. Thus, the time unit is 20.1949 s that is appraised in a workstation 8 core `Intel Xeon` CPU E5-2687W with 64Gb RAM, using the Matlab software environment. Table 1 shows the computational time that each NSC requires for reaching its intrinsical final condition stop. Thus, the second column displays the time that each preconditioning procedure spends. As expected, the LU is very much expensive than the other preconditioners, being the iLU the fastest one. The third column shows the time spent by the solver after preconditioning, where the GMRES employs more time that the BiCG. The total amount of time required by each NSC that is shown in the last column is the sum of the previous two columns. As a result, the most expensive NSC is the BiCG–LU, while the fastest – the BiCG–iLU (in as much as 860 times!). Although there are two NSC providing also small computational (BiCG and BiCG–Fourier Jacoby), they do not fit with needed minimal amount of residual error. As a consequence, the BiCG–iLU is the best NSC in terms of

Table 1. Achieved computational time. Notation * stands for the algorithms reaching the fixed minimal residual error

NSC	Preconditioning	Solver	Total cost
SOR	–	1.0000	1.00
GMRES	–	4.1998	4.20
GMRES–Cholesky	3.4400	1.0826	4.52
GMRES–LU	257.0000	0.5903	258
GMRES–iLU	0.0018	4.2279	4.23
BiCG	–	0.3543	0.35
BiCG–LU*	257.0000	0.4153	258.00
BiCG–Fourier Jacoby	0.0540	0.3223	0.37
BiCG–iLU*	**0.0018**	**0.2991**	**0.30**

computational cost under fixed value of error representation. Further, we test the BiCG–iLU for 10 different values of numerical resolution, where N is the number of divisions in one direction in the cubic fictitious domain (being N^3 the number of rows of the system squared matrix). As seen in Fig. 2a, the highest the resolution – the large the number of needed iterations. This fact may be explained by the conditional number that is computed for the studied range of numerical resolution. As shown in Fig. 2b, the conditional number grows exponentially when increasing the division number N. Therefore, for high resolutions, the coefficient matrix size becomes bigger and is most difficult to solve due to the obtained larger conditional number. However, the BiCG–iLU remains stable even under the largest conditional number 1.63×10^{14}, achieved for $N=61$.

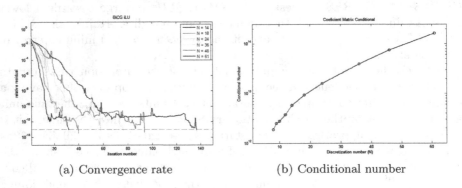

(a) Convergence rate   (b) Conditional number

Fig. 2. Conditional number and convergence rate of BiCG–iLU for different values of numerical resolution

3.2   Validation of BiCG–iLU Numerical Solution

We validate the concordance of the NSC to a suitable analytical solution. To this end, we make use of the spherical head model having $6 \times 6 \times 6, mm$ voxel size. Fig. 3 shows the Potential (μV) versus the electrode channel number computed for the proposed 7-layered spherical head model. The continuous line displays the BiCG–iLU employing aFDM while the circles show the analytical potentials calculated employing the multishell anisotropic spherical model described above. Both, the numerical and analytical, solutions match correctly in computing the potentials for a single source placed at the gray matter area in the normal disposition to the gray matter sphere. Here, the electrodes are placed in 6 different rings covering the entire surface of the scalp surface. Validation

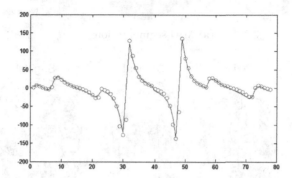

Fig. 3. Validation of BiCG–iLU numerical Vs analytical potentials on the scalp for a spherical head model

of BiCG–iLU is also carried out on realistic head models. Particularly, we the chosen NSC on realistic head models comprising an isotropic representation with a highly anisotropic head model. Image preprocessing is carried out using 3D Slicer built-in modules. The preprocessing steps included: MRI bias correction (N4 ITK MRI bias correction) and registration (BRAINS) for movement correction. We obtain detailed tissue models from the T1-weighted and TOF volumes by using the pipeline described in [10]. The model contain WM, GM, CSF, skull, eyes, muscle, fat, arteries, and skin. The arteries segmentation mask is processed to estimate the direction of blood flow, obtaining a normalized vector map describing the maximum anisotropy inside the arteries. The MRI segmentation holds 9 different tissues with different conductivity values $[S/m]$: (scalp = 0.33; fat = 0.4; muscle = 1.1112; skull = 0.020; eye = 0.0505; CSF = 1.538; GM = 0.3333; WM = 0.14; blood vessels = 0.28). Since the skull and white matter have strongly anisotropic behavior [3], we use a 1:10, radial:tangential anisotropic setup for the skull, based on the volume constraint of the isotropic value. For the anisotropic white matter, DWI are corrected for motion, eddy currents and field inhomogeneities. Diffusion tensor images (DTI) are reconstructed with Diffusion-Toolkit. Finally, registration of DTI images to the anatomical T1 image space is performed using the FSL tool with the preprocessed DWI $b0$ image.

Performance of the BiCG–iLU in highly anisotropic aFDM formulations is carried out for two different realistic head models. The Fig. 4a shows a simplify 5-tissues head model that we set to be completely isotropic. In contrast, we define a 9-tissues head model with anisotropic skull and white matter, including fat,

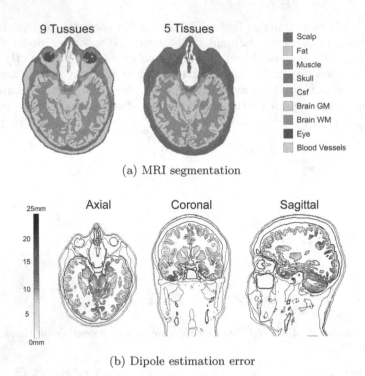

(a) MRI segmentation

(b) Dipole estimation error

Fig. 4. a) Validation of BiCG–iLU numerical Vs analytical potentials on the scalp for a spherical head model b)MRI segmentation. c) Dipole estimation error due to neglect the anisotropic behaviour of skull and white matter tissues in realistic head models.

muscle, eyes and even blood vessels tissues. In accordance to [5], we calculate the dipole estimation errors due to neglect multiple tissue segmentation including the anisotropic skull and white matter. The solver takes about 4 minutes to solve the coefficient matrix for a given source for a 1×1×1mm resolution with N=256. Therefore, the algorithm shows a feasible numerical stability even in the presence of highly anisotropic areas, and larger heterogeneity tissues.

4   Discussion and Concluding Remarks

We discuss the use of the iLU preconditioning within the framework of the Anisotropic-Finite-Difference based Solution for the EEG Forward Problem. To this end, we carry out the comparison of several numerical solver combinations in terms of convergence rate and computational cost. Since there is a need for fast, but accurate forward solver, the minimal error of representation is also included as an additional consideration. From the obtained results of the comparison on real data, we infer that the BiCG solver outperforms the other considered approaches: GMRES and SOR. Furthermore, the BiCG is the only one to reach the fixed amount of residual error. In order to reduce the computational cost significantly, however, a proper choice of the preconditioner for a particular matrix is probably more important than using the optimal outer-level iterative solver. We show that the LU preconditioners is not efficient under the considered settings. On the contrast, the incomplete LU preconditioner improves the solver performance. Thus, the BiCG–iLU results in the best NSC in terms of computational cost within a wide range of numerical resolution. However, if the amount of residual error is relaxed down to $\sim 10^{-10}$, the use of BiCG–Fourier Jacoby leads to a workable NSC.

Validation process comparing numerical and analytical models is carried out and shows that the BiCG–iLU as a numerical approximation accurately matches the potentials over the scalp. Specifically, we select a source placed in the outer gray matter sphere in a 76 artificial electrode configuration covering the entire surface of the sphere with 6 parallel rings. BiCG–iLU performs high accuracy and low computational cost even for realistic head models with high anisotropic areas and heterogeneous tissue conductivity considerations. As a result, the calculation time for a $1\times1\times1mm$ lasts 4 minutes; this amount of time is very competitive in comparison to other aFDM solution techniques reported in literature.

One of the promising approaches for improving the solver performance is the use of the fictitious domain enclosing the data in a redundant low zero voxelization, allowing a straightforward data ordering in the coefficient matrix. Instead of using the fictitious domain, we hypothesize that an inhomogeneous data ordering considering only the non-zero entries of the volumetric model should reduce the size of the coefficient matrix and the number of potential unknowns. As a result, we may improve memory allocation and, at the same time, reduce the whole amount of numerical operations. As future work, the authors plan to test the discussed NSC in realistic head data extracted from EEG recordings, in order to measure the influence of volumetric anisotropic medium on the performance of the source localization problem.

Acknowledgments:. This work was supported by COLCIENCIAS project '*Evaluacion asistida de potenciales evocados cognitivos como marcador del transtorno por deficit de atencion e hiperactividad (TDAH)*' and Programa Nacional de Formacion de Investigadores *"Generacion del Bicentenario"*, 2011.

References

1. Irimia, A., Matthew Goh, S.-Y., Torgerson, C.M., Stein, N.R., Chambers, M.C., Vespa, P.M., Van Horn, J.D.: Electroencephalographic inverse localization of brain activity in acute traumatic brain injury as a guide to surgery, monitoring and treatment. Clinical Neurology and Neurosurgery 115(10), 2159–2165 (2013)
2. Montes, V., van Mierlo, P., Strobbe, G., Staelens, S., Vandenberghe, S., Hallez, H.: Influence of Skull Modeling Approaches on EEG Source Localization. Brain Topography (September 2013)

3. Hallez, H., Vanrumste, B., Grech, R., Muscat, J., De Clercq, W., Vergult, A., D'Asseler, Y., Camilleri, K.P., Fabri, S.G., Van Huffel, S., Lemahieu, I.: Review on solving the forward problem in EEG source analysis. J. Neuroeng. Rehabil. 4, 46 (2007)

4. de Munck, J.C., Peters, M.J.: A fast method to compute the potential in the multiphere model. IEEE Transactions on Bio-Medical Engineering 40(11) (1993)

5. Hallez, H., Vanrumste, B., Van Hese, P., D'Asseler, Y., Lemahieu, I., Van de Walle, R.: A finite difference method with reciprocity used to incorporate anisotropy in electroencephalogram dipole source localization. Physics in Medicine and Biology 50(16), 3787–3806 (2005)

6. Turovets, S., Volkov, V., Zherdetsky, A., Prakonina, A., Malony, A.D.: A 3D finite-difference BiCG iterative solver with the Fourier-Jacobi preconditioner for the anisotropic EIT/EEG forward problem. Computational and Mathematical Methods in Medicine, 2014:426902 (January 2014)

7. Cao, N.: Estimating Extended Brain Sources Using EEG and Diffuse Optical Tomography. ProQuest LLC (2008)

8. Glowinski, R., Pan, T.-W.: A fictitious domain method for dirichlet problem and applications. Computer Methods in Applied Mechanics and Engineering 111(3-4), 283–303 (1994)

9. Saleheen, H.I., Ng, K.T.: New finite difference formulations for general inhomogeneous anisotropic bioelectric problems. IEEE Transactions on Biomedical Engineering 44(9), 800–809 (1997)

10. Hernandez, T., Torrado, C.: Automatic Segmentation Pipeline for Patient-Specific MRI Tissue Models. In: Proc. Intl. Soc. Mag. Reson. Med., vol. (22), pp. 4906–4928 (2014)

EEG Rhythm Extraction Based on Relevance Analysis and Customized Wavelet Transform

L. Duque-Muñoz[1]([✉]), R.D. Pinzon-Morales[2], and G. Castellanos-Dominguez[3]

[1] Universidad de Antioquia, Research group SISTEMIC, Medellin, Colombia
lduquem@unal.edu.co
[2] Chubu University, The Neural Cybernetics laboratory, Kasugai, Jaban
[3] Universidad Nacional de Colombia, Signal Processing and Recognition Group,
Bogotá, Colombia

Abstract. The waveform of physiological signals carries useful information about the brain states. Automated computational algorithms are used in clinical medicine for extracting this information that cannot be read directly by visual inspection. Nonetheless, difficulties arise in the extraction because the intrinsic rhythms of the waveforms vary with the changes in the state of the brain. That is the case for electroencephalogram (EEG) signals from Epileptic seizure events. Here, we address the extraction of information from EEG signals by using a novel methodology that quantitatively measures the intrinsic rhythms of EEG waveforms related to healthy or Epileptic seizure events. In this method, the customized wavelet is used to estimate the EEG rhythms and then the relevance analysis with Fuzzy entropy and Stochastic measure are used to discriminate between seizure free and seizure states. The classification stage is based on classification performance using a support vector machine classifier. The pertinence of the proposed methodology during the Epileptic seizure identification is discussed, and future directions are presented.

Keywords: Customized wavelet · Epileptic seizure detection · Relevant analysis · Brain state classification

1 Introduction

Electroencephalography (EEG) signals reflect the electrical activity of populations of neurons that differ in their characteristic waveforms, termed *physiological rhythms.* These rhythms vary in the frequency of their activity, ranging from slow to fast $(0.5 - 30\ Hz)$ (δ, θ, α, and β, respectively). Physiological rhythms dramatically vary with changes in the state of the brain, and close correspondence has been established with pathological states related to epileptic seizures. In particular, δ and θ rhythms change their characteristics in frequency and amplitude to a larger extent than α waves during epileptic events [1]. These changes in the typical behavior of the rhythms can be used to assist the localization of the epileptogenic zone once the epilepsy is diagnosed [2].

© Springer International Publishing Switzerland 2015
J.M. Ferrández Vicente et al. (Eds.): IWINAC 2015, Part I, LNCS 9107, pp. 419–428, 2015.
DOI: 10.1007/978-3-319-18914-7_44

However, for the automatic epileptic seizure identification, the contribution of each frequency sub-band must be first disassociated carefully. In this line of analysis, digital signal processing techniques applied to EEG signals have provided new insights into the dynamics of the physiological rhythms during the evaluation of different clinical disorders related to epileptic seizures. For evaluating the contribution of physiological rhythms to the EEG signal in discriminating different brain states, there are two stages to be carried out: i) estimation of the time-variant physiological rhythms and ii) relevance evaluation of the estimated rhythm in relation to the state of the brain. Regarding the former item, several time-frequency and time-varying decomposition methods have been proposed to represent the dynamic properties of the EEG signal. Among the most popular methods are the Short-Time Fourier Transform (STFT) [7], the Discrete Wavelet Transform (DWT) [11], the Time-variant Autoregressive Models (TVAR) [10], and the Exponentially Damped Sinusoids (EDS) [3]. In this work, a wavelet approach is followed by using an optimal DWT, which is constructed by an evolutive procedure for creating signal-dependent wavelet analysis [5]. This optimal DWT has been shown to outperform classical wavelets for extraction of discriminant information from physiological signals and different altered brain states such as the ones during Parkinson disease and drowsiness [13].

Regarding the relevance evaluation stage, the analysis of the individual frequency band components in the time-domain has proven to supply useful insight into the physiological rhythms. In particular, the use of relevance weights to determine the optimal frequency band that is subject specific has been presented in [12]. In order to get the relevance weights of the extracted rhythms using DWT, we develop the Fuzzy entropy relevance analysis (FRA). For the sake of comparison, the relevance weights obtained with FRA are contrasted with the ones accomplished by the Stochastic relevance analysis presented in [8]. From the individual rhythms that represent the current brain state, the relevance weights are estimated for improving the analysis/classification performance of the extracted rhythms.

2   Processing Methods

The proposed methodology comprises the following steps: i) Feature extraction from the rhythms of the EEG signal extracted using a customized DWT, ii) Relevance analysis of rhythms for the identification of epileptic and normal brain states.

2.1   Wavelet Customization for Extraction of EEG Rhythms

This work employs the DWT for the EEG rhythm extraction, where selection of a suitable wavelet as well the number of decomposition levels is carried out. The number of decomposition levels is chosen to be five that yields the approximate frequency bands of the EEG physiological rhythms. Lower level decompositions have negligible magnitudes in a standard EEG. Each rhythm is associated with the signal reconstructed (i.e., inverse wavelet transform) from the wavelet coefficients of the level matching the frequency band of each rhythm.

In order to implement the DWT-based time-frequency analysis, the input signal, lasting N samples, $x \in \{x_i = 1, \ldots, N\}$, is projected into an orthogonal (or bi-orthogonal) space, which is created by scaling and translating one single wave-like function (called *wavelet mother*). The successful representation of signals mainly depends on the proper selection of the wavelet mother. Although there are plenty of wavelet prototypes reported in the literature, the existing wavelets may be not adequate for every application. Furthermore, there is not an established rule that states the best wavelet for a given task. Therefore, it is a usual task to test more or less arbitrarily a big amount of different wavelet functions to find one suited for the data at hand [14]. For this reason, the present study employs an evolutionary methodology for the creation of a customized wavelet function based on the use of Genetic Algorithms (GA).

The GA-based procedure for optimization of wavelet functions that has been discussed elsewhere [5] is summarized as follows: Initially, we extract from the measured signal set, $X \in \mathbb{R}^{n \times N}$, a subset of training signals $X_e \subset X$, where $X_e \in \mathbb{R}^{m \times N}$ and $X = \{x_i : i = 1, \ldots, N\}$, being n the number of the whole recorded EEG signals and m the number of recordings selected for training ($m < n$). Then, each input signal is decomposed using the wavelet transform up to level $l = 5$ by using cascade lifting schemes. The use of lifting schemes is conditioned by their fast and flexible implementation of the discrete wavelet transform, comprising a prediction filter $p = \{p_i : i = 1, \ldots, N_p\}$ and an update filter $u = \{u_i : i = 1, \ldots, N_u\}$ with order N_p and N_u, respectively. The former filter, which is associated to the wavelet function in the classical wavelet scheme, is devoted to extract high frequency components from the input signal by eliminating low order polynomials. As a result, we obtain the detail wavelet coefficients $d^{(l)} = x_o^{(l-1)} - p * x_e^{(l-1)}$, where x_o are the odd samples of x (Notation $*$ stands for the convolution). In turn, the update operator is related to the scaling function and the extraction of the approximating coefficients $a^{(l)} = x_e^{(l-1)} - u * d^{(l)}$, where x_e are the even samples of x. Afterward, we generate the feature vector $\phi \in \mathbb{R}^{1 \times k}$ holding the wavelet energy of the detail coefficients at level $l = 5$, and the approximation coefficients at level $l = 5$. For including the frequency band of every rhythm, We fix $k = 6$ as explained below.

Owing to the focus of the present content in discriminating brain states, the GA cost function is stated as a measure evaluating the classifier performance of the feature vector ϕ. For avoiding wrapped measures that may increase considerably the computational burden, rather we use a filter measure without validating any concrete classifier. Namely, we introduce as the cost function the Davies-Bouldin index $J_{DB} \in \mathbb{R}(0, 1)$ that measures the separability among classes. It is worth noting that the J_{DB} assesses, at the same time, both the compactness of each and the global dispersion among classes. This index requires the computation of class–to–class similarity defined as, $r_{ij} = (s_{ii} - s_{jj})/e_{jj}$, where s_{ii} and s_{jj} are the dispersion of the i-th and j-th classes, respectively, and computed as follows:

$$s_{jj} = \sqrt{\mathbb{E}\left\{\sum_{m=1}^{k} \|\phi_{n,m} - \mu_{j,m}\|^2 : \forall n = 1, \ldots, N_j\right\}}, \tag{1}$$

where notation $\mathbb{E}\{\cdot\}$ stands for the expectation operator and e_{ij} is the Euclidean distance between their mean values, given by $e_{i,j}=\|\mu_i - \mu_j\|$, where N_j is the number of signals belonging to the i-th class. $\phi_{n,k}$ is the n-th sample vector of the i-th class over the k-th dimension, and $\mu_{i,k}$ is the mean value of class i over k-th dimension. The fitness function is obtained through determining the worst case of separation for each class and averaging these values, i.e.: $J_{\mathrm{DB}}=r_{ij}/2$, where $i = $ normal, $j = $ epileptic.

The terminating step of the procedure is given by the GA optimization of the filters u and p of the lifting scheme. The GA operates in a guided manner by following symmetrical linear phase and filter normalization constraints [6]. These constrains secure the linear phase, symmetry, compact support, and normalization of lifting filters, in other words, the associated wavelet and scaling functions. Furthermore, the GA procedure only evolves $N_p/2 + N_u/2 - 2$ values because of the introduced constraints.

2.2   Relevance Analysis of Physiological Rhythms

All physiological rhythms are extracted from the measured EEG signal set by using the corresponding inverse wavelet coefficients that correctly reconstruct each i-th waveform (noted as y_i). Namely, $a^5 \to \delta, d^5 \to \theta, d^4 \to \alpha, d^3 \to \beta$. However, we should measure the relevance of each physiological rhythm vector in distinguishing between normal and pathological states of the brain activity. Here, we introduce the Fuzzy entropy Relevance Analysis (FRA) based on an improved estimation of the Shannon entropy using fuzzy sets, defined as follows [9]:

$$\mathbb{H}\{\Xi,y\} = \sum_{c\in C} h(\Xi,y \mid c), \ \mathbb{H}\{\Xi,y\} \in \mathbb{R}(0,1) \tag{2}$$

where Ξ is the fuzzy set of the membership of feature y in class $c\in C$, $h(\Xi,y|c)$ is the fuzzy entropy of class c expressed as:

$$h(\Xi,y \mid c) = -f(\Xi,y \mid c) \log f(\Xi,y \mid c)$$

being $f(\Xi,y|c)$ the summation of membership of feature y in class c, divided by the membership of feature y in all C classes. That is,

$$f(\Xi,y \mid c) = \frac{\sum_{y\subset c} \mu_\Xi(y)}{\sum_{y\subset C} \mu_\Xi(y)}$$

here, $\mu_\Xi(y)$ denotes the membership of a given rhythm y belonging to the fuzzy set Ξ.

The fuzzy entropy measure in Eq. 2 is assumed to evaluate the relevance of the each rhythm. Thus, the higher the value of $\mathbb{H}\{\Xi,y\}$ for a given waveform feature y – the lower its contribution in distinguishing between classes. Consequently, we make the FRA-based relevance weight of the rhythm at hand, $w_{\mathrm{F}}(y_i)\in\mathbb{R}(0,1)$, as follows:

$$w_{\mathrm{F}}(y_i) = (1 - \mathbb{H}\{\Xi,y_i\})/ \max_{\forall i}(1 - \mathbb{H}\{\Xi,y_i\}),$$

For the sake of comparison, we also use the weights derived from the Stochastic relevance analysis (SRA) using latent variable decomposition techniques [8]. Latent variable decomposition finds a transformation mapping of the k–dimensional stochastic waveform set, $\boldsymbol{Y} \in \mathbb{R}^{k \times N}$, into a reduced q–dimensional stochastic set, $\widehat{\boldsymbol{Y}} \in \mathbb{R}^{q \times N}$, where reduction dimension holds whenever $k<q$. So, there is the need for a matrix $\boldsymbol{\Gamma}$ who preserves maximally data information. $\min_{\boldsymbol{\Gamma}^\top} \mathbb{E} \left\{ (\boldsymbol{y} - \boldsymbol{\Gamma}^\top \widehat{\boldsymbol{y}})^\top (\boldsymbol{y} - \boldsymbol{\Gamma}^\top \widehat{\boldsymbol{y}}) \right\}$. The relevant measure w_S obtained with the usual explained variance criterion is:

$$w_S(\boldsymbol{y}) = \mathbb{E} \left\{ |\nu_j^2 V_j| : \forall j = 1, \ldots, p \right\} \tag{3}$$

where ν_j is the ordered eigenvalues of $\boldsymbol{\Gamma}$ and V_j is the set of singular values ranked by the decreasing amplitude. The SRA measure allows capturing the stochastic information embedded in the physiological rhythms.

3 Results

3.1 Electroencephalographic Recording Database

For validating the proposed rhythm extraction methodology, we use the database of epileptic and normal EEG data collected at the *Instituto de Epilepsia y Parkinson del Eje Cafetero, Pereira, Colombia* [8]. The collection includes 160 EEG signals from 20 channels placed on the head according to the International $10 - 20$ standard for EEG electrode placement. Set A holds 80 recordings labeled as normal (seizure-free) from 20 patients each one with four recordings, whereas set E has 80 signals labeled as presenting epileptic activity from another 20 patients; each one with four recordings. A neurologist examined the database to label the epileptic events. Recordings were done under video control to secure an accurate determination of the different seizure stages. EEG signals were sampled at 256 Hz with 12-bit resolution and 2 min duration. Fig. 1 illustrates typical EEG signals with normal and seizure episodes and evidence that measured perturbations obscure the brain activity. EEG signals were acquired in a non-regulated condition. Noise of the EEG recording included the muscle artifacts as well as 60 Hz power line interference. All EEG recordings were digitally band-pass filtered (with a Butterworth filter of order 10, stopband frequencies 0.5–40 Hz) and normalized to the absolute greatest value of each recorded EEG signal.

3.2 Tuning of Customized DWT for Rhythm Extraction

As described above, the reconstruction of the DWT coefficients produces an estimation of the neural activity, where each decomposition level is associated with the respective rhythm as shown in Table 1. The procedure explained above is used to customize the wavelet filter for extraction of the EEG rhythms. In order to perform the training procedure, we randomly select 30% of the whole set of EEG Data. Fig. 2 shows 10 different customized EEG wavelet and scaling

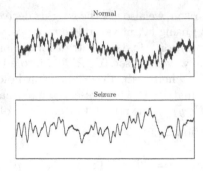

Fig. 1. Typical EEG 10-s segments of the considered subsets, normal (A) and epileptic seizure (E)

Table 1. Used *Daubechies-6* filter wavelet levels in rhythm bandwidth estimation. 256 *Hz* (DS2)

Rhythm	Decomposition	Frequency
β	d_3	16-32
α	d_4	8-16
θ	d_5	4-8
δ	a_5	0-4

functions produced by running 10 times the GA-based optimization procedure fed by various initialization parameters.

Once the customized DWT are calculated, the relevance analysis is carried out relevance analysis on the physiological rhythms. Fig. 3 shows the relevance weights provided by both approaches, FRA and SRA. It is worth noting that the estimated weights should be normalized by their maximum value for handling their comparison. Since the higher the estimated relevance weight – the more relevant the respective rhythm, α proves to the most suitable rhythm for seizure-free EEG signals for either method of computation. This finding confirms several studies of the α rhythm remarking its topographic distribution (maximum amplitude over occipital regions) and high reactivity (have strong attenuation when passing from a normal to alert state). For the seizure–free class, however, both relevance approaches differ in estimating the second weight: the θ rhythm for SRA while β for FRA. In terms of affecting the seizure class, the SRA performs the following weights ranked by decreasing relevance (see Fig. 3): δ, θ, α, β while FRA does δ, α, θ, β. As a result, the δ rhythm gets the largest relevance (FRA or SRA) weight. Furthermore, this low-frequency rhythm may be present in some brain states such as deep sleep or high concentration, its increased activity may be an indication of focal epilepsy on the temporal region of the brain [2]. The second strongest weight is the θ (for SRA) that has been strongly associated with epilepsy [4] while the α rhythm (estimated by FRA) still has not been, meaning that both relevance approaches differ in estimating the second weight again.

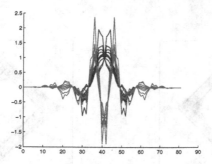

Fig. 2. Customized wavelets (red lines) and correspondind scalling functions (blue lines) for EEG analysis using the procedure described in [5]

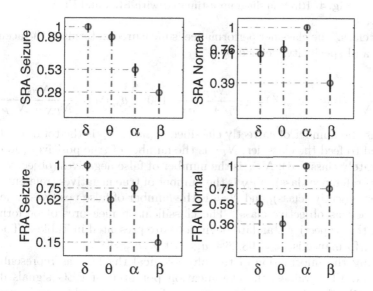

Fig. 3. Estimated weights using SRA and FRA for seizure (left) and normal (right) events of neural activity

The rhythm diagrams introduced in [8] are used to get better interpretation of the assessed feature set Fig. 4. Here for SRA the higher weight is related for $\delta + \theta$ for SRA and $\delta + \alpha$ for FRA.

3.3 Classifier Performance Using Rhythm Relevance Weights

We use a Super vector machine (SVM) classifier for validating the performance in discriminating between the considered classes of neural activity. The SVM classifier is trained under the conventional cross-validation procedure, which consists in dividing the database into 10 folds each one having an equal number of EEG

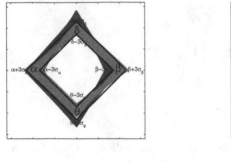
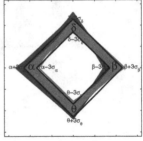

(a) SRA Weights   (b) FRA Weigths

Fig. 4. Rhythm diagram estimated with SRA and FRA

signals per class. The classifier performance is measured in terms of the accuracy, sensitivity and specificity, defined by:

$$a_{ac}(\%) = \frac{N_c}{N_T} * 100; \quad a_{se}(\%) = \frac{N_{TP}}{N_{TP} + N_{FN}} * 100; \quad a_{sp}(\%) = \frac{N_{TN}}{N_{TN} + N_{FP}} * 100$$

where N_c is the number of correctly classified signals, N_T is the total number of signals used to feed the classifier, N_{TP} is the number of true positives (objective class accurately classified), N_{FN} is the number of false negatives (objective class classified as reference class), N_{TN} is the number of true negatives (reference class classified as objective class), and N_{FP} is the number of false positives (reference class classified as objective class). The classification measures of performance along with the respective standard deviations are presented in Table 2. The best classifier performance is (a_{ac}=98.26%, a_{se}=100%, a_{sp}=98.59%).

Combining the information of the physiological rhythms as represented by the SRA and FRA values, the classification potential of EEG signals during normal or epileptic seizures can be improved. Relevance weights can be grouped according to the following four training scenarios for SRA analysis (rhythms by

Table 2. Computed classification accuracy adding their SRA and FRA weights ranked by their decreasing relevance

Rhythms	Classification Performance		
	a_{ac} (%)	a_{se} (%)	a_{sp} (%)
δ	95.21 ± 2.35	94.24 ± 1.46	95.54 ± 2.21
$\delta + \theta$	99.07 ± 0.85	100.00 ± 0.00	99.41 ± 1.12
$\delta + \theta + \alpha$	97.26 ± 1.71	96.63 ± 2.44	97.91 ± 1.92
$\delta + \theta + \alpha + \beta$	96.26 ± 1.78	96.83 ± 2.13	97.48 ± 1.17
δ	95.35 ± 2.21	94.42 ± 1.36	95.68 ± 2.24
$\delta + \alpha$	96.15 ± 1.93	95.36 ± 1.83	97.82 ± 1.41
$\delta + \alpha + \theta$	97.45 ± 1.83	96.21 ± 2.12	97.86 ± 1.95
$\delta + \alpha + \theta + \beta$	96.08 ± 1.55	96.12 ± 1.95	97.16 ± 1.31

decreasing relevance, indicated on the first column on Table 2): 1) δ rhythm, 2) $\delta + \theta$, 3) $\delta + \theta + \alpha$, and 4) $\delta + \theta + \alpha + \beta$. When validation is carried out using the classification measures introduced above, it is evident that low-frequency band rhythms provide the largest discriminant information for the classification task. The starting contribution of the δ rhythm is high and is the greatest when considering both $\delta + \theta$. However, classifier performances decreased when including higher band rhythms. In comparison, FRA suggests that the combination of $\delta + \alpha + \theta$ rhythms provides the highest discrimination of EEG signals 3. However, the combination reaching the best classification scenario is for SRA with $\delta + \theta$ rhythms.

Table 3. Best classification scenario with optimal DWT + SRA, compared with a classical DWT with a Daubechies 6 filter + SRA

Rhythms	Classifier Performance		
	a_{ac} (%)	a_{se} (%)	a_{sp} (%)
$\delta + \theta$	99.07 ± 0.85	100.00 ± 0.00	99.41 ± 1.12
$\delta + \theta$	97.38 ± 1.51	96.29 ± 1.86	98.24 ± 1.32

Lastly, Table 3 shows the best classification scenario achieved with the optimal DWT a_{cc}=99.07% and SRA. For the sake of comparison, the same scenario is repeated with a classical DWT with the Daubechies 6 filter. As shown, the optimal DWT outperforms the classical wavelet.

4   Conclusions

This work discusses a methodology for automatic extraction of discriminant information from physiological rhythms of EEG signals generated during normal and epileptic brain states. Estimation of the physiological rhythms is performed using an optimal discrete wavelet transform that is tailored specifically for the current application.

On other hand, we suggest the use of two different methodologies of rhythm relevance analysis (Fuzzy relevance analysis ans Stochastic relevance analysis), yielding the same outcomes of the strongest weights that enough for distinguishing between seizure–free and seizure events of neural activity. Finding of both classes are in line with other reports in the literature. However, either approach estimates the intermediate rhythm weights distinctly, meaning that further research is to be carried out for improving the knowledge about the EEG rhythms.

We use a Super vector machine classifier for validating the performance of the estimated relevance weights. Results obtained in a real database show that relevance weights provide high accuracy in discriminating between the considered classes of neural activity (seizure–free and seizure events)

Future lines of research include the application of the discussed methodology to analyze other brain activities and to determine the feasibility of seizure prediction.

428   L. Duque-Muñoz et al.

Acknowledgments. This research was supported by COLCIENCIAS project *"Evaluación asistida de potenciales evocados cognitivos como marcador del transtorno por déficit de atención e hiperactividad (TDAH)"*.

References

1. Cantero, J.L., Atienza, M., Salas, R.M.: Human alpha oscillations in wakefulness, drowsiness period, and REM sleep: different electroencephalographic phenomena within the alpha band. Neurophysiologie Clinique Clinical Neurophysiology 32, 54–71 (2002)
2. West, M., Prado, R., Krystal, A.D.: Evaluation and Comparison of EEG Traces: Latent Structure in Nonstationary Time Series. Journal of the American Statistical Association 94, 375–387 (1999)
3. De Clercq, W., Vanrumste, B., Papy, J.-M., Van Paesschen, W., Van Huffel, S.: Modeling common dynamics in multichannel signals with applications to artifact and background removal in EEG recordings. Trans. on Biomed. Eng. 52, 2006–2015 (2005)
4. Gandhi, T., Ketan-Panigrahi, B., Anand, S.: A comparative study of wavelet families for EEG signal classification. Neurocomputing 74, 3051–3057 (2011)
5. Pinzon-Morales, R.D., Orozco-Gutierrez, A.A., Castellanos-Dominguez, G.: Novel signal-dependent filter bank method for identification of multiple basal ganglia nuclei in Parkinsonian patients. J. Neural Eng. 8, 3 (2011)
6. Gouze, A., Antonini, M., Barlaud, M., Macq, B.: Design of signal-adapted multidimensional lifting scheme for lossy coding. IEEE Trans. on Image Processing 13(12), 1589–1603 (2004)
7. Kiymik, M.K., Güler, I., Dizibüyük, A., Akin, M.: Comparison of STFT and wavelet transform methods in determining epileptic seizure activity in EEG signals for real-time application. Computers in Biology and Medicine 35, 603–616 (2005)
8. Duque-Muñoz, L., Espinosa-Oviedo, J., Castellanos-Dominguez, G.: Identification and monitoring of brain activity based on stochastic relevance analysis of short-time EEG rhythms. Biomd Eng Online 13 (2014)
9. Jaganathan, P., Kuppuchamy, R.: A threshold fuzzy entropy based feature selection for medical database classification. Computers in Biology and Medicine 43, 2222–2229 (2013)
10. Li, Y., Wei, H.-L., Billings, S.A., Sarrigiannis, P.G.: Time-varying model identification for time-frequency feature extraction from EEG data. Journal of Neuroscience Methods 196, 151–158 (2011)
11. Subasi, A.: EEG signal classification using wavelet feature extraction and a mixture of expert model. Expert Systems with Applications 32, 1084–1093 (2007)
12. Veluvolu, K., Wang, Y.: Adaptive estimation of EEG-rhythms for optimal band identification in BCI. J. Neurosci. Methods 203, 163–172 (2012)
13. Pinzon-Morales, R.D., Hirata, Y.: Customization of Wavelet Function for Pupil Fluctuation Analysis to Evaluate Levels of Sleepiness. J. Communication and Computer 10, 585–592 (2013)
14. Gandhi, T., Panigrahi, B.K., Anand, S.: A comparative study of wavelet families for EEG signal classification. J. Neurocomputing 17(74), 3051–3057 (2011)

Estimation of M/EEG Non-stationary Brain Activity Using Spatio-temporal Sparse Constraints

J.D. Martínez-Vargas[✉], F.M. Grisales-Franco,
and G. Castellanos-Dominguez

Signal Processing and Recognition Group, Universidad Nacional de Colombia,
Bogotá, Colombia
{jmartinezv,fmgrisalesl,cgcastellanosd}@unal.edu.co

Abstract. Based on the assumption that brain activity appears in localized brain regions that can vary along time, yielding spatial and temporal non-stationary activity, we propose a constrained M/EEG inverse solution, based on the Fused Lasso penalty, that reconstructs brain activity as dynamic small and locally smooth spatial patches. Thus, our main contribution is to provide neural activity reconstruction tracking non-stationary dynamics. We validate the proposed approach in two different ways: i) using simulated MEG data when we have previous knowledge about spatial and temporal signal dynamics, and ii) using real MEG data, particularly we use a faces perception paradigm aimed to examine the M170 response. In the former case of validation, our approach outperforms conventional M/EEG-based imaging algorithms. Besides, there is a high correspondence between brain activities presented on the evaluated real MEG data and the time-varying solution obtained by our approach.

Keywords: Brain mapping · Fused Lasso · M/EEG · Non-stationary activity · Structured sparsity

1 Introduction

The human brain study is an important and exciting area due to its complexity and functionality. Consequently, a better understanding of these facts may lead to the treatment of brain diseases or even to interpret the human cognitive process. Hence, nowadays the discovery of non-invasive brain imaging techniques have become in a field with boosted interest. The most accepted imaging techniques providing high temporal resolution are the Magnetoencephalographic (MEG) and Electroencephalographic (EEG) that have been widely used to study brain dynamics. Both techniques allow identifying and analyzing neural rhythms, evoked potential responses (ERPs), epileptic spikes, among other applications. However, the number of locations measuring Magnetic/electrical activity is relatively small (a couple of hundreds) while the discretized brain activity generators (sources) reaches as much as several thousand. This task, commonly known as

© Springer International Publishing Switzerland 2015
J.M. Ferrández Vicente et al. (Eds.): IWINAC 2015, Part I, LNCS 9107, pp. 429–438, 2015.
DOI: 10.1007/978-3-319-18914-7_45

known as the M/EEG inverse problem, poses a heavily ill-posed problem [1]. Consequently, to ensure a unique and optimal brain activity reconstruction is necessary to provide prior information [2].

A straightforward approach to supply prior information about neuronal activity is to impose some constraints on geometrical or physiological properties of the brain. Mainly, restrictions are considered in the spatial (by inserting an a priori covariance matrix) or temporal (through state space models) domains. In the first case (spatial constraints), most of the state-of-the-art approaches assume brain activity represented by a small/sparse set of spatial basis functions (termed spatial blobs or patches). Thus, brain activity is a linear combination of several predefined spatial patches. The following patch-based approaches are the most representative: Automatic Relevance Determination (ARD) , Greedy Search (GS), Multiple Sparse Priors (MSP) [3,4], Sparse Basis Field Expansion (S-FLEX) [5]. Yet, spatial distribution of those methods states that the active brain patches remain the same throughout the entire solution interval [6]. Such an assumption is far from being totally realistic in many practical tasks where brain activity may have strong spatiotemporal dynamics and non-stationarities [7]. On the other hand (temporal constraints), most of the approaches include dynamic information through state space models [8,9], however, sophisticated tuning along with an increased computational burden make those methods infeasible when the number of brain activity generators becomes large enough. In this paper, our basic assumption is that brain activity can be also represented by a set of small and locally smooth spatial patches varying smoothly over time, in such a way that we can include both spatial and temporal non-stationary dynamics to improve the accuracy of neural activity reconstruction. Namely, based on the fused lasso penalty, first introduced in [10], we encourage the solution to get temporal homogeneity and spatial sparsity, introducing a set of time-varying spatial constraints. This definition leads to a solution with spatial coherence, but considering at the same time the non-stationarity of M/EEG recordings.

2   Methods

This section proceeds as follows: First, we re-introduce the conventional scheme for the M/EEG inverse problem solution. Latter, we describe our contribution that shows how the problem can posed in terms of the spatial basis set, and how the solution can be encouraged to present sparse and time homogeneous activity.

2.1   M/EEG Inverse Problem

In order to represent the electromagnetic field magnitude measured by the scalp, we assume the following linear model [2]:

$$Y = LJ + \Xi, \tag{1}$$

where $Y \in \mathbb{R}^{C \times T}$ is the M/EEG data measured by C sensors at T time samples, $J \in \mathbb{R}^{D \times T}$ is the amplitude of the D current dipoles, distributed through the cortical surface with fixed orientation perpendicular to it, and $L{\in}\mathbb{R}^{C{\times}D}$ (commonly termed *lead field matrix*) is a gain matrix representing the relationship between the dipoles and M/EEG data. Besides, we assume that the M/EEG data are affected by zero mean Gaussian noise $\Xi{\in}\mathbb{R}^{C{\times}T}$ with covariance $\mathrm{cov}\{\Xi\}{=}Q_{\Xi}{=}\lambda_{\Xi}I_{C}{\in}\mathbb{R}^{C{\times}C}$, being $\lambda_{\Xi}{\in}\mathbb{R}^{+}$ the noise variance. Assuming this model, the maximum a-posteriori (MAP) estimate of J can be found by minimizing the following cost function [1]:

$$\underset{J}{\mathrm{argmin}}\{||Y - LJ||_{F}^{2} + \Theta(J, \lambda_{i})\}, \tag{2}$$

where $||.||_{F}$ stands for the frobenius norm, $\Theta(J){\in}\mathbb{R}^{+}$ is a function formalizing the constraints imposed upon the source activity and $\lambda_{i}{\in}\mathbb{R}^{+}$ is a regularization parameter for each constraint.

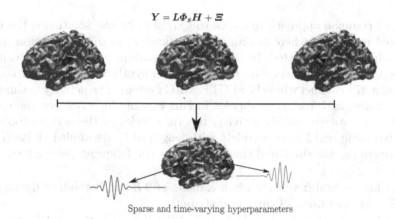

$$Y = L\Phi_{s}H + \Xi$$

Sparse and time-varying hyperparameters

Fig. 1. Illustrative representation of the proposed method. Top: a set of spatial basis Φ_{s} are generated, bottom: solution encouraging sparse and smooth time-varying is obtained.

2.2 Sparsity and Temporal Homogeneity Constraints

In practice, methods encouraging focal solutions have been found to yield better source reconstructions [5]. Consequently, the current density can be expressed as a linear combination of locally smooth, but spatially confined spatial basis functions:

$$J = \Phi_{s}H, \tag{3}$$

where $\Phi_{s}{\in}\mathbb{R}^{D{\times}S}$ holds S spatial basis functions, and $H{\in}\mathbb{R}^{S{\times}T}$ is a matrix of weighting coefficients to be estimated. In order to enforce sparsity and temporal homogeneity, we introduce the following regularization penalty [11]:

$$\Theta(J, \lambda_{s}, \lambda_{t}) = \lambda_{s}||H||_{1} + \lambda_{t}\sum_{t=1}^{T-1}||h_{t+1} - h_{t}||_{1} \tag{4}$$

where λ_s and $\lambda_t \in \mathbb{R}^+$ are the spatial and temporal regularization parameters, and $h_t \in \mathbb{R}^{C \times 1}$ stands for the t-th column of H.

Sparsity is encouraged by the first penalty term, which assigns a large cost to matrices with large absolute values, and thus effectively shrinking elements towards to zero. This situation means that just a few basis of the spatial dictionary will explain the main brain activity. The second penalty term encourages temporal homogeneity by penalizing the difference between consecutive time points, yielding an smooth solution over time. An schematic representation of the proposed algorithm is provided in Fig. 1.

3   Experiments

3.1   Simulated MEG Data

The most common approach to assess the M/EEG inverse solution is the use of simulated recordings where brain activity is known, so that estimation quality can be objectively validated. In this work, simulations are carried out using an MEG system geometry based on the third-order synthetic gradiometer configuration of a 274 channel whole head CTF MEG system. The activity is simulated for one, three, and five active dipoles having random location. For the simulation of non-stationary MEG activity, the time series of the active dipoles are generated using real Morlet wavelets with length of 1.5s, sampled at $120Hz$. As suggested in [6], the simulated time series have the following parameters:

- Random central frequency witsh a mean of $9\,Hz$ and standard deviation of $2\,Hz$, sampled from a Gaussian distribution.
- Random time shift generated by normal distribution with standard deviation of $0.05\,s$ and mean value selected as shown in Table 1.

Table 1. Mean values for simulated Morlet wavelets

# of active sources	Mean value
1	$[0.75]s$
3	$[0.375, 0.75, 1.25]s$
5	$[0.25, 0.5, 0.75, 1, 1.25]s$

Afterwards, each simulated MEG is calculated by multiplying the simulated brain activity by the lead field matrix, as shown in Eqs. 1. Also, measurement noise is added to get SNR levels of $-5, 0, 5, 7, 12$, and $14\,dB$.

For source space modelling, we made use of a tessellated surface of the gray-white matter interface with 8196 vertices (possible source localizations) with source orientations fixed and being perpendicular to the surface. Also, the mean distance between neighborhood vertices is adjusted at $5\,mm$, and the leadfields are computed using a single-SHELL volume conductor. To compare the performance of proposed approach, these simulated data are then source-reconstructed using the proposed, and the Multiple Sparse Prior inversion schemes.

Implementation Issues: The parameters used in the proposed approach are selected as follows:

- As explained in detail in [12], we obtained an spatial projector $U \in \mathbb{R}^{C \times \hat{C}}$, that depends on the forward model. It is obtained from the singular value decomposition of the $LL^\top$ matrix. The default selection in the SPM software framework removes all modes with *eigenvalues* inferior to e^{-16} of the mean. Then, a new lead field matrix $\hat{L} = U^\top L \in \mathbb{R}^{\hat{C} \times D}$ and a new MEG dataset $\hat{Y} = U^\top Y \in \mathbb{R}^{\hat{C} \times T}$ are computed.
- The spatial dictionary comprises $S = 512$ basis covering all the entire cortical surface $\Phi_s = \{q_1, \ldots, q_S\}$, being $q_s \in \mathbb{R}^{D \times 1}$ each element of the dictionary. These functions are locally determined on the basis of brain anatomy with compact spatial support. The spatial extent of a source prior is determined by a smoothing operator that employs a Green's function based on a graph Laplacian that was computed using the vertices and faces provided from a cortical surface mesh derived from a structural MRI [13]:

$$Q_G = e^{\sigma A} \tag{5}$$

being q_s are selected columns of the matrix Q_G, where $A \in \{0, 1\}^{D \times D}$ is a matrix that denotes the neighborhood properties of the vertices. Also, depending on the smoothness parameter σ, the green function connects the patch points from a central vertex up to its 8-th order neighbor. As in the SPM framework, we selected $\lambda = 0.6$ to obtain a trade-off between spatial accuracy and local coherence.
- To solve the high dimensional and large scale problem posed in Eqs. 2 and 4, we used two different approaches: *i*) an extension of the mono-dimensional split Bregman Fused-LASSO (sBFLasso) solver [14], and *ii*) and algorithm based on proximal-gradient method used for optimizing the structured multitask regression, named Graph-guided fused LASSO (GFLasso).
- The most critical issue for solving both the sBFLasso and the GFLasso algorithms are the spatial (λ_s) and temporal (λ_t) regularization parameters. For solving this issue, we empirically tuned both parameters, giving more weight to the spatial term. As a result, we get less parameters different to zero with a not so smooth temporal behavior.

Results of simulated data: As assessment measure, we use the averaged correlation, at the active dipole positions, between the simulated and reconstructed time-series. Thus, the reconstruction is highly penalized if it does not appear in a close neighborhood of any simulated source, yielding an spatio-temporal assessment. For each SNR value and source number (one, three, and five active sources), the experiment is repeated over 100 times. The noise and the simulated signal parameters are drawn randomly for each run. It should be noted that the simulation scenario is designed to imitate an event-related experimental design, in which the activity is time-locked to a given stimulus. Consequently, the higher the number of simulated sources – the more non-stationary the obtained MEG.

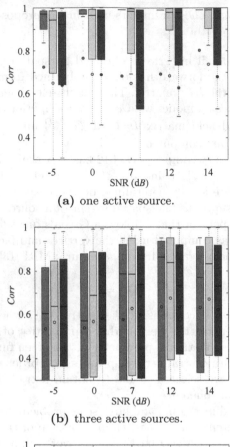

(a) one active source.

(b) three active sources.

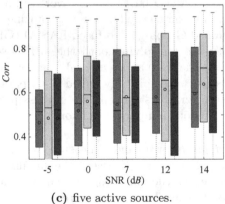

(c) five active sources.

Fig. 2. Assessed quality of achieved activity reconstruction for different estimation methods: –MSP sBFLasso –GFLasso

Fig. 2 shows the measures of activity reconstruction obtained for different estimation methods: sBFLasso, GFLasso, and MSP. As seen in Fig. 2a, the MSP overcomes the proposed approach (greater mean and median values) for one simulated source.

As seen in Fig. 2b Fig. 2c, however, the performance of the proposed approach improves as the number of sources increases (i.e. increasing the non-stationary signal content) for three and five simulated sources, respectively. This situation points out on greater mean and median performance values, and remains for low SNR values, specifically −5 and 0 dB. For three active sources, the performance of the MSP improves as the SNR increases, matching the proposed approach performance. However, our approach gets the best results for all the SNR values in the most demanding scenario in terms of non-stationary activity (that is, five active sources).

3.2   Reconstruction of Brain Neural Activity Using Real MEG Data

Database description: we use the MEG data measured from a single subject while he made symmetry judgements on faces and scrambled faces (for a detailed

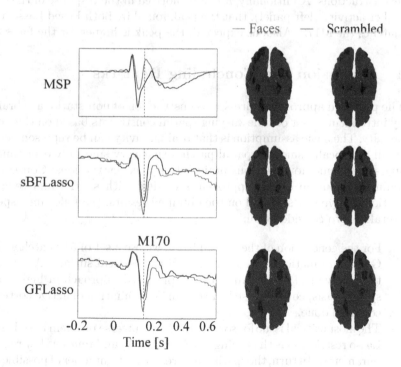

Fig. 3. Source level real data analysis: In the left, the response of the dipole with higher activity at $t=170ms$, and in the right, reconstructed activity mapped over the cortical surface

description of the paradigm see [15][1]). The MEG data were acquired on a 275 channel CTF/VSM system, using second-order axial gradiometers and synthetic third gradient for denoising and sampled at 480 Hz. There are actually 274 MEG channels in this dataset since the system it was recorded on had one faulty sensor. The epochs (168 faces and 168 scrambled faces) were baseline-corrected from $-200\,ms$ to $0\,ms$. Also, data were down-sampled at 200 Hz, and averaged over each condition. For modelling of the source space, we used a tessellated surface of the grey-white matter interface with 8196 vertices (possible source localizations) with source orientations fixed and being orthogonal to the surface. Finally, a single-shell head model was constructed to compute the forward operator L.

Results of real data: The experiment is designed to examine the M170 response (analogous to the N170 response recorded by event-related potentials – ERP), which is a component occurring approximately 170 ms after stimulus onset. The M170 for normal subjects is face-selective with a consistently higher amplitude to faces than to a wide variety of other visual stimulus categories [16].

Fig. 3 shows the obtained results for real data analysis. It can be seen that for both conditions, obtained activity by using the proposed approach corresponds spatially to the one obtained by the MSP framework, as seen in the 3D reconstructions. Additionally, it can be noticed in the response of the dipole with higher activity (left panel), that the peak found by both Fused Lasso approaches matches the M170. Also, as expected, the peak is higher for the faces condition.

4   Discussion and Concluding Remarks

The proposed approach addresses reconstruction of non-stationary brain activity by introducing a set of time-varying spatial constraints based on the Fused Lasso penalty. The basic assumption is that brain activity can be represented by a set of small and locally smooth spatial patches that vary smoothly over time, yielding sparse and time homogeneous brain activity reconstructions. Consequently, the main contribution of our approach is to deal with spatial and temporal non-stationary dynamics. Based on the obtained results, the following aspects are to be taken into consideration:

– For the generation of the spatial basis, we use a set of 512 patches shaped like Gaussian functions extended over all the cortical surface. As an alternative to improve the performance of the proposed approach, a higher number of spatial basis, could be used. Also, a prior estimation of active cortical patches could be done.

– The most critical issue for solving the MEG inverse problem based on the Fused Lasso restriction is the tuning of the spatial λ_s and temporal λ_t regularization parameters. In turn, the spatial term restricts the number of possible active cortical patches (encouraging spatial sparsity). The temporal term, which is based

---

[1] This database can be downloaded from http://www.fil.ion.ucl.ac.uk/spm/data/mmfaces/

on a Markovian restriction, promotes temporal smoothness over the hyperparameters, yielding time-varying solutions that follow non-stationary activity; that is the case of the ERPs. In here, we empirically tune both parameters, however, these can be tuned based on some information criteria as proposed in [11], or based on the residual noise variance [5].

As future work, the authors plan to extend the applicability of the proposed reconstruction method in time-varying brain connectivity analysis.

Acknowledgments. This research was supported by COLCIENCIAS project *Evaluación asistida de potenciales evocados cognitivos como marcador del transtorno por déficit de atención e hiperactividad (TDAH)* and Programa Nacional de Formacion de Investigadores "Generacion del Bicentenario", 2011.

References

1. Grech, R., Cassar, T., Muscat, J., Camilleri, K., Fabri, S., Zervakis, M., Xanthopoulos, P., Sakkalis, V., Vanrumste, B.: Review on solving the inverse problem in eeg source analysis. Journal of NeuroEngineering and Rehabilitation 5(25), 792–800 (2008)
2. Baillet, S., Mosher, J.C., Leahy, R.M.: Electromagnetic Brain Mapping. IEEE Signal Processing Magazine 18, 14–30 (2001)
3. Friston, K., Harrison, L., Daunizeau, J., Kiebel, S., Phillips, C., Trujillo-Barreto, N., Henson, R., Flandin, G., Mattout, J.: Multiple sparse priors for the m/eeg inverse problem. NeuroImage 39(3), 1104–1120 (2008)
4. Wipf, D., Nagarajan, S.: A unified bayesian framework for meg/eeg source imaging. NeuroImage 44(3), 947–966 (2009)
5. Haufe, S., Tomioka, R., Dickhaus, T., Sannelli, C., Blankertz, B., Nolte, G., Muller, K.-R.: Large-scale eeg/meg source localization with spatial flexibility. NeuroImage 54(2), 851–859 (2011)
6. Gramfort, A., Strohmeier, D., Haueisen, J., Hamalainen, M.S., Kowalski, M.: Time-frequency mixed-norm estimates: Sparse m/eeg imaging with non-stationary source activations. NeuroImage 70, 410–422 (2013)
7. Owen, J.P., Wipf, D.P., Attias, H.T., Sekihara, K., Nagarajan, S.S.: Performance evaluation of the champagne source reconstruction algorithm on simulated and real m/eeg data. NeuroImage 60(1), 305–323 (2012)
8. Schmitt, U., Louis, A.K., Wolters, C., Vauhkonen, M.: Efficient algorithms for the regularization of dynamic inverse problems: II. Applications. Inverse Problems 18(3), 659–676 (2002)
9. Barton, M.J., Robinson, P.A., Kumar, S., Galka, A., Durrant-Whyte, H.F., Guivant, J., Ozaki, T.: Evaluating the performance of Kalman-filter-based EEG source localization. IEEE Transactions on Bio-medical Engineering 56(1), 122–136 (2009)
10. Tibshirani, R., Saunders, M., Rosset, S., Zhu, J., Knight, K.: Sparsity and smoothness via the fused lasso. Journal of the Royal Statistical Society Series B, 91–108 (2005)
11. Monti, R.P., Hellyer, P., Sharp, D., Leech, R., Anagnostopoulos, C., Montana, G.: Estimating time-varying brain connectivity networks from functional MRI time series. NeuroImage 103, 427–443 (2014)

12. Belardinelli, P., Ortiz, E., Barnes, G., Noppeney, U., Preissl, H.: Source Reconstruction Accuracy of MEG and EEG Bayesian Inversion Approaches. PloS One 7(12), e51985 (2012)
13. Harrison, L.M., Penny, W., Ashburner, J., Trujillo-Barreto, N., Friston, K.J.: Diffusion-based spatial priors for imaging. NeuroImage 38(4), 677–695 (2007)
14. Ye, G.-B., Xie, X.: Split bregman method for large scale fused lasso. Computational Statistics Data Analysis 55(4), 1552–1569 (2011)
15. Henson, R.N., Wakeman, D.G., Litvak, V., Friston, K.J.: A parametric empirical bayesian framework for the eeg/meg inverse problem: generative models for multisubject and multimodal integration. Frontiers in Human Neuroscience 5(76) (2011)
16. Harris, A.M., Duchaine, B.C., Nakayama, K.: Normal and abnormal face selectivity of the M170 response in developmental prosopagnosics. Neuropsychologia 43(14), 2125–2136 (2005)

Connectivity Analysis of Motor Imagery Paradigm Using Short-Time Features and Kernel Similarities

L.F. Velasquez-Martinez, A.M. Alvarez-Meza,
and G. Castellanos-Dominguez[✉]

Signal Processing and Recognition Group, Universidad Nacional de Colombia,
Manizales, Colombia
{lfvelasquezma,amalvarezme,cgcastellanosd}@unal.edu.co

Abstract. The analysis of coactive regions during a Motor Imagery (MI) task becomes an important issue for revealing the primary neural activity provided by movement intentions. This information should be useful in the design of Brain Computer Interface systems. In this work, a connectivity analysis strategy for the MI paradigm using short-time features and kernel similarities is proposed. Since the imagination and execution of tracking movements are associated with neural rhythm power changes in the μ and β bands, we estimate three representative short-time feature extraction methods (Power spectral density, Hjort, and wavelet parameters). Moreover, a kernel-based pairwise similarity is computed among channels to highlight brain coactive areas during a MI task. In addition, the influence of an EEG preprocessing stage before computing the short-time features and the similarity among channels is studied. The attained results demonstrate that our approach can capture the main brain activity relationships in accordance with the MI paradigm clinic findings.

Keywords: Connectivity analysis · Motor imagery · Short-time analysis · Kernel methods

1 Introduction

Brain Computer Interface (BCI) aims to assess brain activity patterns by analyzing multi-channel time-series extracted from electrical recordings, resulting from neuron interactions, e.g., Electroencephalography signal (EEG). The main BCI assumption is that the neural activity generated by the brain is independent of its normal output pathways of peripheral nerve techniques. Then, the electrical activity of brain function might provide a new non-muscular channel for

Under grants provided by a PhD. scholarship funded by Colciencias and the project "*Convocatoria del programa nacional de apoyo a estudiantes de posgrado para el fortalecimiento de la investigación, creación, e innovación*, Universidad Nacional de Colombia 2013 - 2015 II cohorte 2014".

© Springer International Publishing Switzerland 2015
J.M. Ferrández Vicente et al. (Eds.): IWINAC 2015, Part I, LNCS 9107, pp. 439–448, 2015.
DOI: 10.1007/978-3-319-18914-7_46

sending messages and commands to the external world [1]. Thus, the analysis of the human sensorimotor functions from EEG signals can help people with physical disability or degenerative neuropathologies, where BCI systems are based on the cognitive neuroscience paradigm termed as Motor Imagery (MI) [2]. The MI relies on brain activity patterns of the imagination of a motor action, e.g., the imagination of hand movements [3].

In this sense, the analysis of coactive regions during an MI task becomes an important issue for revealing the primary neural activity provided by movement intentions [4]. However, as there are millions of functionally interconnected neurons, the neural system becomes highly distributed, dynamic, and complex. Therefore, the human brain behaves as a complex network of structural and functional interconnected regions [5]. Several computational methods have been proposed to find the brain regions working together, as both, the coherence and the correlation. The coherence is a simple frequency-dependent measure of association between two processes and the correlation is a measure of linear dependence. Nonetheless, these mentioned measures just capture linear relations in frequency and time domain, respectively, being a strong assumption about the brain region communication process [6]. On the other hand, other measures of dependency are used in the state of the art, for instance, the Generalized Measure of Association (GMA) is used for finding interconnected brain areas on the analysis of others cognitive tasks [7].

We aim to encode brain connectivity across multiple cortical areas by finding spatial similarity among measured EEG channels. In this sense, we introduce a kernel-based pairwise Inter-Channel Similarity (K-ICS). We test the K-ICS method changing two stages: Data enhancement and Feature extraction. The Data enhancement is realized with the purpose of extract primarily MI information using the Empirical Mode Decomposition (EMD) method [8]. Later, the feature vector is estimated using three representative short-time feature extraction methods in order to highlight neural power changes in MI paradigm [9]. We compare the performance of K-ICS against correlation and GMA methods. The results based on short-time features and K-ICS is able to capture the main brain activity relationships in accordance with clinical findings of channel activation for MI task reported in the literature.

The paper is organized as follows. Section 2 describes the theoretical background of the proposed approach. Section 3 provides an overview of the experiments and results from tested methods. Finally, in section 4, we present the conclusions of the work.

2   Materials and Methods

2.1   Feature Vector Extraction

Let $\Psi = \{ \mathbf{Y}^{(r)} : r = 1, \dots, R \in \mathbb{N} \}$ be a set of R raw EEG data trials, where $\mathbf{Y}^{(r)} \in \mathbb{R}^{C \times T}$ is the r-th observed trial with $C \in \mathbb{N}$ channels $\mathbf{y}_c^{(r)} \in \mathbb{R}^T$ ($c \in [1, C]$) and $T \in \mathbb{N}$ time samples. Besides, let $\Upsilon = \{ l^{(r)} \in \{-1, +1\} \}$ be the class label set of Ψ, termed the

MI paradigm condition. Since the imagination and execution of tracking movements are associated with task-related power changes in the μ (8–13 Hz) and β (13–30 Hz) rhythms [10], several parameters characterizing these waveforms are examined [9]. Instead of a long-term parameter set extracted from the input EEG data, however, we incorporate throughout this study a set of short–time parameters, where each EEG signal frame is quantified by a single feature vector.

Power Spectral Density Parameters (PSD): We compute PSD for each EEG channel $y_c \in Y$, noted as $s \in \mathbb{R}^{N_B}$, where $N_B \in \mathbb{N}$ is the number of frequency bins fixed according to the spectral band of each rhythm at hand. Provided the EEG sample frequency $F_s \in \mathbb{R}^+$, the PSD vector $s = \{s_f : f = 1, \ldots, N_B\}$ (with $s_f \in \mathbb{R}$ and $N_B = \lfloor F_s/2 \rfloor$) is computed by means of the nonparametric Welch's method applied to a set of $M \in \mathbb{N}$ overlapping segments, which are split from the input EEG vector. Due to the non–stationary nature of EEG data, the piecewise stationary analysis is carried out over the set of extracted overlapping segments that are further windowed by a smooth-time weighting window $\alpha \in \mathbb{R}^L$ lasting $L \in \mathbb{N}$ ($L < T$). Thus, we accomplish a set of windowed segments $\{v^m \in \mathbb{R}^L : m = 1, \ldots, M\}$, where $v_i^m \in \mathbb{R}$ ($i = 1, \ldots, L$) is the i-th element of v^m. Further, the modified periodogram vector $u = \{u_f \in \mathbb{R}^+ : f = 1, \ldots, N_B\}$, $u \in \mathbb{R}^{N_B}$, is computed based on the Discrete Fourier Transform as follows:

$$u_f = \sum_{m=1}^{M} \left| \sum_{i=1}^{L} v_i^m \exp\left(-j2\pi i f\right) \right|^2 .$$

In the end, each PSD parameter is computed as $s_f = u_f/(M\nu)$, being $\nu = \mathbb{E}\{|\alpha_i|^2 : \forall i \in L\}$. Notation $\mathbb{E}\{\cdot\}$ stands for the expectation operator.

Hjorth Parameters: For each windowed segment v^m, the following time-domain parameters are extracted:

– *Activity*, $\sigma_v^2 \in \mathbb{R}^M$, where each m-th element is directly described by the signal power variance:

$$\sigma_m^2 = \mathrm{var}\left(v^m\right), \tag{1}$$

where var($\cdot$) is the variance operator.
– *Mobility*, $\phi_v \in \mathbb{R}^M$, this parameter measures the signal mean frequency:

$$\phi_m = \sqrt{\mathrm{var}\left(v^{m\prime}\right)/\mathrm{var}\left(v^m\right)}, \tag{2}$$

being v' the derivative of v.
– *Complexity*, $\vartheta_v \in \mathbb{R}^M$, holding parameters measuring frequency variations as the deviation of the signal from the sine shape:

$$\vartheta_m = \phi_m'/\phi_m. \tag{3}$$

Wavelet-Based Parameters: The Continuous Wavelet Transform (CWT) and Discrete Wavelet Transform (DWT) are carried out to capture the spectral dynamics from EEG trials usually having nonstationary spectral components [11]. The former inner-product-based transformation quantifies similarity between one equally sampled time series sampled at time intervals $\delta_t \in \mathbb{R}$ and the base function $\gamma(\eta)$ (termed *mother wavelet*) ruled by the dimensionless parameter vector $\eta \in \mathbb{R}$. Namely, each time element of the CWT vector $\varsigma^g \in \mathbb{C}^T$ is extracted from y_c at scale $g \in \mathbb{R}$ by accomplishing their convolution with the scaled and shifted mother wavelet:

$$\varsigma_t^g = \sum_{\tau=1}^{T} y_\tau \gamma^* ((\tau - t)\delta_t/g), \tag{4}$$

where notation (*) stands for the complex conjugate. For building a picture showing amplitude variations through time in eq. (4), both procedures are used: the Wavelet scaling g and translating through the localized time index $t \in T$. In turn, the DWT adequately addresses the trade-off between time and frequency resolution for nonstationary signal analysis. DWT also provides multi-resolution and non-redundant representation by decomposing the considered time-series into a number of sub-bands at different scales, yielding a more precise time-frequency information about y_c [12]. Aiming to extract suitable time-frequency information from the DWT, the detail parameter vector $b^j \in \mathbb{C}$ at level j is defined as follows:

$$b_t^j = \sum_{k \in \mathbb{Z}} a_{j,k} \psi_{j,k}(t), \tag{5}$$

where $a_{j,k} = \sum_{t \in T} y_t h_{j,k}(t)$, with $a_{j,k} \in \mathbb{C}$, $h_{j,k}(t) \in \mathbb{C}$ is the impulse response of a given wavelet filter. Then, provided the wavelet $\psi(\cdot)$, the DWT-based decomposition of y^c is computed as $y_t = \sum_{j \in \mathbb{Z}} \sum_{k \in \mathbb{Z}} a_{j,k} \psi_{j,k}(t)$.

Lastly, once the short-time parameters mentioned above are computed for every channel y_c, several of their statistical measures are applied to extract the feature vector $x_c \in \mathbb{R}^Q$, with $Q \in \mathbb{N}$.

2.2   Kernel-Based Connectivity Analysis

Let $\varphi : \mathbb{R}^Q \to \mathcal{H}$ be a nonlinear mapping function from the original feature space, $\mathbb{R}^D$, to a Reproducing Kernel Hilbert Space (RKHS), $\mathcal{H}$. Regarding this, a connectivity measure between channels $\{x_i, x_j\}$, $\forall i, j \in C$, can be computed as a pairwise kernel-based similarity measure $\kappa(x_i, x_j) = \langle \varphi(x_i), \varphi(x_j) \rangle$, where $\kappa(\cdot, \cdot)$ is a positive definite kernel function. In addition, the so-called "*kernel trick*" avoids the need for computing directly $\varphi(\cdot)$. Moreover, due to the universal approximation ability, the well-known Gaussian kernel is employed commonly to estimate the pair-wise sample relationship as follows:

$$g(x_i, x_j; \sigma) \triangleq \exp\left(-\|x_i - x_j\|_2^2/(2\sigma^2)\right), \tag{6}$$

where $\sigma \in \mathbb{R}^+$ is the kernel band-width and $\| \cdot \|_2$ stands for the 2-norm. So, based on eq. (6), a Kernel-based Inter-Channel Similarity (K-ICS) matrix $\boldsymbol{K} \in \mathbb{R}^{C \times C}$ can be computed as follows [13]: $k_{ij} = g(\boldsymbol{x}_i, \boldsymbol{x}_j; \sigma)$. Hence, the matrix $\boldsymbol{K}$ encodes the information about the main brain activity connections for each EEG trial.

3 Experiments and Results

EEG database: We carry out experimental testing using the well-known Motor Imagery (MI) database provided by the Berlin Brain-Computer Interface group[1]. These data, employed in the BCI competition IV (2008), were collected under the cognitive neuroscience paradigm of imagination of hand movements. The database (noted as D1) holds EEG signals recorded from seven subjects belonging either to the left or right-hand class [3]. From each subject, the recordings were measured in 59 EEG positions, being the sensorimotor area the most densely covered by the electrodes. All signal set was preprocessed as follows: Firstly, band-pass filtered between 0.05 and 200 Hz, then digitized at 1000 Hz, and lastly, down-sampled at F_s=100 Hz, but previously an order 10 low-pass Chebyshev II filter had been employed having stop-band ripple 50 dB down and stopband edge frequency 49 Hz. For each person, the whole MI session was conducted without feedback, performing 100 repetitions per class. The EEG data was recorded for 4 s while a cue (an arrow) was pointing either side on one screen. Recordings were interleaved by a blank screen pause and a fixation cross shown to the screen center (either pause lasting 2 s). Overall, the set of EEG signals $\Psi = \{ \boldsymbol{Y}^{(r)} : r=1, \ldots, 200 \}$ (with $\boldsymbol{Y}^{(r)} \in \mathbb{R}^{59 \times 400}$) was acquired for each subject going under evaluation.

In order to assess the proposed methodology as a suitable tool to support connectivity analysis tasks, we carry out the following three stages: *i)* EEG data enhancement, *ii)* feature extraction, and *iii)* brain connectivity estimation. In the beginning, signal representation is enhanced by using the Empirical Mode Decomposition (EMD) providing adaptive extraction of the main MI information [8,9]. The EMD iteratively estimates each zero-mean Intrinsic Mode Function (IMF) amplitude based on the first order derivative criterion until the residual value asymptotically becomes a small constant, when no more IMF terms can be further decomposed. All first $N_I < N$ IMF terms are calculated, where we fix N_I=4 for the purpose of accentuating MI information concentrated in the μ and the β rhythms [8].

During the next stage, the short-time parameter set is initially calculated, where the window size is set as $L > F_r / F_s$ for computing the PSD and the Hjort parameters. As suggested in [14], the smallest considered frequency is F_r=8 Hz. In the case of the CWT analysis, we use a couple of Morlet wavelets [15]: one centered at 10 Hz (to extract the μ band) and another at 22 Hz (β band). For DWT analysis, the Symlet wavelet (Sym-7) is used to compute the detail parameter vector b^j as to include both rhythms, resulting in the second and third DWT levels [9]. From the computed short-time parameter set, we extract the

---

[1] http://bbci.de/competition/iv/desc_1.html

feature vector as follows: The norm and two first statistical moments of the PSD s for both brain activity bands, while the maximum value and two first statistical moments of the Hjorth and WT parameters. As a result, we get the feature vector sizing $x \in \mathbb{R}^{27}$ ($Q=27$) computed for each channel.

In the last stage of the brain connectivity estimation, the Gaussian kernel estimates the pairwise relationships among channels as in eq. (6), fixing the kernel bandwidth based on the maximization of the information potential variability [16]. For the sake of comparison, we also estimate the connectivity between channels by using the Pearson's correlation measure [17] and the Generalized Measure of Association (GMA) [7]. The latter allows computing each pair-wise inter-channel dependence preserving time resolution and generalizes the concept of relationship by reflecting the distance between realizations rather than their absolute locations. Due to temporal restrictions, GMA is estimated on the basis of the raw EEG time-series as well as on the enhanced EEG time-series. The Pearson's correlation is also calculated from the feature vectors.

For illustrative purposes, we show the assessed values of brain connectivity related to just the subjects numbered as S1, S2, and S7 in D1. The mean inter-channel dependencies per MI condition, i.e., imagination of left or right hand, are shown in figs. 1 to 3, where all maps of the coactive brain areas are computed taking as reference the C_z channel location of the 10–20 system.

figs. 1a to 1f show the obtained Pearson's correlation computed from the raw temporal representation. As seen, we get high values of connectivity dependency between channels that are spread over the entire head for both MI conditions. Almost the same values of connectivity are assessed for the enhanced EEG signal, meaning that the inclusion of this stage barely influences on the estimated measure. Nonetheless, the Pearson's correlation computed from the feature vectors leads to a notorious activation over the C-electrodes as seen in figs. 1g to 1l. The observed activation should be related to the Homunculus, i.e., the Motor Function Area (MF). Such behavior is supported by the MI paradigm clinical findings stating that the primary brain active regions must be related to the motor system [18].

In contrast, the GMA-based connectivity computed from raw EEG data clearly shows that the joint activity of the MI task gets more localized over the scalp as seen in figs. 2a to 2c. In fact, high values of the association are accomplished close to the C_z channel (central brain region over the Homunculus and the MF). Yet, a joint activation also appears around the Primary Motor Cortex (PCM - FC electrodes) that is assumed not to play a fundamental role in the MI process. Likewise, the observed activity in Parietal Cortex (CP electrodes) area most likely reflects the movement mode but not the imagery mode [19]. Here, the signal enhancement makes the GMA-based inter-channel highlight areas that are coactive during the MI task, especially, around the MF region (see figs. 2d to 2f). However, the attained connectivity representation is not in accordance with the modern findings of the clinical MI paradigm.

As seen in figs. 3a to 3c showing the connectivity values of K-ICS obtained from the raw data, the main brain connectivity are mapped over the MC central

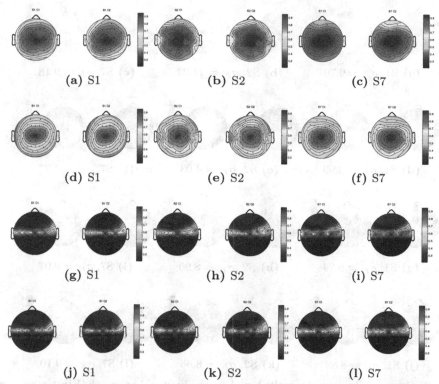

Fig. 1. Connectivity analysis results based on Pearson's correlation measure. **First row**- raw time-series representation. **Second row**- enhanced time-series representation. **Third row**- feature vector extracted from raw data. **Fourth row**- feature vector extracted from enhanced EEG data.

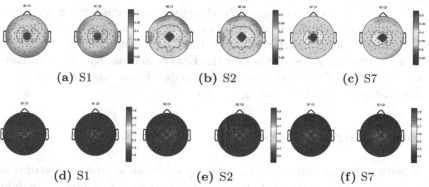

Fig. 2. Connectivity analysis results for GMA. **First row** - raw time-series representation. **Second row**- Enhanced EEG data.

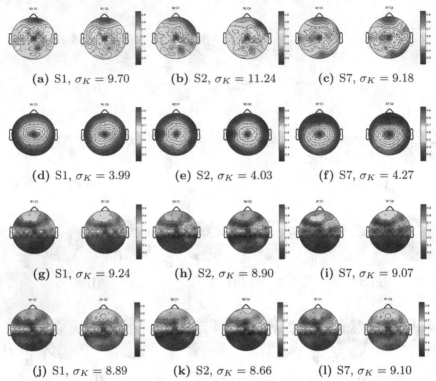

(a) S1, $\sigma_K = 9.70$ (b) S2, $\sigma_K = 11.24$ (c) S7, $\sigma_K = 9.18$

(d) S1, $\sigma_K = 3.99$ (e) S2, $\sigma_K = 4.03$ (f) S7, $\sigma_K = 4.27$

(g) S1, $\sigma_K = 9.24$ (h) S2, $\sigma_K = 8.90$ (i) S7, $\sigma_K = 9.07$

(j) S1, $\sigma_K = 8.89$ (k) S2, $\sigma_K = 8.66$ (l) S7, $\sigma_K = 9.10$

Fig. 3. Connectivity analysis results based on the proposed approach. **First row-** raw time-series representation. **Second row-** enhanced time-series representation. **Third row-** Feature vector extracted from raw EEG data. **Fourth row-** Feature vector extracted from enhanced EEG data.

region (around the C_z electrode) having some spotted activations in the PCM and the PC areas. However, those PCM activations vanish when the K-ICS measure is applied to the enhanced signal as shown in figs. 3d to 3e. An improved connectivity estimation is performed when the K-ICS measure is calculated from the feature vector regardless of whether the EEG recordings have been enhanced (see figs. 3g to 3l). These results of the estimated brain connectivity resemble the state-of-the-art MI clinic findings.

4   Concluding Remarks

In this work, a connectivity analysis strategy based on short-time features and kernel similarities is proposed. Our approach is tested as a tool to highlight brain areas that are coactive during a MI paradigm task. Due to the imagination and execution of tracking movements are associated with neural rhythm power changes in the μ and β bands, we consider three representative short-time feature extraction methods that have been widely applied in many MI studies:

PSD, Hjorth, and Wavelet methods. Thus, each EEG channel is represented as a feature vector encoding some statistical measures from the calculated short-time parameters. Afterwards, a kernel-based pairwise similarity is computed among channels to highlight brain areas that are coactive during a MI paradigm task. Moreover, the influence of a EEG preprocessing stage before computing the feature vector and the brain connectivity is studied. In addition, the well-known Pearson's correlation and the GMA are tested as baselines for estimating the inter-channel relationships.

We consider the use of EMD for enhancement of EEG data. Grounded on the obtained results, we show that this case of enhancement stage applied on raw data may facilitate the estimation of relevant brain activity regions. Moreover, a consistent localization of brain active regions increases when we compute the described short-time features from the EEG records. Indeed, a suitable brain connectivity behavior is achieved when the proposed feature extraction is applied over the raw EEG data. In this regard, the introduced short-time features highlight the main power variations in the μ and β rhythms allowing to identify the underlying neurological user mechanism during the MI task [20]. Now, with respect to the Person's correlation technique, it is possible to notice that a suitable brain connectivity representation is achieved when short-time features are computed from the EEG signals. Besides, the GMA-based connectivity is not able to highlight the main brain areas that are coactive during the MI task. In contrast, attained results demonstrate that our approach, which includes short-time feature extraction and K-ICS, can capture the main brain activity relationships in accordance with the MI paradigm clinic findings.

As future work, authors plan to test the introduced approach over different MI paradigms and other Brain Computer Interface tasks. Furthermore, an online extension of the introduced connectivity analysis strategy can be proposed to include directly the temporal variations of the inter-channel relationships.

References

1. Wolpaw, J.R., Birbaumer, N., McFarland, D.J., Pfurtscheller, G., Vaughan, T.M.: Brain–computer interfaces for communication and control. Clinical Neurophysiology 113(6), 767–791 (2002)
2. Yao, L., Meng, J., Zhang, D., Sheng, X., Zhu, X.: Combining motor imagery with selective sensation towards a hybrid-modality bci (2013)
3. Allison, B.Z., Wolpaw, E.W., Wolpaw, J.R.: Brain-computer interface systems: progress and prospects. Exp. Rev. of Med. Dev. 4(4), 463–474 (2007)
4. Heger, D., Terziyska, E., Schultz, T.: Connectivity based feature-level filtering for single-trial eeg bcis. In: ICASSP, pp. 2064–2068. IEEE (2014)
5. Lin, L., Il Memming, P., Seth, S., Sanchez, J., Príncipe, J.: Functional connectivity dynamics among cortical neurons: a dependence analysis. IEEE Transactions on Neural Systems and Rehabilitation Engineering 20(1), 18–30 (2012)
6. Sakkalis, V.: Review of advanced techniques for the estimation of brain connectivity measured with eeg/meg. Computers in Biology and Medicine 41(12), 1110–1117 (2011)

7. Fadlallah, B., Seth, S., Keil, A., Principe, J.: Quantifying cognitive state from eeg using dependence measures. IEEE Transactions on Biomedical Engineering 59(10), 2773–2781 (2012)
8. He, W., Wei, P., Wang, L., Zou, Y.: A novel emd-based common spatial pattern for motor imagery brain-computer interface. In: IEEE EMBC (2012)
9. Alvarez-Meza, A.M., Velasquez-Martinez, L.F., Castellanos-Dominguez, G.: Time-series discrimination using feature relevance analysis in motor imagery classification. Neurocomputing 151, 122–129 (2015)
10. Zhang, H., Guan, C., Ang, K.K., Wang, C.: Bci competition iv–data set i: learning discriminative patterns for self-paced eeg-based motor imagery detection. Frontiers in Neuroscience 6 (2012)
11. Corralejo, R., Hornero, R., Álvarez, D.: Feature selection using a genetic algorithm in a motor imagerybased brain computer interface. In: IEEE EMBC (2011)
12. Carrera-Leon, O., Ramirez, J.M., Alarcon-Aquino, V., Baker, M., D'Croz-Baron, D., Gomez-Gil, P.: A motor imagery bci experiment using wavelet analysis and spatial patterns feature extraction. In: 2012 Workshop on Engineering Applications (WEA), pp. 1–6. IEEE (2012)
13. García-Vega, S., Álvarez-Meza, A.M., Castellanos-Domínguez, G.: Neural decoding using kernel-based functional representation of ECoG recordings. In: Bayro-Corrochano, E., Hancock, E. (eds.) CIARP 2014. LNCS, vol. 8827, pp. 247–254. Springer, Heidelberg (2014)
14. Teixeira, A.R., Tomé, A.M., Boehm, M., Puntonet, C., Lang, E.: How to apply nonlinear subspace techniques to univariate biomedical time series. IEEE Trans. on Instrument. and Measur. 58(8), 2433–2443 (2009)
15. Lemm, S., Schafer, C., Curio, G.: Bci competition 2003-data set iii: probabilistic modeling of sensorimotor μ rhythms for classification of imaginary hand movements. IEEE Transactions on Biomedical Engineering 51(6), 1077–1080 (2004)
16. Álvarez-Meza, A.M., Cárdenas-Peña, D., Castellanos-Dominguez, G.: Unsupervised kernel function building using maximization of information potential variability. In: Bayro-Corrochano, E., Hancock, E. (eds.) CIARP 2014. LNCS, vol. 8827, pp. 335–342. Springer, Heidelberg (2014)
17. Bonita, J.D., Ambolode II, L.C.C., Rosenberg, B.M., Cellucci, C.J., Watanabe, T.A.A., Rapp, P.E., Albano, A.M.: Time domain measures of inter-channel eeg correlations: a comparison of linear, nonparametric and nonlinear measures. Cognitive Neurodynamics 8(1), 1–15 (2014)
18. Hanakawa, T., Immisch, I., Toma, K., Dimyan, M.A., Van Gelderen, P., Hallett, M.: Functional properties of brain areas associated with motor execution and imagery. Journal of Neurophysiology 89(2), 989–1002 (2003)
19. Velásquez-Martínez, L.F., Álvarez-Meza, A.M., Castellanos-Domínguez, C.G.: Motor imagery classification for BCI using common spatial patterns and feature relevance analysis. In: Ferrández Vicente, J.M., Álvarez Sánchez, J.R., de la Paz López, F., Toledo Moreo, F. J. (eds.) IWINAC 2013, Part II. LNCS, vol. 7931, pp. 365–374. Springer, Heidelberg (2013)
20. Mason, S.G., Birch, G.E.: A general framework for brain-computer interface design. IEEE Transactions on Neural Systems and Rehabilitation Engineering 11(1), 70–85 (2003)

Robust Linear Longitudinal Feedback Control of a Flapping Wing Micro Air Vehicle

Lidia María Belmonte[1], R. Morales[1], Antonio Fernández-Caballero[1(✉)],
and José A. Somolinos[2]

[1] Escuela de Ingenieros Industriales, Universidad de Castilla-La Mancha,
02071-Albacete, Spain
antonio.fdez@uclm.es
[2] Escuela Técnica Superior de Ingenieros Navales, Universidad Politécnica de
Madrid, 28040-Madrid, Spain

Abstract. This paper falls under the idea of introducing biomimetic
miniature air vehicles in ambient assisted living and home health applications. The concepts of active disturbance rejection control and flatness
based control are used in this paper for the trajectory tracking tasks in
the flapping-wing miniature air vehicle (FWMAV) time-averaged model.
The generalized proportional integral (GPI) observers are used to obtain accurate estimations of the flat output associated phase variables
and of the time-varying disturbance signals. This information is used in
the proposed feedback controller in (a) approximate, yet close, cancela-
tions, as lumped unstructured time-varying terms, of the influence of the
highly coupled nonlinearities and (b) the devising of proper linear output
feedback control laws based on the approximate estimates of the string
of phase variables associated with the flat outputs simultaneously provided by the disturbance observers. Numerical simulations are provided
to illustrate the effectiveness of the proposed approach.

1 Introduction

The creation of flapping wing micro air vehicles (FWMAV) is a challenging problem. The potential benefits for insect-like flapping wing micro air vehicles are
numerous [1]. The hovering ability of insects, coupled with the ability for a quick
transition to forward flight, provide an ideal indoor/outdoor reconnaissance platform for search and rescue, reconnaissance and surveillance and ambient assisted
living and home health, among others [2]-[8]. Indeed, this paper falls within a
project called "Improvement of the Elderly Quality of Life and Care through
Smart Emotion Regulation". The long-term objective of the project is to find
solutions for improving the quality of life and care of ageing adults at home by
using emotion detection and regulation techniques. We believe that miniature
air vehicles at home settings are capable of including some sensors that capture
the mood of the ageing adults.

Different control methods have been found in the literature. Deng *et al.* developed in [9] a nominal state-space linear time-invariant model in hover through

© Springer International Publishing Switzerland 2015
J.M. Ferrández Vicente et al. (Eds.): IWINAC 2015, Part I, LNCS 9107, pp. 449–458, 2015.
DOI: 10.1007/978-3-319-18914-7_47

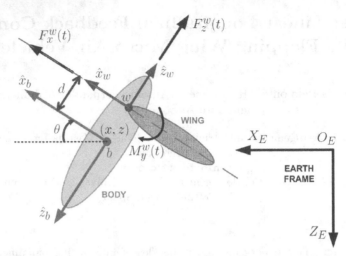

Fig. 1. Coordinate systems and longitudinal motion of FWMAV with respect to the earth frame

linear estimation. Also, a LQG controller was designed and compared with a PD controller. A state feedback attitude controller control scheme using the sensor output as feedback was designed by Schenato *et al.* [10]. Campolo *et al.* realized in [11] a geometric approach to robust attitude estimation, derived from multiple and possibly redundant bio-inspired navigation sensors, for attitude stabilization of a micromechanical flying insect.

The use of time-averaging theory has been used within the control of FWMAV because it helps to simplify the complex aerodynamics associated to the flapping wings [12]-[13] because the aerodynamic forces and torques, generated by the wings, affect the behavior of the FWMAV only by their mean values since the dynamics of the body are much slower than the flapping wings ones. Deng *et al.* provided a methodology to approximate the time-varying body dynamics caused by the aerodynamic forces with time-invariant dynamics using averaging theory and a biomimetic parametrization of wing trajectories [14]. Also, a Linear Quadratic Gaussian (LQG) controller was designed which does not require the knowledge of an accurate model for the insect morphological parameters, such as moment of inertia and mechanical part's sizes, nor an accurate model of the aerodynamics. Rifaï *et al.* developed in [15] a bounded state feedback control of the forces and torques and takes into account the saturation of the actuators driving the flapping wings and Khan *et al.* realized in [16] a differential flatness based non-linear controller based on the time-averaging theory for the control of the longitudinal dynamics of FW-MAV.

Taking into consideration the highly nonlinear nature of the FWMAV, active disturbance rejection control (ADRC) appears as an excellent methodology for the control of uncertain linear and nonlinear systems (see the work of Han [17] for the initial theoretical aspects of this new area of research). The objective of ADRC stems in the accurate estimation of the unknown part of the controlled system dynamics and proceed to cancel its effects in the feedback control law. Gao and its coworkers have proposed new advances in controllers, including practical applications, in a similar manner to that of Han [18], [19].On the other hand, Sira-Ramírez and its coworkers have contribute to the area emphasizing the use of *generalized proportional integral (GPI) observers* [20]-[22].

In this line of action, in this article, we propose a robust observer-based linear output feedback control scheme for the trajectory tracking tasks in the flapping-wing miniature air vehicle time-averaged model. The linear observer-based controller design approach rests on using highly simplified models of the inputs differential parameterizations, provided by the flatness property. Within the simplification task proposed, only the order of integration of the subsystems and the control inputs, along with their associated matrix gains, are retained in full detail. All the additive nonlinearities, including their state couplings and complexities, are regarded as, unstructured, time-varying signals that need to be online estimated, and canceled, at the controller specification within an Active Disturbance Rejection Control Scheme. After input gain matrix cancelation, the resulting system consists of pure integration (linear) perturbed systems with time-varying additive disturbances. A set of linear extended observers, here denominated as *GPI observers*, are capable of accurate on-line estimations of: (1) the output related phase variables; (2) the, state dependent, additive perturbation input signal itself; and (3) the estimation of a certain number of the perturbation input time derivatives. This last feature facilitates the task of perturbation input prediction as GPI observers are the most naturally applicable to the control of perturbed differentially flat nonlinear systems [23]-[25].

The remainder of the article is structured as follows: Section 2 presents the flapping-wing miniature air vehicle time-averaged model and its flatness property. Additionally, this section proposes a simplified model of the system and formulates the problem to be solved. Section 3 describes the active disturbance rejection controller design and the results are applied for the stabilization and trajectory tracking problem of the time-averaged model for the flapping-wing miniature air vehicle. Section 4 presents the obtained simulation results and, finally, Section 5 is devoted for the conclusions of this study and future works.

2   Problem Formulation and Its Flatness Property

2.1   System Dynamics

Consider the following time-averaged model for the flapping-wing miniature air vehicle (FWMAV) based on Newtonian approach derived in [16]:

$$\dot{x} = v_x C_\theta + v_z S_\theta \tag{1}$$

$$\dot{z} = -v_x S_\theta + v_z C_\theta \tag{2}$$

$$\dot{\theta} = \omega \tag{3}$$

$$\dot{v}_x = -g S_\theta - \omega v_z + F_x \tag{4}$$

$$\dot{v}_z = -g C_\theta + \omega v_x - F_z \tag{5}$$

$$\dot{\omega} = -\frac{F_x}{E} \tag{6}$$

where $S_\theta = \sin\theta, C_\theta = \cos\theta, g$ is the gravity acceleration, (x, z) are the coordinates of the center of mass in the earth frame, θ represents the pitch angle, (v_x, v_z) express the velocity of the body of the FWMAV in the body frame and ω is the angular velocity of the body and (F_x, F_z) represent the aerodynamic forces. The constant $E = \frac{I_b}{md}$, being I_b the moment of inertia of the body about the y axis of the body frame, m is the mass and d denotes the distance from the axis of oscillation to the center of mass of the body. Fig. 1 shows the coordinate systems and longitudinal motion of the FWMAV with regard to the earth frame.

2.2   Flatness of the System

According to the theory of differential flatness [21], a dynamic system, $\dot{\mathbf{x}} = \mathbf{f}(\mathbf{x}, \mathbf{u})$, with $\mathbf{x} \in \mathbb{R}^n$ and $\mathbf{u} \in \mathbb{R}^m$, is said to be differentially flat if there exist, m, differentially independent variables called *flat outputs* (differentially independent meaning that they are not related by differential equations), which are functions of the state vector and, possibly, of a finite number of time derivatives of the state vector (i.e., derivatives of the inputs may be involved in their definition), such that *all* system variables (states, inputs, outputs, and functions of these variables) can, in turn, be expressed as functions of the flat outputs and of a finite number of their time derivatives. This parameterization establishes a one-to-one mapping from the states and the inputs to the flat outputs.

The proposed system is differentially flat with flat outputs given by the coordinates of the *Huygens center of oscillation* [22] given by:

$$F = x + ES_\theta, \quad L = z + EC_\theta \tag{7}$$

Proposition 1. *The flapping-wing miniature air vehicle given in (1)-(6) is differentially flat, with flat outputs given by F and L, i.e., all system variables in (1)-(6) can be differentially parameterized solely in terms of F, L, and a finite number of their time derivatives.*

Proof. If the equations given in (7) are differentiated with regard to time, we obtain the first and second derivatives of the flat outputs:

$$\dot{F} = \dot{x} + E\omega C_\theta = v_x C_\theta + v_z S_\theta + E\omega C_\theta \tag{8}$$
$$\dot{L} = \dot{z} - E\omega S_\theta = -v_x S_\theta + v_z C_\theta - E\omega S_\theta \tag{9}$$
$$\ddot{F} = \xi S_\theta \tag{10}$$
$$\ddot{L} = g + \xi C_\theta \tag{11}$$

where $\xi = -\left(F_z - E\omega^2\right)$ is defined as a new virtual input vector. Upon operating with (10) and (11) we achieve:

$$\xi = \sqrt{\ddot{F}^2 + \left(\ddot{L} - g\right)^2}; \quad \theta = \arctan\left(\frac{\ddot{F}}{\ddot{L} - g}\right) \tag{12}$$

$$S_\theta = \frac{\ddot{F}}{\sqrt{\ddot{F}^2 + \left(\ddot{L} - g\right)^2}}; \quad C_\theta = \frac{\ddot{L} - g}{\sqrt{\ddot{F}^2 + \left(\ddot{L} - g\right)^2}} \tag{13}$$

If the expressions (10) and (11) are differentiated with regard to time, it is obtained

$$F^{(3)} = \dot{\xi} S_\theta + \xi \omega C_\theta; \quad L^{(3)} = \dot{\xi} C_\theta - \xi \omega S_\theta \tag{14}$$

Rearranging terms in (14) yields

$$\dot{\xi} = \frac{\ddot{F} F^{(3)} + \left(\ddot{L} - g\right) L^{(3)}}{\sqrt{\ddot{F}^2 + \left(\ddot{L} - g\right)^2}}; \quad \omega = \dot{\theta} = \frac{F^{(3)} \left(\ddot{L} - g\right) - L^{(3)} \ddot{F}}{\ddot{F}^2 + \left(\ddot{L} - g\right)^2} \tag{15}$$

Now, operating with (4) and (5) one obtains

$$v_x = \dot{x} C_\theta - \dot{z} S_\theta = \dot{F} C_\theta - \dot{L} S_\theta - E\omega \tag{16}$$
$$v_z = \dot{x} S_\theta + \dot{z} C_\theta = \dot{F} S_\theta + \dot{L} C_\theta \tag{17}$$

Combining (16) and (17) with (13) and (15), we conclude that v_x and v_z are also functions of $(\dot{F}, \dot{L}, \ddot{F}, \ddot{L}, F^{(3)}, L^{(3)})$. On the other hand, differentiating expressions (14) with regard to time and rearranging terms

$$F^{(4)} = S_\theta \ddot{\xi} - \frac{\xi C_\theta}{E} F_x - \omega^2 \xi S_\theta + 2\dot{\xi} \omega C_\theta \tag{18}$$

$$L^{(4)} = C_\theta \ddot{\xi} + \frac{\xi S_\theta}{E} F_x - \omega^2 \xi C_\theta - 2\dot{\xi} \omega S_\theta \tag{19}$$

Similarly, upon operating with (18), it is achieved

$$\ddot{\xi} = S_\theta F^{(4)} + C_\theta L^{(4)} + \omega^2 \xi \tag{20}$$

$$F_x = \frac{-EC_\theta}{\xi} F^{(4)} + \frac{ES_\theta}{\xi} L^{(4)} + 2\frac{E\omega \dot{\xi}}{\xi} \tag{21}$$

Finally, substituting (12), (13) and (15) into (20) shows that all the system variables can be expressed as a function of (F, L) and their derivatives, proving that the flat output vector composed by (F, L) constitute a flat output vector for system (1)-(6).

2.3   Simplified Model and Problem Formulation

On the basis of (20), we adopt the following simplified perturbed model for the underlying FWMAV (18):

$$\begin{bmatrix} F^{(4)} \\ L^{(4)} \end{bmatrix} = \underbrace{\begin{bmatrix} S_\theta & -\frac{\xi C_\theta}{E} \\ C_\theta & \frac{\xi S_\theta}{E} \end{bmatrix}}_{\mathcal{N}(\theta, \xi)} \begin{bmatrix} \ddot{\xi} \\ F_x \end{bmatrix} + \underbrace{\begin{bmatrix} \varphi_F \\ \varphi_L \end{bmatrix}}_{\varphi(t)} \tag{22}$$

where $\varphi(t) = [\varphi_F, \varphi_L]^T$ involves state dependent expressions, the possibly unmodeled dynamics and external unknown disturbances affecting the system. We lump all these uncertain terms into an unknown but uniformly absolutely bounded disturbance input that needs to be on-line estimated by means of an observer and, subsequently, canceled from the simplified system dynamics via feedback in order to regulate the flat output vector, $[F, L]^T$, towards the desired reference trajectories $[F^*, L^*]^T$. Finally, the formulation of the problem is stated as follows: *Given a desired flat output vector of reference trajectories $[F^*, L^*]^T$, devise a linear multi-input output feedback controller for system (22) such that the flat output vector $[F, L]^T$ is forced to track the given reference flat output vector $[F^*, L^*]^T$. This objective must be achieved even in the presence of unknown disturbances and coupling nonlinearities, represented by $[\varphi_F, \varphi_L]^T$.*

3   GPI Observer-Based Active Disturbance Rejection Controller

A GPI observer including a reasonable, self-updating, time-polynomial model is considered for each unknown component disturbance input vector $\varphi(t)$. For this internal model, we use for each component of $\varphi(t)$ an unspecified element of a fifth order family of time-polynomials, denoted by $\varphi_1^{(6)}(t) = [\varphi_{1F}^{(6)}, \varphi_{1L}^{(6)}]^T = \mathbf{0}$. The GPI observer based flat output feedback controller is devised as follows:

$$\begin{bmatrix} \dot{\xi} \\ F_x \end{bmatrix} = \underbrace{\begin{bmatrix} S_{\hat{\theta}_s} & C_{\hat{\theta}_s} \\ \dfrac{EC_{\hat{\theta}_s}}{xi_s} & \dfrac{ES_{\hat{\theta}_s}}{xi_s} \end{bmatrix}}_{\mathcal{N}^{-1}(\theta, \xi)} \begin{bmatrix} \nu_F \\ \nu_L \end{bmatrix} \tag{23}$$

with

$$\nu_F = -\hat{\varphi}_{1Fs} + [F^*(t)]^{(4)} - \sum_{i=0}^{3} k_i^F \left(\hat{F}_s^{(i)} - [F^*]^{(i)} \right)$$

$$\nu_L = -\hat{\varphi}_{1Ls} + [L^*(t)]^{(4)} - \sum_{i=0}^{3} k_i^L \left(\hat{L}_s^{(i)} - [L^*]^{(i)} \right) \tag{24}$$

where the quantities with subindex s are smoothing observer variables which are carried out by means of the following *clutching function*, avoiding possible large peaks in their high gain induced responses:

$$s_f(t) = \begin{cases} 1 \text{ for } t > \varepsilon \\ \sin^8 \left(\frac{\pi t}{2\varepsilon} \right) \text{ for } t \leq \varepsilon \end{cases} \tag{25}$$

with $\epsilon = 2\,[s]$. The design coefficients k_i^F and k_i^L, $i = 0, 1, 2, 3$, are chosen so that the dominant characteristic polynomials are 4th-degree Hurwitz polynomials, i.e.,

$$p_F(s) = s^4 + k_3^F s^3 + k_2^F s^2 + k_1^F s + k_0^F \in \text{Hurwitz}_4(s)$$

$$p_L(s) = s^4 + k_3^L s^3 + k_2^L s^2 + k_1^L s + k_0^L \in \text{Hurwitz}_4(s) \tag{26}$$

render an asymptotically, exponentially convergence of the flat output error vector, $[e_F, e_L]^T = [F - F^*, L - L^*]^T$, towards a small vicinity of the origin of the tracking error phase space. Furthermore, the variables $\hat{F}^{(j)} = F_j$ and $\hat{L}^{(j)} = L_j$, $j = 0, 1, \ldots, 3$ are generated by:

$$\dot{F}_0 = F_1 + \lambda_8^F(F - F_0)$$
$$\dot{F}_1 = F_2 + \lambda_7^F(F - F_0)$$
$$\dot{F}_2 = F_3 + \lambda_6^F(F - F_0)$$
$$\dot{F}_3 = S_\theta\ddot{\xi} - \frac{\xi C_\theta}{E}F_x + \varphi_{1F} + \lambda_5^F(F - F_0)$$
$$\dot{\varphi}_{1F} = \varphi_{2F} + \lambda_4^F(F - F_0)$$
$$\dot{\varphi}_{2F} = \varphi_{3F} + \lambda_3^F(F - F_0)$$
$$\dot{\varphi}_{3F} = \varphi_{4F} + \lambda_2^F(F - F_0) \qquad (27)$$
$$\dot{\varphi}_{4F} = \varphi_{5F} + \lambda_1^F(F - F_0)$$
$$\dot{\varphi}_{5F} = \lambda_0^F(F - F_0)$$

$$\dot{L}_0 = L_1 + \lambda_8^L(L - L_0)$$
$$\dot{L}_1 = L_2 + \lambda_7^L(L - L_0)$$
$$\dot{L}_2 = L_3 + \lambda_6^L(L - L_0)$$
$$\dot{L}_3 = C_\theta\ddot{\xi} + \frac{\xi S_\theta}{E}F_x + \varphi_{1L} + \lambda_5^L(L - L_0)$$
$$\dot{\varphi}_{1L} = \varphi_{2L} + \lambda_4^L(L - L_0)$$
$$\dot{\varphi}_{2L} = \varphi_{3L} + \lambda_3^L(L - L_0)$$
$$\dot{\varphi}_{3L} = \varphi_{4L} + \lambda_2^L(L - L_0)$$
$$\dot{\varphi}_{4L} = \varphi_{5L} + \lambda_1^L(L - L_0) \qquad (28)$$
$$\dot{\varphi}_{5L} = \lambda_0^L(L - L_0)$$

where the design coefficients λ_i^F and λ_i^L, $i = 0, 1, \ldots, 8$, are chosen so that the reconstruction error dynamics dominant characteristic polynomials are 9th-degree Hurwitz polynomials, i.e.,

$$p_{F_0}(s) = s^9 + \lambda_8^F s^8 + \lambda_7^F s^7 + \ldots + \lambda_1^F s + \lambda_0^F \in \text{Hurwitz}_9(s)$$
$$p_{L_0}(s) = s^9 + \lambda_8^L s^8 + \lambda_7^L s^7 + \ldots + \lambda_1^L s + \lambda_0^L \in \text{Hurwitz}_9(s) \qquad (29)$$

and their roots are located sufficiently far from the imaginary axis, in the left half of the complex plane, then the trajectories of the flat output estimation error vector, $[\tilde{e}_F, \tilde{e}_L]^T = [F - F_0, L - L_0]^T$, and of its time derivatives, will converge to a small neighborhood of the origin of the phase space of the observer estimation error. The further away the roots are located in the left half of the complex plane, the smaller the radius of the disk representing the neighborhood around the origin of the estimation error phase space will be.

4   Numerical Simulations

Numerical simulations were carried out in order to verify the efficiency of the proposed approach in terms of quick convergence of the tracking errors to a small neighborhood of zero and smooth transient responses. The system parameters are: $m = 2.5 \cdot 10^{-3} \, [kg]$ and $I_b = 8.125 \cdot 10^{-7} \, [kg \, m^2]$. The flat output vector $[F, L]^T$ has been designed to track the following reference trajectories:

$$F^* = R \sin(At) + E \sin(\alpha(t)) \tag{30}$$
$$L^* = R \left[\cos(At) - 1 \right] - z_0 + E \cos(\alpha(t)) \tag{31}$$

where $R = 7 \, [m]$, $z_0 = 0.5 \, [m]$, $A = 2\pi/30 \, [rad/s]$ and $\alpha(t) = B_1 \sin(B_2 t)$ being $B_1 = \pi/180 \, [rad]$ and $B_2 = 2\pi/30 \, [rad/s]$.

The time sampling used in all the simulations is $T = 0.001 \, [s]$. The observer gains, $\{\lambda_8^F, \dots, \lambda_0^F\}$ and $\{\lambda_8^L, \dots, \lambda_0^L\}$ were selected by identifying, term by term, the coefficients of the polynomials given in expression (29) with those of a desired Hurwitz polynomial given by $p_{obs}(s) = \left(s^2 + 2\zeta_o \omega_{no} s + \omega_{no}^2 \right)^4 \cdot (s + p_o)$, with $\omega_{no} = 15, \zeta_o = 1.5$ and $p_o = 15$. On the other hand, the controller gains, $\{k_3^F, \dots, k_0^F\}$ and $\{k_3^L, \dots, k_0^L\}$, governing the dominant dynamics, were set by identifying, term by term, the coefficients of the polynomials given in expression (26) with the Hurwitz polynomial $p_{cont}(s) = \left(s^2 + 2\zeta_c \omega_{nc} s + \omega_{nc}^2 \right)^2$, with $\omega_{nc} = 2, \zeta_c = 1$. Fig. 4a and Fig. 4b illustrate the path tracking and the closed loop trajectories for the coordinates of the center of mass (x, z) in the earth frame showing that the system follows the desired trajectory in an accurate manner. On the other hand, Fig. 4c

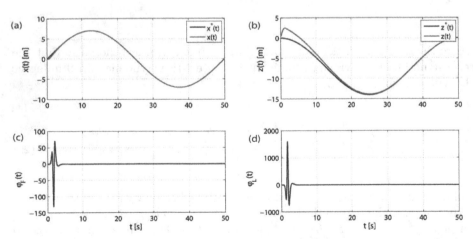

Fig. 2. Evolution of: (a) Coordinate x of the center of mass in the earth frame; (b) Coordinate z of the center of mass in the earth frame; (c) State-dependent estimated disturbance φ_F and; (d) State-dependent estimated disturbance φ_L

and Fig. 4d depict the evolution of the GPI observer state dependent disturbance estimation.

5   Conclusions and Future Work

This paper is related to the introduction of biomimetic miniature air vehicles in ambient assisted living and home health applications. Indeed, the proposal described falls within the complete project "Improvement of the Elderly Quality of Life and Care through Smart Emotion Regulation". The long-term objective of the project is to find solutions for improving the quality of life and care of the elderly who can or wants to continue living at home by using emotion detection and regulation techniques. We believe that miniature air vehicles at home settings can carry some fundamental sensors to capture the mood of the ageing adult.

In this way, this particular work has explored, within the context of the trajectory tracking problem, the use of approximate, yet accurate, total active disturbance rejection schemes, based on linear GPI observers, for the flapping-wing miniature air vehicle time-averaged model. Numerical simulations were provided where the efficiency of the proposed control method is assessed. Finally, in future work, we try to extend this control scheme to the full 6 DOF flight dynamics.

Acknowledgments. This work was partially supported by Spanish Ministerio de Economía y Competitividad / FEDER under TIN2013-47074-C2-1-R grant.

References

1. Ellington, C.P.: The novel aerodynamics of insect flight: Applications to microair vehicles. The Journal of Experimental Biology 202, 3439–3448 (1999)
2. de Clerq, K.M.E., de Kat, R., Remes, B., van Oudheusden, B.W., Bijl, H.: Aerodynamic experiments on delfly ii: unsteady lift enhancement. In: Proceedings of the 2009 European Micro-air Vehicle Conference and Competition 2009, pp. 255–262 (2009)
3. Conn, A., Burgess, S., Hyde, R., Ling, C.S.: From natural flyers to the mechanical realization of a flapping wing micro air vehicle. In: Proceedings of the 2006 IEEE International Conference on Robotics and Biomimetics, pp. 439–444 (2006)
4. Fenelon, M.A.A.: Biomimetic flapping wing aerial vehicle. In: Proceedings of the 2008 IEEE International Conference on Robotics and Biomimetics, pp. 1053–1058 (2009)
5. Fernández-Caballero, A., Latorre, J.M., Pastor, J.M., Fernández-Sotos, A.: Improvement of the elderly quality of life and care through smart emotion regulation. In: Pecchia, L., Chen, L.L., Nugent, C., Bravo, J. (eds.) IWAAL 2014. LNCS, vol. 8868, pp. 348–355. Springer, Heidelberg (2014)
6. Castillo, J.C., Carneiro, D., Serrano-Cuerda, J., Novais, P., Fernández-Caballero, A., Neves, J.: A multi-modal approach for activity classification and fall detection. International Journal of Systems Science 45, 810–824 (2014)
7. Carneiro, D., Castillo, J.C., Novais, P., Fernández-Caballero, A., Neves, J.: Multimodal behavioral analysis for non-invasive stress detection. Expert Systems with Applications 39, 13376–13389 (2012)

8. Oliver, M., Montero, F., Fernández-Caballero, A., González, P., Molina, J.P.: RGB-D assistive technologies for acquired brain injury: description and assessment of user experience. Expert Systems (2014), doi:10.1111/exsy.12096

9. Deng, X., Schenato, L., Sastry, S.: Model identification and attitude control scheme for a micromechanical flying insect. In: International Conference on Control, Automation, Robotic and Vision, pp. 1007–1012 (2002)

10. Schenato, L., Wu, W.-C., Sastry, S.: Attitude control for a micromechanical flying insect via sensor output feedback. IEEE Transaction on Robotics and Automation 20, 93–106 (2004)

11. Campolo, D., Barbera, G., Schenato, L., Pi, L., Deng, X., Guglielmelli, E.: Attitude stabilization of a biologically inspired robotic housefly via dynamic multimodal attitude estimation. Advanced Robotics 23, 955–977 (2009)

12. Guckenheimer, J., Holmes, P.: Nonlinear Oscillations, Dynamical Systems, and Bifurcations of Vector Fields. Springer (1990)

13. Sanders, J.A., Verhulst, F.: Averaging Methods in Non-Linear Dynamical Systems. Springer (1985)

14. Deng, X., Schenato, L., Wu, W.-C., Sastry, S.: Flapping flight for biomimetic robotic insects: part II–flight control design. IEEE Transactions on Robotics 22, 789–803 (2006)

15. Rifaï, H., Marchand, N., Poulin-Vittrant, G.: Bounded control of an underactuated biomimetic aerial vehicle–Validation with robustness tests. Robotics and Autonomous Systems 60, 1165–1178 (2012)

16. Khan, Z.A., Agrawal, S.K.: Control of longitudinal flight dynamics of a flapping-wing micro air vehicle using time-averaged model and differential flatness based controller. In: Proceedings of the 2007 American Control Conference, pp. 5284–5289 (2007)

17. Han, J.: From PID to active disturbance rejection control. IEEE Transactions on Industrial Electronics 56, 900–906 (2009)

18. Radke, A., Gao, Z.: A survey of state and disturbance observers for practicioners. In: Proceedings of the 2006 American Control Conference, pp. 5183–5188 (2006)

19. Qing, Z., Gao, Z.: On practical applications of active disturbance rejection control. In: Proceedings of the 29th Chinese Control Conference, pp. 6095–6100 (2010)

20. Luviano-Juárez, A., Cortés-Romero, J., Sira-Ramírez, H.: Synchronization of chaotic oscillators by means of generalized proportional integral observers. International Journal of Bifurcation and Chaos 20, 1509–1517 (2010)

21. Fliess, M., Lévine, J., Martin, P., Rouchon, P.: Flatness and defect of nonlinear systems: Introductory theory and examples. International Journal of Control 61, 1327–1361 (1995)

22. Sira-Ramírez, H., Agrawal, S.: Differentially Flat Systems. Marcel Dekkert Inc. (2004)

23. Sira-Ramírez, H., Núñez, C., Visairo, N.: Robust sigma-delta generalized proportional integral observer based on a 'buck' converter with uncertain loads. International Journal of Control 83, 1631–1640 (2009)

24. Morales, R., Sira-Ramírez, H., Somolinos, J.A.: Robust control of underactuated wheeled mobile manupulators using GPI disturbance observers. Multibody Systems Dynamics 32, 511–533 (2014)

25. Morales, R., Sira-Ramírez, H., Somolinos, J.A.: Linear active disturbance rejection control of the hovercraft vessel model. Ocean Engineering 96, 100–108 (2015)

Use and Adoption of a Touch-Based Occupational Therapy Tool for People Suffering from Dementia

René F. Navarro[1(✉)], Marcela D. Rodríguez[2], and Jesús Favela[3]

[1] Industrial Engineering Department, Universidad de Sonora, Mexico, USA
rnavarro@industrial.uson.mx
[2] School of Engineering, UABC, Mexucali, Mexico
marcerod@uabc.mx
[3] Department of Computer Science, CICESE, Mexico, Ensenada
favela@cicese.mx

Abstract. Even in its early stages, the cognitive deficits in persons with dementia (PwD) can produce significant functional impairment. Dementia is characterized by changes in personality and behavioral functioning that can be very challenging for caregivers and patients. This paper presents results on the use and adoption of a cognition assistive system to support occupational therapy to address psychological and behavioral symptoms of dementia. During 6 months we conducted an in situ system evaluation with a caregiver-PwD dyad to evaluate the adoption and effectiveness of the system to ameliorate challenging behaviors. Evaluation results indicate that intervention personalization and touch-based systems interfaces encouraged the adoption and the positive effect in reducing challenging behaviors in PwD and decreases caregiver burden.

Keywords: Non-pharmacological interventions · Occupational therapy · Cognitive assistive systems · Ambient assisted living

1 Introduction

Dementia is characterized by the loss of intellectual functions to the extent that it interferes with daily activities [1]. For instance, memory difficulties in PwD can impact on self-confidence, may lead to withdrawal from day-to-day activities, anxiety, and depression. Family caregivers are also affected due to the practical impact of memory problems on everyday life and to the strain of frustration that can result from it [2]. Besides cognitive decline, PwD presents Behavioral and psychological symptoms of dementia (BPSD), which are defined as "symptoms of disturbed perception, thought content, mood, behavior frequently occurring in patients with dementia [3]. Psychological symptoms of dementia relate to anxiety, depression, and psychosis whereas behavioral symptoms include aggression, apathy, agitation, disinhibited behaviors, wandering, nocturnal disruption, and vocally disruptive behaviors. Such behaviors are typically identified by observation of the PwD and only considered challenging when they affect other people

© Springer International Publishing Switzerland 2015
J.M. Ferrández Vicente et al. (Eds.): IWINAC 2015, Part I, LNCS 9107, pp. 459–468, 2015.
DOI: 10.1007/978-3-319-18914-7_48

or cause self-injury. It is estimated that behavioral symptoms occur in as many as 90% of PwDs [4]. These challenging behaviors are associated with high levels of distress in both PwD and their caregivers.

1.1   Occupational Therapy for Dementia Treatment

The care for a PwD is a complex and challenging task since the natural evolution of dementia is one of progressive decline, requiring increasing degrees of care, and gradually deteriorating individuals cognitive, physical, and social functions. There is a growing agreement that dementia treatment should initiate with non-pharmacological interventions to ameliorate challenging behaviors such as those aforementioned because they address the psychosocial/environmental causing the behavior [4]. Non-pharmacological interventions have been classified as: a) cognitive/emotion-oriented interventions; b) sensory stimulation interventions; c) behavior management techniques; and d) other psychosocial interventions such as Occupational Therapy. To maximize the effect of these interventions, requires an individualized intervention planning and execution according to the unique needs and strengths of the PwD [3]. There is evidence that occupational therapy (OT) is effective in dementia treatment [5,6]. The focus of OT is to improve PwD ability to perform activities of daily living promoting their independence and participation in social activities and to reduce the burden on the caregiver. Typical OT intervention involves the assessment of PwD abilities, training family caregivers in skills such as problem solving and coping strategies, and to implement environmental and compensatory strategies to assist the PwD to engage in meaningful activities. multicomponent psycho-social interventions that are tailored and focused on the patientcaregiver dyad are the most effective in dementia [6]. Occupational therapists can play an important role in the care of the PwD, given their expertise at understanding the complex relationships between person, environment and occupation required for successful activity execution. Therapists accompany PwD during the course of the disease, providing education and skills training, and supporting the caregivers.

1.2   Assistive Technologies for Dementia Treatment

Ambient Intelligence (AmI) has been recognized, as a promising approach for improving home and community-based care, aiming at mitigating dementia effects on individuals and families [7]. Ambient intelligence is a new paradigm in information technology aimed at empowering peoples capabilities by the means of digital environments that are sensitive, adaptive, and responsive to human needs [8]. The development of AmI systems to support dementia treatment, should consider that non-pharmacological interventions not only need to be adapted to the particular PwD and caregiver needs, but also may need to evolve or change as the dementia progresses. An Ambient-assisted Intervention System (AaIS) uses AmI to improve PwDs quality of life by identifying the presence of BPSDs, deciding on an appropriate intervention, and either modifying the environment or persuading the PwD or the caregiver to act on the systems advice [9]. This paper presents results on the use and adoption of an AaIS to support personalized occupational therapy interventions to address psychological and behavioral symptoms of dementia.

2   Personalized Ambient-Assisted Interventions

Assistive technologies have the potential to assist occupational therapist in gathering assessment data, executing interventions, and monitoring responses to therapy. For example, the Engaging Platform for Art Development (ePAD) enable PwD independent access to art creation [10]. ePAD is customizable such that an art therapist can choose themes and tools that they feel reflect PwD needs and preferences.

The AaIS approach consists on a set of autonomous and collaborative agents that implements the services depicted in Figure 1. Thus, a behavior analysis agent identify BPSD episodes, which can be explicitly observed and reported by the caregiver, or alternatively, they can be inferred by analyzing the information perceived from agents attached to sensors located in the environment or worn by the PwD. Agitation, for instance, is manifested via repetitive movement and verbal expressions such as shouting or continuous talk. Finally, once there is evidence that the PwD is exhibiting a BPSD, a decision model agent is used to select on a behavioral intervention, which will be enacted through ambient actuator agents that: a) Intervene directly to change the configuration of the physical environment; b) Communicate with the caregiver to recommend an action to perform; or, c) Communicate with the PwD to suggest an activity or provide him with information that could change his current behavior. Our model for tailoring the AaIS services is supported by an ontology, which is described in [9] . The system user interface is based on two components: AnswerBoard and AnswerPad. AnswerBoard is a public ambient display implemented on a touch screen LCD computer. Located in a common area within the PwDs home, it provides information of their activities for the current day, the current date, and time of day. Reminder messages are displayed on the AnswerBoard to prompt the patient on relevant events on his agenda, such as medication. The caregiver may create reminder notes from scratch or select one of the predefined templates completing the required information.

AnswerPad is an application running on an Android mobile phone with touch screen. It includes different widgets aiming to offer the PwD time and place awareness, reminder notes, cues on his/her current activity, and to maintain the connection with his/her social network. AnswerPad collects data from the mobile phones sensors to feed the intervention engine. Additionally, caregivers may use AnswerPad to manage elders daily activities, keep track of his/her whereabouts, create reminder notes, and keep a diary of patient's.

3   Study Design

This section describes a case of study in which the effectiveness of the AaIS to support occupational therapy interventions is evaluated. The participants in the

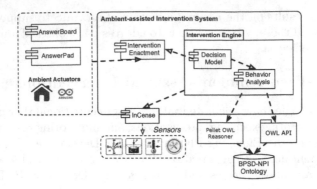

Fig. 1. Ambient-assisted intervention system architecture and applications

study were Jose a 70 years male with Alzheimers disease (MMSE=17), their primary caregivers Ana (wife, 66 years old) and Sonia (daughter, 43 years old). The study involved the visits of a therapist to the participants home during a period of 20 weeks. The therapist applied non-pharmacological intervention to address BPSD as suggested in clinical guidelines. Variables observed during the study were:

- The presence and severity of challenging behaviors estimated by the scores of the Cumming's neuropsychiatric inventory questionnaire (NPI-Q) to evaluate the effect of the intervention on the behavior in the PwD [11].
- The occurrence of apathy measured by the apathy evaluation scale (AES), developed to measure the apathy resulting from neurological diseases [12].
- Caregiver burden is the psychological state resulting from the combination of physical, emotional job and social restrictions associated with caring for a sick person. In the study the outcome of the Zarit burden interview (ZBI) was used to observe variations in the subjective burden reported by the caregiver [13].
- Caregiver Self-efficacy, which refers to a subjective belief that a person has about his/her ability to successfully carry out certain kinds of behavior. This was observed using the revised scale for caregiving self-efficacy (RSCSE) [14].

Additionally, caregivers kept a diary of PwD behavior to document incidents they considered problematic, unusual changes in behavior, health status, mood or memory problems. Caregivers reported each incident by describing the incident, the response of relatives to the incident, and the context in which it occurred (date and time). This information was reviewed on interviews with caregivers during follow-up visits, which enabled us to assess the effect of the intervention on the behavior of the participants. The study was divided in two stages.

3.1 Stage A: Intervention with Traditional Artifacts

In this first stage, and for a period of 4 weeks, the therapist implemented a combination of strategies using traditional means, which included the use of external memory aids, cognitive training, reminiscence therapy, and techniques to enhance communication. In assessment interviews, caregivers expressed a particular concern about the lack of interest of Jose for any type of activity during the day. They noted that Jose spends most of the day asleep in his room. To address this behavior, and in agreement with caregivers, the therapist defined a weekly schedule of activities that could be attractive for the PwD. For instance, Jose enjoys solving crossword and Sudoku puzzles prior the dementia onset. So, one the activities required solve crossword puzzle using pencil and paper. Visits were performed three times a week and during these visits the primary caregiver was involved. In each visit, the therapist guided the session consisting in the execution of three activities scheduled for the day. To perform each activity a maximum time of 30 minutes was allocated in order to avoid fatigue of the participants. During rest periods of 10 minutes between each activity the therapist promoted communication between the PwD and his caregiver. Another concern for caregivers was the refusal of Mr. Jose to take his medication. They need to remind him the medication schedule and monitor intake. Often Jose hid the pills in his pocket or mouth for later disposal. Due this situation, an external memory aid based intervention was implemented trough a whiteboard (40cm x 30cm) placed in the kitchen for displaying his medication schedule, and the use of paper cards with written instructions contained in a labeled pill organizer.

3.2   Stage B: Intervention Supported with AaIS

In the second stage of the study the AaIS supported the intervention. The AaIS services were tailored according the particular needs of the participants. As in the first stage, the therapist conducted the sessions alternating the execution activities supported by the AaIS with activities using traditional artifacts. As part of the deployment of the AaIS, AnswerBoard was installed on a 20-inch touchscreen computer over a table in the living room. On the agenda of activities AnswerBoard the weekly schedule of predefined activities are added. AnswerBoard displays the agenda of activities previously defined in stage A. To complete the deployment of the AaIS, two mobile phones running AnswerPad were given to Mr. Jose and their caregivers. Using AnswerPad the caregivers created and delivered medication reminders and other activities prompts, which Mr. Jose would receive in his AnswerPad or in the AnsswerBoards screen. AnswerBord included of two games:

a) **Memorama:** A card game in which all the cards are laid face down on the touchscreen display and two cards are flipped face up over each turn (Fig. 2a). The object of the game is to turn over pairs of matching cards. The user chooses two cards touching the screen to turns them face up. If cards show the same picture, the cards disappear displaying part of the background image. If they dont show the same picture, they are turned face down again. The game ends when the last pair of cards has been picked up.

b) **Alphabet soup:** In this activity, the participant is presented with a list of words that must be found in a grid of letters showing in the touchscreen display (Fig. 2b). Each list has words of one of the following categories: a) animals; b) objects in the house; c) months of the year; d) names of family members; e) fruit; and f) countries. Each list has a maximum of 20 and a minimum of 10 words. Words can appear horizontally, vertically or diagonally in the grid. Participants mark each found word dragging their finger over the word. If the word matches a word on the list, the word is highlighted and removed from the list.

Fig. 2. Activities implemented on AnswerBoard a) Memorama; b) Alphabet soup

Research has proposed matching activities to individual interests and retained skills to engage persons with dementia and maintain involvement [15]. The selection of games implemented was stirred by the adoption of similar activities that Mr. Jose found enjoyable. During the first week was observed that Mr. Jose had difficulties to identify and select the words in Alphabet soup activity. The game interface was customized increasing the font size and the dimensions of the grid in which the letters are shown. Likewise, caregivers pointed out that a hear impairment in Mr. Jose some-times prevented him for hearing the reminders audio notification in AnswerPad, so it was configure to vibrate whenever it received the reminder.

4   Evaluation Results

In this section we present the results obtained from the application of assessment instruments, interviews with caregiver and system usage data obtained from the logs generated by the AaIS.

4.1   Results on Adoption and Usability

The average daily running time of AnswerBoard, was 11:30 h (σ=5:53 h). The average number of days used in a week was 6.7 days. Figure 3 shows the average hours of daily use for each week of the stage B of the study.

Figure 4 shows the number of reminders received by the participant through Answer-Board classified into four categories: a) Medications: Reminders for medication; b) Activities: Reminders about the activities on the agenda; c) Prompting: Directions to support an activity; d) Orientation: Reminder for temporal or

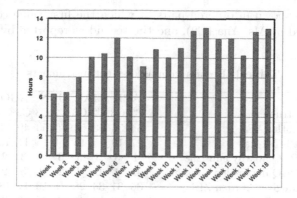

Fig. 3. AnswerBoard average daily usage

spatial orientation. All the reminders were created and delivered through An-swerPad by the caregivers. As shown in fig 4, the predominant type of reminder is for medication, except for the month of May, since only the last week of the month the system was deployed. The decrease in the number of medication re-minders from August is associated with the removal of medications prescribed to Mr. Jose by his family physician.

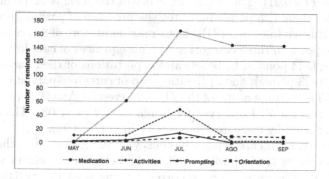

Fig. 4. Reminders delivered to the PwD using AnswerPad

With respect to the activities implemented through AnswerBoard. For activity Memorama an average daily use of 29 minutes ($\sigma=42$ m) was observed, and on average the activity was performed 5 times a day ($\sigma=4$). The average time to complete the Memorama was 6 minutes ($\sigma= 6$ m). The Alphabet Soup activity average daily use was 37 minutes ($\sigma=26$ m). The average daily usage was 6 ($\sigma=4$) times.

4.2   Results on Challenging Behaviors and Caregiver Burden

This section presents the results obtained from the application of assessment instruments. Table 1 summarizes the results obtained in 20 weeks of the study.

The results are shown in ordered pairs in which the first value corresponds to results re-ported by Ms. Ana (C2), and the second as reported by Sonia (C3).

Table 1. Results from assessment instruments

Instrument	APR		MAY		JUN		AGO	
	C2	C3	C2	C3	C2	C3	C2	C3
Apathy AES	64	54	63	54	60	50	49	51
NPI-Q Total	27	19	15	16	14	15	14	13
NPI-Q apathy subscale	4	3	3	3	3	3	3	2
NPI-Q depression subscale	0	4	0	3	2	3	3	2
Caregiver burden ZBI	36	31	34	28	21	16	17	18
Caregiver self-efficacy RSCSE	81	39	83	41	83	45	90	49

Throughout the study Ana reported incidents in 7 of the 12 NPI-Qs subscales ($\mu=17.5$, $\sigma=5.5$). Sonia reported incidents in 4 of the 12 NPI-Qs subscales ($\mu=15.75$, $\sigma=2.17$). The results of the apathy evaluation scale (AES) show a slight variation in the scores reported by both caregivers Ana ($\mu=58.20$, $\sigma=5.56$) and Sonia ($\mu=53.20$, $\sigma=2.48$). The correlation of the scores reported by caregivers is weak ($\Upsilon=0.21$). The maximum score in the scale is 72 points. In PwD a score greater than 41.5 points indicates the presence of pathological apathy.

The Zarit Burden Interview (ZBI) has a maximum score of 88 points and questions are grouped into three categories: a) Consequences of care in the caregiver (11 questions, 0-44 points), b) beliefs and expectations of their caregiving skills (7 questions, 0-28 points), and c) Relationship of caregiver-patient (4 questions, 0-16 points). The average rating of the ZBI observed for Ana is 24 ($\sigma=9.44$). The average rating of the ZBI observed for Sonia is 21.2 ($\sigma=7.03$). The correlation ZBI scores reported by the caregivers is very high ($\Upsilon=0.97$).

The revised scale for caregiving self-efficacy (RSCSE) measures the perceived ability of caregivers to deal with challenging behaviors of elders with cognitive impairment. The mean scores reported by Ana is 84.13 ($\sigma=3.05$), which is considerably high compared to the average scores observed in Sonia ($\mu=47.47$, $\sigma=8.98$).

5   Discussion

The analysis of the study focuses on the effects of the intervention on issues such as medication, apathy, overload/efficacy of caregivers and adoption system. The logs generated by the AaIS show that at least 12 hours a day the system was running. Also, the system logs show that nearly seven days a week participants used the system. One of the basic functions of AaIS evaluated in the study was the intervention based on external memory aids. From the analysis of data from the system log can be observed a reduced response time to reminders. To remove the reminder note from the AnswerBoard screen the user has to touch over the reminder. It was observed that typically Mr. Jos, after reading the reminder note, went to the kitchen for a glass of water, took the medication and deactivated the reminder. On average, the reminders were visible in AnswerBoard for 5 minutes ($\sigma = 13$ minutes) before being removed, and 88.22% of the reminders were eliminated in less than 5. The skill to remove the reminders improved throughout the study. In the first four weeks 28.95% of the reminders were removed within 2 minutes after being displayed on the AnswerBoard. After 18 weeks, 49.07% were removed in less than 2 minutes. Further evidence of the adoption of the system is derived from qualitative data gathered from interviews with the caregivers and the therapists log. Ana: *I feel good because now I know hes taking his pills, not like before. He hid the pills and now he doesnt I feel safer now because already taking the pills.* Sonia: *Very helpful. It really is. Because now, even when we are busy. He asks us: "Do I have to take the pill right now?" And that helped us.*

During home visits it was observed that normally when Mr. Jos read the reminder, either from AnswerBoard or AnswerPad, he went to the kitchen, grabbed a glass of water, took the medication and deactivated the reminder. Given the positive response to medication reminders, it was possible to implement an automated reminders strategy. Reminders were scheduled to automatically deliver to AnswerBoard and AnswerPad, relieving the caregivers responsibility of creating the reminders.

As noted in the previous section, quantitative variation in the results of apathy sub-scale of the NPI-Q shows a decrease during the study. Likewise, the assessment of apathy by AES shows a reduction in scores reported by the caregivers. Although scores remain above the cutoff (41.5) suggesting the presence of apathy in PwD, the results of the interviews provide evidence of an increase in the motivation of Jose in performing activities. Ana: *I see him more awake at all. More cheerful too. Better mood, because before he was sleep most of the day. He was very quiet, not talking, and now he ask questions.* Sonia: *He is more active now. He can last 2 to 3 hours playing. He is entertained and no longer sleeps. For instance, yesterday I ask him to take a nap in the afternoon and he said: "No, I'll be here for a while".*

6   Conclusions and Future Work

We have presented an ambient intervention system to support occupational therapy aimed at addressing psychological and behavioral symptoms of people suffering from dementia. The approach leverages the cognitive remnants resources of PwD and allows them to take some active role to help themselves. Since dementia has very broad effects on the person and their caregiver, the proposal has a comprehensive approach that takes into account their needs to individualize treatment. The system relieved the caregiver of the task of continually supporting the PwD in medication and to maintain PwD engagement in pleasurable activities. Our findings suggest that PwD might expect greater benefits from personalized intervention.

References

1. Galvin, J.E., Sadowsky, C.H.: Practical Guidelines for the Recognition and Diagnosis of Dementia. J. of the Am. Board of Family Medicine 25(3), 367–382 (2012)
2. Rockwood, K.: Treatment expectations in Alzheimer's disease. The Canadian Review of Alzheimer's Disease and Other Dementias 9, 4–8 (2007)
3. Desai, A.K., Galliano Desai, F.: Management of Behavioral and Psychological Symptoms of Dementia. Current Geriatrics Reports 3(4), 259–272 (2014)
4. Sadowsky, C.H., Galvin, J.E.: Guidelines for the Management of Cognitive and Behavioral Problems in Dementia. J. of the Am. Board of Family Medicine 25(3), 259–272 (2012)
5. Gitlin, L.N., et al.: The Cost-Effectiveness of a Nonpharmacologic Intervention for Individuals With Dementia and Family Caregivers: The Tailored Activity Program. The A. J. of Geriatric Psychiatry 18(6), 510–519 (2010)
6. Van't Leven, et al.: Barriers to and facilitators for the use of an evidence-based occupational therapy guideline for older people with dementia and their carers. Int. J. of Geriatric Psychiatry 27(7), 742–748 (2011)
7. Rashidi, P., Mihailidis, A.: A Survey on Ambient-Assisted Living Tools for Older Adults. IEEE J. Biomed. Health Inform. 17(3), 579–590 (2013)
8. Sadri, F.: Ambient intelligence: A Survey. ACM Computing Surveys 43(4), 1–66 (2011)
9. Navarro, R.F., et al.: Intervention tailoring in augmented cognition systems for elders with dementia. IEEE J. Biomed. Health Inform. 18(1), 361–367 (2014)
10. Leuty, V., et al.: Engaging Older Adults with Dementia in Creative Occupations Using Artificially Intelligent Assistive Technology. Assistive Technology 25(2), 7–79 (2013)
11. Cummings, J.L., et al.: The Neuropsychiatric Inventory: comprehensive assessment of psychopathology in dementia. Neurology 44(12), 2308–2314 (1994)
12. Clarke, D.E., et al.: Apathy in dementia: an examination of the psychometric properties of the apathy evaluation scale. The Journal of Neuropsychiatry and Clinical Neuroscience 19(1), 57–64 (2007)
13. Zarit, S.H., et al.: Subjective burden of husbands and wives as caregivers: a longitudinal study. The Gerontologist 26(3), 260–266 (1986)
14. Steffen, A.M., et al.: The revised scale for caregiving self-efficacy reliability and validity studies. The Journals of Gerontology Series B: Psychological Sciences and Social Sciences 57(1), 74–86 (2002)
15. Brodaty, H., Burns, K.: Nonpharmacological management of apathy in dementia: a systematic review. Am. J. Geriatr. Psychiatry 20(7), 549–564 (2012)

Multisensory Treatment of the Hemispatial Neglect by Means of Virtual Reality and Haptic Techniques

Miguel A. Teruel[1], Miguel Oliver[1], Francisco Montero[2], Elena Navarro[2(✉)], and Pascual González[2]

[1] LoUISE Research Group, Research Institute of Informatics,
02071, Albacete, Spain
{MiguelAngel.Teruel,Miguel.Oliver}@uclm.es
[2] LoUISE Research Group, Computing Systems Department,
University of Castilla-La Mancha, 02071, Albacete, Spain
{Francisco.MSimarro,Elena.Navarro,Pascual.Gonzalez}@uclm.es

Abstract. The syndrome of hemispatial neglect is usually associated to a lesion of the brain and is characterized by a reduced or lack of awareness of one side of space, even though there may be no sensory loss. Although it is extremely common, it has proven to be a challenging condition both to understand and to treat. This paper focuses on reviewing this syndrome and proposing new therapies based on multisensory feedback in a virtual environment. These therapies have been designed to improve the awareness of the neglected side by using visual, auditory and haptic feedback.

1 Introduction

After a hemisphere of the brain sustain a damage, a deficit of both attention to and awareness of one side of space can be observed. This is a syndrome that has been called *Hemispatial neglect* and has been defined as a failure to attend to the "contralesional" side of space, that is, to the opposite side of the damaged hemisphere [8]. Usually, this syndrome is the result of a damage to the right cerebral whose consequence is a visual neglect of the left-hand side of space [23]. Right-sided spatial neglect seldom appears because there is a redundant processing of the right space by both the left and right cerebral hemispheres. However, most of the left-dominant brains only process the left space by using the right cerebral hemisphere [26].

Although it mostly affects visual perception ('visual neglect') [7], neglect in other forms of perception, such as auditory [2] and tactile [10], can also be found, either alone or in combination with visual neglect. As Mattingley and Bradshaw stated [18], from a clinical perspective, tactile neglect may be the most salient manifestation of a lateralized attentional impairment. In addition to this sensory neglect, the patient can also suffer other problems such as motor neglect due to a reduced use of the contralesional hemisphere. Therefore, this syndrome greatly

© Springer International Publishing Switzerland 2015
J.M. Ferrández Vicente et al. (Eds.): IWINAC 2015, Part I, LNCS 9107, pp. 469–478, 2015.
DOI: 10.1007/978-3-319-18914-7_49

lessen people's ability to carry out some simple tasks of daily life such as washing up, dressing, eating and even strolling.

This syndrome is extremely frequent among people with Acquired Brain Injury (ABI)[19] and among older people after suffering a stroke, one of the main causes of hemispatial neglect. A recent study [21] states that the Total European 2010 cost of brain disorders was 798 billion euros, among which 37% was direct health care cost, 23% was direct non-medical cost, and 40% was indirect cost. Particularly the cost associated with stroke and traumatic brain injury was 64.1 and 33.0, respectively. Thus, it can be stated that this is a relevant problem not only due to the difficulties it can cause to the people that suffer this syndrome but also to its associated costs.

In order to treat this brain disorder some techniques for diagnosis and rehabilitation have been proposed [23]. Related to diagnosis, several simple bedside screening tests have been developed. The assessment of hemispatial neglect is done by using just pencil and paper tests that consist in asking the patient drawing tasks such as line bisection, target cancellation or copies [1]. Patients with hemispatial neglect make incorrect bisections, fail to cancel the targets on the left side, etc. Regarding treatment, the most common techniques are behavioral intervention and drug treatments. The treatments do not have the same effectiveness for every patient because everyone has a different combinations of cognitive deficits. Also, it is important to keep in mind that those patients who have damaged certain areas of their brain, could be unresponsive to the treatment.

Hemispatial neglect is not a unitary deficit; instead, it involves neglect of different portions of space and hence different functions. In addition, as Bonato stated [6] the performance of brain-damaged patients is negatively affected by increased task demands, which can result in the emergence of severe awareness deficits for contralesional space even in patients who perform normally on paper-and-pencil tests. Finally, although most of the researchers have prevalently attended to the visual symptoms of neglect, others [13][12] have highlighted that this syndrome has an impact on non-visual sensory modalities as well. In particular, neglect-related symptoms have been described in the haptic, tactile and auditory modalities. This conception fits well with the fact that areas usually affected by lesions causing neglect are known to contain multisensory neurons responsible for the convergence and integration of information coming from different senses to build multiple multisensory representations of space. Taking into account all of these characteristics of the hemispace neglect we have developed a novel system that allows the therapists to design their own therapies so that they can be adapted to each specific patient. In particular, these therapies not only stimulate the visual sense but also include other sensory stimuli, mainly auditory and haptic ones. Thus, we can talk about multisensory therapies applied to hemispace neglect patients.

This paper is organized as follows. First, in the next section, we give an overview of the application of virtual reality to the evaluation and treatment of hemispace neglect. Section 3 describes the new tool that allows the therapists to design their own therapies. Finally, last section provides some final conclusion and future works.

2   Related Work

Virtual reality (VR) environment is a computer-simulated environment that can simulate physical presence of individuals in places of the real or imagined worlds. VR can be either non-immersive or immersive, depending on the device used to provide visual feedback [22]. Non-immersive systems use a monitor to provide visual feedback, while fully immersive systems commonly use head mounted displays.

Several VR-based therapy methods for hemispatial neglect have been proposed up to date [25] [11]. For instance, the eye-patching technique was integrated in a VR environment [3] to hide parts of the virtual world. This environment also included auditory stimulation, both lateralized and spatial, as well as a variant of optokinetic stimulation implemented as a flow of dots superimposed over the virtual world. This environment was extended later on to simulate the hemispatial neglect [4]. This simulation showed to be useful not only to determine the potential of a particular rehabilitation technique but to help the rehabilitation staff and the relatives to get a better understanding of the patient's condition. VR and auditory stimulation were also employed in others applications such as [20] and [15]. Myers et al. [20] integrated several aids in their application, including partial patching of the right hemispace and auditory stimulation. Another interesting aid was a variation of optokinetic stimulation that consisted in a virtual dog moving from right to left in one of the rooms of the virtual house. On the other hand, Kim et al. [15] developed several aids to assist the patients in tracking a visual target. They incorporated auditory stimulation in their application in the form of a bilateral alerting tone. Visual aid to direct the attention of the participant towards the target was also provided.

There are already experimental data documenting the effectiveness of using VR-based therapy methods. For instance, a recent study [16] suggests that VR training may be a beneficial therapy for patients with hemispatial neglect after stroke. For instance, Castiello et al. [9] used a PC screen for visual display combined with a DataGlove for hand-motion tracking. They suggested that the virtual hand was incorporated as part of patients' body so that the space representation was extended to include that virtual space. Katz et al. [14] analyzed the efficiency of a non-immersive virtual environment training for safe street crossing of right hemisphere stroke patients with regard to computer based visual scanning tasks. After nine hours of training (distributed over 4 weeks), both control and experimental groups of participants with hemispatial neglect improved on standard hemispatial neglect measures. However, experimental group achieved better results on some measures of the real street crossing. Smith et al. [24] also carried out a single-subject experiment with 4 patients with hemispatial neglect who carried out several tests, before and after training in a VR environment with developed VR environment. The main conclusion was that patients show slight

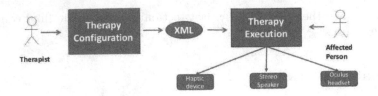

Fig. 1. System's parts

improvements on their test and also reported some improvements on everyday tasks such as reading.

However, we propose that training of patients with Hemispatial neglect using both VR and haptic feedback may be a potential new intervention option because it can stimulate new learning and foster an accelerated recovery. The haptic human-computer interaction is has been defined as the interaction between a human computer user and the computer user interface by means of the powerful human sense of touch. Haptic hardware has been discussed and exploited for some time, particularly in the context of computer games. However, so far, little attention has been paid to the general principles of haptic HCI in neglect rehabilitation. As far as we know, only a system [5] combining VR and haptic feedback has been developed to overcome the lack of proper quantification of neglect that occurs when using traditional the Behavioural Inattention Test. In this work, we propose a new environment that provides therapists with facilities to customize the therapy according to the special needs of patients with hemispatial neglect.

3   A Tool for Multisensory Rehabilitation of Hemispace Neglect

The main goal of the system developed is to aid experts as well as patients in the process of hemispatial neglect rehabilitation by using multisensorial feedback in VR environment. On the one hand, the system proposed enable experts to design therapies customized according to the specific characteristics of each patient, and monitor the results of each exercise done by the patients. On the other hand, patients only have to deal with their exercise, without worrying about the configuration of the application. Furthermore, the exercises have been designed as games in order to take advantage of their playful and engaging aspect, encouraging patients to keep on their rehabilitation process.

Figure 1 shows the functional parts of the system developed and their relationships with the system's roles. As can observed, the therapist creates an XML configuration file to define the patient's environment and the therapies to perform. Moreover, thanks to this configuration file, each patient always have his/her rehabilitation environment properly configured. As soon as a configuration file is created, the therapist can try and assess the designed game immediately.

3.1   Senses Stimulated

The system has been developed to enable therapists to define which patient's senses they want to stimulate and test. The senses that can be stimulated in a game are:

- *Visual*: A virtual world where the user can interact and move will be presented by means of a computer monitor or a VR system. This sense is critical, since patients suffering from hemispatial neglect do not respond properly to visual stimuli.
- *Aural*: To support the rehabilitation of eyesight, aural stimuli will be used to help the patient to determine when a hazard is approaching and proceed consequently.
- *Haptic*: The users can feel vibrations on both sides of their bodies, depending on which the origin of the hazards is in the virtual worlds. The haptic system has been implemented by means of the VitaKi prototype [17], which can send haptic stimuli to a set of vibrators located on the patients' skin or clothes.

3.2   Therapies Supported

The system provides the therapist with support to create the following three types of games, therapies indeed:

1. *Path.* The main goal of this game is to train the patients in a real-world situation, namely to walk through an environment full of obstacles which they must dodge in order not to collide with them. While playing this game, the patients try to select the right path, so that their avatars start to move forwards while a series of obstacles appear. The goal is to avoid such obstacles by moving right or left (see Figure 2).
   In this game, the patient, by means of its avatar, can move along six roads or lanes. If an obstacle appears in the lane where the patient is currently, the audio system starts to play a sound to warn him/her. This sound is higher, as closer the patient is to the obstacle. Similarly, the haptic system starts to transmit a harder-and-harder vibration to the user. If the obstacle is in an adjacent lane, the system's behavior is similar, but the stimuli are weaker and they are applied only to the patient's body part closer to the location of the obstacle in the virtual world (i.e. if the obstacle is in the user's right lane, he/she would feel a vibration on his/her right arm).
   If the patient's avatar collide with an obstacle in the virtual world, he/she will lose one live and, if he/she changes to a different lane without a collision hazard, that is, there is no near obstacle in the current lane, an unnecessary change of lane will be accounted. The game will end satisfactorily when all the obstacles set by the therapist have been dodged. If the user loses all the lives, reach the unnecessary changes number, the game would end but in a not satisfactory manner.
   The user's movements, the number and position of the obstacles which were dodged successfully and unsuccessfully, as well as the number of lives and

Fig. 2. From leftmost top to bottommost right: (1) Path game, (2) Shield game and (3) Shield while walking game

unnecessary movements are saved along with their timestamp in a log file. This log file can be always consulted and used by the therapist to virtually recreate the patient's results. Hence, this log files are a comprehensive log of the patient's evolution over time.

2. *Shield.* The goal of this game is, once again, to train the patient in a possible real-world situation where several moving objects can collide with him/her. In this game, patients will have to defend themselves against those objects coming towards them. Thus,the system throws several virtual objects to the patient's avatar, then the patient will have to move down that of the two walls that enables him/her to avoid their collision with the avatar (see Figure 2). If the patient moves down the wall on time and the object collide with it, a correct protection will be accounted. However, if the object collides with the avatar, the patient will lose a live. Alternatively, if the patient anticipates and defends a position too early, it will be accounted as an error. The game will end when the patient avoids all the objects, loses all the lives or reaches the maximum number of errors, winning in the first case and losing in the last two. The number of lives, points, objects avoided and errors will be logged as in the previous game, thus enabling the therapist to access to previous patient's results.

   As in the previous game, the audio and haptic system will warn the patient that an object has been thrown, being these stimuli increasingly stronger depending on the hazard distance. In the event that an object is thrown, but it will not collide with the avatar, the user will be warn as well, but with less intensity.

3. *Shield while walking.* Following the same principles of the previous game, this one includes the user's movement in order to make more difficult to identify which objects could hit the avatar (see Figure 2). The goal still is to train the patient in real-world situations, like crossing a street.

3.3 Therapist Environment

The therapist's environment is the most complex and configurable of the system, since it provides support for the configuration of the system as well as for the development of personalized therapies. Those options which are common to all the games can be configured by using the form shown in Figure 3(a). The keys that the patient should press to activate the left and right action are set here. This enables the therapist to decide whether the rehabilitation procedure will be performed by using either one or two hands. Another configuration option is to establish whether the image will be shown by means of a computer screen or a VR headset. Finally, another option enables the therapist to pair the haptic device with the system by using Bluetooth.

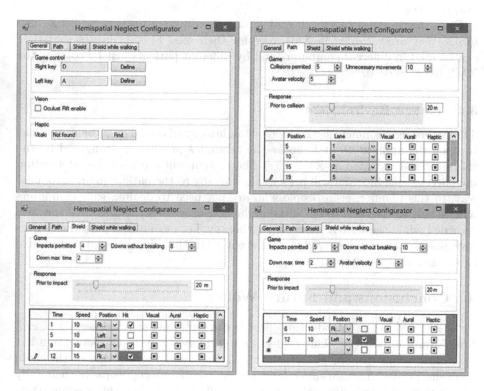

Fig. 3. From leftmost top to bottommost right: (a) General configuration form, (b) "Path" game configuration form, (c) "Shield" game configuration form and (d) "Shield while walking" game configuration form

In order to enable the therapist to configure the "Path" game, the form shown in Figure 3(b) has been created. It can be used to set the number of collisions, the number of unnecessary movements a patient can use to pass a level, the speed that he/she will walk at (in meters per second) as well as the maximum distance the patient can move within a lane without causing an unnecessary movement (in meters). Finally, the obstacles that the patient will find along his/her path are established here. With this aim, the therapist has to create each obstacle by setting its position along the path (in meters), the lane where it will be located, and the stimuli to activate (visual, aural or haptic).

The form shown in Figure 3(c) can be used to configure the "Shield" game. The therapist will set the options number of lives as well as the maximum number of errors and times that walls can be move down. It is also used to set the number and configuration of each object to be thrown to the avatar (at which time to throw it, its speed and direction, whether it is supposed to hit the avatar, and the stimuli to activate).

For the third game, "Shield while walking", the form shown in Figure 3(d) can be used to set the same options than in the previous game as well as the avatar's speed.

Once the therapist has finished of setting the different options by using these forms, they are recorded in XML files for their use. This facilitates that they can be used in different computers and by different patients just by copying the XML file as needed.

3.4   Patient Environment

The patient environment is the more straightforward one. The patient only has to open the application and choose the therapies previously configured by the therapist and recorded as XML files. The system will use with the information included in such files and will present the therapy to the patient in an automatic way. In this manner, even people who are not used to dealing with computers will be able to use this system without the assistance of a therapist. All the data recorded while the user is executing a therapy, i.e. movements performed, lives lost, points, etc., will be saved in a different XML that the therapist will be able to see and analyse afterwards.

4   Conclusions and Future Works

As was stated in the introduction, people can suffer hemispatial neglect after a brain injury or a stroke. This neglect can affect not only visual perception but also other forms of perception such as auditory and tactile. One of the main problems derived from this disorder is that it greatly affects people in their daily life as they can find difficulties even to stroll or dress. For this reason, every effort to help these people to overcome this disorder is welcome.

For this aim, a system has been developed focusing mainly on one of the special needs when treating people with hemispatial neglect: customization. As was presented in the previous section, the therapists can configure totally how the therapy will be carried by establishing senses to stimulate, obstacles, movements, etc. They can even establish which key of the keyboard will be used by the patient in order to require him/her to use one or two hand or whether the patient will use a VR headset. This enables therapist to carry out motor rehabilitation as well.

Moreover, the literature analysis showed that the existing tools focus just on visual neglect and use just visual stimuli by means of VR environments. For this reason, the system was developed to be a step forward by facilitating the system supporting haptic and aural stimuli in addition to the visual ones. Finally, thanks to the use of a VR environment and the presentation of therapies as games, therapist can improve patient engagement.

Therefore, the experience of the VR application approach on neglect rehabilitation suggests that this element seems a promising approach for motor and cognitive rehabilitation, with a wide range of applicability. As ongoing work, we are conducting several experiments with both patients and therapies to evaluate the usability of the system.

Acknowledgements. This work was partially supported by Spanish Ministerio de Economía y Competitividad / FEDER under TIN2012-34003 grant and through the FPU scholarships (AP2010-0259 and FPU13/03141) from the Spanish Government. It is also relevant the collaboration of the specialists of the ADACE association.

References

1. Albert, M.L.: A simple test of visual neglect. Neurology 23(6), 658–664 (1973)
2. Altman, J.A., Balonov, L.J., Deglin, V.L.: Effects of unilateral disorder of the brain hemisphere function in man on directional hearing. Neuropsychologia 17, 295–301 (1979)
3. Baheux, K., Yoshizawa, M., Tanaka, A., Seki, K., Handa, Y.: Diagnosis and rehabilitation of patients with hemispatial neglect using virtual reality technology. In: 26th Annual Int. Conf. Engineering in Medicine and Biology Society (IEMBS 2004), vol. 2, pp. 4908–4911 (2004)
4. Baheux, K., Yoshizawa, M., Tanaka, A., Seki, K., Handa, Y.: Virtual reality pencil and paper tests for neglect: A protocol. Cyberpsychology and Behaviour 9(2), 192–195 (2006)
5. Baheux, K., Yoshizawa, M., Yoshida, Y.: Simulating hemispatial neglect with virtual reality. Journal of Neuroengineering and Rehabilitation 4, 27 (2007)
6. Bonato, M.: Neglect and Extinction Depend Greatly on Task Demands: A Review. Frontiers in Human Neuroscience 6, 195 (2012)
7. Brain, W.R.: Visual disorientation with special reference to lesions of the right cerebral hemisphere. Brain 64, 244–272 (1941)
8. Buxbaum, L.J., Ferraro, M.K., Veramonti, T., Farne, A., Whyte, J., Ladavas, E., Frassinetti, F., Coslett, H.B.: Hemispatial neglect: Subtypes, neuroanatomy, and disability. Neurology 62, 749–756 (2004)
9. Castiello, U., Lusher, D., Burton, C., Glover, S., Disler, P.: Improving left hemispatial neglect using virtual reality. Neurology 62, 1958–1962 (2004)
10. De Renzi, E., Faglioni, P., Scotti, G.: Hemispheric contribution to exploration of space through the visual and tactile modality. Cortex; A Journal Devoted to the Study of the Nervous System and Behavior 6, 191–203 (1970)
11. Fasotti, L., van Kessel, M.: Novel insights in the rehabilitation of neglect. Frontiers in Human Neuroscience 7, 780, 192–199 (2013)
12. Jacobs, S., Brozzoli, C., Farnè, A.: Neglect: A multisensory deficit? Neuropsychologia 50, 1029–1044 (2012)

13. Karnath, H.O., Dieterich, M.: Spatial neglect - A vestibular disorder? Brain. A Journal of Neurology 129, 293–305 (2006)

14. Katz, N., Ring, H., Naveh, Y., Kizony, R., Feintuch, U., Weiss, P.L.: Interactive virtual environment training for safe street crossing of right hemisphere stroke patients with unilateral spatial neglect. Disability and Rehabilitation 27, 1235–1243 (2005)

15. Kim, K., Kim, J., Ku, J., Kim, D.Y., Chang, W.H., Shin, D.I., Lee, J.H., Kim, I.Y., Kim, S.I.: A virtual reality assessment and training system for unilateral neglect. Cyberpsychology & Behavior: The Impact of the Internet, Multimedia and Virtual Reality on Behavior and Society 7, 742–749 (2004)

16. Kim, Y.M., Chun, M.H., Yun, G.J., Song, Y.J., Young, H.E.: The Effect of Virtual Reality Training on Unilateral Spatial Neglect in Stroke Patients. Annals of Rehabilitation Medicine 35(3), 309–315 (2011)

17. Martinez, J., Garcia, A., Oliver, M., Molina, J., González, P.: Vitaki: A vibrotactile prototyping toolkit for virtual reality and video games. International Journal of Human-Computer Intraction 30(11), 855–871 (2014)

18. Mattingley, J.B., Bradshaw, J.L.: Can tactile neglect occur at an intra-limb level? Vibrotactile reaction times in patients with right hemisphere damage. Behavioural Neurology 7, 67–77 (1994)

19. Montero, F., López-Jaquero, V., Navarro, E., Sánchez, E.: Computer-aided relearning activity patterns for people with acquired brain injury. Computers & Education 57(1), 1149–1159 (2011)

20. Myers, R.L., Bierig, T.A.: Virtual Reality and Left Hemineglect: A Technology for Assessment and Therapy. CyberPsychology & Behavior 3, 465–468 (2000)

21. Olesen, J., Gustavsson, A., Svensson, M., Wittchen, H.U., Jönsson, B.: The economic cost of brain disorders in Europe. European Journal of Neurology 19, 155–162 (2012)

22. Parsons, T.D., Trost, Z.: Virtual Reality Graded Exposure Therapy as Treatment for Pain-Related Fear and Disability in Chronic Pain. In: Ma, M., Jain, L.C., Anderson, P. (eds.) Virtual, Augmented Reality and Serious Games for Healthcare 1. ISRL, vol. 68, pp. 523–546. Springer, Heidelberg (2014)

23. Parton, A., Malhotra, P., Husain, M.: Hemispatial neglect. Journal of Neurology, Neurosurgery and Psychiatry 75(1), 13–21 (2004)

24. Smith, J., Hebert, D., Reid, D.: Exploring the effects of virtual reality on unilateral neglect caused by stroke: Four case studies. Technology and Disability. IOS Press 19(1), 29–40 (2007)

25. Tsirlin, I., Dupierrix, E., Chokron, S., Coquillart, S., Ohlmann, T.: Uses of virtual reality for diagnosis, rehabilitation and study of unilateral spatial neglect: review and analysis. Cyberpsychology & Behavior: The Impact of the Internet, Multimedia and Virtual Reality on Behavior and Society 12, 175–181 (2009)

26. Weintraub, S., Daffner, K.R., Ahern, G.L., Price, B.H., Mesulam, M.M.: Right sided hemispatial neglect and bilateral cerebral lesions. Journal of Neurology, Neurosurgery and Psychiatry 60(3), 342–344 (1996)

Evaluation of Color Preference
for Emotion Regulation

Marina V. Sokolova[1], Antonio Fernández-Caballero[1(✉)],
Laura Ros[2], José Miguel Latorre[2], and Juan Pedro Serrano[2]

[1] Instituto de Investigación en Informática de Albacete, Universidad de Castilla-La
Mancha, 02071-Albacete, Spain
antonio.fdez@uclm.es
[2] Instituto de Investigación en Discapacidades Neurológicas, Universidad de
Castilla-La Mancha, 02071-Albacete, Spain

Abstract. This paper introduces a study on the relationship between
emotion regulation and color preference. In the described pilot study,
participants are asked to label uniform color images by using opposite
meaningful words belonging to four semantic scales, namely "Tension"
(ranging from *Relax* to *Stress*), "Temperature" (*Coldness* to *Warmness*),
"Amusement" (*Boredom* to *Fun*) and "Attractiveness" (*Pleasantness* to
Unpleasantness). Simultaneously, the participants have to indicate if they
feel certain emotions while observing each colored image, as well as to
rate the intensity of the feeling. The labeled emotions are "Joy", "Happiness", and "Sadness". The results demonstrate that people generally
perceive color emotions for one-colored images in similar ways, though
showing some variations for males and females. Several conclusions about
the relations between color and emotions are presented.

Keywords: Color emotion · Emotion regulation · Color preference

1 Introduction

Emotional well-being has become paramount for computer based systems as well
as applications and gadgets related to health care, and nursing, among others [1].
This article is based on the assumption of the power of color to change mood. The
described proposal falls within the project "Improvement of the Elderly Quality
of Life and Care through Smart Emotion Regulation" [2], [3]. The objective of
the project is to find solutions for improving the quality of life and care of the
elderly who can or wants to continue living at home by using emotion detection
and regulation means. Cameras and body sensors are used for monitoring the
ageing adults' facial and gestural expression, activity and behavior, as well as
relevant physiological data. This way the older people's emotions are inferred
and recognized. Music, color and light are the stimulating means to regulate
their emotions towards a positive and pleasant mood. This article introduces
the first steps in the use of color to regulate affect.

© Springer International Publishing Switzerland 2015
J.M. Ferrández Vicente et al. (Eds.): IWINAC 2015, Part I, LNCS 9107, pp. 479–487, 2015.
DOI: 10.1007/978-3-319-18914-7_50

Besides of the typical popular beliefs that tend to construct some unsophisti-
cated ideas as "red is stimulating" and "blue is tranquil" [4], there is a considerable number of studies which do present empirical results on this topic. So, we
believe that color is strongly related to emotional state, and it brings information about emotions as well as can influence on them. So, assuming that color
affects the emotional state of a person (and specifically ageing adults), a new
approach to the evaluation of the influence between color and mood is proposed.
The evaluation is based on a color emotion test. Firstly, the computerized test
is aimed to discover and evaluate the participants' preference to color. Then,
relations between color and emotion are established with the objective of facilitating emotion regulation. We have to highlight that a pilot experimentation
with adult people is performed in this initial work. An exhaustive experimentation with ageing adults is foreseen in a close future.

2   On Color and Emotion Regulation

It has been demonstrated that color characteristics like chroma, hue or lightness
produce an impact on emotions [5], [6]. It has also been proved that chromatic
images transmit emotionally charged information on contrary to the achromatic
ones [7]. Contrary to chroma and lightness, the hue of a color does not affect the
emotional state [5]. Another work [8] points out that the older observers show
strong preference to colors with higher chroma, and the younger prefer achromatic colors. A study on color-emotion associations reveals that the principal
hues comprise the highest number of positive emotions, and are followed by the
intermediate hues and achromatic colors [9]. Based on the obtained responses,
the green color attained the highest number of positive emotions, followed with
yellow, blue, red and purple. Among the intermediate hues the most preferred
include blue-green, red-purple, yellow-red and purple-blue. For the achromatic
colors, the most attractive color is white in a great measure, followed by black
and, finally, gray.

An emotion manifests psychological and physiological changes. It can be detected through identification of facial expressions, bodily and behavioral changes
[10]. The use of emotionally charged words allows studying changes in feelings and
moods in and indirect manner [11]. For example, semantic scales "warm/cool",
"heavy/light", "active/passive", and "like/dislike" are used to evaluate color emotion for a two-color model [8]. Also, twelve emotion variables represented with
word pairs have been studied [8]. In [12] semantic word pairs are used as indicators
for emotion expression factors: "bright-dark" (activity factor), "like-dislike"
(evaluation factor) and "strong-weak" (potency factor).

Nevertheless, evaluation with semantic scales has often been claimed to be
more generalized than self-report instruments. The latter are more subjective
as they do rely on the participants both rational and irrational own assessment
of the emotional experience. For instance, contrary to the indirect measurement
approach, a self-reported evaluation of emotions is presented for the case of
emotion elicitation using films [13]. This approach has been used in psychological

practice, where participants directly rate their feelings [13] or indicate them from the available sets [14].

3   Color Emotion Evaluation

As it has been seen previously, many attempts have been undertaken so far to evaluate the potential relations between color and emotion. Factor analysis, correlation analysis and ANOVA are the most widely used classification methods for color emotion. These have been used in a great number of research (e.g. [8], [15], [12], [16], [11], [17], and [18]). RGB-histograms for measuring statistical properties of color images are related to color emotion vectors [11].

Identification and clustering are the two procedures that are commonly used to study the relationship between emotion and color. Among these methods, one of the most visible and understandable interpretation is decision trees. [19] considers the application of this method through studying dependencies of hue and given brightness level choice for geographically different groups of participants. Also, a cluster analysis for web site designers to evaluate color preference has been described [18]. With the purpose of studying color emotion, [12] uses factor analysis, and, if it fails, it changes to independent component analysis with the intrinsic statistical properties of data.

In the current study, some considerations are made to properly approaching the evaluation of the relation between emotions and colors. The first consideration refers to the nature of emotion measurement. As it has been discussed before, emotions can both be detected and evaluated with semantic words and through reporting about the own emotional state. Reasonably, the optimal solution emerges from a combination of these two approaches, as shown in Fig. 1. This approach has been accepted for this study as it offers the following advantages: (a) possibility of a comprehensive assessment of changes in the subject's reactions, (b) mutual confirmation at the coincidence of the responses, and, (c) indirect proof of the sincerity of answers.

The second consideration is that the subsystem, which includes all information on each participant and color, has to be seen as a complex system. Next, in order to model an emotion modulation system for the experiment, the relationship between the dependent and independent variables is presented as black box model (see Fig. 2).

The four input variables of the subsystem are: age, gender, initial emotional state, and color. Here, only the "color" variable is independent. It is the only one that can be changed within the given subsystem. After the test is carried out and all the answers are stored into a database, the information is processed within a black box, and the final output dependency $F(x)$ is calculated. Although different methods can be applied for classification, it has been decided to base on decision trees. One of the advantages of their use is that they allow building tree-like structures of dependencies between variables. Such structures are easily understandable and can be well processed. Another advantage is their possibility to cut off unnecessary leaves and branches, thereby simplifying the structure of the model.

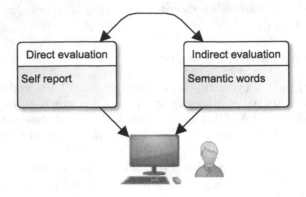

Fig. 1. Emotion evaluation model

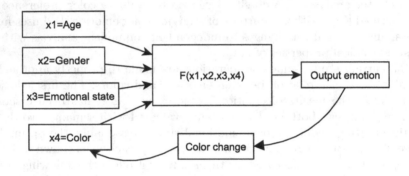

Fig. 2. The model of the complex subsystem for emotion modulation

3.1 Description of the Experiment

The experimentation is carried out in a specially organized room with white colored walls, where each participant is placed in front of a computer. Evaluation of color preference with the aim of regulating emotions is performed with a software application. The graphical user interface of the test is shown in Fig. 3.

The test lasts up to 30 minutes, depending on the test participant who may answer the complete questionnaire quicker or slower. Each participant is asked to judge about each of the 32 images which are included in the image set. The set of images consists, first, of uniform or textured one-color pictures, and, second, of images taken form nature, mostly landscapes, where one dominating color is prevailing. The colors used in the pilot study are gray, light blue, pink, dark blue, light brown, brown, violet, green, light green, red, orange, and yellow. A sample of the image set is provided in Fig. 4.

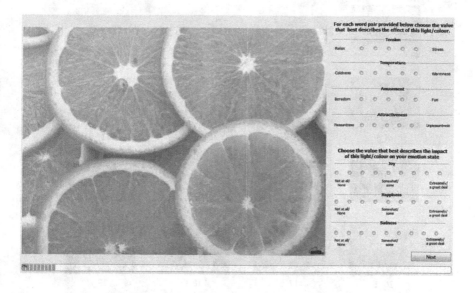

Fig. 3. Graphical user interface for the experimentation

Since the purpose of this study is to explore the possibility of "improving" the emotional state, the number of emotions under study have been limited to the following "basic" ones: sad and happy. When the test starts, the participant is asked to label the randomly appearing images for the following semantic scales, where each scale is represented with two opposite semantic words:

1. "Tension": *Relax / Stress*
2. "Temperature": *Coldness / Warmness*
3. "Amusement": *Boredom / Fun*
4. "Attractiveness": *Pleasantness / Unpleasantness*

Next, each participant reports if he/she felt one of emotions "Joy", "Happiness", and "Sadness". The choice of these emotions is based on the premise that the practical purpose of this study is to find mechanisms to improve mood. It is therefore important to check whether participants experienced the feelings of joy, happiness and sadness. The participant also indicates a value corresponding to the intensity of each feeling.

Each of the semantic scales can be evaluated with a value from the set $[-2, -1, 0, 1, 2]$. Referring to the self-reported emotions, they also should be ranked similarly with a a discrete integer value within the interval $[-9, \ldots, 9]$ (see Fig. 5).

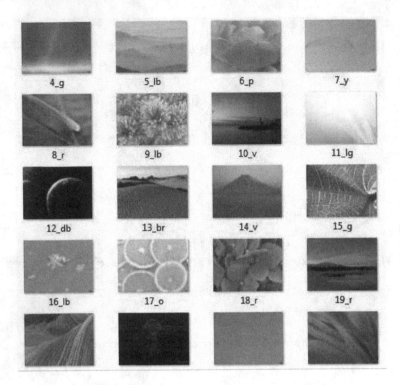

Fig. 4. Some pictures from the image set

3.2   Description of the Results

Sixteen graduate males and females aged from 25 to 36 years old have taken part in the pilot test. The results show marked similarities in the opinions about colors for the participants of the pilot test. Thus, the outcomes demonstrate that people generally perceive color emotions for one-colored images in a similar way. The highest values for the semantic scale "Temperature" are for red and orange colored images. The highest value for "Tension" (*Relax*) is for light brown or sandy color, which is also associated with "Attractiveness" (*Pleasantness*). The semantic scale "Amusement" is characterized with green, dark blue, gray, and, finally, violet for *Boredom*, and with pink, red, yellow and orange for *Fun*.

When the test participants report about their emotions, both men and women indicate that brown and light blue images produce no feeling of "Joy". The findings reveal that the major part of participants rate as maximum their feeling of "Happiness" for light brown (sand). The pictures that receive the next intensive rates are those colored in pink and violet colors for women, and in orange for men. With respect to "Joy", pink, red, orange and yellow are indicated. "Sadness" correlates with dark brown, gray, and, in a lesser extent, violet. For a small part of participants feeling of "Joy" is strongly related with "Happiness".

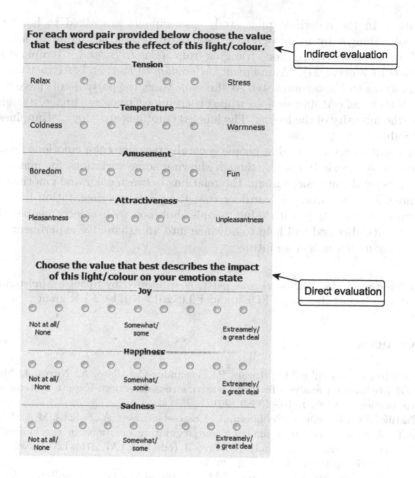

Fig. 5. Graphical user interface with questions

4 Conclusions

This article has described the first steps in the use of color to regulate affect. The proposal is based on the assumption of the power of color to change mood. It belongs to a running project denominated "Improvement of the Elderly Quality of Life and Care through Smart Emotion Regulation". The objective of the project is to find solutions for improving the quality of life and care of ageing adults living at home by using emotion elicitation.

After assuming that color affects the emotional state of a person, a new approach to the evaluation of the influence of color in mood has been proposed in this paper. The evaluation is based on a color emotion test. Firstly, the computerized test evaluates the participants' preference to color. Then, relations between color and emotion are established with the objective of facilitating emotion regulation. In the described pilot study, participants are asked to label uniform color images by using opposite meaningful words belonging to four semantic scales, namely "Tension" (ranging from *Relax* to *Stress*), "Temperature" (*Coldness* to *Warmness*), "Amusement" (*Boredom* to *Fun*) and "Attractiveness" (*Pleasantness* to *Unpleasantness*). At the same time, the participants have to indicate if they feel certain emotions while observing each colored image, as well as to rate the intensity of the feeling. The labeled emotions are "Joy", "Happiness", and "Sadness".

The results demonstrate that people generally perceive color emotions for one-colored images in similar ways, though showing some variations for males and females. Several conclusions about the relations between color and emotions are presented. It is important to consider that the pilot experimentation has been performed at this stage with adult people, but not with elderly people. The initial results obtained will help to advance into an exhaustive experimentation with ageing adults in a close future.

Acknowledgements. This work was partially supported by Spanish Ministerio de Economía y Competitividad / FEDER under TIN2013-47074-C2-1-R grant.

References

1. Carneiro, D., Castillo, J.C., Novais, P., Fernández-Caballero, A., Neves, J.: Multimodal behavioral analysis for non-invasive stress detection. Expert Systems with Applications 39(18), 13376–13389 (2012)
2. Castillo, J.C., Fernández-Caballero, A., Castro-González, Á., Salichs, M.A., López, M.T.: A framework for recognizing and regulating emotions in the elderly. In: Pecchia, L., Chen, L.L., Nugent, C., Bravo, J. (eds.) IWAAL 2014. LNCS, vol. 8868, pp. 320–327. Springer, Heidelberg (2014)
3. Fernández-Caballero, A., Latorre, J.M., Pastor, J.M., Fernández-Sotos, A.: Improvement of the elderly quality of life and care through smart emotion regulation. In: Pecchia, L., Chen, L.L., Nugent, C., Bravo, J. (eds.) IWAAL 2014. LNCS, vol. 8868, pp. 348–355. Springer, Heidelberg (2014)
4. O'Connor, Z.: Colour psychology and colour therapy: caveat emptor. Color Research & Application 36(3), 229–234 (2011)
5. Xin, J.H., Cheng, K.M., Taylor, G., Sato, T., Hansuebsai, A.: Cross-regional comparison of colour emotions Part I: Quantitative analysis. Color Research & Application 29(6), 451–457 (2004)
6. Xin, J.H., Cheng, K.M., Taylor, G., Sato, T., Hansuebsai, A.: Cross-regional comparison of colour emotions Part II: Qualitative analysis. Color Research & Application 29(6), 458–466 (2004)
7. Jue, J., Kwon, S.M.: Does colour say something about emotions?: Laypersons' assessments of colour drawings. The Arts in Psychotherapy 40(1), 115–119 (2013)
8. Ou, L.C., Luo, M.R., Sun, P.L., Hu, N.C., Chen, H.S.: Age effects on colour emotion, preference, and harmony. Color Research & Application 37(2), 92–105 (2012)
9. Kaya, N., Epps, H.: Color-emotion associations: Past experience and personal preference. In: Proceedings of the AIC 2004 Color and Paints, Interim Meeting of the International Color Association, vol. 5, p. 31 (2004)

10. Picard, R.W.: Emotion research by the people, for the people. Emotion Review 2(3), 250–254 (2010)
11. Solli, M., Lenz, R.: Color emotions for multi-colored images. Color Research & Application 36(3), 210–221 (2011)
12. Hanada, M.: Analyses of color emotion for color pairs with independent component analysis and factor analysis. Color Research & Application 38(4), 297–308 (2013)
13. Rottenberg, J., Ray, R.D., Gross, J.J.: Emotion elicitation using films. In: Handbook of Emotion Elicitation and Assessment, pp. 9–28. Oxford University Press, New York (2007)
14. Desmet, P.: Measuring emotion: Development and application of an instrument to measure emotional responses to products. Funology, 111–123 (2005)
15. Gao, X.P., Xin, J.H., Sato, T., Hansuebsai, A., Scalzo, M., Kajiwara, K., Guan, S.S., Valldeperas, J., Lis, M.J., Billger, M.: Analysis of cross-cultural color emotion. Color Research & Application 32(3), 223–229 (2007)
16. Küller, R., Mikellides, B., Janssens, J.: Color, arousal, and performance - A comparison of three experiments. Color Research & Application 34(2), 141–152 (2009)
17. Choi, C.J., Kim, K.S., Kim, C.M., Kim, S.H., Choi, W.S.: Reactivity of heart rate variability after exposure to colored lights in healthy adults with symptoms of anxiety and depression. International Journal of Psychophysiology 79(2), 83–88 (2011)
18. Bonnardel, N., Piolat, A., Le Bigot, L.: The impact of colour on Website appeal and users' cognitive processes. Displays 32(2), 69–80 (2011)
19. Lechner, A., Simonoff, J.S., Harrington, L.: Color-emotion associations in the pharmaceutical industry: understanding universal and local themes. Color Research & Application 37(1), 59–71 (2012)

Elicitation of Emotions through Music:
The Influence of Note Value

Alicia Fernández-Sotos[1], Antonio Fernández-Caballero[2]([⊠]),
and José Miguel Latorre[3]

[1] Facultad de Educación de Albacete, Universidad de Castilla-La Mancha,
02071-Albacete, Spain & Conservatorio Profesional de Música Maestro Gómez Villa,
30530-Cieza (Murcia), Spain
[2] Instituto de Investigación en Informática de Albacete,
Universidad de Castilla-La Mancha, 02071-Albacete, Spain
antonio.fdez@uclm.es
[3] Instituto de Investigación en Discapacidades Neurológicas,
Universidad de Castilla-La Mancha, 02071-Albacete, Spain

Abstract. This article is based on the assumption of the power of music to change the listener's mood. The proposal studies the participants' changes in emotional states through listening different auditions. This way it is possible to answer to the question if music is able to induce positive and negative emotions in the listener. The present research focuses on the musical parameter of note value through its four basic components of the parameter note value, namely, beat, rhythm, harmonic rhythm and rhythmic accompaniment to detect the individual preferences of the listeners. The initial results prove that the influence of beat in music for eliciting emotions is dependent of the personality of each participant in terms of neuroticism and extraversion.

Keywords: Emotion regulation · Music · Note value · Beat · Rhythm

1 Introduction

This article is based on the assumption of the power of music to change the listener's mood. The proposal falls within the project "Improvement of the Elderly Quality of Life and Care through Smart Emotion Regulation" [1], [2]. The objective of the project is to find solutions for improving the quality of life and care of the elderly who can or wants to continue living at home by using emotion elicitation techniques. Cameras and body sensors are used for monitoring the ageing adults' facial and gestural expression, activity and behavior, as well as relevant physiological data [3], [4]. This way the older people's emotions are inferred and recognized. Music, color and light are the stimulating means to regulate their emotions towards a positive and pleasant mood. This article introduces the first steps in the use of music to regulate affect.

The aim of the article is to study the listener's changes in emotional state through playing different auditions. This way, it is possible to come to the conclusion if music is able to induce positive and negative emotions in the listener.

J.M. Ferrández Vicente et al. (Eds.): IWINAC 2015, Part I, LNCS 9107, pp. 488–497, 2015.
DOI: 10.1007/978-3-319-18914-7_52

The present research focuses on one principal musical parameter, namely, the note value. In this sense, the article introduces a series of tests aimed at detecting the individual preferences related to the four basic components of the parameter note value, that is, beat, rhythm, harmonic rhythm and rhythmic accompaniment. In our opinion, the information obtained from the experimentation will be crucial to understand the mood baseline of the listener before entering into the task of detecting and regulating his/her emotional state.

The tests described below are based on the belief that there is a custom internal criterion in the perception of the beat of a particular music piece. When listening to a piece that maintains a constant beat, although this tempo is not altered at all, listeners use to experience changes in tempo stability. They may perceive music as faster as or slower than the initial tempo [5]. Therefore, it is believed that there exists an inter-individual parameter to judge the tempo of a hearing, Each individual senses tempo (or beat) in a particular way. Moreover, the conclusion is that there is no absolute notion of tempo. The conclusion cones after making a theoretical review of the structural relationships that affect the perception of musical tempo [6]. The intensity, harmony and harmonic rhythm directly influence the perception of tempo by listeners. The previously cited research addresses the individual preference of the listeners from a varying age (preteens, teens and adults) perspective. The participants listen six musical fragments of different styles where the beat is modified. The listeners report about their beat preferences and they compare the perceived tempos in comparison with the original tempos of the performances. The study concludes with the assertion that the human ear tries to supply its own tempo to the perceived one (tempo thought to be the right one) in order to ensure a meaningful coordination.

Lastly, this proposal combines the previously mentioned studies with a recent research [7] which is centered in the analysis of harmonies to demonstrate their influence in the induction of emotions. We attempt to formulate a categorization of different beats, rhythms, harmonic rhythms and rhythmic accompaniments which cause changes in the emotional state of a music listener through conducting a series of tests. The idea is to associate emotional terms to the listened auditions. Also, it is relevant to categorize musical chords and scales so powerful as to cause specific emotional states in the audience.

2   Preference and Musical Tests

Firstly, a test is performed to discover individual preferences towards beat, rhythm, harmonic rhythm and rhythmic accompaniment. Secondly, a series of experiment, consisting of four different musical tests, are carried out for the sake of studying the changes in the participants' emotions from the musical parameter note value. Thus, it will be possible to prove the capacity of inducing emotions through the several variations of the note value parameter.

2.1   Preference Test

The first test studies and evaluates the personality factors of each listener (participant). We start from the hypothesis that the personality is a determinant factor in music perception. So, before the participants perform the four musical tests described below, they have to respond to sixty questions which enable evaluating these personality factors. The questions are extracted from the NEO-PI-R ("Revised Neo Personality Inventory") test belonging to the well-known "Big Five Personality Factors". In this particular case, its reduced version "NEO Five-Factor Inventory" (NEO-FFI) is used. This test analyzes human factors in relation to his/her levels of neuroticism, extraversion, openness, kindness and responsibility. However, although the sixty questions apply, this study focuses on the factors of neuroticism and extraversion to establish prototypes of people. The prototypes are subsequently converted into prototypes of listeners, allowing to obtain individualized information from the tests contained in the next part of the overall experimentation.

2.2   First Musical Test: The Pulse

There is no doubt that the beat is the essential element of note value. Indeed, the rhythm is based around the beat. The beat enables perceiving music in an organized manner. It forms the basis on which the melodic-harmonic lines are built. The promotion in children of perception, acquisition and reproduction of the beat is a widely advocated topic. This practice has a positive effect on reading assignments, learning vocabulary, math and motor coordination of the younger [8]. Children perceive better the responses that they receive from the exterior through a constant beat, allowing giving logical sense to their world. This element is present in daily actions as observed in speech and body movements made by the human being [9]. On the one hand, there is a social synchrony between human movements; the tempo is an underlying social interaction organizer [10]. On this basis, Norris shows that two individuals in contact tend to synchronize their movements and they reach to establish a common beat pattern. Tempos are also observed in verbal discourse, for example when a question is posed and an answer is provided. This fact is noted in the gestures and movements associated with the discourse. Moreover, this situation also occurs in listening to background music. It is worth highlighting that the listener synchronizes his/her movements with the beat perceived in music.

Thus, the first musical test that is proposed here focuses on the evaluation of three tunes by the listener. The three melodies are really the same one, but it is varied on two occasions by altering the beat. The piece is titled "Walking on the Street", framed in a suite called "Three Little Bar Songs Suite" (see Fig. 1). It has been written by the contemporary composer Juan Francisco Manzano Ramos. We wanted to start with this little piece in non-classical style to bring variety to the experimentation. The different experiments combine both classical and contemporary elements of music. The only requirement is that both music pieces share a tonal harmonic language, with a harmonic rhythm of classical music and repetitive rhythmic parameters. This enables to highlight each of

Three little bar songs suite

Fig. 1. "Walking on the Street" theme and 3 variations

the auditions to categorize them correctly. So, in this way, we have a piece which rhythm uses constantly alternating dotted notes (providing a touch of swing) and syncopated notes in prominent places. Then, changes are provided to the harmonic rhythm used.

The order of appearance of each melody is random for each listener in the computer program. However, notice that the three melodies are available to the listener. The beats to be listened are 90, 120 and 150 beats per minute, respectively. The listener presses on each of the buttons of the melodies and labels them in the way he/she considers them more suited. The labeling lists have been formulated through following a list of antonyms. The following ones have been considered: "Relaxing - Stressing", "Expressionless - Striking", "Boring - Funny" and "Pleasant - Unpleasant". It is important to highlight that the listener can select the same word for more than one melody. Finally, on a second program screen, the listener checks on each melodic version the type "Like", "Dislike", "Irrelevant" to categorize his/her tastes related to the beats.

2.3   Second Musical Test: The Rhythm

In relation to the rhythm, typically the basic rhythmic figures are addressed in duple measure, or triple measure present in the continuous movements of the human being as, for instance, walking. Jaques Dalcroze emphasizes the importance of implementing the rhythmic movements, perceived in music and represented through the human body in its rhythmic part, in the right balance of the nervous system. Dalcroze stresses that the rhythm is movement and all motion is material. Therefore, any movement is in need of space and time. Thus, Dalcroze starts from the binary rhythm in his teaching, associating it to freely walking. Thus, one of the basic methodologies used is the association of black rhythmic figure (basic beat in measure two by four) with walking, quavers with running, and quavers with dotted note and semiquavers with jumping.

In this sense, the second musical test is geared to the variation of the rhythm parameter. In other words, it is the variation of the rhythm of the melody without altering the melodic line or harmonic rhythm or beat. To do this, from the main melody of the symphony "Surprise" by Haydn, three rhythmic variations are established (see Fig. 2). The listener hears and labels what the melody suggests to him from the list of antonyms used in the first musical test. The listener sorts the melodies from lowest to highest in order of personal preference. The theme is characterized by the use of rhythmic black and white figures. For this, the original melody is slightly modified, especially in the last four bars where a cadence amendment is made. Variation number 1 is characterized by the predominance of the rhythmic formula of two quavers. A slight variation is introduced in the sixth bar in order to bring interest to the resolution of the theme. The second variation is characteristic for the use of simple combinations of the representation of semiquavers, that is, rhythmic formulas of four semiquavers, two semiquavers-quaver and quaver-two semiquavers. Finally, the third variation uses syncopated notes, dotted notes and triplets, all in value of a beat.

2.4   Third Musical Test: The Harmonic Rhythm

Thirdly, there is the harmonic rhythm. In this musical test of individual preference, participants are exposed to listening two movements of two sonatas. Physiological data are captured to see the listener's response to the harmonic rhythm. The two sonata movements have chosen because one remains a clear quadrature and another not, breaking the periods of traditional phrase constructed from sequences composed of four and eight bars. Thus, in addition to studying the behavior before the rhythmic-harmonic changes, it is judged whether the quadrature influences the way music is perceived. Orff, along with other contemporary educators, stresses the importance of organizing the musical rhythmic patterns in sentences of two, four or eight bars, proportionally. This arrangement provides adequate mental models that facilitate the understanding of the works. Musical phrases are equated to oral language, also highlighting the use of musical elements in their natural form without modification. Thus it can be seen that natural and spontaneous songs hold a quadrature system. Moreover, notice that

Fig. 2. "Surprise" theme and 3 variations

Martenot argues that a musical phrase must be the principle for the realization of rhythm. Therefore, in the different experiments the intentionality that a melodic line supports each of the proposals is always maintained, even when what is really intended is to measure rhythmic accompaniments to the central proposal.

Thus, the third musical test provides the listener with two sonatina movements. The first of them is the first movement of the sonatina in sonata form named "Sonatina for clarinet and piano in SibM" from Wolfgang Amadeus Mozart (see Fig. 3a). The second one is the first movement of the "Sonata no. 1 for trumpet and piano in MibM" by James Hook in rondo form (see Fig. 3b). Both sonatas are performed in a version for clarinet and piano, so that the change of instrumental timbre does not affect the listener. For ease of listening and understanding of both works a basic sheet music indicating color and squares is used. It is marked with lighting on the computer screen during the performance. Each measure is lit so that the listener understands the several parts of the piece, repetitions or variations thereof, and listening is comprehensive. After listening each of the pieces, the participant writes a phrase that describes his/her feeling. The physiological data from listeners are considered for evaluation of the musical test. The labels described in the previous tests in relation to their physiological data are used. This is to analyze what happens when listening to these pieces.

Fig. 3. Rhythmic schemes. Left: First movement of the Mozart sonatina. Right: First movement of the Hook sonatina.

There is a special focus on the possible variations in acceleration or deceleration of harmonic rhythm. To this end, a scheme of harmonic rhythm has been provided for each of the pieces.

2.5   Fourth Musical Test: The Rhythmic Accompaniment

Finally, the fourth musical test intends to associate some types of rhythmic accompaniments to emotional states. To do this, listeners hear different rhythmic fragments on an equal harmonic basis, altering only the rhythms of accompaniments in each case. The listener is offered the theme and seven variants based on both classical and modern music. In this case, the variant executed by the clarinet is excluded in the main melody (theme). It remains unchanged, so that the rhythmic variation does not influence the response of the listener. Therefore, all variations in the rhythm accompaniment (but never in the harmonic rhythm) are performed by the piano. For this, a specific composition that adheres to the above parameters is used. It is entitled "Theme and seven rhythmic variations (Eight versions of accompaniment to an unaltered melody harmony)" by the composer Juan Francisco Manzano Ramos.

The theme features a chord accompaniment (basic links from a chord) in white, with the use of black to mark the end of every sentence, thus adapting the harmonic rhythm. Variation I is characterized by the use of triplets in the right hand of the piano in a headless manner (always beginning with a silence),

Fig. 4. An excerpt of "Theme and seven rhythmic variations"; theme and variation I

which the left hand completes with a shared chord (see Fig. 4). The use of passing notes in the right hand should be emphasized.

Variation II is characterized by the use in the left hand of an Alberti bass in semiquavers, with a tessitura focused on the central "do". Moreover, the right hand supplements the above with blacks in a tessitura an octave higher (with repetition of chords through white). Variation III is characterized by the use of demisemiquavers. The first beat starts with an upward deployment of the chord (arpeggio), which the left hand completes in the second beat. The fourth beat always supposes a return to the initial situation with the deployment of the chord in the left hand. Variation IV uses the "pasodoble" rhythm with use of the characteristic contretemps. Thus, the left hand takes the bass (alternating the fundamental note and the fifth of the chord), and the right hand makes the harmonic filler with the use of contretemps. Variation V uses an example of blues, alternating swing in its accompaniment (right hand), while the left hand holds the bass in quaver notes, alternating with strange notes the proper chord. Variation VI is ragtime, with the characteristic rhythm in both right and left hands. Variation VII is an example of using rhythmic asymmetry in the accompaniment. Thus, the use of octaves with chords without thirds in rhythms that do not follow a particular pattern is alternating with a variety of tessituras.

3 Data and Results

At this moment of research some basic and limited experimentation has been performed. This study has involved ten people from different ages. The initial results prove that the influence of note value in music for eliciting emotions is dependent of the personality of each participant in terms of neuroticism and extraversion. Although the number of participants is too little to draw clear conclusions, we believe that there is some evidence about some general tendencies related to the results obtained from the influence of beat.

The most neurotic people appreciate less difference between "Stressing" and "Relaxing" when the beat is varied. People with greater differentiation between levels of neuroticism and extraversion (standing out in introversion or neuroticism, but not the other) show a greater perception in the increase of expressiveness (from "Expressionless" to "Striking") with an increasing beat. Extroverted listeners show a more proportional correlation to the description of "Pleasant" vs. "Unpleasant" regarding increased or decreased beat correlation.

The results for the other tests draw no clear conclusions, as there are too few participants. The number of possibilities grows a lot from musical test 1 to 4, and, obviously there is a need of working with a much higher number of individuals.

4 Conclusions

This article has described the first steps in the use of music to regulate affect. The proposal is based on the assumption of the power of music to change mood. It belongs to a running project denominated "Improvement of the Elderly Quality of Life and Care through Smart Emotion Regulation". The objective of the project is to find solutions for improving the quality of life and care of ageing adults living at home by using emotion elicitation.

The proposal has studied the participants' changes in emotional states through listening different auditions. The present research has focused on the musical parameter of note value through its four basic components of the parameter note value, namely, beat, rhythm, harmonic rhythm and rhythmic accompaniment to detect the individual preferences of the listeners. The initial results prove that the influence of beat in music for eliciting emotions is dependent of the personality of each participant in terms of neuroticism and extraversion.

It is important to consider that the experimentation has been performed on a limited number of subjects. It is foreseen to engage into an exhaustive experimentation with ageing adults in a close future.

Acknowledgements. This work was partially supported by Spanish Ministerio de Economía y Competitividad / FEDER under TIN2013-47074-C2-1-R grant.

References

1. Castillo, J.C., Fernández-Caballero, A., Castro-González, Á., Salichs, M.A., López, M.T.: A framework for recognizing and regulating emotions in the elderly. In: Pecchia, L., Chen, L.L., Nugent, C., Bravo, J. (eds.) IWAAL 2014. LNCS, vol. 8868, pp. 320–327. Springer, Heidelberg (2014)
2. Fernández-Caballero, A., Latorre, J.M., Pastor, J.M., Fernández-Sotos, A.: Improvement of the elderly quality of life and care through smart emotion regulation. In: Pecchia, L., Chen, L.L., Nugent, C., Bravo, J. (eds.) IWAAL 2014. LNCS, vol. 8868, pp. 348–355. Springer, Heidelberg (2014)
3. Costa, A., Castillo, J.C., Novais, P., Fernández-Caballero, A., Simoes, R.: Sensor-driven agenda for intelligent home care of the elderly. Expert Systems with Applications 39(15), 12192–12204 (2012)
4. Fernández-Caballero, A., Castillo, J.C., Rodríguez-Sánchez, J.M.: Human activity monitoring by local and global finite state machines. Expert Systems with Applications 39(8), 6982–6993 (2012)
5. Dahl, S.: On the beat: human movement and timing in the production and perception of music. Doctoral Thesis. KTH Royal Institute of Technology, Stockholm, Sweden (2005)
6. Lapidaki, E.: Young people's and adults' large-scale timing in music listening. In: Proceedings of the Sixth International Conference on Music Perception and Cognition, pp. 278–290 (2000)
7. Willimek, D.: Music and Emotions – Research on the Theory of Musical Equilibration (die Strebetendenz-Theorie) (2013)
8. Weikart, P.S.: Value for learning and living – Insights on the value of music and steady beat. Child Care Information Exchange 153, 86–88 (2003)
9. Norris, S.: Tempo, auftakt, levels of actions, and practice: rhythm in ordinary interactions. Journal of Applied Linguistics 6, 333–355 (2009)
10. Scollon, R.: The rhythmic integration of ordinary talk. In: Analyzing Discourse: Text and Talk, pp. 335–349 (1982)

Towards Emotionally Sensitive Conversational Interfaces for E-therapy

David Griol[1(✉)], José Manuel Molina[1], and Zoraida Callejas[2]

[1] Group of Applied Artificial Intelligence (GIAA)
Computer Science Department
Carlos III University of Madrid, Madrid, Spain
{david.griol,josemanuel.molina}@uc3m.es
[2] Spoken and Multimodal Dialogue Systems Group (SISDIAL)
Dept. of Languages and Computer Systems
University of Granada, Granada, Spain
zoraida@ugr.es

Abstract. In this paper, we enhance systems interacting in healthcare domains by means of incorporating emotionally sensitive spoken conversational interfaces. The emotion recognizer is integrated in these systems as an intermediate phase between natural language understanding and dialog management in the architecture of a spoken dialog system. The prediction of the user's emotional state, carried out for each user turn in the dialog, makes it possible to adapt the system dynamically selecting the next system response taking into account this valuable information. We have applied our proposal to develop an emotionally sensitive conversational system adapted to patients suffering from chronic pulmonary diseases, and provide a discussion of the positive influence of our proposal in the perceived quality.

Keywords: Conversational Interfaces · Dialog Systems · Emotion Recognition · E-therapy · Adaptation · Spoken Interaction · Mobile Interfaces

1 Introduction

Conversational interfaces [17] have been proven useful for providing the general public with access to telemedicine services, promoting patients' involvement in their own care, assisting in health care delivery, and improving patient outcome [4]. Bickmore and Giorgino defined these systems as being "those automated systems whose primary goal is to provide health communication with patients or consumers primarily using natural language dialog" [4].

During the last two decades, these interfaces have been increasingly used in healthcare and E-therapy providing services such as interviews [16], counseling [12], chronic symptoms monitoring [14], medication prescription assistance and adherence [5], changing dietary behavior [8], promoting physical activity [9], helping cigarette smokers quit [18], or speech therapy [19].

The proposal that we present in this paper is focused on the design of healthcare systems in which speech is the only modality used as input and output for

© Springer International Publishing Switzerland 2015
J.M. Ferrández Vicente et al. (Eds.): IWINAC 2015, Part I, LNCS 9107, pp. 498–507, 2015.
DOI: 10.1007/978-3-319-18914-7_52

the system. On the one hand, speech and natural language technologies allow users to access applications in which traditional input interfaces cannot be used (e.g. in-car applications, access for disabled persons, etc.). Also speech-based interfaces work seamlessly with small devices and allow users to easily invoke local applications or access remote information. For this reason, conversational agents are becoming a strong alternative to traditional graphical interfaces, which might not be appropriate for all users and/or applications domains [17].

Health dialog systems must confront social, emotional and relational issues in order to enhance patients satisfaction. However, although emotion is receiving increasing attention from the dialog systems community, most research described in the literature is devoted exclusively to emotion recognition. For example, a comprehensive and updated review can be found in [20,3].

Emotions affect the explicit message conveyed during the interaction and is frequently mentioned in the literature as the most important factor in establishing a working alliance in healthcare applications [5]. They change people voices, facial expressions, gestures, and speech speed. Emotions can also affect the actions that the user chooses to communicate with the system. Emotions have also been recently considered as a very important factor of influence in decision making processes.

Despite its benefits, the recognition of emotions in dialog systems presents important challenges which are still unresolved. The first challenging issue is that the way a certain emotion is expressed generally depends on the speakers, their culture and environment. Another problem is that some emotional states are long-term (e.g. sadness), while others are transient and do not last for more than a few minutes. Thus, it is not trivial to select the categories being analyzed and classified by an automatic emotion recognizer. Also there is not a clear agreement about which speech features are most powerful in distinguishing between emotions.

In this paper, we describe a proposal that address these important issues by developing affective dialog models for healthcare conversational systems, which take into account both emotions and the dialog acts in the users' utterances to select the next system action. Our approach for emotion recognition is focused on recognizing negative emotions that might discourage users from employing the system again or even lead them to abort an ongoing dialog. The dialog manager of the system tailors the next system answer to the user emotional state by changing the help providing mechanisms, the confirmation strategy, and the interaction flexibility.

2   Our Proposal to Develop Emotionally Sensitive Conversational Interfaces

A spoken dialog system integrates five main tasks to deal with user's spoken utterances in natural language: automatic speech recognition (ASR), natural language understanding (NLU), dialog management (DM), natural language generation (NLG), and text-to-speech synthesis (TTS). We propose to add an emotion recognizer in this architecture to process the users' emotional state during the interaction, which is considered as an additional valuable input for the dialog manager to select the next system action.

Our proposal to develop an emotion recognizer is based solely in acoustic and dialog information because in most application domains the user utterances are not long enough for the linguistic parameters to be significant for the detection of emotions. Our recognition method, based on the previous work described in [7], firstly takes acoustic information into account to distinguish between the emotions which are acoustically more different, and secondly dialog information to disambiguate between those that are more similar. We are interested in recognizing negative emotions that might discourage users from employing the system again or even lead them to abort an ongoing dialog. Concretely, we have considered three negative emotions: anger, boredom, and doubtfulness, where the latter refers to a situation in which the user uncertain about what to do next).

Following the proposed approach, our emotion recognizer employs acoustic information to distinguish anger from doubtfulness or boredom and dialog information to discriminate between doubtfulness and boredom, which are more difficult to discriminate only by using phonetic cues.

This process is shown in Figure 1. As can be observed, the emotion recognizer always chooses one of the three negative emotions under study, not taking neutral into account. This is due to the difficulty of distinguishing neutral from emotional speech in spontaneous utterances when the application domain is not highly affective. This is the case of most spoken dialog systems, in which a baseline algorithm which always chooses "neutral" would have a very high accuracy, which is difficult to improve by classifying the rest of emotions, that are very subtlety produced.

The first step for emotion recognition is feature extraction. The aim is to compute features from the speech input which can be relevant for the detection of emotion in the users' voice. We extracted the most representative selection from the list of 60 features shown in Table 1. The feature selection process is carried out from a corpus of dialogs on demand, so that when new dialogs are available, the selection algorithms can be executed again and the list of representative features can be updated. The features are selected by majority voting of a forward selection algorithm, a genetic search, and a ranking filter using the default values of their respective parameters provided by the Weka toolkit.

The second step of the emotion recognition process is feature normalization, with which the features extracted in the previous phase are normalized around the user neutral speaking style. This enables us to make more representative classifications, as it might happen that a user 'A' always speaks very fast and loudly, while a user 'B' always speaks in a very relaxed way. Then, some acoustic features may be the same for 'A' neutral as for 'B' angry, which would make the automatic classification fail for one of the users if the features are not normalized.

Once we have obtained the normalized features, we classify the corresponding utterance with a multilayer perceptron (MLP) into two categories: *angry* and

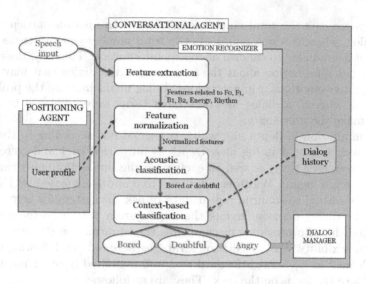

Fig. 1. Schema of the proposed emotion recognizer

doubtful_or_bored. The precision values obtained with the MLP are discussed in detail in [7], where we evaluated the accuracy of the initial version of this emotion recognizer. If an utterance is classified as angry, the emotional category is passed to the dialog manager of the system. If the utterance is classified as *doubtful_or_bored,* it is passed through an additional step in which it is classified according to two dialog parameters: depth and width. Dialog context is considered for emotion recognition by calculating these parameters.

Table 1. Features defined for emotion detection from the acoustic signal [11,22,15]

Groups	Features	Physiological changes related to emotion
Pitch	Minimum value, maximum value, mean, median, standard deviation, value in the first voiced segment, value in the last voiced segment, correlation coefficient, slope, and error of the linear regression.	Tension of the vocal folds and the sub glottal air pressure.
First two formant frequencies and their bandwidths	Minimum value, maximum value, range, mean, median, standard deviation and value in the first and last voiced segments.	Vocal tract resonances.
Energy	Minimum value, maximum value, mean, median, standard deviation, value in the first voiced segment, value in the last voiced segment, correlation, slope, and error of the energy linear regression.	Vocal effort, arousal of emotions.
Rhythm	Speech rate, duration of voiced segments, duration of unvoiced segments, duration of longest voiced segment and number of unvoiced segments.	Duration and stress conditions.

Depth represents the total number of dialog turns up to a particular point of the dialog, whereas width represents the total number of extra turns needed throughout a subdialog to confirm or repeat information. This way, the emotion recognizer has information about the situations in the dialog that may lead to certain negative emotions, e.g. a very long dialog might increase the probability of boredom, whereas a dialog in which most turns were employed to confirm data can make the user angry.

The computation of depth and width is carried out according to the dialog history, which is stored in log files. Depth is initialized to 1 and incremented with each new user turn, as well as each time the interaction goes backwards (e.g. to the main menu). Width is initialized to 0 and is increased by 1 for each user turn generated to confirm, repeat data or ask the system for help.

Then, the dialog manager tailors the next system answer to the user state by changing the help providing mechanisms, the confirmation strategy, and the interaction flexibility. The conciliation strategies adopted are, following the constraints defined in [6], straightforward and well delimited in order not to make the user loose the focus on the task. They are as follows:

- If the recognized emotion is doubtful and the user has changed his behavior several times during the dialog, the dialog manager changes to a system-directed initiative and adds at the end of each prompt a help message describing the available options.
- In the case of anger, if the dialog history shows that there have been many errors during the interaction, the system apologizes and switches to DTMF (Dual-Tone Multi-Frequency) mode. If the user is assumed to be angry but the system is not aware of any error, the system's prompt is rephrased with more agreeable phrases and the user is advised that they can ask for help at any time.
- In the case of boredom, if there is information available from other interactions of the same user, the system tries to infer from those dialogs what the most likely objective of the user might be. If the detected objective matches the predicted intention, the system takes the information for granted and uses implicit confirmations.
- In any other case, the emotion is assumed to be neutral, and the next system prompt is decided only on the dialog history.

3   Practical Application: Patients with Domiciliary Oxygen Therapy

Domiciliary oxygen therapy has been used during the last five decades to alleviate reduced arterial oxygenation (hypoxemia) and its consequences [13,2]. It is considered to be the only therapeutic approach that can prolong survival in patients with chronic pulmonary diseases. This therapy is also aimed at relieving dyspnea and improving exercise capacity and sleep quality. Patients have portable cylinders, concentrators and portable liquid systems as well a pulse oximeter that monitors the oxygen saturation of a patient's blood and changes in blood volume in the skin. The pulse oximeter is usually incorporated into a multiparameter patient monitor, which also monitors and displays the pulse rate and blood pressure [21].

We have applied our proposal to develop and evaluate an adaptive system that provides functionalities oriented to these patients. The system is able of greeting the patient, conducting a chat, assessing the patient's behavior since the last conversation, collecting data to monitor the patients' current state, providing feedback on this behavior, setting new behavioral goals for the patient to work towards before the next conversation, promoting medication adherence, providing personalized tips or relevant educational material, creating a self-report survey with questions assessing the patient's attitude towards the agent, providing nearest pharmacies on duty, and personalized farewell exchanges. The information offered to the patient is extracted from different web pages. Several databases are also used to store this information and automatically update the data that is included in the application.

The greeting and farewell functionalities have been designed to achieve the personalization of the system right from the beginning of the interaction, modifying the structure of the initial and ending prompts to incorporate not only the name of the patient, but also additional functionalities like encouraging them to follow personalized advices.

Given that continuous control and monitoring is a key factor for these diseases, this is one of the main functionalities of the system. The data collected by the system are the patient's oxygen saturation level, heart rate, and blood pressure (systolic and diastolic values). The system validates and analyzes the data, providing some immediate feedback to the patients regarding their current progress as well as communicating the results to doctors at the hospital who are able to review the patient's progress graphically and deal with alerts generated by the system concerning abnormal developments.

The evolution of the patient is also taken into account in the personalized tips functionality (e.g., "drink often to avoid dehydration", "keep a varied and balanced diet", "try to keep in the same weight", "avoid caffeine and salty food", "eat in a relaxed environment without hurry", "visit the doctor at the first evidence of cold or influenza", etc.). Also for these patients it is important to receive support, as they sometimes suffer from anxiety and diminished self-esteem because the illness deeply affects their social life.

The chat functionality extends this goal by means of personalized forms related to educational hints explaining details of their illness (e.g., how the respiratory system works and what are the consequences of their treatment so that they can better face them). The medication adherence functionality emulates previous works [5,1] to remind patients to take all their medications as prescribed in the medical profile. Finally, the pharmacies functionality is based on dynamic information automatically provided by the system and related to the current location of the terminal and the daily updated list of pharmacies on duty.

A previously developed automatic user simulation technique [10] has been employed to generate the dialog corpus required for learning the neural networks for the emotion recognizer. To generate the emotion label for each turn of the simulated user, we employ a rule-based approach based on dialog information similar to the threshold method employed as a second step in the emotion recognizer described in the previous section. In each case, the method chooses randomly (0.5 probabilities) between an emotion (doubtful, bored, or angry) and neutral. The probability of choosing the emotion rises to 0.7 when the same emotion was chosen in the previous turn, which allows simulating moderate changes of the emotional state. Although the simulated users resemble the behavior of the real users in the initial corpus acquired for the task (the changes in the emotional state correspond to the same transitions observed in the dialog states), they are more emotional, as the probability of neutral in this corpus was 0.85. This way, it is possible to obtain different degrees of emotional behavior with which evaluate the benefits of our proposal.

4   Preliminary Evaluation

For comparison purposes, we have developed two systems providing the functionalities described in the previous section. The baseline system does not carry out any adaptation to the user, while the emotionally sensitive system integrates an emotion recognizer developed following our proposal. A total of 30 recruited users participated in the evaluation, aged 51 to 69 (mean 57.2), 67% male, five with chronic pulmonary diseases. Although not all users suffered from them, they were recruited taking into account the age range which is more affected by these disorders. Additionally, the design of the application and its functionalities was carried out with the continuous feedback of several patients and the medical personal that treats them.

A total of 90 dialogs was recorded from the interactions of the recruited users, 15 users employed the emotionally sensitive system, and 15 users employed the baseline version of the system. The users were provided with a brochure describing the scenarios that they were asked to complete and main functionalities of the system. A total of 45 scenarios was defined to consider the different queries that may be performed by users. Each scenario specified a set of objectives that had to be fulfilled by the user at the end of the dialog and they were designed to include and combine the complete set of functionalities previously described for the system.

We asked the recruited users to complete a questionnaire to assess their subjective opinion about system performance. The questionnaire had eight questions: i) Q1: How well did the system understand you?; ii) Q2: How well did you understand the system messages?; iii) Q3: Was it easy for you to get the requested information?; iv) Q4: Was the interaction rate adequate?; v) Q5: If the system made errors, was it easy for you to correct them?; vi) Q6: How much did you feel that the system cares about you?; vii) Q7: How much did you trust the system?; viii) Q8: With which frequency would you continue working with the system?

The possible answers for each one of the questions were the same: Never/Not at all, Seldom/In some measure, Sometimes/Acceptably, Usually/Well, and Always/Very Well. All the answers were assigned a numeric value between one and five (in the same order as they appear in the questionnaire). The following subsections present the results obtained for the four types of evaluation metrics previously described.

Table 2 shows the average results obtained with respect to the subjective evaluation carried out by the recruited users. As can be observed, the two systems correctly understand the different user queries and obtain a similar evaluation regarding the user observed easiness in correcting errors made by the ASR module. However, the emotionally sensitive system has a higher evaluation rate regarding the user observed easiness in obtaining the data required to fulfill the complete set of objectives defined in the scenarios, as well as the suitability of the interaction rate during the dialog. Ratings of satisfaction, ease of use, trust, and desire to continue using the system were also improved by the emotionally sensitive system. Together, these results indicate that the conversational system represents a viable and promising medium for helping patients with the described diseases.

The following main conclusions can also be extracted from the analysis of the results obtained for the different questions and systems. With regard questions Q1 and Q2 (users understanding system responses and system understanding users responses), the analysis of the results showed that there were not significant differences between the two systems. This might be because of both systems integrated the same ASR, NLU and TTS modules. A similar conclusion can be extracted from the analysis of the facility of correcting errors (question Q5).

Regarding the easiness of obtaining information (question Q3) and the adequacy of the interaction rate (question Q4), the emotionally sensitive system improves the results obtained with the baseline system. The same conclusion can be extracted from the analysis of the users' perception about the credibility and concern transmitted by the system (questions Q6 and Q7). Users also significantly prefer to continue working with the emotionally sensitive system, which obtained the highest mean and lowest standard deviation for question 8. In our opinion, this can be explained by the user's adaptation achieved by the introduction of our proposal in the emotionally sensitive system.

Table 2. Results of the subjective evaluation with real users (For the mean value M: 1=worst, 5=best evaluation)

	Baseline	Emotionally sensitive System
Q1	M = 4.62, SD = 0.37	M = 4.82, SD = 0.34
Q2	M = 3.65, SD = 0.24	M = 3.93, SD = 0.27
Q3	M = 3.84, SD = 0.56	M = 4.36, SD = 0.34
Q4	M = 3.43, SD = 0.28	M = 4.24, SD = 0.29
Q5	M = 3.27, SD = 0.59	M = 3.34, SD = 0.57
Q6	M = 3.71, SD = 0.41	M = 4.30, SD = 0.35
Q7	M = 4.22, SD = 0.46	M = 4.51, SD = 0.26
Q8	M = 3.80, SD = 0.42	M = 4.47, SD = 0.36

5   Conclusions and Future Work

Emotions are frequently mentioned in the literature as the most important factor in establishing a working alliance in healthcare applications. In this paper, we contribute a proposal to develop emotionally sensitive spoken conversational interfaces for healthcare applications. Our proposal is focused on recognizing negative emotions that might discourage users from employing the system again or even lead them to abort an ongoing dialog. The recognized emotion is used as an additional valuable information to select and adapt the next system response.

We have provided a practical application of our proposal by means of a system that provides personalized services for patients suffering from chronic pulmonary diseases. From a set of dialogs acquired with recruited users we have studied the influence of the emotional adaptation on the quality of the services that are provided by the system.

As future work, we want to carry out a detailed study with a large number of patients in a continuous use of the system during several months. We are also interested in extending and evaluating our proposal for emotion recognition considering its combination with sentiment analysis approaches analyzing the text transcription hypothesis provided by the ASR module.

Acknowledgements. This work was supported in part by Projects MINECO TEC2012-37832-C02-01, CICYT TEC2011-28626-C02-02, CAM CONTEXTS (S2009/TIC-1485).

References

1. Allen, J., Ferguson, G., Blaylock, N., Byron, D., Chambers, N., Dzikovska, M., Galescu, L., Swift, M.: Chester: towards a personal medication advisor. Journal of Biomedical Informatics 39(5), 500–513 (2006)
2. Antoniu, S.: Outcomes of adult domiciliary oxygen therapy in pulmonary diseases. Expert Review of Pharmacoeconomics and Outcomes Research 6(1), 9–66 (2006)
3. El Ayadi, M., Kamel, M., Karray, F.: Survey on speech emotio nrecognition: Features, classification schemes, and databases. Pattern Recognition 44, 572–587 (2011)
4. Bickmore, T., Giorgino, T.: Health dialog systems for patients and consumers. Journal of Biomedical Informatics 39(5), 556–571 (2006)
5. Bickmore, T., Puskar, K., Schlenk, E., Pfeifer, L., Sereika, S.: Maintaining reality: Relational agents for antipsychotic medication adherence. Interacting with Computers 22, 276–288 (2010)
6. Burkhardt, F., van Ballegooy, M., Engelbrecht, K., Polzehl, T., Stegmann, J.: Emotion detection in dialog systems - Usecases, strategies and challenges. In: Proc. ACII 2009, pp. 1–6 (2009)
7. Callejas, Z., López-Cózar, R.: Influence of contextual information in emotion annotation for spoken dialogue systems. Speech Communication 50(5), 416–433 (2008)
8. Delichatsios, H., Friedman, R., Glanz, K., Tennstedt, S., Smigelski, C., Pinto, B.: Randomized trial of a talking computer to improve adults eating habits. American Journal of Health Promotion 15, 215–224 (2000)

9. Farzanfar, R., Frishkopf, S., Migneault, J., Friedman, R.: Telephone-linked care for physical activity: A qualitative evaluation of the use patterns of an information technology program for patients. Biomedical Informatics 38, 220–228 (2005)
10. Griol, D., Hurtado, L., Sanchis, E., Segarra, E.: Acquiring and evaluating a dialog corpus through a dialog simulation technique. In: Proc. SIGdial 2007, pp. 29–42 (2007)
11. Hansen, J.: Analysis and compensation of speech under stress and noise for environmental robustness in speech recognition. Speech Communication 20(2), 151–170 (1996)
12. Hubal, R., Day, R.: Informed consent procedures: An experimental test using a virtual character in a dialog systems training application. Journal of Biomedical Informatics 39, 532–540 (2006)
13. Jindal, S.: Oxygen therapy: important considerations. Indian Journal of Chest Disease and Allied Science 50(1), 97–107 (2008)
14. Migneault, J.P., Farzanfar, R., Wright, J., Friedman, R.: How to write health dialog for a talking computer. Journal of Biomedical Informatics 39(5), 276–288 (2006)
15. Morrison, D., Wang, R., DeSilva, L.: Ensemble methods for spoken emotion recognition in call-centres. Speech Communication 49(2), 98–112 (2007)
16. Pfeifer, L., Bickmore, T.: Designing Embodied Conversational Agents to Conduct Longitudinal Health Interviews. In: Proc. IVA 2010, pp. 4698–4703 (2010)
17. Pieraccini, R.: The Voice in the Machine: Building Computers That Understand Speech. MIT Press (2012)
18. Ramelson, H., Friedman, R., Ockene, J.: An automated telephone-based smoking cessation education and counseling system. Patient Education and Counseling 36, 131–143 (1999)
19. Saz, O., Yin, S.C., Lleida, E., Rose, R., Vaquero, C., Rodríguez, W.R.: Tools and Technologies for Computer-Aided Speech and Language Therapy. Speech Communication 51(10), 948–967 (2009)
20. Schuller, B., Batliner, A., Steidl, S., Seppi, D.: Recognising realistic emotions and affect in speech: state of the art and lessons learnt from the first challenge. Speech Communication 53(9-10), 1062–1087 (2011)
21. Shah, N., Ragaswamy, H., Govindugari, K., Estanol, L.: Performance of three new-generation pulse oximeters during motion and low perfusion in volunteers. Journal of Clinical Anesthesia 24(5), 385–391 (2012)
22. Ververidis, D., Kotropoulos, C.: Emotional speech recognition: resources, features and methods. Speech Communication 48, 1162–1181 (2006)

Automatic Drawing Analysis of Figures Included in Neuropsychological Tests for the Assessment and Diagnosis of Mild Cognitive Impairment

M. Rincón[1]([✉]), S. García-Herranz[2], M.C. Díaz-Mardomingo[2], R. Martínez-Tomás[1], and H. Peraita[2]

[1] Department of Inteligencia Artificial. E.T.S.I. Informática, UNED, Madrid, Spain
[2] Department Psicología Básica I. Facultad de Psicología, UNED, Madrid, Spain
mrincon@dia.uned.es

Abstract This proposal is framed within the group's general working line of applying artificial intelligence techniques to advance in early mild cognitive impairment diagnosis. If impairment in semantic production was studied in previous works, now we rely on the reduced ability to reproduce or copy simple figures, part of standardized neuropsychological tests designed to assess mild cognitive impairment. Although the long-term goal of this project is to work with all figures from these tests, in this paper we will focus on the automatic analysis of the alternating graphs figure. We develop a quantitative descrition of different features that appear to be very abstract in the test norms and define new features that are not considered so far. Results with just one figure are quite promising (77.7% precision and 77.1 recall).

Keywords: Drawing analysis · Mild Cognitive Impairment · Alzheimer disease · Drawings in neuropsychological tests · Test Barcelona

1 Introduction

It is known that mild cognitive impairment (MCI) is detected in the very early stages of Alzheimer disease (AD) and other neurodegenerative dementias, which is essential for achieving maximum effectiveness in pharmacological treatments and cognitive therapies. One part of standardized neuropsychological tests designed to assess MCI include reproducing or copying figures. The main idea of this work is to analyse, using artificial intelligence techniques, from a typical pattern or standard of each one of the figures with which we have worked in an investigation on early detection of mild cognitive impairment [1,2,3], the extent to which certain distortions from the standard may indicate of different profiles and degrees of MCI. These figures come from the Mini Examen Cognoscitivo (MEC) [4], the Test Barcelona [5], and the Rey complex-figure Test [6].

The processes and cognitive functions —cognitive domains— that are supposedly assessed by the execution of these figures are: the executive function, visual and/or visuospatial perception, motor skills, and spatial memory. In each

© Springer International Publishing Switzerland 2015
J.M. Ferrández Vicente et al. (Eds.): IWINAC 2015, Part I, LNCS 9107, pp. 508–515, 2015.
DOI: 10.1007/978-3-319-18914-7_53

of these functions or processes, the traits or components (control, inhibition, planning, etc.) involved in the patterns to be analyzed can be defined, for example, following the indications of Dr. Peña Casanova with regard to the drawings of alternating graphs and loops [5].

An important complementary aspect is to be able to obtain not only the distortion with regard to the standard figure at any given time, but also to be able to determine how this evolves over the years, an aspect that can be analyzed in the future, as it constitutes the data of our longitudinal research.

Although the long-term goal of this project is to work with all figures from the above-mentioned tests, we present herein only the results of the figure alternating graphs, which is one of the subtests of the Barcelona Test. The diagnostic value of this test, selected for this automatic exploratory analysis, lies in the fact that in its execution are involved some executive function components, such as seriation, planning, flexibility, inhibition, as well as praxic capacity.

The task consisted of copying one figure, one in which peaks and plateaus should alternate. Following the test norms, scoring is done according to the quality of the copy, assigning 0 to 2 points for each one of them. The distortions that can emerge in the execution of these figures can be of different types: variations in the size of the execution of the figure, alteration of features by addition or deletion, scribbling, perseverations, rotations, etc. These errors or alterations in the reproduction of the drawing may be markers of more severe dysfunctions, which can be of an apraxic type, in which the executive functions are also involved, as in Alzheimer's type dementia [7].

Whereas in other types of standardized tests that assess episodic memory, verbal fluency, etc. it is much easier to obtain normative data that allow scoring free from subjectivity, in these types that involve reproducing and copying figures, it is much more difficult, imposing the subjectivity and some discretionality by the evaluator. Although scoring criteria exist, there is a large component of subjectivity, that can undermine the reliability if discrepancies among the evaluators are not corrected [8]. As a result of this lack of agreement in the correction, there may be an important margin of error in the detection of certain problems of motor skills, visual and visuospatial perception, etc. within the framework of a general plan for early detection of MCI, either prior or not, to AD. The method proposed herein is a method of Artificial Intelligence (A.I), inexpensive, easy to apply and implement and that provides a convergence of criteria of both methods: the manual one and the automatic one. This could lead to discovering certain aspects of figure execution that may be significant for the early detection of alterations within the framework of specific cognitive domains.

In previous publications, our group has worked to define an economic procedure for MCI diagnosis by analysing cognitive alterations affecting declarative semantic memory [9,10]. As in this work, our goal there was to objectify and automate the analysis of a test that, because of its low cost, it could be used for routine clinical evaluations or screenings that could lead to more expensive and selective tests that confirm or rule out the disease accurately. We confirmed that, in this context, Bayesian networks are the most appropriate tool for this purpose because they allow us to combine previous knowledge with case data (the network structure, the qualitative part of the model, is obtained from psychology experts and epidemiological studies, and the network parameters, the quantitative part of the model, are learnt automatically from epidemiological studies and a linguistic corpus of oral definitions [11].

Other non-conventional diagnostic methods proposed to evaluate the cognitive state of patients or even to detect motor deficiencies caused by a brain haemorrhage from monitoring daily activity. In particular, Matic et al. analyse the act of getting dressed [12] and Kearns et al. analyse the tortuosity in movement paths—irregular movements—of elderly people with cognitive impairment [13].

The remainder of the paper is organised as follows: Section 2 describes the materials and methods, i.e., the sociodemographic and clinical data of the participants in the study and purpose specific automatic method for alternating graphs drawing analysis. In section 3 we present the experimental results. Finally, section 4 presents our concluding remarks as well as the future lines of research.

2   Materials and Methods

2.1   Participants

The sample (N = 40 participants) was recruited from a larger sample of participants in an ongoing longitudinal study (ref. SEJ 2004-04233 and SEJ 2007-63325) focused on determining the prevalence the different MCI subtypes [1,3]. The participants were recruited in the Autonomous Community of Madrid, Spain. They were assessed longitudinally with a neuropsychological battery during an average period of 3 years. MCI was defined as having a score of 1.5 SD below the mean in at least 2 of the tests applied. Depending on the data obtained through the different neuropsychological assessments, the participants were classified in one of the following cognitive profiles: healthy individuals (n=16)—expected performance according to references scales—or MCIs (n=24). Of the MCI group, after a 3-year follow-up, 10 of them had a diagnosis compatible with an initial phase of probable AD (see Table 1).

2.2   Methods

We have defined a purpose specific automatic method for alternating graphs drawing analysis. The general approach considered in this paper consists of the following steps: 1) drawing digitalisation, 2) drawing segmentation, 3) line extraction, 4) pattern matching and characterization and 5) MCI diagnosis. Figure 1 shows an example with the intermediate results of each one of these steps.

Drawing Segmentation. After digitalisation with a standard scanner, a grayscale image is obtained (Fig. 1.a). The image contains two figures, the pattern

Table 1. Baseline sociodemographic and clinical descriptive data of healthy, MCI and MCI-converters

	Healthy n = 16	MCI n = 14	MCI-Converters n = 10
	Mean (SD)	Mean (SD)	Mean (SD)
Gender (female)	13 (81.25%)	11.00 (4.24)	4 (40%)
Age (years)	69.75 (4.85)	72.57 (5.31)	71.60 (5.08)
Formal education (years)	11.00 (4.24)	5.50 (6.51)	8.30 (5.53)
Geriatric Depression Scale (GDS) (Yesavage scale)[14]	3.44 (2.78)	3.93 (2.99)	4.20 (3.39)
Functional level (Blessed scale) [15]	0.59 (0.45)	0.67 (0.57)	1.00 (1.00)
Cognitive status (MEC 0-35)[4]	33.50 (2.47)	29.14 (3.78)	30.30 (2.00)

and the manual drawing. Histogram analysis is performed in order to segment the objects from the background. It is assumed that background is white and that the drawing is a continuous black line that comprises a small amount of the image pixels. Due to the fact that posterior pattern analysis is simplified if segmentation obtains a thin object, an iterative threshold selection is used that evaluates the results and stops when more than two big objects are obtained (Fig. 1.b). After that, the region of interest is rescaled in x and y to a standard size. The segmented line quality depends tremendously on the conditions under which the drawing is performed, the pen type, and the scanning process.

Line Extraction. Line thickness is various pixels wide and we therefore need to thin it down. We have used mathematical morphology for extracting the skeleton, which is very noisy and contains many small branches that must be eliminated (Fig. 1.c) Assuming that both line ends are in the x-coordinate extremes, the rest of braches are eliminated by deleting iteratively the endpoints. If small loops are detected, they are broken and the intermediate branches are eliminated again. The final result is a continuous line with no loops. Then, we use the recursive Douglas-Peucker line simplification algorithm to approximate the curve with line segments to a specified tolerance (Fig. 1.d).

Pattern Matching and Characterization. In this point, we have to make clear that our interest is not limited to recognise the alternating graph pattern or to assess if the pattern is copied correctly or not, instead, we are interested in wider assessment metrics that serve us to discriminate between groups of people. Because the alternating graph consists of repeating five times the peak-plateau

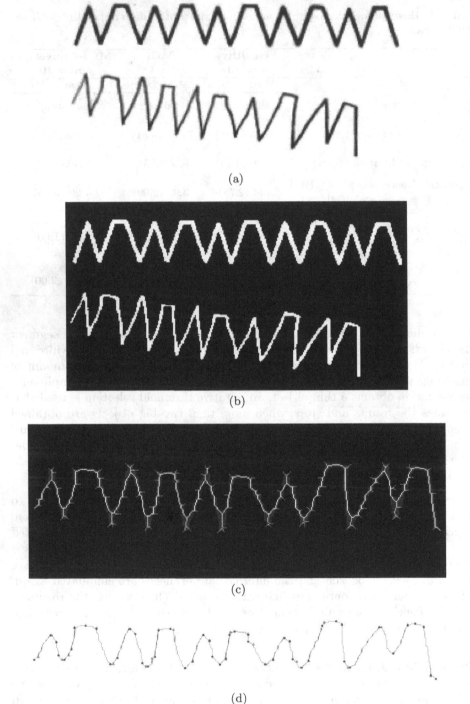

Figure 1. Peak-Plateau drawing segmentation and segment decomposition. a) scanned image; b) segmented image; c) skeleton extraction and d) segment approximation

pattern, which in turn consists of a sequence of various segments with different orientations (ascending, descending and horizontal segments), we analyse whether the segments found in the drawing match the pattern or not. The result is a measure of the number of valid and invalid segments found in the drawing. Table 2 summarises the list of features used for characterising the alternating graph that allows us to compare it with the model.

Table 2. List of features used for characterising the peak-plateau drawing

Feature	Description
height_pattern_diff	height difference between pattern and drawing
width_pattern_diff	width difference between pattern and drawing
drawing_tilt	tilt of the drawing with respect to the sample.
vertical_dist_to_model	vertical distance between drawing and model
X_scale_factor	scale factor in X coordinate used for drawing normalization
Y_scale_factor	scale factor in Y coordinate used for drawing normalization
alternance	variable associated with a perfect match between drawing and model in the five peak-plateau pattern
#valid_patterns	number of valid peak-plateau patterns
phase_diff	average x coordinate difference between pattern and drawing
plateau_width	average width of the horizontal segment in plateau subpatterns
#segments	total number of detected segments
#valid_segments	number of drawing segments that match the pattern segments
#invalid_segments	number of drawing segments that does not match the pattern
average_segment_tilt	average tilt of the drawing segments
last_segm_match	last consecutive valid segment

DCL Diagnosis. We have built a supervised machine learning classifier using the dataset described in subsection 2.1. In this stage of the study, we are interested in the analysis of the discrimination power of the different features. Therefore, we have used a J48 decision tree to implement the classifier because of ease of understanding. Due to the small sample size, we have used leave-one-out cross-validation for evaluating the classifier, which allows to use the largest possible training sample while keeping a reliable performance estimation.

3   Experimental Results

The final features included in the decision tree are #invalid_segments, X_scale_factor and vertical_dist_to_model. It is worth noting that #valid_segments is not very discriminative and it is highly correlated with #invalid_segments. Besides, the most discriminative measures, vertical_dist_to_model and X_scale_factor, are not considered in the neuropsicological tests. The results are quite promising (0.777 precision / 0.771 recall) having in mind the limitations imposed by the small sample size and the fact that the analysis is based on just one item of one test. As mentioned in the introduction, the expert's score in this test is quite generic and only distinguishes three possible values: good performance (2), good performance with some defects (1) and mistake (0).

4 Conclusions y Further Research

In this paper we have demonstrated that AI techniques can offer solutions to support the automatic analysis of drawings included in neuropsychological tests for the assessment and diagnosis of MCI. The main advantage of the automatic analysis is that it includes a larger amount of metrics for characterising the drawings, which makes it more quantitative, robust and user independent.

We have implemented an automatic system alternating graphs analysis and we have found discriminative metrics for diagnosing MCI. The fundamental problem in machine learning AD diagnosis, as in most neurological studies, is the absence of a training dataset large enough to build a reliable AD diagnosis system using supervised learning. We have to recognise that with this small sample size, we can only conclude that these proposed features are good candidates for MCI diagnosis, but they alone can not distinguish between different MCI types. In future works we will broaden the sample and will combine features from different figures to improve the classification performance. Therefore, this work intends to be a pilot study of a much broader work in which, from the longitudinal data already available, we can make different types of analysis:

- Longitudinal analysis of healthy controls: comparing, over a series of years, the execution patterns of healthy control subjects, to verify their stability in the execution of the figures.
- Longitudinal analysis of the stable MCIs: the same type of comparative approach but in subjects with stable MCI, to attempt to answer the following questions: Is the execution of the figures stable? Of all of them? For how long?
- Longitudinal analysis of the MCIs that evolve to AD: the same in subjects with an evolutionary MCI to Alzheimer's disease or another dementia.
- Transversal analysis of the different types of MCI: the same in subjects with various types of MCI: amnestic, multidomain and nonamnestic.
- Relationship between the figure drawing and ideomotor praxia: to study the relationship between the execution of the figures and the ideomotor praxia obtained by the same subjects in other tests.
- Socio-demographic analysis: to analyze the relationships between socio-demographic variables, such as age, gender and level of education, and the execution patterns of the above-mentioned figures.

Acknowledgments. This research has been supported by project No. 018-ABEL-CM-2013 of NILS Science and Sustainability program coordinated by Universidad Complutense (Spain), and UNED projects SEJ-2004-04233 and SEJ-2007-63325.
We thank all voluntary participants of las Rozas, Pozuelo and Madrid for taking part in the longitudinal study whose data have led to this work for their selfless collaboration, as well as the municipalities and institutions that facilitated contact details and provided an adequate logistics to carry out evaluations.

References

1. Díaz, C., Peraita, H.: Detección precoz del deterioro cognitivo ligero de la tercera edad. Psicothema 20(3), 438–444 (2008)
2. Díaz-Mardomingo, M.C., García-Herranz, S., Peraita, H.: Detección del DCL y conversión a la EA: un estudio longitudinal de casos. Psicogeriatría 2(2), 105–111 (2010)
3. Peraita, H., García-Herranz, S., Díaz-Mardomingo, M.C.: Evolution of Specific Cognitive Subprofiles of Mild Cognitive Impairment in a Three-Year Longitudinal Study. Current Aging Science 4, 171–182 (2011)
4. Lobo, A., Ezquerra, J., Gómez, F., Sala, J.M., Seva, A.: El mini-examen cognoscitivo. Un test sencillo, práctico, para detectar alteraciones intelectivas en pacientes médicos. Actas Luso Españolas de Neurolología, Psiquiatría y Ciencias Afines 3, 189–202 (1979)
5. Peña-Casanova, J.: Programa integrado de exploración neuropsicológica "test Barcelona". Normalidad, semiología y patología neuropsicológica. Masso, Barcelona (1991)
6. Rey, A.: Rey. Test de copia y de reproducción de memoria de figuras geométricas complejas. TEA, Madrid (2003)
7. Freeman, R.Q., Giovannetti, T., Lamar, M., Cloud, B.S., Stern, R.A., Kaplan, E., Libon, D.J.: Visuoconstructional problems in dementia: contribution of executive systems functions. Neuropsychology 14(3), 414–426 (2000)
8. Urbina, S.: Claves para la evaluación con tests psicológicos (Kaufman, A.S., Kaufman, N.L (trad.)). TEA Ediciones (publicado originalmente en 2004), Madrid (2007)
9. Guerrero Triviño, J.M., Martínez-Tomás, R., Peraita Adrados, H.: Bayesian Network-Based Model for the Diagnosis of Deterioration of Semantic Content Compatible with Alzheimer's Disease. In: Ferrández, J.M., Álvarez Sánchez, J.R., de la Paz, F., Toledo, F.J. (eds.) IWINAC 2011, Part I. LNCS, vol. 6686, pp. 419–430. Springer, Heidelberg (2011)
10. Guerrero, J.M., Martínez-Tomás, R., Rincón, M., Peraita, H.: Bayesian network model to support diagnosis of cognitive impairment compatible with an early diagnosis of Alzheimer's disease. Methods of Information in Medicine (in revision, 2015)
11. Peraita, H., Grasso, L.: Corpus lingüístico de definiciones de categorías semánticas de personas mayores sanas y con la enfermedad de Alzheimer. Una investigación transcultural hispano-argentina. Fundación BBVA (2010), http://www.fbbva.es/TLFU/dat/ DT%203_2010_corus%20linguistico_peraita_web.pdf
12. Matic, A., Mehta, P., Rehg, J.M., Osmani, V., Mayora, O.: Monitoring Dressing Activity Failures through RFID and Video. Methods Inf. Med. 51(1), 45–54 (2012)
13. Kearns, W.D., Nams, V.O., Fozard, J.L.: Tortuosity in Movement Paths Is Related to Cognitive Impairment. Methods Inf. Med. 6(49), 592–598 (2010)
14. Yesavage, J.A., Brink, T.L., Rose, T.L., Lum, O.: Development and validation of a geriatric depression scale: A preliminary report. Journal of Psychiatry Research 17(1), 37–49 (1983)
15. Blessed, G., Tomlinson, B.E., Roth, M.: The association between quantitative measures of dementia and of senile changes in the cerebral grey matter of elderly subjects. British Journal of Psychiatry 114, 797–811 (1968)

Identification of Loitering Human Behaviour in Video Surveillance Environments

Héctor F. Gómez A.[1]([✉]), Rafael Martínez Tomás[2], Susana Arias Tapia[1],
Antonio Fernández Caballero[3], Sylvie Ratté[4], Alexandra González Eras[1],
and Patricia Ludeña González[1]

[1] Instituto de Ciencias de la Computación. Escuela de Ciencias de la Computación,
Universidad Técnica Particular de Loja, Marcelino Champagnat S/N, 1101608, Loja,
Ecuador
hfgomez@utpl.edu.ec
[2] Dpto. Inteligencia Artificial. Escuela Técnica Superior de Ingeniería Informática,
Universidad Nacional de Educación a Distancia, Juan del Rosal 16, 28040 Madrid,
Spain
[3] Departamento de Sistemas Informáticos e Instituto de Investigación en Informática,
Universidad de Castilla-La Mancha, 02071, Albacete, Spain
[4] Department of Software and IT Engineering. Université du Québec - École de
Technologie Supérieure. 1100 Notre-Dame West, Montreal, Québec, H3C 1K3,
Canada

Abstract. Loitering is a common behaviour of the elderly people.
We goal is develop an artificial intelligence system that automatically
detects loitering behaviour in video surveillance environments. The first
step to identify this behaviour was used a Generalized Sequential Patterns that detects sequential micro-patterns in the input loitering video
sequences. The test phase determines the appropriate percentage of inclusion of this set of micro-patterns in a new input sequence, namely
those that are considered to form part of the profile, and then be identified as loitering. The system is dynamic; it obtains micro-patterns on a
repetitive basis. During the execution time, the system takes into account
the human operator and updates the performance values of loitering in
shopping mall. The profile obtained is consistent with what has been documented by experts in this field and is sufficient to focus the attention
of the human operator on the surveillance monitor.

1 Introduction

Modelling and automatic identifying human behaviour is an area that has been
developed significantly over the last few years in artificial intelligence and artificial vision. The aim of this type of investigations corresponds to the social need
for more security in particular, but also in general, in the form of automatic
observation of behaviour, such as in the health sector. We worked under the assumption that it was possible to develop a system that emulated the ability of an
expert in recognizing loitering behaviour by considering a set of repeated actions
(micro-patterns) that are part of the loitering profile. This system updates the

© Springer International Publishing Switzerland 2015
J.M. Ferrández Vicente et al. (Eds.): IWINAC 2015, Part I, LNCS 9107, pp. 516–525, 2015.
DOI: 10.1007/978-3-319-18914-7_54

profile with the contribution of a human operator, with the aim of covering the widest possible positive cases. The system is interactive and dynamic, because it enables interaction between the system and the human operator and because it facilitates the updating of the loitering profile.

We worked with the Generalized Sequential Patterns (GSP) algorithm to obtain the micro-patterns (patterns comprised of a small number of sufficiently repeated events). Srikant and Agrawal [1] use GSP to obtain sequential patterns based on data about consumer shopping habits at supermarkets. In our study, we had to change the input sequences of shopping behaviour to input sequences of labelled loitering activities from video surveillance to obtain micro-patterns that characterized the target behaviour. These micro-patterns constituted the loitering profile for identifying loitering behaviour in video surveillance domains. These micro-patterns are initially identified by positive sequences and loitering characteristics. Afterwards, a sensitivity analysis is performed on new cases of loitering sequences. Another step is the testing phase, which enables us to determine the appropriate percentage of inclusion for this set of micro-patterns in a new input stream. Since the system is dynamic, it obtains micro-patterns on a repetitive basis; the sensitivity analysis is continually updated too. During the execution time, the system takes into account the human operator annotations and updates the performance values.

The next section presents a review of works related to the learning and sequential micro-pattern recognition of human behaviour. After, we describe the proposed system based on micro-pattern matching with GSP and the selection of those micro-patterns that best characterized (profile) the target situation. Section 4 provides details of the experiments and the last section of the article consists of our conclusions and also proposes new areas of related research.

2 Related Work

Park, et al. [2] use a probabilistic scoring function to calculate the temporal similarity of event sequences with behavioural patterns that are defined as a priori, that is, they identify Daily Living Activities (DLA) of people at home, such as reading, listening to music, etc. Their approach consists of identifying previously known behaviour using more explicit knowledge.

In our research, we examined the repetition of the occurrence of an event, or several events, that led to the expected behaviour. Robertson, et al. [3] for example, use rules of behaviour with a probabilistic algorithm, namely the Hidden Markov Model (HMM), which identifies the behaviour of pedestrians crossing a street in various situations such as when there is a lot of traffic, or when the traffic lights change, etc. Other studies (e.g., see [4–6]) identify the behaviour of people in video images based on the recognition of human movements. For example, a sequential analysis of events was used with HMM to detect domestic accidents, and to identify health problems such as feinting and cardiac arrests, etc.

Chikhaoui et al. [7] use GSP to search behavioural patterns of persons during their daily routines with the objective of distinguishing individual behaviour.

The results show that there are clear differences between individual and typical behaviour of people in activity day live. Moreover, it is relatively easy to model normal behaviour. For example, the daily activity of a person at home in the morning is often the same, and can thus be modelled a priori [8]. We believe this proposed approach is innovative and has the potential of opening investigation into adjacent domains of research, such as in healthcare and psychology. However, this proposal is not an attempt to replace the human operator who monitors peoples behaviour; instead, it should be viewed as a practical alternative for preventing delinquent behaviour using state of the art surveillance technology (GSP and sensitivity analysis).

3   Loitering Behavior Identification Based on Sequential Micro-Patterns

It shows the methodological structure of both stages or scenarios of our proposal: the a priori training/learning process and the identification of patterns (Fig.1), and the stage when the system is in operation (Fig. 2).

In the training stage (Fig. 1), the first step (1) is to find micro-patterns using GSP. As mentioned in the introduction, micro-patterns are small patterns comprised of several sufficiently repeated events of loitering behaviour. In order to obtain these, we must have a series of positive case sequences. Each sequence represents the behaviour of a person and is obtained by labelling the individual activities of the monitored person for each second of video surveillance, e.g. walk, walk, stop, stop, walk, walk, stop, stop. It may be assumed that vision algorithms can recognize these events (see [6, 8–10]), or that they can be labelled manually or semi-manually.

To obtain the micro-patterns, GSP searches all frequent sequences in the database. Frequent sequences are those whose frequency exceeds a threshold value known as minimum support. In first stage, GSP searches for these frequent

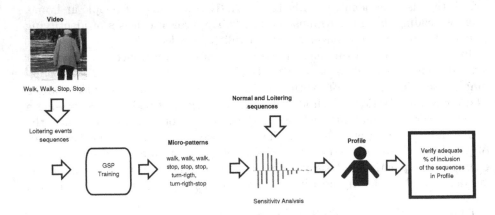

Fig. 1. Identify loitering human behavior: Training stage

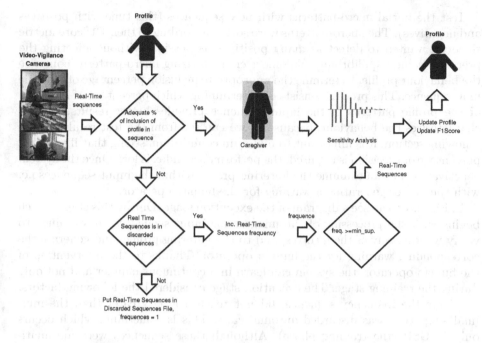

Fig. 2. Identify loitering human behavior: Test stage

sequences in the database (using hashing tree algorithm) from the sequences of size 1 (a sequence that contains 1 item), with 1-sequence frequent (candidate sequences); and from these, GSP builds sequences with size 2 (a sequence composed of 2 items) and select the frequent 2-sequence. Frequent 2-sequences (candidate sequences) are joined with frequent 1-sequences in order to form sequences of size 3. With these sequences, other sequences of higher orders are generated. GSP search ends when there are sequences of a desired length that appears more frequently in the database. Finally, in the second stage, GSP removes non-candidate sequences and as such obtains frequent sequences, known in this study how micro-patterns . A sensitivity analysis is applied to all the micro-patterns. Then, the most reliable ones for the target situation are selected, which form a more representative behaviour profile (see Fig. 1). The sensitivity analysis of the micro-patterns is carried out as follows:

Recognize (match) a p micro-pattern in a new sequence (s) implies extracting an s sub-sequence from s, which is of the same length as p, and calculate the Levenshtein distance (L) [11, 12] between p and s. If L is less than the threshold α, there is therefore an occurrence with a positive result. This is repeated for all the s that form part of s. We defined the matching threshold (α) as in [13] (In this case, it is related to football strategies). This is done to determine when a micro-pattern appears in a sequence. The use of this threshold is justified since it is difficult for an entire micro-pattern to appear exactly in the new sequence, given the variability of behaviour (see [14, 15]).

Test the initial micro-patterns with new sequences (this time with positives and negatives). The micro-patterns were sorted according to their F1Score metric since we wanted to detect as many positives as possible without affecting the precision value (equilibrium). The most characterizing micro-patterns comprise the behaviour profile. Determine the appropriate inclusion percentage of a profile in a sequence. This process consists of determining which percentage of inclusion of the profile patterns in the input sequence provides better results, or best characterizes the behaviour, because as we can see from the test results in the following section, it is important to maintain equilibrium, seeing that if too high percentage of precision is required, the performance index is low. Once the system is activated, we can examine the loitering profile within the input sequences i.e. with the aim of generating a warning for the human operator.

In Fig. 2, we can see a diagram of the execution stage. During this stage, which begins with the processing of the images, the input sequences are obtained to verify whether any of them correspond to the loitering profile, and generate the corresponding warning for the human operator. Thanks to the intervention of the human operator, the system can learn in a continuous manner and not only during the training stage. The execution stage considered the following factors:

Where the frequency sequence did not reach a level greater than the minimal support it was discarded automatically. This is something, which occurs only in GSP (the training phase). Although these sequences were automatically discarded by the system, there is also a way of retrieving them. On the other hand, if the new sequences that are inputted into the system contain the discarded sequences, their frequency will increase. Moreover, if this sequence frequency reached a higher level than the minimal support, this sequence was determined to be a micro-pattern. Consequently, we needed to repeat the entire training phase and the sensitivity analysis again. Therefore, it could be concluded that whenever there was a new micro-pattern, there would also be other micro-patterns.

During the updating of the sensitivity analysis, we observed that it was necessary for the human operator to determine whether true positive or false negative cases were needed, or if, by default, neither of these options were required. Afterwards, the human operator updated the sensitivity analysis of micro-patterns, namely those that constituted the profile.

4   Experimentation

The experimentation dataset, training and testing, is comprised of the following: 35 loitering sequences from CAVIAR-Project [16] test-bed contain footage from a camera situated in a shopping centre alley (outside a shopping mall).

Loitering video observation and manual labelling of events: 100 video recordings of loitering behaviour, recorded by video surveillance systems, were analyzed. As with other previous examples, a security assistant observed each of the video recordings for 40 seconds (timestamp). Then, the observations were manually registered in a software program specially designed for this investigation.

The 135 positive sequences were obtained from a single file of input sequences for the GSP. Group (b) was created in the same way. We generated 100 sequences by labelling negative events. Finally, we obtained a dataset of 235 mixed sequences. By using the 135 positives sequences to train the GSP, we were able to obtain micro-patterns. The results showed that with the value MS = 0.4 (where 40% of the sequences include the micro-pattern) and $\alpha = 2$ (where the distance from the micro-pattern is less than or equal to 2), we were able to obtain the required micro-pattern data. The values that were obtained during the testing phase provided the most accurate results.

For the sensitivity analysis of loitering behaviour (see Table 1), we used 135 positive sequences with their respective number (100) of negative sequences. This procedure helped us to obtain micro-patterns in order to make the desired profile. Finally, we used the same sequences (235 sequences) to determine the optimal percentage of micro-patterns. By also having the micro-patterns appear in a sequence we were able to ensure that the sequence contained the profile of loitering behaviour, thus obtaining new F1Score:

Table 1. Sensitivity analysis of micro-patterns obtained with *GSP*

Micro-patterns	Precision	Recall	F1 Score
walks, walks, walks, stops, stops, stops, walks, walks, walks	0,85	0,94	0,89
stops, stops, stops, stops, stops, walks, turns-right, walks, walks, turns-right, walks	0,73	0,96	0,82
stops, stops, stops, walks, walks, walks, walks, stops, stops, turns-left, walks	0,71	0,96	0,81
stops, stops, stops, stops, walks, walks, walks, turns-right, turns-right, browses, browses	0,81	0,96	0,87
walks, walks, walks, stops, stops, stops, walks, walks, walks	0,85	0,94	0,89

As can be seen from the Table 2, the results with the highest values are observed when the sequences contain 75 % of the profile micro-patterns (see row highlighted in bold). These results are considered valid (see [15, 16]) for this study as they provide a high recall value and because the level of precision does not severely decrease, but instead gradually increases based on the fact that the input sequences contain the optimum number of micro-patterns. The optimum percentage (75 %) of inclusion of the micro-patterns in the input sequences is thus determined by the highest value of the F1Score (0.91).

It is worth highlighting the trend, that where there is a precision level of 0.64, this indicates that a greater number of false alerts are generated compared with the values of the last three rows of the table (0.93). This theory can be sustained when we examine what happens with a precision level of 1.00. Where there is an optimum precision level of 1.00, there will be minimal false alerts. The precision value and the recall value rise and fall alternately, i.e. where one value increases the other value decreases. The precision and recall values are mutually dependent and the F1Score shows the relationship between these values.

Table 2. Sensitivity analysis to determine the optimal percentage of inclusion in the profile

Percentage of inclusion in the profile	Precision	Recall	F1 Score
35	**0,64**	0,98	0,77
40	0,64	0,976	0,77
45	0,67	0,97	0,79
50	0,71	0,95	0,81
55	0,74	0,943	0,82
60	0,74	0,94	0,82
65	0,74	0,94	0,82
70	0,76	0,94	0,84
75	**0,89**	**0,94**	**0,91**
80	0,9	0,919	0,90
85	0,9	0,87	0,88
90	**0,93**	0,865	0,89
95	**0,93**	0,84	0,88
100	**0,93**	0,77	0,84

To explain this point further from a theoretical perspective, the system identifies the maximum number of loitering sequences (recall) in relation to the minimum number of false alerts (precision). Therefore, the equilibrium between precision and recall can be found where the F1Score is 0.91.

The experimentation found that if there is no equilibrium between these values (i.e. when the number of input sequences is too small to generate representative micro-patterns of loitering behaviour), we must increase the number of input sequences by labelling more video recordings. For this reason, it is essential to achieve equilibrium between precision and recall values.

Table 3. Performance results of micro-patterns (new test)

Micro-patterns	Precision	Recall	F1 Score
walks, walks, walks, stops, stops, stops, walks, walks, walks	0,64	0,94	0,76
stops, stops, stops, stops, stops, walks, turns-right, walks, walks, turns-right, walks	0,67	0,94	0,78
stops, stops, stops, walks, walks, walks, walks, stops, stops, turns-left, walks	0,73	0,91	0,81
stops, stops, stops, stops, walks, walks, walks, turns-right, turns-right, browses, browses	0,77	0,9	0,82
walks, walks, walks, stops, stops, stops, walks, walks, walks	0,64	0,94	0,76
stops, stops, stops, stops, stops, walks, turns-right, walks, walks, turns-right, walks	0,67	0,94	0,78
stops, stops, stops, walks, walks, walks, walks, stops, stops, turns-left, walks	0,73	0,91	0,81

To test whether our online system worked, we performed a final test to check the learning capacity of the system with 100 new positive and 100 new negative sequences. In this case (see Table 3), the sensitivity analysis did not generate new micro-patterns, as there were not any sequences that reached the minimum support level. Therefore, there did not exist sufficient changes in the sensitivity analysis to produce new micro-patterns. Although the sensitivity analysis could be updated, we still used the same micro-patterns. As we can see below, the F1 score remains high, and as with the example of loitering behaviour, the results confirmed the micro-pattern percentage inclusion of 75% (see Table 4).

Table 4. Sensitivity analysis to determine the optimal percentage of inclusion in the profile (new test)

Percentage of inclusion in the profile	Precision	Recall	F1 Score
35	0,56	0,66	0,60
40	0,62	0,64	0,62
45	0,66	0,67	0,66
50	0,67	0,67	0,67
55	0,67	0,67	0,67
60	0,67	0,72	0,69
65	0,67	0,72	0,09
70	0,7	0,74	0,71
75	**0,7**	**0,77**	**0,73**
80	0,7	0,7	0,7
85	0,7	0,68	0,68
90	0,66	0,61	0,63
95	0,74	0,53	0,61
100	0,87	0,53	0,65

The results from Table 1 and Table 4 (i.e. after updating the sensitivity analysis of the selected micro-patterns) show that the obtained profile is used to distinguish between normal and potential theft behaviour. Furthermore, with the sequences that were used in the experimentation stage, the proposed system is capable of distinguishing between normal and loitering behaviour, a problem that was not, however, resolved in [17]. Indeed, this finding constitutes another important contribution to our study.

5   Conclusions

In this paper, we aimed to test the hypothesis that there is a common denominator in loitering behaviour which human experts are capable of identifying, namely in determining target scenarios, but which, at the same time, may result in difficulties when actually defining them. Our subsequent approach to this problem consisted of generating automatic alerts for human operators based on the creation of a pre-selected loitering profile and the implementation of a sensitivity analysis.

There were occasional occurrences of false alerts during the testing phase. These false alerts may be reduced by obtaining the optimum percentage of inclusion of micro-patterns in the input sequences and by repeating the sensitivity analysis. However, the proposed system is designed in such a way that, theoretically once it is fully installed and operational, it would work on its own by automatically generating message alerts for the human operator, who, in turn, would take the necessary security action.

This entire process is based on the identification and labelling of what we call elementary or basic activities, namely events that are recognizable by artificial vision algorithms or intelligent sensory monitoring techniques (segmentation, targeting, tracking and classification). By using these labelled sequences, i.e. where loitering behaviour usually occurs, we can obtain sequential micro-patterns with the GSP algorithm. After doing a sensitivity analysis with sequences showing normal (negative) and loitering (positive) behaviour, the most characteristic micro-patterns were selected, thereby confirming the loitering behaviour profile.

During runtime, i.e. when an input sequence contains the optimal percentage of the profile, a message alert is raised for the human operator. The human operator would then confirm the true positives and mark the false negatives. This human interaction with the system therefore helps to update the sensitivity analysis of the profile. Moreover, in real time the results that are originally discarded can likewise be recovered if their frequency of occurrences reaches the minimum required level.

To test our hypothesis, we carried out an experiment on the identification of loitering behaviour in a shopping mall. This scenario was chosen because they represented situations that fulfilled the conditions of the main areas of gerontology i.e Alzheimer. In this case, we manually labelled the video recordings, thus facilitating the sequencing of event labels.

Our results strongly suggest, therefore, that the implementation of a micro-pattern profile in video surveillance situations helps in the prediction and prevention of loitering activity, thereby serving as a fundamental tool for the human operator.

Acknowledgements. The authors are grateful to the Spanish Ministerio de Economía y Competitividad / FEDER for their financial contribution through the projects TIN2010-20845-C03-01 and TIN2010-20845-C03-02.

References

1. Srikant, R., Agrawal, R.: Mining sequential patterns: Generalizations and performance improvements. In: Apers, P.M.G., Bouzeghoub, M., Gardarin, G. (eds.) EDBT 1996. LNCS, vol. 1057, pp. 3–17. Springer, Heidelberg (1996)
2. Park, K., Lin, Y., Metsis, V., Le, Z., Makedon, F.: Abnormal human behavioural pattern detection in assisted living environmets. In: 3rd International Conference on Pervasive Technologies Related to Assistive Environments (2010)

3. Robertson, N., Reid, I., Brady, M.: Automatic human behaviour recognition and explanation for CCTV video surveillance. Security Journal 21(3), 173–188 (2008)
4. Quian, H., Ou, Y., Wu, X., Meng, X., Xu, Y.: Support vector machine behaviour-based driver identification system. Journal of Robotics, 173–188 (2008)
5. Leo, M., Spagnolo, P., D'Orazio, R., Mazzeo, P., Distante, A.: Real-Time smart surveillance using motion analysis. Expert Systems 25(5), 314–337 (2010)
6. Cuchiara, R., Prati, A., Vezzani, R.: A multi-camera vision system for fall detection and alarm generation. Expert Systems 24(5), 334–345 (2007)
7. Chikhaoui, B., Wang, S., Pigot, H.: A new Algorithm Based on Sequential Pattern for person identification in ubiquitous environments. In: Proceedings of the Fourth International Workshop on Knowledge Discovery from Sensor Data, pp. 20–28 (2010)
8. Fern, X., Komireddy, C., Burnnet, M.: Mining Interpretable Human Strategies: A Case Study. In: 7th IEEE International Conference on Data Mining, pp. 475–480 (2007)
9. Chun, L.: Definition, detection, and evaluation of meeting events in airport surveillance videos. In: TRECVID Workshop (2008)
10. Charkraborty, B., Bagdanov, A., González, J., Roca, X.: Human action recognition using an ensemble of body-part detectors. Expert Systems (2011), doi:10.1111/j.1468-0394.2011.00610.x
11. Levenshtein, V.: Binary codes capable of correcting deletions, insertions, and reversals. Soviet Physics Doklady 10(8) (1966)
12. Hernández-Vela, A., Bautista, M., Perez-Sala, X., Ponce-López, V., Escalera, S., Baró, X., Angulo, C.: Probability-based dynamic time warping and bag-of-visual-and-depth-words for human gesture recognition in rgb-d. Pattern Recognition Letters 50, 112–121 (2014)
13. Iglesias, A.: Modelado Automático del Comportamiento de Agentes Inteligentes. PhD dissertation Universidad Carlos III Madrid, Spain (2010)
14. Pillai, J., Vyas, O.P.: Overview of itemset utility mining and its applications. International Journal of Computer Applications 5(11), 9–13 (2010)
15. Manganaris, S., Christensen, M., Zerkle, D., Hermiz, K.: A data mining analysis of RTID alarms. Computer Networks 34(4), 571–577 (2000)
16. CAVIAR-Project: CAVIAR Project (May 20, 2009), http://home-pages.inf.ed.ac.uk/rbf/CAVIAR
17. Williem, A., Vamsi, M., Boles, K., Wageeh, W.: Detecting Uncommon Trajectories. In: Digital Image Computing: Techniques and Applications, pp. 398–404 (2008)

Stress Detection Using Wearable Physiological Sensors

Virginia Sandulescu[1], Sally Andrews[2], David Ellis[2], Nicola Bellotto[3],
and Oscar Martínez Mozos[3(✉)]

[1] Politehnica University of Bucharest, Bucharest, Romania
s_virg@yahoo.com
[2] School of Psychology, University of Lincoln, UK
{s.andrews,dellis}@lincoln.ac.uk
[3] School of Computer Science, University of Lincoln, UK
{nbellotto,omozos}@lincoln.ac.uk

Abstract. As the population increases in the world, the ratio of health carers is rapidly decreasing. Therefore, there is an urgent need to create new technologies to monitor the physical and mental health of people during their daily life. In particular, negative mental states like depression and anxiety are big problems in modern societies, usually due to stressful situations during everyday activities including work. This paper presents a machine learning approach for stress detection on people using wearable physiological sensors with the final aim of improving their quality of life. The presented technique can monitor the state of the subject continuously and classify it into "stressful" or "non-stressful" situations. Our classification results show that this method is a good starting point towards real-time stress detection.

Keywords: Stress detection · Wearable physiological sensors · Assistive technologies · Signal classification · Quality of life technologies

1 Introduction

As the population increases in the world, the ratio of health carers is rapidly decreasing. Actually, the Organisation for Economic Co-operation and Development (OECD) warns about future shortages of available health workers and doctors [3]. Therefore, there is an urgent need to create new technologies to monitor the health of people, both physical and mental, during their daily life with the aim of supporting health workers, caregivers, and doctors in their tasks. These technologies, also known as Quality of Life Technologies (QoLTs), have emerged as the concept of applying findings from different technological areas to assist people and improve their quality of life.

An emerging research topic inside QoLTs is their application to psychology and self-therapy to improve the mood of people and thus, their quality of life. Although there exist several technologies to support the health of people at the physiological level, the technologies that are able to provide similar support at the mental level are almost inexistent.

© Springer International Publishing Switzerland 2015
J.M. Ferrández Vicente et al. (Eds.): IWINAC 2015, Part I, LNCS 9107, pp. 526–532, 2015.
DOI: 10.1007/978-3-319-18914-7_55

Treating negative mental states in people is becoming a priority in our new societies. In particular, stress is a big problem in modern populations due to the increment of stressful situations during everyday activities including work. Stress is a natural reaction of the human body to an outside perturbing factor. The physiological responses to stress are correlated with variations in heart rate, blood volume pulse, skin temperature, pupil dilation, electro-dermal activity [18,17,13]. Stress may have beneficial effects on fighting the stress factor, like increasing reflexes, but it was determined that long term stress is correlated with various health problems like depression and premature ageing [16], [9].

Stress is creating new problems that have a great impact in our societies and economies. For example, according to the Mental Health Foundation in UK [2], around 12 million adults in the UK visit their general practitioner doctor (GP) each year with mental health problems, most of which are related to stress. As a consequence, 13.3 million working days are lost per year due to stress problems. Moreover, according to the World Health Organization [4], stress has a cost of around 8.4 million to UK enterprises. Finally, current appointments for national health mental services in UK, such as Cognitive Behavioural Therapy (CBT) [5] are taking 3-6 months to be processed, with the subsequent danger for the patient because cumulative stress may have broad negative consequences on societal well-being and costs [15]. Thus, the research of this paper emerges as a necessity to create new wearable technologies to monitor stress on people during their daily life.

This paper presents a machine learning approach for stress detection on people using wearable physiological sensors with the final aim of improving their quality of life. Moreover, the presented technique monitors the state of the subject continuously and classifies it into "stressful" or "non-stressful" situations. Finally, our classification results shows that our approach is a good starting point towards real-time stress detection and treatment.

2   Wearable Physiological Sensors

In this paper we aim to detect stress in people using wearable sensors that measure physiological responses. In particular, we have used the BioNomadix module from Biopac, model BN-PPGED [1] as shown in Figure 1.

The BN-PPGED is worn as a wristband on the non-dominant hand of a subject with two electrodes situated on two fingers that measure the electro-dermal activity (EDA) and the pulse plethysmograph (PPG) signals. EDA, sometimes measured as electrodermal response, skin conductance activity, or galvanic skin response, is an indication of skin sweating activity. PPG, also known as Blood Volume Pulse (BVP), is obtained using a pulse oximeter which illuminates the skin and measures the differences in light absorption. The amount of light that returns to the PPG sensor is proportional to the volume of blood in the tissue [14].

In our experiments the EDA and PPG physiological signals were acquired at a 1000 Hz sampling frequency. After the acquisition the signals were down-sampled to 10 Hz. Afterwards, a filtering and artefact removal approach was applied by using

Fig. 1. BioNomadix model BN-PPGED and MP150 station by biopac [1]

the routines included in the AcqKnowledge software [1]. In adittion, AcqKnowledge was used to extract the PPG autocorrelation signal and the Heart Rate Variability (HRV). HRV represents the beat-to-beat variability over a given period of time and is computed by calculating the standard deviation of the average of normal-to-normal heartbeats [14].

The BN-PPGED connects though wireless to a Biopac MP150 communication station as shown in Figure 1. The MP150 station directly connects to a computer that runs AcqKnowledge 4 software for real-time data acquisition [1]. In this way, the subject wearing the sensors can move freely while the experiments and the different signals are send through wireless to a computer.

3   Classification of Physiological Signals

In our approach we classify the state of each person at 0.1 seconds intervals. Each state is composed of four measurements: PPG value (ppg), PPG autocorrelation value ($ppgau$), HRV value (hrv), and EDA value (eda). Thus, we represent each sample at time t as the feature vector $x_t = \{ppg_t, ppgau_t, hrv_t, eda_t\}$, where t is sampled at 0.1 seconds intervals.

Each sample x_t was labelled according to the state of the person at that time, i.e. *stressed*, or *not stressed*. Thus, our dataset was composed of the measurements obtained at each time interval together with their corresponding label as $D = \{(x_t, l_t)\}$, with $l_t \in L = \{stressed, not\_stressed\}$. The state of the person l_t was defined by the activity that person was performing at time t during the experiment (see Section 4).

The classification of the sampled meassurements was done using a support vector machine (SVM) [8,6]. Support vector machines take as input a set of n feature vectors x_i together with their labels $y_i \in Y = \{1, -1\}$. The idea behind SVMs is to find the hyperplane that maximizes the distance between the examples of the two classes $\{1, -1\}$. This is done by finding a solution to the optimization problem

$$\min_{w,b,\xi} \quad C \sum_{i=1}^{n} \xi_i + \frac{1}{2} \|w\|^2, \tag{1}$$

subject to the condition

$$y_i \left(w^T \phi(x_i) + b \right) \geq 1 - \xi_i \,, \tag{2}$$

where w is the normal to the hyperplane, and $\xi_i \geq 0$ are slack variables that measure the error in the misclassification of x_i. In addition, we use a radial basis function (RBF) kernel

$$K(x_i, x_j) = \exp\left(-\gamma \|x_i - x_j\|^2\right), \gamma > 0 \tag{3}$$

In our case, we map our original labels $L = \{stressed, not\_stressed\}$ into $Y = \{1, -1\}$ so that our examples could be used in a SVM.

4   Experimental Setup

To check the validity of our stress detector we prepared an experimental setup where different subjects experimented different stressful situations. In this section we will describe the complete experimental setup and protocol.

In the study presented in [9], more than 200 stress experiments are reviewed in terms of activities involved in the experiments and the cortisol responses measured on the subjects performing these activities. According to the same source, the most effective tasks for inducing stress are public speaking and cognitive tasks, because during these tasks the highest increases in cortisol levels are measured. This is why our designed experiment contained both a public speaking task and a cognitive task.

Our final designed experiment is based on the Trier Social Stress Test (TSST) [12]. This is a very popular experimental setup and it has been used in more than 4000 sessions during the last decades [9]. The TSST consists of a neutral task followed by a public speaking task, a cognitive task and another neutral task in the end. Each neutral task consists of 2 minutes of predefined neutral questions like: "How do you find the weather today" or "How did you get here?". The public speaking is a 5 minute interview for a desired job. After this, the participant is asked to count back in steps of 13, starting from 1022. This is the cognitive task. All the previous tasks are performed in front of a live audience and a video camera. The camera is only used to induce the stress more reliably [9], so the recordings are not stored. The neutral tasks are thought as *non-stressful* situations, while the speaking and cognitive tasks are considered *stressful* situations.

In more detail, our protocol for the TSST was as follows. When the participants enter the experiment room, they are given verbal and written information about the procedures involved in the experiment. The participants are asked to fill in a consent form and to confirm that they do not suffer from any cardiovascular or anxiety disorder that might be affected by experiencing stress or that might affect the results of the experiment.

After being briefed, the participants are asked to fill in a State Trait Anxiety Inventory (STAI) [10] to estimate the current level of stress. They are then fitted with the sensors. There is a 2 minutes period of time when the participants are asked predefined neutral questions, in order to determine the baseline, which we will used as neutral state. Afterwards, the participants are asked to sit at a desk and prepare a presentation for an job interview job during 3 minutes. They are given a pen and a paper for this. When the 3 minutes time expires, they are asked to hand out the sheet of paper and stand up in a predefined square on the ground and begin their presentation. During the 5 minutes of the presentation, the participants are encouraged to speak continuously. If the participants stop during the presentation, at the first pause, they are told about the remaining time and asked to continue. At the next pause, they are asked a set of predefined typical interview questions like: "What are your strengths/weaknesses?", "Where do you see yourself in 5 years?" and so on. After the 5 minutes presentations. The participants are explained a cognitive task and the 5 minutes timer is started. Whenever the participants say the wrong number, they are asked to start again from 1022. At the end of the cognitive task, the participants are given a short time to relax, while given another debrief. Then another two minutes of neutral questions are recorded. Finally, the participants are then asked to fill in the STAI questionnaire to estimate the current level of stress and the general level of stress.

5   Experimental Results

The previous TSST session was conducted on 5 participants $\{P1, P2, P3, P4, P5\}$, that were volunteering students from the School of Psychology, at the University of Lincoln, aged 18 to 39, both males and females.

The goal of these experiments is to check the performance of our approach to create a personalized stress detector for each participant. For this reason we created independent datasets of measurements for each participant $\mathcal{D} = \{D_1, D_2, D_3, D_4, D_5\}$. The size of datasets D_k, i.e. number of feature vectors (c.f. Section 3), were $|D_1| = 11620, |D_2| = 13450, |D_3| = 13740, |D_4| = 13740$, and $|D_5| = 13000$.

We trained a SVM_k for each participant using the corresponding dataset D_k and evaluated the classifier according to it. For each SVM_k we used 75% of the corresponding dataset D_k for training and the remaining 25% for testing. To create this sets we used a stratified selection to ensure the same class distribution in the subset as in the original set. Then the training and set data were scaled to have values in the $[-1, 1]$ range.

In our experiments we used the LIBSVM library [7]. Moreover, following the method in [11], the parameters C and γ for each SVM_k were selected by grid-search using cross-validation.

The results of the different detectors are shown in Table 1. We can see that we obtain very good detection results in all the participants, with accuracies over 82% in two cases, and precissions over 80% in the majority of the patients.

The individual confusion matrices for each participant are shown in Table 2. The results suggest a bias to classify non-stressful states as *stressed*. We think this is due to the fact that people remained stress during short periods of times during the transitions to neutral tasks, since they need time to relax. However, this transition time was not taken into account in these results.

Table 1. Classification accuracy and precision

Participant No.	Accuracy [%]	Precision [%]
P1	78.90	80.19
P2	73.26	73.61
P3	83.08	83.87
P4	82.82	83.20
P5	76.83	76.67

Table 2. Individual confusion matrices

Participant No.	P1		P2		P3		P4		P5	
	stressed	not stressed	stressed	not stressed	stressed	not stressed	stressed	not stressed	not stressed	stressed
stressed	94.04	5.95	88.21	11.79	90.50	9.50	91.55	8.44	91.62	8.38
not stressed	60.62	39.38	50.73	49.27	29.49	70.51	32.53	67.47	49.36	50.64

6   Conclusions

In this paper we have presented an approach for stress detection using wearable physiological sensors. Our approach is able to analyse the state of the subject at any instant an decide about his/her stress situation. Detection results in our experiment demonstrate that our approach is a good starting point towards real-time mental mood detection and treatment on people to improve their quality of life.

References

1. Biopac, http://www.biopac.com
2. Mental Health Foundation in UK, http://www.mentalhealth.org.uk
3. The Organisation for Economic Co-operation and Development (OECD), http://www.oecd.org
4. World Health Organization (WHO), http://www.who.org
5. Beck, A.T.: Cognitive therapy and the emotional disorders. International Universities Press, Inc., Madison (1975)
6. Bishop, C.M.: Pattern Recognition and Machine Learning. Springer (2006)
7. Chang, C.-C., Lin, C.-J.: LIBSVM: A library for support vector machines. ACM Transactions on Intelligent Systems and Technology 2, 27:1–27:27 (2011), Software available at http://www.csie.ntu.edu.tw/~cjlin/libsvm
8. Cortes, C., Vapnik, V.: Support-vector network. Machine Learning 20, 273–297 (1995)
9. Dickerson, S.S., Kemeny, M.E.: Acute stressors and cortisol responses: A theoretical integration and synthesis of laboratory research. Psychological Bulletin 130(3), 355–391 (2004)

10. Elwood, L.S., Wolitzky-Taylor, K., Olatunji, B.O.: Measurement of anxious traits: a contemporary review and synthesis. Anxiety Stress Coping 25(6), 647–666 (2012)
11. Hsu, C.-W., Chang, C.-C., Lin, C.-J.: A practical guide to support vector classification (2010), http://www.csie.ntu.edu.tw/~cjlin/papers/guide/guide.pdf
12. Kirschbaum, C., Pirke, K.M., Hellhammer, D.H.: The 'Trier Social Stress Test' – A tool for investigating psychobiological stress responses in a laboratory setting. Neuropsychobiology, 76–81 (1993)
13. Wikgren, M., Maripuu, M., Karlsson, T., Nordfjäll, K., Bergdahl, J., Hultdin, J., Del-Favero, J., Roos, G., Nilsson, L.G., Adolfsson, R., Norrback, K.F.: Short telomeres in depression and the general population are associated with a hypocortisolemic state. Biological Psychiatry 71(4), 294–300 (2012)
14. Peper, E., Harvey, R., Lin, I.-M., Tylova, H., Moss, D.: Is there more to blood volume pulse than heart rate variability, respiratory sinus arrhythmia, and cardiorespiratory synchrony? Biofeedback 35(2), 54–61 (2007)
15. Perkins, A.: Saving money by reducing stress. Harvard Business Review 72(12) (1994)
16. Rai, D., Kosidou, K., Lundberg, M., Araya, R., Lewis, G., Magnusson, C.: Psychological distress and risk of long-term disability: population-based longitudinal study. Journal of Epidemiology and Community Health 66(7), 586–592 (2011)
17. Sun, F.-T., Kuo, C., Cheng, H.-T., Buthpitiya, S., Collins, P., Griss, M.: Activity-aware mental stress detection using physiological sensors. In: Griss, M., Yang, G. (eds.) MobiCASE 2010. LNICST, vol. 76, pp. 211–230. Springer, Heidelberg (2012)
18. Sung, M., Pentland, A.: PokerMetrics: Stress and Lie Detection through Non-invasive Physiological Sensing. PhD thesis, MIT Media Laboratory (2005)

An Embedded Ground Change Detector for a "Smart Walker"

Viviana Weiss[1], Aleksandr Korolev[1], Guido Bologna[1(✉)],
Séverine Cloix[1,2], and Thierry Pun[1]

Computer Science Department, University of Geneva, Geneva, Switzerland
{viviana.weiss,Aleksandr.Korolev,guido.bologna,thierry.pun}@unige.ch
Centre Suisse d'Electronique de de Microtechnique, CSEM, Neuchâtel, Switzerland
severine.cloix@csem.ch

Abstract. Millions of elderly people around the world use the walker for their mobility; nevertheless, these devices may lead to an accident. One of the cause of these accidents is misjudge the terrain. The main objective of this work is the implementation of a ground change detector in real time on a small and light embedded system that can be clipped on a rollator. As a long-term goal, this device will allow users to anticipate entering dangerous situations. We implemented an algorithm to detect ground changes based on color histograms and texture descriptor given as inputs to multi-layer perceptrons. Experiments were performed both off-line and with an embedded system. The obtained results indicated that it is possible to have an accurate detector which is able to distinguish ground changes in real-time.

Keywords: Ground change detector · Embedded system · Artificial Neural Network (ANN) · Elderly care · Gerontechnology

1 Introduction

The world is facing a situation without precedents. The proportion of old people and the expectation of life increase everywhere. The number of elederly people is projected to grow up to more than 2 billions, by 2050 [1]. Reducing severe disability from disease and health conditions is one key to holding down health and social costs.

The rollator (Fig. 1), widely spread among elderly, aims at helping users keep their independence and mobility. However, these tools can lead to falls, especially in urban areas and buildings. They occur when the user misjudges the nature of ground, which can happen in any kind of familiar or unknown environments. Approximately, 87% of elderly people falls are attributable to walkers use [2].

Various prototypes of "intelligent walkers" are motorized, equipped with route planning and obstacle detection, relying on active sensing (laser, sonar, IR ligth), or passive sensing (RFID tags, visual signs) [3,4,5]. Such aids are complex, thus expensive and exist only at prototype level. In practice, their use is limited to indoor situations due to their short battery life.

© Springer International Publishing Switzerland 2015
J.M. Ferrández Vicente et al. (Eds.): IWINAC 2015, Part I, LNCS 9107, pp. 533–542, 2015.
DOI: 10.1007/978-3-319-18914-7_56

Fig. 1. Typical walker with four wheels, handles with breaks and a seat

In this work, we present the *"Eye Walker"* project which aims at developing a low-cost, ultra-light computer vision-based prototype for users with mobility problems. This device is meant to be small and lightly embedded; it would be easily fixed on a standard rollator. One of the goals is to warn users before they enter in a dangerous ground. It has to operate in indoor and outdoor environments. The users initially targeted by this project are elderly persons that still live independently.

In this work, our goal is to asses the accuracy of our ground change detector and to determine at which image resolution it could be used in a real time implementation. Our key idea is based on the estimation of changes of brightness, color and textures under real environmental conditions and on the reduction of image resolution, in order to reduce the time processing.

This paper is organized as follows: Section 2 describes the material and the methodology to detect ground changes. Section 3 presents the experiments involving off-line and embedded systems at different resolutions, before concluding remarks.

2 System Design

This section briefly describes the methodology to detect the ground change, the different hardware set-up used for the off-line and embedded systems and the datasets used to compare both systems.

2.1 Detection Process

We would like to prevent in real time the falls related to the loss of balance caused by the ground change. As a result, the ground change detection is based on the comparison between the current frame and the average of k previous frames.

The procedure is divided in two different steps. Firstly, we extract the image descriptor from the current frame and the average of the k previous frames. Secondly, we use an Artificial Neural Network (ANN) trained on color histograms

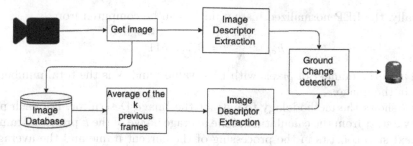

Fig. 2. General block diagram to detect ground changes

and a texture descriptor to decide whether to warn the user. The block diagram to detect ground changes is illustrated in Fig.2.

Image Descriptor

From each video frame, we calculate an image descriptor based on colors and textures. Specifically, a similarity measure between the current image and the k previous image is determined and provided to a neural network.

For the color feature, different types of color space were tested in [6]. As a result, the HSV color space demostrated to be the most suitable. From the image, we obtain the normalized color histogram h_c for each color channel using the following equation:

$$h_{c_i}^{H,S,V} = \frac{n_i}{N}, i = 0, \ldots, 255 \tag{1}$$

where n_i is the number of pixels with color label i and N is the total number of pixels in the image for each channel.

As a texture feature, the Local Edge Pattern (LEP) was used. LEP describes the spatial structure of the local texture according to the organization of edge pixels. To compute the LEP histogram, an edge image must be obtained first. The edge image is obtained by applying the Sobel edge detector to intensity gray level. The binary values are then multiplied by the corresponding binomial weights in a LEP mask, and the resulting values are summed to obtain the LEP value.

The LEP value is defined as [7],

$$LEP(n, m) = \sum_{i,j \in I} K_e(i, j) \times I_e(n, m) \tag{2}$$

where $I_e(n, m)$ denotes the binary image, K_e is the LEP mask and $LEP(n, m)$ is the LEP value for the pixel (n, m). The LEP mask is given by:

$$K_e = \begin{pmatrix} 1 & 2 & 4 \\ 128 & 256 & 8 \\ 64 & 32 & 16 \end{pmatrix} \tag{3}$$

Finally, the LEP normalized histogram h_e can be computed from

$$h_{e_i} = \frac{n_i}{N}, i = 0, \ldots, 511 \tag{4}$$

where n_i is the number of pixels with LEP value i and N is the total number of pixels in the image.

Fig.3 shows the methodology to extract the Image Descriptor. For each new frame we start from the calculation of the average image of the k previous frames. The next step consists in the processing of the current frame and the averaged image. Specifically:

- we apply the blur filter to reduce image noise and reduce detail;
- we convert the image from the RGB to the HSV color space;
- we calculate the histograms for H, S and V color variables;
- we transform the RGB image into a gray level image;
- we apply the Sobe filter to detect the edges;
- we use the LEP filter and calculate the LEP histogram.

We take into account the values of each bin in the histograms (H,S,V channels and LEP) as an image descriptor. Since we have three color channels, each channel having 256 bins and a texture channel represented by 512 bins, the final Image Descriptor contains 2560 bins (Fig. 4). Principal Component Analysis (PCA) is used to reduce the dimensionality of the image descriptor to 200 bins.

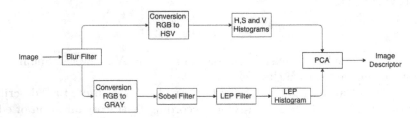

Fig. 3. Block diagram to extract Image Descriptor

Ground Change Detector

The color's distribution and LEP are used to obtain a distance measure between two images characterising the inhomogeneity of a surface. Specifically, this is calculated between the current image and the average of the k previous images. To measure the similarity between frames, several methods were presented in [8]. Here, to distinguish the ground change we implemented an Artificial Neural Network (ANN).

We used a multilayer perceptron with n neurons in the input layer, m hidden layers and 2 neurons in the output layer. Values for n and m were determined

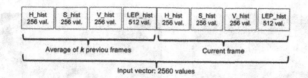

H_hist 256 val.	S_hist 256 val.	V_hist 256 val.	LEP_hist 512 val.	H_hist 256 val.	S_hist 256 val.	V_hist 256 val.	LEP_hist 512 val.

Average of *k* previou frames   Current frame

Input vector: 2560 values

Fig. 4. Image Descriptor configuration

empirically in [6]. The output layer has two neurons, one represents the detection of the ground change and the other indicates unchanged ground. In this neural network, the input vector theoretically is represented as a vector of 2560 neurons (see Fig. 4). In order to fulfil the real time requirements, the size of the input vectors was reduced to 200 neurons by means of PCA.

2.2 Set-Up

A set of outdoor video sequences were collected from a campus path at Geneva University using the walker shown in Fig.1. This data set was recorded with two set-ups, the first set-up (video set-up I) is a walker equipped with a color webcam (Logitech HD Webcam C510, 8 Mpixels) with a resolution of 640x480 pixels. The second set-up (video set-up II) is based on the camera sensor "Caspa VL" (Fig.7(a)) with a image resolution of 752x480 pixels. The cameras were located 60 cm from the ground. The covered visual field region is about 130 cm long, as shown in Fig.5. We use a total of six videos recorded at different times for the first set-up, each video containing between 188 and 313 frames and two to four ground change transitions. For the second set-up, 25 videos were recorded with 1687 frames in total. At the Fig.6 shows an example of both video sets. Note that videos were recorded at an approximative speed of 0.6 m/s (2.3 km/h) with a frame rate of $25\,fps$.

60 cm

130 cm

Fig. 5. Walker set-up for the detection of ground changes

The final system is composed by the camera sensor and the computer. In the first set-up, the color webcam is connected to a Dell computer with an Intel(R) Core(TM) i7-2600 CPU (3.4 GHz). For the second set-up, the detector was implemented in the Linux operating system with the use of the OpenCV library [9]. Specifically, the camera sensor "Caspa VL" [10] is connected to a "Tobi" platform (Fig.7(b)) [11] with Overo® (Computer On-Module) coard [12]. Tobi

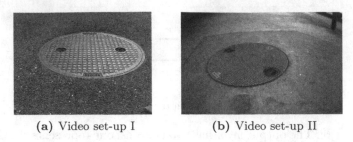

(a) Video set-up I   (b) Video set-up II

Fig. 6. Video sets image examples

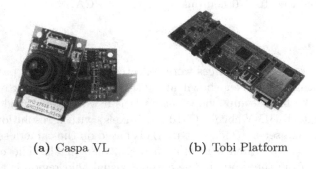

(a) Caspa VL   (b) Tobi Platform

Fig. 7. Hardware embedded system

and Overo® work with a ARM Cortex-A8 (Texas Instruments DaVinci DM3730 up to 1GHz capable) CPU with 512MB RAM.

3   Experiments

3.1   Implementation Comparisons

One of the goals of this research was to investigate the plausibility of migrating from the off-line system (Matlab) to embedded system (C++).

During the calculation of the histograms a substantial difference between the results obtained in MatLab and OpenCV was discovered. For the color feature, differences in the HSV conversion between both systems were found. On one hand, the biggest difference is observed in the S component and the smallest is for the V component.

On the other hand, differences in the implementation of the Sobel edge detection operator and gray scale images in MatLab and OpenCV are the main reasons of the discrepancy in the textural descriptor. One of the first major problems encountered during the migration process was the different implementation of the Sobel Filter. The implementation of the standard method for the Sobel edge detection in MatLab and OpenCV is quite different. In MatLab the result is a binary image, whereas in OpenCv it is a gray level image. Moreover, the standard Sobel function does not give any binary image and requires the additional applying of the threshold and "skeleton" in OpenCV. Because of this, there is a significant difference that is characterized by a Peak Signal to Noise Ratio (PSNR) equal to $64.46dB$. PSNR is an unit to measure the image distortion.

Other source of difference is the sensor camera. The camera Caspa VL is not a camera of full value. The pictures produced by the Caspa sensor are the results of applying the Bayer filter (see example at Fig. 8).

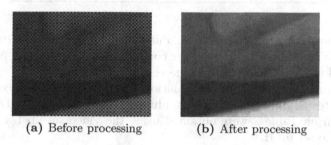

(a) Before processing   (b) After processing

Fig. 8. Bayer filter result

The Bayer filter mosaic is a color filter array (CFA) for arranging RGB color filters on a square grid of photo sensors. The raw output of Bayer-filter sensors is referred to as a Bayer pattern image. Since each pixel is filtered to record only one of three colors, the data from each pixel cannot fully specify each of the red, green, and blue values on its own.

3.2 Evaluation

To assess the performance of our approach, an extensive and systematic evaluation in terms of accuracy and processing time of image resolution were conducted on a data set labelled manually.

The two classes, ground change and no change, are defined as follows: a frame is labelled positive as soon as a ground change enters the visual field and it remains positive until the user is completely on the new terrain; the frame is labelled negative, otherwise [6]. To compare the methods, we use the confusion matrix shown in Table 1, where accuracy is defined as:

$$Accuracy = \frac{tp + tn}{tp + tn + fp + fn} \tag{5}$$

In our system, we have two critical values. The first one is the missing alarm because it can generate an accident. We however must minimize the false positive rate to ensure user acceptance. And the second one is the processing time.

We tested our detector in both videos set-ups. To evaluate the classifiers for each video set-up, we performed a ten-fold cross-validation by creating 10 different training/validation pairs by sliding the training data window by 10% each

Table 1. Terminology use for the evaluation

		Condition	
		Ground change	No change
System result	Ground change	True positive "tp" (Correct alarm)	False Positive "fp" (Unexpected alarm)
	No change	False negative "fn" (Missing alarm)	True negative "tn" (Correct absence of alarm)

time. Then, for each of the training/validation pair, we performed 10 classificatory runs. We trained each run using the corresponding training set. Afterwards, we evaluated the classification using the rest of the dataset.

To determinate how the image resolution affects the performance of our algorithm, we tested different scales of resolution reduction between 1 to 16. The results shown on Table 2 were obtained using the video set-up I and Table 3 with the video set-up II.

Table 2. Comparison of accuracy, false alarm, missing alarm using different image resolutions on the video set-up I

Image resolution	Accuracy (%)	False Alarm (%)	Missing Alarm (%)
640x480	99.59	0.16	1.12
320x240	99.52	0.17	1.37
160x120	99.17	0.41	2.06
80x60	99.11	0.67	2.41
40x30	98.17	0.77	4.93

Table 3. Comparison of accuracy, false alarm, missing alarm using different image resolutions on the video set-up II

Image resolution	Accuracy (%)	False Alarm (%)	Missing Alarm (%)
752x480	98.28	1.02	2.78
376x240	98.06	1.12	3.19
188x120	97.86	1.32	3.40
94x60	97.41	1.47	4.29
47x30	96.32	2.08	6.126

The purpose of this experiment is to see the influence of the resolution in terms of accuracy, false alarm and missing alarm rate. For the first video set-up, we have a difference of 1.42% in terms of accuracy between the biggest and the smallest image resolution. With the second video set-up, this difference is around 2%. Fig. 9 shows the performance of our detector and the impact of the scaling.

An evaluation criterion is the processing time between the input image and the instant of detection. In these experiments, the average of processing time for

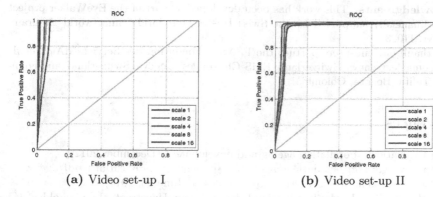

(a) Video set-up I   (b) Video set-up II

Fig. 9. Receiver operating characteristic (ROC) curves of two videos set-up with different images resolutions

a full image is around 7 s. This time is not acceptable on a real-time system. Table 4 shows the impact of scaling in the execution time. The main idea is to determine a good compromise between accuracy and execution time. Hence by reducing the resolution eight times for each dimension, the processing time is close to 0.3 s with fairly good accuracy.

Table 4. Comparison of time execution using different image resolutions on the video set-up II.

Image resolution	Execution Time (ms)
752x480	7097.10
376x240	1855.71
188x120	602.03
94x60	295.92
47x30	227.41

4 Conclusion

In this paper, we presented an embedded ground change detector aiming at warning walker's users before entering dangerous situations. This detector was based on a multilayer perceptron processing the current frame and a number of preceding frames.

The obtained results demonstrated the possibility to implement a ground change detector in real-time on an embedded system. Moreover, image resolution reduction showed that image resolution reduction has small impact on accuracy loss, but with a high reduction factor of time.

Finally, the very promising results allow us to advocate the view that our detector could be possibly embedded in a device showing high accuracy in realistic situations.

Acknowledgments. This work has been developed as part of the EyeWalker project that is financially supported by the Swiss Hasler Foundation Smartworld Program, grant Nr. 11083.

We thank our end-users partner the FSASD, Fondation des Services dAide et de Soins Domicile, Geneva, Switzerland; EMS-Charmilles, Geneva, Switzerland; and Foundation Tulita, Bogota, Colombia.

References

1. United Nations. Population ageing and development (December 2012), http://www.un.org/en/development/desa/population/publications/pdf/ageing/ 2012PopAgeingandDev_WallChart.pdf (accessed June 28, 2014)
2. Bleijlevens, M., Diederiks, J., Hendriks, M., van Haastregt, J., Crebolder, H., van Eijk, J.: Relationship between location and activity in injurious falls: An exploratory study. BMC Geriatrics 10(1), 40 (2010)
3. Rentschler, A.J., Simpson, R., Cooper, R.A., Boninger, M.L.: Clinical evaluation of guido robotic walker. Journal of Rehabilitation Research and Development 45(9), 1281–1293 (2008)
4. Frizera, A., Ceres, R., Pons, J.L., Abellanas, A., Raya, R.: The smart walkers as geriatric assistive device. The simbiosis purpose. Gerontechnology 7(2) (2008)
5. Dubowsky, S., Genot, F., Godding, S., Kozono, H., Skwersky, A., Yu, H., Yu, L.S.: Pamm - a robotic aid to the elderly for mobility assistance and monitoring: A "helping-hand" for the elderly. In: IEEE International Conference on Robotics and Automation, pp. 570–576 (2000)
6. Weiss, V., Cloix, S., Bologna, G., Hasler, D., Pun, T.: A robust, real-time ground change detector for a smart walker. In: 9th Int. Conf. on Computer Vision Theory and Applications, VISAPP 2014, Lisbon, Portugal (2014)
7. Kumar, P., Gupta, V.: Learning based obstacle detection for textural path. International Journal of Emerging Technology and Advanced Engineering 2, 436–439 (2012)
8. Penatti, O.A.B., Valle, E., da S. Torres, R.: Comparative study of global color and texture descriptors for web image retrieval. Journal of Visual Communication and Image Representation 23, 359–380 (2012)
9. OpenCV. Opencv (March 2015), http://opencv.org/ (accessed March 20, 2015)
10. Gumstix. Caspa vl (March 2015), https://store.gumstix.com/index.php/products/260/ (accessed March 20, 2015)
11. Gumstix. Tobi (March 2015), https://store.gumstix.com/index.php/products/230/ (accessed March 20, 2015)
12. Gumstix. Overo waterstorm com (March 2015), https://store.gumstix.com/index.php/products/265/ (accessed March 20, 2015)

Author Index

Printed in the United States
By Bookmasters